T0178482

Introduction to Queueing Systems
with Telecommunication Applications

László Lakatos • László Szeidl • Miklós Telek

Introduction to Queueing Systems with Telecommunication Applications

Second Edition

László Lakatos
Eotvos Lorant University
Budapest, Hungary

László Szeidl
Obuda University
Budapest, Hungary

Miklós Telek
Technical University of Budapest
Budapest, Hungary

ISBN 978-3-030-15144-7   ISBN 978-3-030-15142-3 (eBook)
https://doi.org/10.1007/978-3-030-15142-3

Library of Congress Control Number: 2019936306

Mathematics Subject Classification: 60J10, 60J27, 60K25, 60K30

This Springer imprint is published by the registered company Springer Nature Switzerland AG.
The registered company address is: Gewerbestrasse 11, 6330 Cham, Switzerland

Preface to the Second Edition

In early 2017, we were approached by the editor of the first edition of this book, Donna Chernyk, for potential interest in revising and extending the book for a second edition. After some rounds of discussions, we agreed to sign up for a new project to renew the book. Due to the fact that structure of the book was given and we had a clear plan on new contents to add, the preparation of this second edition was easier and faster than the preparation of the first edition, while we added a significant amount of new materials. Apart from the extensions to the formerly existing sections, we have added several new parts to this second edition:

- Section 2.7.5: Campbell's and Displacement Theorems
- Section 3.4: Birth–Death and Birth–Death-Immigration Processes
- Section 3.4.3: Restricted or Taboo Process
- Section 5.2.4: Properties of Batch Markov Arrival Processes
- Section 5.2.5: Marked Markov Arrival Process
- Section 5.3.3: Analysis of the Level Process of Quasi-Birth–Death Processes
- Section 5.3.4: Discrete-Time Quasi-Birth–Death Process
- Section 5.4: G/M/1-Type Process
- Section 5.5: M/G/1-Type Process
- Section 8.4: Queue Length in Discrete-Time System
- Section 9.4: Queues with Underlying G/M/1 and M/G/1-Type Processes
- Section 11.7.4: Queue Length for the Discrete Cyclic Waiting System
- Solution manual for each exercises.

Section 2.7.5 consists of the Campbell's and displacement theorems which establish specific properties related to a Poisson point process with deterministic and random transformation. In general form, the Campbell's theorem characterizes the stochastic behavior of the sum of random variables defined by a deterministic function on the jump times of a Poisson process. The displacement theorem establishes that under given conditions, randomly and independently displaced points of a Poisson point process to some new location form a new Poisson point process.

Section 3.4 extends the results of the (time-)homogeneous birth–death process to the case of (time-)inhomogeneous Kendall (or in other word, linear) birth–death process with immigration.

Section 8.4 studies the discrete-time version of the M/G/1 system with some restriction on the service time; it may be interesting in the applications. In Sect. 11.7, we added the discrete-time variant of cyclic-waiting systems both for the queue length and waiting time; there one can allow the case of general service time as well. Section 11.8 demonstrates the idea of queues with negative arrivals introduced by Gelenbe.

The extensions of Chaps. 5 and 9 are closely related. Most of the extensions in Chap. 5 are introduced in order to provide necessary materials on the behavior of QBD, M/G/1-type, and G/M/1-type processes in order to analyze queues with such structures. The analysis of some queues based on M/G/1-type and G/M/1-type processes is provided in Sect. 9.4. In case of BMAP/MAP/1 and MAP/BMAP/1 queues, the underlying process is directly G/M/1-type and an M/G/1-type process, respectively. In case of BMAP/G/1 and G/BMAP/1 queues the analysis goes through an embedded Markov chain approach, where the embedded Markov chain is M/G/1 type in the earlier and G/M/1 type in the later case. It is interesting to see that some essential results are obtained utilizing the fact that some characterizing matrices commute.

To be more specific about the extensions in Chap. 5, apart from the introduction and analyses of G/M/1-type and M/G/1-type processes, we provided a more detailed discussion on the level process of QBD processes, which results in quadratic equations for the transient behavior of the characterizing matrices ($R(s)$ and $G(s)$) in Laplace domain. Additionally, we introduced and analyzed the discrete-time versions of QBD processes and, as a kind of generalization of BMAPs, introduced the concept of marked Markov arrival process, whose analysis resembles to the one of BMAPs.

Acknowledgment

Thanks to OTKA K-123914 and MTA-BME Information Systems Research Group.

Budapest, Hungary   László Lakatos
Budapest, Hungary   László Szeidl
Budapest, Hungary   Miklós Telek
April 30, 2018

Preface to the First Edition

The development of queueing theory dates back to more than a century. Originally the concept was examined for the purpose of maximizing performance of telephone operation centers; however, it was realized soon enough that issues in that field that were solvable using mathematical models might arise in other areas of everyday life as well. Mathematical models, which serve to describe certain phenomena, quite often correspond with each other, regardless of the specific field for which they were originally developed, be that telephone operation centers, the planning and management of emergency medical services, the description of the computer operation, banking services, transportation systems, or other areas. The common feature in these areas is that demands and services occur (also at an abstract level) with various contents depending on the given questions. In the course of modeling, irrespective of the meaning of demand and service in the modeled system, one is dealing with only moments and time intervals. Thus it can be concluded, despite the diversity of the problems, a common theoretical background and a mathematical toolkit can be relied upon that ensures the effective and multiple application of the theory. It is worth noting as an interesting aspect that the beginning of the development of queueing theory is closely connected to the appearance of telephone operation centers more than a century ago, as described previously; nevertheless, it still plays a significant role in the planning, modeling and analyzing of telecommunication networks supplemented by up-to-date simulation methods and procedures.

The authors of this book have been conducting research and modeling in the theoretical and practical field of queueing theory for several decades and teaching in both bachelor's, master's and doctoral programs in the Faculty of Informatics, Eötvös Loránd University, Faculty of Engineering Sciences, Széchenyi István University, Faculty of Electrical Engineering and Informatics, Budapest University of Technology and Economics (all located in Hungary).

The various scientific backgrounds of the authors complement each other; therefore, both mathematical and engineering approaches are reflected in this book. The writing of this book was partly triggered by requests from undergraduate students and Ph.D. students and by the suggestions of supportive colleagues of whom all expressed the necessity to write a book that could be directly applied to informatics, mathematics, and applied mathematics education as well as in other fields. In considering the structure of the book, the authors tried to briefly summarize the necessary theoretical basis of probability theory and stochastic processes, which provide a uniform system of symbols and conventions to study and master the material presented here. At the end of Part I., the book provides a systematic and detailed treatment of Markov chains, renewal and regenerative processes, and Markov chains with special structures. Following the introductory chapters on probability theory and stochastic processes, we will disregard the various possible interpretations concerning the examples to emphasise terms, methodology, and analytical skills; therefore we will provide the proofs for each of the given examples. We think that this structure help readers to study the material more effectively since they may have different backgrounds and knowledge concerning this area. Regarding the basics of probability theory, we refer the interested reader to the books [8, 11, 15, 28]. With respect to general results of stochastic processes and Markov chains, we refer the reader to the following comprehensive literature: [9, 10, 13, 14, 19, 20, 22, 25].

In Part II., the book introduces and considers the classical results of Markov and non-Markov queueing systems. Then queueing networks and applied queueing systems (performance analysis of switching fabric with fixed size packets, conflict resolution methods of random access protocols, queueing systems with priorities and repeated orders queueing systems) are analyzed. For more on the classical results of queueing theory, we refer the reader to [3, 7, 16, 21, 23, 24, 26] whereas in connection with modern theory of queueing and telecommunication systems the following books may be consulted: [1, 2, 4–6, 12, 17, 18, 27], as well as results published mainly in journals and conference papers. The numerous exercises at the end of chapters ensure a better understanding of the material.

A short appendix appears at the end of the book that sums up those special concepts and ideas that are being used in the book and that help the reader to understand the material better.

This work was supported by the European Union and co-financed by the European Social Fund under grant TÁMOP 4.2.1/B-09/1/KMR-2010-0003, and by the OTKA grant No. K-101150. The authors are indebted to the Publisher for the encouragement and the efficient editorial support.

This book is recommended for students and researchers studying and working in the field of queueing theory and its applications.

Budapest, Hungary   László Lakatos
Györ, Hungary   László Szeidl
Budapest, Hungary   Miklós Telek
June 30, 2012

References

1. Artalejo, J.R., Gomez-Corral, A.: Retrial Queueing Systems: A Computational Approach. Springer, Berlin (2008)
2. Asmussen, S.: Applied Probability and Queues. Springer, Berlin (2003)
3. Baccelli, F., Brémaud, P.: Elements of Queueing Theory, Applications of Mathematics. Springer, Berlin (2002)
4. Borovkov, A.A.: Stochastic Processes in Queueing Theory. Applications of Mathematics. Springer, Berlin (1976)
5. Borovkov, A.A.: Asymptotic Methods in Queueing Theory. Wiley, New York (1984)
6. Breuer, L., Baum, D.: An Introduction to Queueing Theory and Matrix-Analytic Methods. Springer, Berlin (2005)
7. Chen, H., Yao, D.D.: Fundamentals of Queueing Networks: Performance, Asymptotics, and Optimization. Springer, Berlin (2001)
8. Chow, Y., Teicher, H.: Probability Theory. Springer, Berlin (1978)
9. Chung, K.: Markov Chains with Stationary Transition Probabilities. Springer, Berlin (1960)
10. Daley, D.J., Vere-Jones, D.: An Introduction to the Theory of Point Process, 2nd edn. Springer, Berlin (2008)
11. Feller, W.: An Introduction to Probability Theory and its Applications, vol. I. Wiley, New York (1968)
12. Giambene, G.: Queuing Theory and Telecommunications: Networks and Applications. Springer, Berlin (2005)
13. Gikhman, I., Skorokhod, A.: The Theory of Stochastic Processes, vol. I. Springer, Berlin (1974)
14. Gikhman, I.I., Skorokhod, A.V.: The Theory of Stochastic Processes, vol. II. Springer, Berlin (1975)
15. Gnedenko, B.V.: Theory of Probability, 6th edn. Gordon and Breach Publishers, London (1997)
16. Gnedenko, B.V., Kovalenko, I.N.: Introduction to Queueing Theory, 2nd edn. Birkhäuser, Boston (1989)
17. Gross, D., Shortle, J.F., Thompson, J.M., Harris, C.M.: Fundamentals of Queueing Theory, 4th edn. Wiley, London (2008)
18. Kalashnikov, V.: Mathematical Methods in Queueing Theory. Kluwer, Dordecht (1994)
19. Karlin, S., Taylor, H.M.: A First Course in Stochastic Processes. Academic Press, New York (1975)
20. Karlin, S., Taylor, H.M.: A Second Course in Stochastic Processes. Academic Press, New York (1981)
21. Kelly, F.P.: Reversibility and Stochastic Networks. Wiley, London (1979)
22. Kingman, J.F.C.: Poisson Processes. Clerendon Press, Oxford (1993)
23. Kleinrock, L.: Queuing Systems, Volume 1: Theory. Wiley Interscience, New York (1975)

24. Matveev, V.F., Ushakov, V.G.: Queueing Systems. MGU, Moscow (1984)
25. Meyn, S., Tweedie, R.: Markov Chains and Stochastic Stability. Springer, Berlin (1993)
26. Saaty, T.: Elements of Queueing Theory. McGraw-Hill, New York (1961)
27. Serfozo, R.: Introduction to Stochastic Networks. Springer, Berlin (1999)
28. Shiryaev, A.N.: Probability. Springer, Berlin (1994)

Contents

Part I
Introduction to Probability Theory and Stochastic Processes

Chapter 1
Introduction to Probability Theory

1.1 Summary on Basic Notions of Probability Theory

In this chapter we summarize the most important notions and facts of probability theory that are necessary for elaboration of our topic. In the present summary, we will apply the more specific mathematical concept and facts—mainly measure theory and analysis—only to a necessary extent while, however, maintaining mathematical precision. Readers interested in more detailed introduction to probability theory might consult [1, 4].

Random Event We consider experiments whose outcomes are uncertain, where the totality of the circumstances that are or can be considered does not determine the outcome ω of the experiment. A set consisting of all possible outcomes is called a **sample space**. We define **random events** (**events** for short) as certain sets of outcomes (subsets of sample space). It is assumed that the set of events is closed under countable set operations, and we assign probability to events only; they characterize the quantitative measure of the degree of uncertainty. Henceforth countable means finite or countably infinite.

Denote the sample space by $\Omega = \{\omega\}$. If Ω is countable, then the space Ω is called **discrete**. In a mathematical approach, events can be defined as subsets $A \subset \Omega$ of the possible outcomes Ω having the properties (σ-algebra properties) defined subsequently.

A given event A occurs in the course of an experiment if the outcome of the experiment belongs to the given event, that is, if an outcome $\omega \in A$ exists. An event is called simple if it contains only one outcome ω. It is always assumed that the whole set Ω and the empty set $\varnothing$ are events that are called a **certain** event and an **impossible event**, respectively.

Operation with Events; Notion of σ-Algebra Let A and B be two events. The **union** $A \cup B$ of A and B is defined as an event consisting of all elements $\omega \in \Omega$ belonging to either event A or B, i.e., $A \cup B = \{\omega : \omega \in A \text{ or } \omega \in B\}$.

© Springer Nature Switzerland AG 2019
L. Lakatos et al., *Introduction to Queueing Systems with Telecommunication Applications*, https://doi.org/10.1007/978-3-030-15142-3_1

The **intersection (product)** $A \cap B$ (AB) of events A and B is defined as an event consisting of all elements $\omega \in \Omega$ belonging to both A and B, i.e.,

$$A \cap B = \{\omega : \omega \in A \text{ and } \omega \in B\}.$$

The **difference** $A \backslash B$, which is not a symmetric operation, is defined as the set of all elements $\omega \in \Omega$ belonging to the event A but not to event B, i.e.,

$$A \backslash B = \{\omega : \omega \in A \text{ and } \omega \notin B\}.$$

The **complementary event** $\overline{A}$ of A is defined as a set of all elements $\omega \in \Omega$ that does not belong to A, i.e.,

$$\overline{A} = \Omega \backslash A.$$

If $A \cap B = \emptyset$, then sets A and B are said to be **disjoint** or **mutually exclusive**.

Note that the operations $\cup$ and $\cap$ satisfy the associative, commutative, and distributive property

$$(A \cup B) \cup C = A \cup (B \cup C), \quad \text{and} \quad (A \cap B) \cap C = A \cap (B \cap C),$$

$$A \cup B = B \cup A, \quad \text{and} \quad A \cap B = B \cap A,$$

$$A \cap (B \cup C) = (A \cap B) \cup (A \cap C), \quad \text{and} \quad A \cup (B \cap C) = (A \cup B) \cap (A \cup C).$$

DeMorgan identities are valid also for the operations union, intersection, and complementarity of events as follows:

$$\overline{A \cup B} = \overline{A} \cap \overline{B}, \qquad\qquad \overline{A \cap B} = \overline{A} \cup \overline{B}.$$

With the use of the preceding definitions introduced, we can define the notion of σ-algebra of events.

Definition 1.1 Let Ω be a nonempty (abstract) set, and let $\mathscr{A}$ be a certain family of subsets of the set Ω satisfying the following conditions:

(1) $\Omega \in \mathscr{A}$,
(2) If $A \in \mathscr{A}$, then $\overline{A} \in \mathscr{A}$,
(3) If $A_1, A_2, \ldots \in \mathscr{A}$ is a countable sequence of elements, then

$$\bigcup_{i=1}^{\infty} A_i \in \mathscr{A}.$$

The family $\mathscr{A}$ of subsets of the set Ω satisfying conditions (1)–(3) is called a σ-**algebra**. The elements of $\mathscr{A}$ are called **random events**, or simply **events**.

Remark 1.1 The pair $(\Omega, \mathscr{A})$ is usually called a **measurable space,** which forms the general mathematical basis of the notion of probability.

Probability Space, Kolmogorov Axioms of the Probability Theory Let Ω be a nonempty sample set, and let $\mathscr{A}$ be a given σ-algebra of subsets of Ω, i.e., the pair $(\Omega, \mathscr{A})$ is a measurable space. A nonnegative number $\mathbf{P}(A)$ is assigned to all events A of σ-algebra satisfying the axioms as follows.

A1. $0 \le \mathbf{P}(A) \le 1, A \in \mathscr{A}$.
A2. $\mathbf{P}(\Omega) = 1$.
A3. If the events $A_i \in \mathscr{A}, i = 1, 2, ...$, are disjoint (i.e., $A_i A_j = \emptyset, i \ne j$), then

$$\mathbf{P}\left(\bigcup_{i=1}^{\infty} A_i\right) = \sum_{i=1}^{\infty} \mathbf{P}(A_i).$$

The number $\mathbf{P}(A)$ is called the **probability** of event A, the axioms A1, A2, and A3 are called the **Kolmogorov axioms**, and the triplet $(\Omega, \mathscr{A}, \mathbf{P})$ is called the **probability space**. As usual, axiom A3 is called the σ-additivity property of the probability. The probability space characterizes completely a random experiment.

Remark 1.2 In the measure theory context of probability theory, the function $\mathbf{P}$ defined on $\mathscr{A}$ is called a probability measure because conditions A1–A3 ensure that $\mathbf{P}$ is nonnegative, σ-additive set function on $\mathscr{A}$ and $\mathbf{P}(\Omega) = 1$, i.e., a normed measure on $\mathscr{A}$. Our discussion basically does not require the direct use of measure theory, but some assertions cited in this work essentially depend on this theory.

Main Properties of Probability Let $(\Omega, \mathscr{A}, \mathbf{P})$ be a probability space. The following properties of probability are valid for all probability spaces.
   Elementary properties:

(a) The probability of an impossible event is zero, i.e.,
   $\mathbf{P}(\emptyset) = 0$.
(b) $\mathbf{P}(\overline{A}) = 1 - \mathbf{P}(A)$ for all $A \in \mathscr{A}$.
(c) If the relationship $A \subseteq B$ is satisfied for given events $A, B \in \mathscr{A}$, then
   $\mathbf{P}(A) \le \mathbf{P}(B), \mathbf{P}(B \setminus A) = \mathbf{P}(B) - \mathbf{P}(A)$.

Definition 1.2 A collection $\{A_i, \ i \in I\}$ of a countable set of events is called **complete system** of events if $A_i, \ i \in I$ are disjoint (i.e., $A_i \cap A_j = \emptyset$ if $i \ne j$, $i, j \in I$) and $\bigcup_{i \in I} A_i = \Omega$.

Remark 1.3 If the collection of events $\{A_i, \ i \in I\}$ forms a complete system of events, then

$$\mathbf{P}\left(\bigcup_{i \in I} A_i\right) = 1.$$

Probability of Sum of Events, Poincaré Formula For any events A and B it is true that

$$\mathbf{P}(A \cup B) = \mathbf{P}(A) + \mathbf{P}(B) - \mathbf{P}(AB).$$

Using this relation, a more general formula, called the **Poincaré formula**, can be proved. Let n be a positive integer number; then, for any events $A_1, A_2, ..., A_i \in \mathscr{A}$,

$$\mathbf{P}\left(\bigcup_{i=1}^{n} A_i\right) = \sum_{k=1}^{n} (-1)^{k-1} S_k^{(n)},$$

where $S_k^{(n)} = \sum_{1 \le i_1 \le ... \le i_k \le n} \mathbf{P}\left(\bigcup_{j=1}^{k} A_{i_j}\right)$.

Subadditive Property of Probability For any sequence of events $A_1, A_2, ...$ the inequality

$$\mathbf{P}\left(\bigcup_{i=1}^{\infty} A_i\right) \le \sum_{i=1}^{\infty} \mathbf{P}(A_i)$$

is true.

The proof is simple. Let us denote $B = \bigcup_{i=1}^{\infty} A_i$, $B_1 = A_1$, $B_i = A_i \setminus \bigcup_{n=1}^{i-1} A_n$, $i \ge 2$ then $B_i \subset A_i$; B_i, $i \ge 1$ are disjoint events and $B = \bigcup_{i=1}^{\infty} B_i$, consequently, $\mathbf{P}(B) = \sum_{i=1}^{\infty} \mathbf{P}(B_i) \le \sum_{i=1}^{\infty} \mathbf{P}(A_i)$.

Continuity Properties of Probability Continuity properties of probability are valid for monotonically sequences of events, each of which is equivalent to axiom A3 of probability. A sequence of events $A_1, A_2, ...$ is called monotonically increasing (resp. decreasing) if $A_1 \subset A_2 \subset ...$ (resp. $A_1 \supset A_2 \supset ...$).

Theorem 1.1 *If the sequence of events $A_1, A_2, ...$ is monotonically decreasing, then*

$$\mathbf{P}\left(\bigcap_{i=1}^{\infty} A_i\right) = \lim_{n \to \infty} \mathbf{P}(A_n).$$

If the sequence of events $A_1, A_2, ...$ is monotonically increasing, then

$$\mathbf{P}\left(\bigcup_{i=1}^{\infty} A_i\right) = \lim_{n \to \infty} \mathbf{P}(A_n).$$

Conditional Probability and Its Properties, Independence of Events In practice, the following obvious question arises: if we know that event B occurs (i.e., the outcome is in $B \in \mathscr{A}$), what is the probability that the outcome is in $A \in \mathscr{A}$? In other words, how does the occurrence of an event B influence the occurrence of another event A? This effect is characterized by the notion of the conditional probability $\mathbf{P}(A|B)$ as follows.

Definition 1.3 Let A and B be two events, and assume that $\mathbf{P}(B) > 0$. The quantity

$$\mathbf{P}(A|B) = \mathbf{P}(AB)/\mathbf{P}(B)$$

is called the **conditional probability of A given B.**

It is easy to verify that the conditional probability possesses the following properties:

1. $0 \leq \mathbf{P}(A|B) \leq 1$.
2. $\mathbf{P}(B|B) = 1$.
3. If the events $A_1, A_2, \ldots$ are disjoint, then

$$\mathbf{P}\left(\sum_{i=1}^{\infty} A_i \Big| B\right) = \sum_{i=1}^{\infty} \mathbf{P}(A_i|B).$$

4. The definition of conditional probability $\mathbf{P}(A|B) = \mathbf{P}(AB)/\mathbf{P}(B)$ is equivalent to the so-called theorem of multiplication

$$\mathbf{P}(AB) = \mathbf{P}(A|B)\mathbf{P}(B) \text{ and } \mathbf{P}(AB) = \mathbf{P}(B|A)\mathbf{P}(A).$$

Note that these equations are valid in the cases $\mathbf{P}(B) = 0$ and $\mathbf{P}(A) = 0$ as well.

One of the most important concepts of probability theory, the independence of events, is defined as follows.

Definition 1.4 We say that the events A and B are **independent** if the equation

$$\mathbf{P}(AB) = \mathbf{P}(A)\mathbf{P}(B)$$

is satisfied.

Remark 1.4 If A and B are independent events and $\mathbf{P}(B) > 0$, then the conditional probability $\mathbf{P}(A|B)$ does not depend on the event B since

$$\mathbf{P}(A|B) = \frac{\mathbf{P}(AB)}{\mathbf{P}(B)} = \frac{\mathbf{P}(A)\mathbf{P}(B)}{\mathbf{P}(B)} = \mathbf{P}(A).$$

This relation means that knowing that an event B occurs does not change the probability of another event A.

The notion of independence of an arbitrary collection A_i, $i \in I$ of events is defined as follows.

Definition 1.5 A given collection of events A_i, $i \in I$ is said to be **mutually independent (independent** for short) if, having chosen from among them any finite number of events, the probability of the product of the chosen events equals the product of the probabilities of the given events. In other words, if $\{i_1, ..., i_k\}$ is any subcollection of I, then one has

$$\mathbf{P}\left(A_{i_1} \cap ... \cap A_{i_k}\right) = \mathbf{P}\left(A_{i_1}\right)...\mathbf{P}\left(A_{i_k}\right).$$

This notion of independence is stricter when pairs are concerned since it is easy to create an example where pairwise independence occurs but mutual independence does not.

Example 1.1 We roll two dice and denote the pair of results by

$$(\omega_1, \omega_2) \in \Omega = \{(i, j), 1 \le i, j \le 6\}.$$

The number of elements of the set Ω is $|\Omega| = 36$, and we assume that the dice are standard, that, is $P\{(\omega_1, \omega_2)\} = 1/36$ for every $(\omega_1, \omega_2) \in \Omega$. Events A_1, A_2, and A_3 are defined as follows:

  $A_1 = \{$the result of the first dice is even$\}$,
  $A_2 = \{$the result of the second dice is odd$\}$,
  $A_3 = \{$both the first and second dice are odd or both of them are even$\}$.

We check that the events A_1, A_2, and A_3 are pairwise independent, but they are not (mutually) independent. It is clear that

$$A_1 = \{(2, 1), \dots, (2, 6), (4, 1), \dots, (4, 6), (6, 1), \dots, (6, 6)\},$$
$$A_2 = \{(1, 1), \dots, (6, 1), (1, 3), \dots, (6, 3), (1, 5), \dots, (6, 5)\},$$
$$A_3 = \{(1, 1), (1, 3), (1, 5), (2, 2), (2, 4), (2, 6), (3, 1), (3, 3),$$
$$(3, 5), \dots, (6, 2), (6, 4)(6, 6)\},$$

thus

$$|A_1| = 3 \cdot 6 = 18, \quad |A_2| = 6 \cdot 3 = 18, \quad |A_3| = 6 \cdot 3 = 18.$$

We have, then, $\mathbf{P}(A_i) = \frac{1}{2}$, $i = 1, 2, 3$, and the relations

$$\mathbf{P}\left(A_i A_j\right) = \frac{1}{4} = \mathbf{P}(A_i)\mathbf{P}\left(A_j\right), \ 1 \le i, j \le 3, \ i \ne j,$$

which means events A_1, A_2, and A_3 are pairwise independent. On the other hand,

$$\mathbf{P}(A_1 A_2 A_3) = 0 \ne \frac{1}{8} = \mathbf{P}(A_1)\mathbf{P}(A_2)\mathbf{P}(A_3);$$

consequently, the mutual independence of events A_1, A_2, and A_3 does not follow from the pairwise independence.

Formula of Total Probability, Bayes' Rule Using the theorem of multiplication for conditional probability we can easily derive the following two theorems. Despite the fact that the two theorems are not complicated, they represent quite effective tools in the course of the various considerations.

Theorem 1.2 (Formula of Total Probability) *Let the sequence* $\{A_i, \ i \in I\}$ *be a complete system of events with* $\mathbf{P}(A_i > 0), \ i \in I$; *then for all events B*

$$\mathbf{P}(B) = \sum_{i \in I} \mathbf{P}(B|A_i)\mathbf{P}(A_i)$$

is true.

Theorem 1.3 (Bayes' Rule) *Under the conditions of the preceding theorem, the following relation holds for all indices* $n \in I$ *and* $\mathbf{P}(B) > 0$:

$$\mathbf{P}(A_n|B) = \frac{\mathbf{P}(B|A_n)\mathbf{P}(A_n)}{\sum_{i \in I} \mathbf{P}(B|A_i)\mathbf{P}(A_i)}.$$

Concept of Random Variables Let $(\Omega, \mathscr{A}, \mathbf{P})$ be a probability space that is to be fixed later on. In the course of random experiments, the experiments usually result in some kind of value. This means that the occurrence of a simple event ω results in a random $X(\omega)$ value. Different values might belong to different simple events; however, the function $X(\omega)$, depending on the simple event ω, will have a specific property. We must answer basic questions like, what is the probability that the result of the experiment will be smaller than or equal to a certain given value x, needs to be answered. We have only determined probabilities of events (only for elements of the set $\mathscr{A}$) in connection with the definition of probability space; therefore, it has the immediate consequence that we may only consider the probability of the set if the set $\{\omega : X(\omega) \leq x\}$ is an event, which means that the set belongs to σ-algebra $\mathscr{A}$:

$$\{\omega : X(\omega) \leq x\} \in \mathscr{A}.$$

This fact led to one of the most important notions of probability theory.

Definition 1.6 The real valued function $X : \Omega \to \mathbb{R}$ is called **a random variable** if the relationship

$$\{\omega : X(\omega) \leq x\} \in \mathscr{A}$$

is valid for all real numbers $x \in \mathbb{R}$. A function satisfying this condition is called $\mathscr{A}$ **measurable**.

A property of random variables should be mentioned here. Define by $\mathscr{B} = \mathscr{B}_1$ the σ-algebra of Borel sets of $\mathbb{R}$ as the minimal σ-algebra containing all intervals of $\mathbb{R}$; the elements of $\mathscr{B}$ are called the Borel sets of $\mathbb{R}$. If X is $\mathscr{A}$ measurable, then for all Borel sets D of $\mathbb{R}$ the set $\{\omega : X(\omega) \in D\}$ is also an element of $\mathscr{A}$, i.e., $\{\omega : X(\omega) \in D\}$ is an event. Thus the probability $\mathbf{P}_X[D] = \mathbf{P}(\{\omega : X(\omega) \in D\})$, and so $\mathbf{P}(\{\omega : X(\omega) \leq x\})$ are well defined. An important special case of random variables is the so-called **indicator variables** defined as follows. Let $A \in \mathscr{A}$ be an event, and let us introduce the random variable $\mathscr{I}_{\{A\}}$, $A \in \mathscr{A}$:

$$\mathscr{I}_{\{A\}} = \mathscr{I}_{\{A\}}(\omega) = \begin{cases} 1, & \text{if } \omega \in A, \\ 0, & \text{if } \omega \notin A. \end{cases}$$

Distribution Function Let $X = X(\omega)$ be a random variable; then the probability $\mathbf{P}(X \leq x)$, $x \in \mathbb{R}$, is well defined.

Definition 1.7 The function $F_X(x) = \mathbf{P}(X \leq x)$ for all real numbers $x \in \mathbb{R}$ is called a **cumulative distribution function** (CDF) of random variable X.

Note that the CDFs F_X and function $\mathbf{P}_X$ determine each other mutually and unambiguously. It is also clear that if the real line $\mathbb{R}$ is chosen as a new sample space, and $\mathscr{B}$ is a σ-algebra of Borel sets as the σ-algebra of events, then the triplet $(\mathbb{R}, \mathscr{B}, \mathbf{P}_X)$ determines a new probability space, where $\mathbf{P}_X$ is referred to as a probability measure induced by the random variable X.

The CDF F_X has the following properties.

(1) In all points of a real line $-\infty < x_0 < \infty$ the function $F_X(x)$ is continuous from the right, that is,

$$\lim_{x \to x_0+0} F_X(x) = F_X(x_0).$$

(2) The function $F_X(x)$, $-\infty < x < \infty$ is monotonically increasing function of the variable x, that is, for all $-\infty < x < y < \infty$ the inequality $F_X(x) \leq F_X(y)$ holds.

(3) The limiting values of the function $F_X(x)$ exist under the conditions $x \to -\infty$ and $x \to \infty$ as follows:

$$\lim_{x \to -\infty} F_X(x) = 0 \quad \text{and} \quad \lim_{x \to \infty} F_X(x) = 1.$$

(4) The set of discontinuity points of function $F_X(x)$, that is, the set of points $x \in \mathbb{R}$ for which $F_X(x) \neq F_X(x - 0)$, is countable.

Remark 1.5 It should be noted in connection with the definition of the CDF that the literature is not consistent. The use of $F_X(x) = \mathbf{P}(X < x)$, $-\infty < x < \infty$ as a CDF is also widely applied. The only difference between the two definitions lies within property (1) (see preceding discussion), which means that in the latter case the CDF is continuous from the left and not from the right, but all the other properties remain the same. It is also clear that if the CDF is continuous in all $x \in \mathbb{R}$, then there is no difference between the two definitions.

Remark 1.6 From a practical point of view, it is sometimes useful to allow that the CDF F_X does not satisfy property (3), which means that, instead, one or both of the following relations hold:

$$\lim_{x \to -\infty} F_X(x) = \alpha > 0 \qquad \text{or} \qquad \lim_{x \to \infty} F_X(x) = \beta < 1.$$

In this case $\mathbf{P}\left(|X| < \infty\right) < 1$, and the CDF of random variable X has a **defective distribution function**.

Let a and b be two arbitrary real numbers for which $-\infty < a < b < \infty$; then we can determine the probability of some frequently occurring events with the use of the CDF of X as follows:

$$\mathbf{P}\left(X = a\right) = F_X(a) - F_X(a - 0),$$

$$\mathbf{P}\left(a < X < b\right) = F_X(b - 0) - F_X(a),$$

$$\mathbf{P}\left(a \le X < b\right) = F_X(b - 0) - F_X(a - 0),$$

$$\mathbf{P}\left(a < X \le b\right) = F_X(b) - F_X(a),$$

$$\mathbf{P}\left(a \le X \le b\right) = F_X(b) - F_X(a - 0).$$

These equations also determine the connection between the CDF F_X and the distribution $\mathbf{P}_X$ for special Borel sets of real number line.

Discrete and Continuous Distribution, Density Function We distinguish two important types of distributions in practice, the so-called discrete and continuous distributions. There is also a third type of distribution, the so-called singular distribution, in which case the CDF is continuous everywhere and its derivative (with respect to the Lebesgue measure) equals 0 almost everywhere; however, we will not consider this type. This classification follows from the Jordan decomposition theorem of monotonically functions, that is, an arbitrary CDF F can always be decomposed into the sum of three functions—the monotonically increasing absolutely continuous function, the step function with finite or countably infinite sets of jumps (this part corresponds to discrete distribution), and the singular function.

Definition 1.8 Random variable X is **discrete** or has a **discrete distribution** if there is a finite or countably infinite set of values $\{x_k, \ k \in I\}$ such that $\sum_{k \in I} p_k = 1$, where $p_k = \mathbf{P}\left(X = x_k\right), \ k \in I$. The associated function

$$f_X(x) = \begin{cases} p_k, & \text{if } x = x_k, \ k \in I \\ 0, & \text{if } x \neq x_k, \ k \in I \end{cases}, \quad x \in \mathbb{R}$$

is termed a **probability density function** (PDF) or **probability mass function** (PMF).

It is easy to see that if the random variable X is discrete with possible values $\{x_k, k = 0, 1, ...\}$ and with distribution $\{p_k, k = 0, 1, ...\}$, then the relationship between CDF F_X and PMF can be given as

$$F_X(x) = \sum_{x_k \leq x} p_k, \quad -\infty < x < \infty.$$

Definition 1.9 A random variable X is **continuous** or has a **continuous distribution** if there exists a nonnegative integrable function $f_X(x)$, $-\infty < x < \infty$ such that for all real numbers a and b, $-\infty < a < b < \infty$,

$$F_X(b) - F_X(a) = \int_a^b f_X(x)\mathrm{d}x$$

holds. *The function* $f_X(x)$ *is called the PDF of random variable* X, *or just the* *density function* *of* X.

Remark 1.7 The notion of the integral in Definition 1.9 is taken in ordinary (Riemann) sense. If we are interested in calculating the probability $\mathbf{P}(X \in S) = \int_S f(x)dx$, where S is a general subset of the real line or in a case when the integrand is not "well behaved," then the use of Lebesgue integration is a possible way of defining the integral in terms of measure theory (see, e.g., [1]).

It is clear that

$$F_X(x) = \int_{-\infty}^x f_X(u)\mathrm{d}u, \quad -\infty < x < \infty,$$

and it is also true that the PDF is not uniquely defined since if we take instead of $f_X(u)$ the function $f_X(u) + g(u)$, where the function $g(u)$ is nonnegative, integrable, and $\int_{-\infty}^x g(u)\mathrm{d}u = 0$, then the function $f_X(u) + g(u)$ is also a PDF of random variable X, which can naturally differ from the original f_X.

An arbitrary PDF $f_X(x)$ is nonnegative and integrable,

$$\int_{-\infty}^{\infty} f_X(x)\mathrm{d}x = 1,$$

and almost everywhere in $\mathbb{R}$ (with respect to the Lebesgue measure) the equation $F_X'(x) = f_X(x)$ is true.

Distribution of a Function of a Random Variable Let $X = X(\omega)$ be a random variable. Let $h(x)$, $x \in \mathbb{R}$ be a real-valued function, and let us define it as $Y = h(X)$.

The equation $Y = h(X)$ determines a random variable if for all $y \in \mathbb{R}$ the set $\{\omega : Y(\omega) = h(X(\omega)) \leq y\}$ is an event that is an element of σ algebra $\mathscr{A}$. If h is a continuous function or, more generally, is a Borel-measurable function (h is Borel measurable if for all x the relationship $\{u : h(u) \leq x\} \in \mathscr{B}$ is true), then Y, which is determined by the equation $Y = h(X)$, is a random variable. The question is how the CDF and the density function (if the latter exists) of random variable Y can be determined. It is usually true that

$$F_X(y) = \mathbf{P}(Y \leq y) = \mathbf{P}(h(X) \leq y) = \mathbf{P}_X[\{x : h(x) \leq y\}], \quad -\infty < y < \infty.$$

If h is a strictly monotonically increasing function, then this formula can be given in a simpler form. Let us denote by h^{-1} the inverse function of h, which in this case must exist. Then

$$F_X(y) = \mathbf{P}(h(X) \leq y) = \mathbf{P}\left(X \leq h^{-1}(y)\right) = F_X(h^{-1}(y)), \quad -\infty < y < \infty.$$

If h is a strictly monotonically decreasing function, then

$$F_X(y) = \mathbf{P}(h(X) \leq y) = \mathbf{P}\left(X \geq h^{-1}(y)\right) = 1 - F_X(h^{-1}(y) - 0), \quad -\infty < y < \infty.$$

With these relations, a formula can be given to PDF of Y in special cases.

Theorem 1.4 *Let us suppose that the random variable X has a PDF f_X and h is a strictly monotonically, differentiable real function. Then*

$$f_Y(y) = f_X(h^{-1}(y)) \left| \frac{d}{dy} h^{-1}(y) \right|, \quad -\infty < y < \infty.$$

Remark 1.8 If h is a linear function, that is, $h(y) = ay + b$, $a \neq 0$, and X has a PDF f_X, then the random variable $Y = h(X)$ also has a PDF and the formula

$$f_Y(y) = \frac{1}{|a|} f_X\left(\frac{y - a}{b}\right), \quad -\infty < y < \infty.$$

is true.

Joint Distribution and Density Function of Random Variables, Marginal Distributions In the majority of problems arising in practice, we have not one but several random variables, and we examine the probability of events where random variables simultaneously satisfy certain conditions.

Let $(\Omega, \mathscr{A}, \mathbf{P})$ be a probability space, and let there be two random variables X and Y on that space. The joint statistical behavior of the two random variables can be determined by a **joint CDF**. We should note that the joint analysis of the random variables X and Y corresponds to the examination of two-dimensional random vector variables such as (X, Y) that have random variable coordinates.

Definition 1.10 The function

$$F_{XY}(x, y) = \mathbf{P}(X \le x, Y \le y), \quad -\infty < x, y < \infty,$$

is called the **joint CDF** of random variables X and Y.

From a practical point of view, the two most important types of distributions are the discrete and the continuous ones, as in the one-dimensional case.

Definition 1.11 The joint distribution function of random variables X and Y is called **discrete**; in other words, the random vector (X, Y) has a **discrete distribution** if random variables X and Y are discrete. If we denote the values of random variables X and Y by $\{x_i, \ i \in I\}$ and $\{y_j, \ j \in J\}$, respectively, then the function

$$f_{X,Y}(x, y) = \begin{cases} p_{i,j}, & \text{if } x = x_i, \ y = y_j, \ i \in I, \ j \in J, \\ 0, & \text{if } x \ne x_i, \ y \ne y_j, \ i \in I, \ j \in J, \end{cases}$$

is called **joint PMF** or **joint PDF**.

It is clear that in the discrete case the joint distribution function is

$$F_{XY}(x, y) = \sum_{x_i \le x, \ y_j \le y} p_{ij}.$$

The case of joint continuous distribution is analogous to the discrete one.

Definition 1.12 The joint distribution of random variables X and Y is called **continuous**; in other words, the random vector (X, Y) has a **continuous distribution** if there exists a nonnegative, real-valued integrable function on the plane $f_{XY}(x, y)$ $-\infty < x, y < \infty$, for which the relation

$$F_{XY}(x, y) = \int\limits_{-\infty}^{x} \int\limits_{-\infty}^{y} f_{XY}(u, v) du dv$$

holds for all $-\infty < x, y < \infty$.

Definition 1.13 If F_{XY} denotes the joint CDF of random variables X and Y, then the CDFs

$$F_X(x) = \lim_{y \to \infty} F_{XY}(x, y),$$

$$F_Y(y) = \lim_{x \to \infty} F_{XY}(x, y)$$

are called **marginal distribution functions.**

It is not difficult to see that the marginal distribution functions do not determine the joint CDF. It is also clear that if the joint PDF $f_{XY}(x, y)$ of random variables X and Y exists, then the marginal PDFs can be given in the form

$$f_X(x) = \int_{-\infty}^{\infty} f_{XY}(x, y)dy,$$

$$f_Y(y) = \int_{-\infty}^{\infty} f_{XY}(x, y)dx,$$

for $-\infty < x < \infty$ and $-\infty < y < \infty$.

If there are more than two random variables $X_1, \ldots, X_n, n \geq 3$, i.e., in the case of n-dimensional random vector $(X_1, \ldots, X_n)$, then the definitions of joint distribution functions and density functions can be given analogously to the case of two random variables, so there is no essential difference. We will return to this question when we introduce the concept of stochastic processes.

Conditional Distributions   Let A be an arbitrary event, with $P(A) > 0$, and X an arbitrary random variable. Using the notion of conditional probability, we can define the **conditional distribution** of random variable X given event A as the function

$$F_X(x|A) = \mathbf{P}(X \leq x|A), \quad x \in \mathbb{R}.$$

The function $F_X(x|A)$ has all the properties of a distribution function mentioned previously.

The function $f_X(x|A)$ is called **conditional density function** of random variable X given event A if a nonnegative integrable function $f_X(x|A)$ exits for which the equation

$$F_X(x|A) = \int_{-\infty}^{x} f_X(u|A)du, \quad -\infty < x < \infty$$

holds.

The result for distribution function $F_X(x)$ can be easily proved in the same way as the theorem of total probability. If the sequence of events $A_1, A_2, \ldots$ is a complete system of events with property $\mathbf{P}(A_i) > 0$, $i = 1, 2, \ldots$, then

$$F_X(x) = \sum_{i=1}^{\infty} F_X(x|A_i)\mathbf{P}(A_i), \quad -\infty < x < \infty.$$

A similar relation holds for the conditional PDFs $f_X(x|A_i)$, $i \geq 1$, if they exist:

$$f_X(x) = \sum_{i=1}^{\infty} f_X(x|A_i)\mathbf{P}(A_i), \quad -\infty < x < \infty.$$

A different approach is required to define the conditional distribution function $F_{X|Y}(x|y)$ of random variable X given $Y = y$, where Y is another random variable. The difficulty is that if a random variable Y has a continuous distribution function, then the probability of the event $\{Y = y\}$ equals zero, and therefore the conditional distribution function $F_{X|Y}(x|y)$ cannot be defined with the help of the notion of conditional probability. In this case the conditional distribution function $F_{X|Y}(x|y)$ is defined as follows:

$$F_{X|Y}(x|y) = \lim_{\Delta y \to 0+} \mathbf{P}(X \leq x|y \leq Y < y + \Delta y)$$

if the limit exists.

Let us assume that the joint density function $f_{XY}(x, y)$ of random variables X and Y exists. In such a case random variable X has the conditional CDF $F_{X|Y}(x|y)$ and conditional PDF $f_{X|Y}(x|y)$ given $Y = y$. If the joint PDF exists and $f_X(y) > 0$, then it is not difficult to see that the following relation holds:

$$\begin{aligned}
F_{X|Y}(x|y) &= \lim_{\Delta y \to 0+} \mathbf{P}(X \leq x|y \leq Y < y + \Delta y) \\
&= \lim_{\Delta y \to 0+} \frac{\mathbf{P}(X \leq x, y \leq Y < y + \Delta y)}{\mathbf{P}(y \leq Y < y + \Delta y)} \\
&= \lim_{\Delta y \to 0+} \frac{\frac{F_{XY}(x, y+\Delta y)-F_{XY}(x, y)}{\Delta y}}{\frac{F_Y(y+\Delta y)-F_Y(y)}{\Delta y}} = \frac{1}{f_Y(y)}\frac{\partial}{\partial y}F_{XY}(x, y).
\end{aligned}$$

From this relation we get the conditional PDF $f_{X|Y}(x|y)$ as follows:

$$f_{X|Y}(x|y) = \frac{\partial}{\partial x}F_{X|Y}(x|y) = \frac{1}{f_Y(y)}\frac{\partial^2}{\partial x \partial y}F_{XY}(x, y) = \frac{f_{XY}(x, y)}{f_Y(y)}. \tag{1.1}$$

Independence of Random Variables Let X and Y be two random variables. Let $F_{XY}(x, y)$ be the joint distribution function of X and Y, and let $F_X(x)$ and $F_Y(y)$ be the marginal distribution functions.

Definition 1.14 The random variables X and Y are called independent of each other, or just independent, if the identity

$$F_{XY}(x, y) = F_X(x)F_Y(y)$$

holds for any x, y, $-\infty < x, y < \infty$.

In other words, random variables X and Y are independent if and only if the joint distribution function of X and Y equals the product of their marginal distribution functions.

The definition of independence of two random variables can be easily generalized to the case where an arbitrary collection of random variables $\{X_i, \ i \in I\}$ is given, analogously to the notion of the independence of events.

Definition 1.15 A collection of random variables $\{X_i, \ i \in I\}$ is called **mutually independent** (or just **independent**), if for any choice of a finite number of elements $X_{i_1}, ..., X_{i_n}$, the relation

$$F_{X_{i_1}, ..., X_{i_n}}(x_1, ..., x_n) = F_{X_{i_1}}(x_1) \cdot ... \cdot F_{X_{i_n}}(x_n), \ x_1, ..., x_n \in \mathbb{R}$$

holds.

Note that from the **pairwise independence** of random variables $\{X_i, \ i \in I\}$, which means that the condition

$$F_{X_{i_1}, X_{i_2}}(x_1, x_2) = F_{X_{i_1}}(x_1) F_{X_{i_2}}(x_2), \ x_1, x_2 \in \mathbb{R}, \ i_1, i_2 \in I,$$

is satisfied, mutual independence does not follow.

Example 1.2 Consider Example 1.1 given earlier and preserve the notation. Denote by $X_i = \mathscr{I}_{\{A_i\}}$ the indicator variables of the events $A_i, \ i = 1, 2, 3$. Then we can verify that the random variables X_1, X_2, and X_3 are pairwise independent, but they do not satisfy mutual independence. The pairwise independence of random variables X_i can be easily proved. Since the events A_1, A_2, A_3 are pairwise independent and

$$\{X_i = 1\} = A_i \quad \text{and} \quad \{X_i = 0\} = \overline{A}_i,$$

then, using the relation proved in Example 1.1, we obtain for $i \neq j$

$$\mathbf{P}\left(X_i = 1, X_j = 1\right) = \mathbf{P}\left(A_i A_j\right) = \mathbf{P}(A_i)\mathbf{P}\left(A_j\right) = \frac{1}{4},$$

$$\mathbf{P}\left(X_i = 1, X_j = 0\right) = \mathbf{P}\left(A_i \overline{A}_j\right) = \mathbf{P}(A_i)\mathbf{P}\left(\overline{A}_j\right) = \frac{1}{4},$$

$$\mathbf{P}\left(X_i = 0, X_j = 0\right) = \mathbf{P}\left(\overline{A}_i \overline{A}_j\right) = \mathbf{P}\left(\overline{A}_i\right)\mathbf{P}\left(\overline{A}_j\right) = \frac{1}{4},$$

while, for example,

$$\mathbf{P}\left(X_1 = 1, X_2 = 1, X_3 = 1\right) = \mathbf{P}\left(A_1 A_2 A_3\right) = 0 \neq \frac{1}{8}$$

$$= \mathbf{P}(A_1)\mathbf{P}\left(A_2\right)\mathbf{P}\left(A_3\right) = \mathbf{P}\left(X_1 = 1\right)\mathbf{P}\left(X_2 = 1\right)\mathbf{P}\left(X_3 = 1\right).$$

Consider how we can characterize the notion of independence for two random variables in the discrete and continuous cases (if more than two random variables are given, we may proceed in a similar manner).

Firstly, let us assume that the sets of values of discrete random variables X and Y are $\{x_i, i \geq 0\}$ and $\{y_j, j \geq 0\}$, respectively. If we denote the joint and marginal distributions of X and Y by

$$\{p_{ij} = \mathbf{P}\left(X = x_i, Y = y_j\right), \ i, j \geq 0\}, \ \{q_i = \mathbf{P}\left(X = x_i\right), \ i \geq 0),$$

$$\text{and } \{r_j = \mathbf{P}\left(Y = y_j, \ j \geq 0\right), \},$$

then the following assertion holds. Random variables X and Y are independent if and only if

$$p_{ij} = q_i r_j, \ i, j \geq 0.$$

Now assume that random variables X and Y have joint density $f_{XY}(x, y)$ and marginal densities $f_X(x)$ and $f_Y(y)$. Thus, in this case, random variables X and Y are independent if and only if the joint PDF takes product form, that is,

$$f_{XY}(x, y) = f_X(x) f_Y(y), \ -\infty < x, y < \infty.$$

Convolution of Distributions Let X and Y be independent random variables with distribution functions $F_X(x)$ and $F_Y(y)$, respectively, and let us consider the distribution of random variable $Z = X + Y$.

Definition 1.16 The distribution (CDF, PDF) of the random variable $Z = X + Y$ is called the **convolution** of distribution (CDF, PDF), and the equations expressing the relation among them are called convolution formulas.

Definition 1.17 Let $X_1, X_2, \ldots$ be independent identically distributed random variables with the common CDF F_X. The CDF F_X^{*n} of the sum $Z_n = X_1 + \ldots + X_n$ ($n \geq 1$) is uniquely determined by F_X and it is called the n-**fold convolution** of the CDF of F_X.

Note that the CDF $F_Z(z)$ of the random variable $Z = X + Y$, which is called the **convolution** of CDFs $F_X(x)$ and $F_Y(y)$, can be given in the general form

$$F_Z(z) = \mathbf{P}\left(Z \leq z\right) = \mathbf{P}\left(X + Y \leq z\right) = \int\limits_{-\infty}^{\infty} F_X(z - y)\, dF_Y(y).$$

This formula gets a simpler form in cases where the discrete random variables X and Y take only integer numbers, or if the PDFs $f_X(x)$ and $f_Y(y)$ of X and Y exist.

Let X and Y be independent discrete random variables taking values in $\{0, \pm 1, \pm 2, \ldots\}$ with probabilities $\{q_i = \mathbf{P}\left(X = i\right)\}$ and $\{r_j = \mathbf{P}\left(Y = j\right)\}$, respectively. Then the random variable $Z = X + Y$ takes values in $\{0, \pm 1, \pm 2, \ldots\}$, and its distribution satisfies the identity

$$s_k = \sum_{n=-\infty}^{\infty} q_{k-n} r_n, \ k = 0, \pm 1, \pm 2, \ldots.$$

If the independent random variables X and Y have a continuous distribution with the PDFs $f_X(x)$ and $f_Y(y)$, respectively, then the random variable Z is continuous and its PDF $f_Z(z)$ can be given in the integral form

$$f_Z(z) = \int_{-\infty}^{\infty} f_X(z-y) f_Y(y) dy.$$

Mixture of Distributions Let $F_1(x), \ldots, F_n(x)$ be a given collection of CDFs and let $a_1, \ldots, a_n$ be nonnegative numbers with the sum $a_1 + \ldots + a_n = 1$. The function

$$F(x) = a_1 F_1(x) + \ldots + a_n F_n(x), \ -\infty < x < \infty,$$

is called a **mixture** of CDFs $F_1(x), \ldots, F_n(x)$ with weights $a_1, \ldots, a_n$.

Remark 1.9 Any CDF can be given as a mixture of discrete, continuous, and singular CDFs, where the weights can also take a value 0.

Clearly, the function $F(x)$ possesses all the properties of CDFs; therefore it is also a CDF. In practice, the modeling of mixture distributions plays a basic role in stochastic simulation methods. A simple way to model mixture distributions is as follows.

Let us assume that the random variables $X_1, \ldots, X_n$ with distribution functions $F_1(x), \ldots, F_n(x)$ can be modeled. Let Y be a random variable taking values in $\{1, \ldots, n\}$ and independent of $X_1, \ldots, X_n$. Assume that Y has distribution $P(Y = i) = a_i$, $1 \le i \le n$ ($a_i \ge 0$, $a_1 + \ldots + a_n = 1$). Let us define random variable Z as follows:

$$Z = \sum_{i=1}^{n} \mathscr{I}_{\{Y=i\}} X_i,$$

where $\mathscr{I}_{\{\}}$ denotes the indicator variable. Then the CDF of random variable Z equals $F(z)$.

Proof Using the formula of total probability, we have the relation

$$\mathbf{P}(Z \le z) = \sum_{i=1}^{n} \mathbf{P}(Z \le z | Y = i) \mathbf{P}(Y = i) = \sum_{i=1}^{n} \mathbf{P}(X_i \le z) a_i = F(z).$$

Concept and Properties of Expectation A random variable can be completely characterized in a statistical sense by its CDF. To define a distribution function $F(x)$, one needs to determine its values for all $x \in \mathbb{R}$, but this is not possible in many cases.

Fortunately, there is no need to do so because in many cases it suffices to give some values that characterize the CDF in certain sense depending on concrete practical considerations. One of the most important concepts is expectation, which we define in general form, and we give the definition for discrete and continuous distributions as special cases.

Definition 1.18 Let X be a random variable and let $F_X(x)$ be its CDF. The **expected value** (or **mean value**) of the random variable X is defined as

$$\mathbf{E}(X) = \int_{-\infty}^{\infty} x \, dF_X(x)$$

if the expectation exists.

Note that the finite expected value $\mathbf{E}(X)$ exists if and only if $\int_{-\infty}^{\infty} |x| dF_X(x) < \infty$. It is conventional to denote the expected value of the random variable X by μ_X.

Expected Value of Discrete and Continuous Random Variables Let X be a discrete valued random variable with countable values $\{x_i, i \in I\}$ and with probabilities $\{p_i = \mathbf{P}(X = x_i), i \in I\}$. The finite expected value $\mathbf{E}(X)$ of random variable X exists and equals

$$\mathbf{E}(X) = \sum_{i \in I} p_i x_i$$

if and only if the sum is absolutely convergent, that is, $\sum_{i \in I} p_i |x_i| < \infty$. In the case of continuous random variables, the expected value can also be given in a simple form. Let $f_X(x)$ be the PDF of a random variable X. If the condition $\int_{-\infty}^{\infty} |x| f_X(x) dx < \infty$ holds (i.e., the integral is absolutely convergent), then the finite expected value of X exists and can be given as

$$\mathbf{E}(X) = \int_{-\infty}^{\infty} x f_X(x) dx.$$

From a practical point of view, it is generally enough to give two special, discrete, and continuous cases. Let X be a random variable that has a mixed CDF with discrete and continuous components $F_1(x)$ and $F_2(x)$, respectively, and with weights a_1 and a_2, that is

$$F(x) = a_1 F_1(x) + a_2 F_2(x), \ a_1, a_2 \geq 0, \ a_1 + a_2 = 1.$$

Assume that the set of discontinuities of $F_1(x)$ is $\{x_i,\ i \in I\}$ and denote $p_i = F_1(x_i) - F_1(x_i-),\ i \in I$. In addition, we assume that the continuous CDF $F_2(x)$ has the PDF $f(x)$. Then the expected value of random variable X is determined as follows:

$$\mathbf{E}(X) = a_1 \sum_{i \in I} p_i x_i + a_2 \int_{-\infty}^{\infty} x f(x) dx$$

if the series and the integral on the right-hand side of the last formula are absolutely convergent. The expected values related to special and different CDFs are to be given later in this chapter.

The operation of expectation can be interpreted as a functional

$$\mathbf{E} : X \rightarrow \mathbf{E}(X)$$

that assigns a real value to the given random variable. We enumerate the basic properties of this functional as follows.

1. If random variable X is finite, i.e., there are constants x_1 and x_2 for which the inequality $x_1 \leq X \leq x_2$ holds, then

$$x_1 \leq \mathbf{E}(X) \leq x_2.$$

   If random variable X is nonnegative and the expected value $\mathbf{E}(X)$ exists, then

$$\mathbf{E}(X) \geq 0.$$

2. Let us assume that the expected value $\mathbf{E}(X)$ exists; then the expected value of random variable cX exists for an arbitrary given constant c, and the identity

$$\mathbf{E}(cX) = c\mathbf{E}(X)$$

   is true.
3. If the random variable X satisfies the condition $\mathbf{P}(X = c) = 1$, then

$$\mathbf{E}(X) = c.$$

4. If the expected values of random variables X and Y exist, then the sum $X + Y$ has an expected value and the equality

$$\mathbf{E}(X + Y) = \mathbf{E}(X) + \mathbf{E}(Y)$$

   holds. This relation can usually be interpreted in such a way that the operation of expectation on the space of random variables is an additive functional.

5. The preceding properties can be expressed in a more general form. If there are finite expected values of random variables $X_1, \ldots, X_n$ and $c_1, \ldots, c_n$ are constants, then the equality

$$\mathbf{E}\,(c_1 X_1 + \ldots + c_n X_n) = c_1 \mathbf{E}\,(X_1) + \ldots + c_n \mathbf{E}\,(X_n)$$

holds. This property means that the functional $\mathbf{E}\,()$ is a linear one.

6. Let X and Y be independent random variables with finite expected value. Then the expected value of the product of random variables $X \cdot Y$ exists and equals the product of expected values, i.e., the equality

$$\mathbf{E}\,(XY) = \mathbf{E}\,(X) \cdot \mathbf{E}\,(Y)$$

is true.

Expectation of Functions of Random Variables, Moments, and Properties Let X be a discrete random variable with finite or countable values $\{x_i, i \in I\}$ and with distribution $\{p_i, i \in I\}$. Let $h(x)$, $x \in \mathbb{R}$ be a real valued function for which the expected value of random variable $Y = h(X)$ exists; then the equality

$$\mathbf{E}\,(Y) = \mathbf{E}\,(h(X)) = \sum_{i \in I} p_i h(x_i)$$

holds.

If the continuous random variable X has a PDF $f_X(x)$ and the expected value of random variable $Y = h(X)$ exists, then the expected value of Y can be given in the form

$$\mathbf{E}\,(Y) = \int_{-\infty}^{\infty} h(x) f_X(x) \mathrm{d}x.$$

In cases where the expected value of functions of random variables (functions of random vectors) are investigated, analogous results to the one-dimensional case can be obtained. We give the formulas in connection with the two-dimensional case only. Let X and Y be two random variables, and let us assume that the expected value of random variable $Z = h(X, Y)$ exists. With the appropriate notation, used earlier, for the cases of discrete and continuous distributions, the expected value of random variable Z can be given in the forms

$$\mathbf{E}\,(Z) = \sum_{i \in I} \sum_{j \in J} h(x_i, y_j) \mathbf{P}\,\big(X = x_i, Y = y_j\big),$$

$$E(Z) = \int\limits_{-\infty}^{\infty} \int\limits_{-\infty}^{\infty} h(x, y) f_{XY}(x, y) dx dy.$$

Consider the important case where h is a power function, i.e., for given positive integer number k, $h(x) = x^k$. Assume that the expected value of X^k exists. Then the quantity

$$\mu_k = E\left(X^k\right), \ k = 1, 2, ...,$$

is called the k-th moment of random variable X. It stands to reason that the **first moment** $\mu = \mu_1 = E\left(X^1\right)$ is the expected value of X and the frequently used **second moment** is $\mu_2 = E\left(X^2\right)$.

Theorem 1.5 *Let j and k be integer numbers for which $1 \le j \le k$. If the kth moment of random variable X exists, then the jth moment also exists.*

Proof From the existence of the kth moment it follows that $E\left(|X|^k\right) < \infty$. Since $k/j \ge 1$, then the function $x^{k/j}$, $x \ge 0$, is convex, and by the use of Jensen's inequality we get the relation

$$\left[E\left(|X|^j\right)\right]^{k/j} \le E\left(\left(|X|^j\right)^{k/j}\right) = E\left(|X|^k\right) < \infty.$$

The kth central moment $E\left((X - E(X))^k\right)$ is also used in practice; it is defined as the kth moment of the random variable centered at the first moment (expected value). The kth central moment $E\left((X - E(X))^k\right)$ can be expressed by the noncentral moments μ_i, $1 \le i \le k$ of random variable X as follows:

$$E\left((X - E(X))^k\right) = E\left(\sum_{i=0}^{k} \binom{k}{i} X^i (-E(X))^{k-i}\right)$$

$$= \sum_{i=0}^{k} \binom{k}{i} E\left(X^i\right)(-E(X))^{k-i}.$$

In the course of a random experiment, the observed values fluctuate around the expected value. One of the most significant characteristics of the quantity of fluctuations is the variance. Assume that the second moment of random variable X is finite. Then the quantities

$$\mathbf{Var}(X) = E\left((X - E(X))^2\right)$$

are called the **variance** of random variable X. The **standard deviation** of a random variable X is the square root of its variance:

Remark 1.10 Let X be a random variable. If $h(x)$ is a convex function and $\mathbf{E}(h(X))$ exists, then the Jensen inequality $\mathbf{E}(h(X)) \geq h(\mathbf{E}(X))$ is true. Using this inequality we can obtain some other relations, similar to the case of the Markov inequality.

Example 1.3 As a simple application of the Chebyshev inequality, let us consider the average $(X_1 + \dots + X_n)/n$, where the random variables $X_1, \dots, X_n$ are independent identically distributed with finite second moment. Let us denote the joint expected value and variance by μ and σ^2, respectively. Using the property (1.2) of variance and the Chebyshev inequality and applying $(n\varepsilon)$ instead of ε, we get the inequality

$$\mathbf{P}(|X_1 + \dots + X_n - n\mu| \geq n\varepsilon) = \mathbf{P}\left((X_1 + \dots + X_n - n\mu)^2 \geq n^2\varepsilon^2\right)$$

$$\leq \frac{n\sigma^2}{(n\varepsilon)^2} = \frac{\sigma^2}{n\varepsilon^2},$$

then

$$\mathbf{P}\left(\left|\frac{X_1 + \dots + X_n}{n} - \mu\right| \geq \varepsilon\right) \leq \frac{\sigma^2}{n\varepsilon^2}.$$

As a consequence of the last inequality, for every fixed positive constant ε the probability $\mathbf{P}\left(\left|\frac{X_1 + \dots + X_n}{n} - \mu\right| \geq \varepsilon\right)$ tends to 0 as n goes to infinity. This assertion is known as the **weak law of large numbers**.

Generating and Characteristic Functions So far, certain quantities characterizing the distribution of random variables have been provided. Now such transformations of distributions will be given where the distributions and the functions obtained by the transformation uniquely determine each other. The investigated transformations provide effective tools for determining, for instance, distributions and moments and for proving limit theorems.

Definition 1.19 Let X be a random variable taking values in $\{0, 1, \dots\}$, with probabilities $p_0, p_1, \dots$. Then the power series

$$G_X(z) = \mathbf{E}\left(z^X\right) = \sum_{i=0}^{\infty} p_i z^i$$

is convergent for all $z \in [-1, 1]$ and the function $G_X(z)$ is called the **probability generating function** (or just **generating function**) of the discrete random variable X.

In engineering practice the power series defining the generating function is applied in a more general approach instead of in the interval $[-1, 1]$, and the generating function is defined on the closed complex unit circle $z \in \mathbb{C}$, $|z| \leq 1$, which is usually called z-**transform** of the distribution $\{p_i, i = 0, 1, \dots\}$. This notion is also applied if, instead of distribution, a transformation is made to an arbitrary sequence of real numbers.

It should be noted that $|G_X(z)| \leq 1$ if $z \in \mathbb{C}$ and the function $G_X(z)$ is differentiable on the open unit circle of the complex plane $z \in \mathbb{C}$, $|z| < 1$ infinitely many times and the kth derivative of $G_X(z)$ equals the sum of kth derivative of the members of the series.

It is clear that

$$p_k = G_X^{(k)}(0)/k!, \ k = 0, 1, \dots.$$

This formula makes it possible to compute the distribution if the generating function is given. It is also true that if the first and second derivatives $G'_X(1-)$ and $G''_X(1-)$ exist on the left-hand side at $z = 1$, then the first and second moments of random variable X can be computed as follows

$$\mathbf{E}(X) = G'_X(1-) \quad \text{and} \quad \mathbf{E}(X^2) = \left. \left(z G'_X(z) \right)' \right|_{z=1} = G''_X(1-) + G'_X(1-).$$

From this we can obtain the variance of X as follows:

$$\mathbf{D}^2(X) = G''_X(1-) + G'_X(1-) - (G'_X(1-))^2.$$

It can also be verified that if the nth derivative of the generating function $G_X(z)$ exists on the left-hand side at $z = 1$, then

$$\mathbf{E}(X(X-1)\dots(X-m+1)) = \sum_{k=m}^{\infty} k(k-1)\dots(k-m+1)p_k$$

$$= G_X^{(m)}(1-), \ 1 \leq m \leq n.$$

Computing the expected values on the left-hand side of these identities, we can obtain linear equations between the moments $\mu_k = \mathbf{E}(X^k)$, $1 \leq k \leq m$, and the derivatives $G_X^{(m)}(1-)$ for all $1 \leq m \leq n$. The moments μ_m, $m = 1, 2, \dots, n$ can be determined in succession with the help of the derivatives $G_X^{(k)}(1-)$, $1 \leq k \leq m$. The special cases of $k = 1, 2$ give the preceding formulas for the first and second moments.

Characteristic Function

Definition 1.20 The complex valued function

$$\varphi_X(s) = \mathbf{E}\left(e^{\Im s X}\right) = \mathbf{E}(\cos(s X)) + \Im \mathbf{E}(\sin(s X)), \ s \in \mathbb{R}$$

is called the **characteristic function** of random variable X, where $\Im = \sqrt{-1}$.

Note that a characteristic function can be rewritten in the form

$$\varphi_X(s) = \int\limits_{-\infty}^{\infty} e^{\Im s x} dF_X(x),$$

which is the well-known Fourier–Stieltjes transform of the CDF $F_X(x)$.

Using conventional notations, in discrete and continuous cases we have

$$\varphi_X(s) = \sum_{k=0}^{\infty} p_k e^{\Im s x_k}, \quad \text{and} \quad \varphi_X(s) = \int\limits_{-\infty}^{\infty} e^{\Im s x} f_X(x) dx.$$

The characteristic function and the CDFs determine each other uniquely. Now some important properties of characteristic functions will be enumerated.

1. The characteristic function is real valued if and only if the distribution is symmetric.
2. If the kth moment $\mathbf{E}\left(X^k\right)$ exists, then

$$\mathbf{E}\left(X^k\right) = \frac{\varphi_X^{(k)}(0)}{\Im^k}.$$

3. If the derivative $\varphi_X^{(2k)}(0)$ is finite for a positive integer k, then the moment $\mathbf{E}\left(X^{2k}\right)$ exists. Note that from the existence of the finite derivative $\varphi_X^{(2k+1)}(0)$ only the existence of the finite moment $\mathbf{E}\left(X^{2k}\right)$ follows.
4. Let $X_1, \ldots, X_n$ be independent random variables; then the characteristic function of the sum $X_1 + \ldots + X_n$ equals the product of the characteristic functions of the random variables X_i, that is,

$$\varphi_{X_1 + \ldots + X_n}(s) = \mathbf{E}\left(e^{\Im s(X_1 + \ldots + X_n)}\right) = \mathbf{E}\left(e^{\Im s X_1} \ldots e^{\Im s X_n}\right)$$

$$= \mathbf{E}\left(e^{\Im s X_1}\right) \cdot \ldots \cdot \mathbf{E}\left(e^{\Im s X_n}\right) = \varphi_{X_1}(s) \ldots \varphi_{X_n}(s).$$

5. If there exists the kth absolute moment $\mathbf{E}\left(|X|^k\right) < \infty$, then the characteristic function φ_X possesses the following asymptotic relation

$$\varphi_X(s) = \sum_{j=0}^{k} \varphi_X^{(k)}(0)\frac{s^k}{k!} + o(s^k) = \sum_{j=0}^{k} \Im^j \mu_j \frac{s^j}{j!} + o(s^j), \quad s \to 0.$$

Note that properties 4 and 5 play an important role in the limit theorems of probability theory.

6. The following so-called inversion formula is valid between the CDF F_X and the characteristic function φ_X (see, p. 264 in [1]):

$$F_X(x_2) - F_X(x_1) = \frac{1}{2\pi} \lim_{C \to \infty} \int_{-C}^{C} \frac{e^{-\Im s x_1} - e^{-\Im s x_2}}{\Im s} \varphi_X(s) ds,$$

where x_1, x_2 are arbitrary continuity points of the CDF F_X and $x_1 < x_2$.

The CDF F_X has at most countable set of discontinuity points and it is continuous from right, consequently, the characteristic function φ_X uniquely determines the CDF F_X at all points of the real line.

It is an important special case when the characteristic function φ_X satisfies the condition $\int_{-\infty}^{\infty} |\varphi_X(s)| ds < \infty$, then the CDF $F_X(x)$ has the continuous PDF $f_X(x)$ and it is determined by the so-called inverse Fourier transform as follows

$$f_X(x) = \frac{1}{2\pi} \int_{-\infty}^{\infty} e^{-\Im s x} \varphi_X(s) ds.$$

Laplace–Stieltjes and Laplace Transforms If, instead of the CDFs, the Laplace–Stieltjes and Laplace transforms were used, the problem could be solved much easier in many practical cases and the results could additionally often be given in more compact form. Let X be a nonnegative random variable with CDF $F(x)$ $(F(0) = 0)$. Then the real or, in general, complex varying function

$$F^{\sim}(s) = \mathbf{E}\left(e^{-sX}\right) = \int_{0}^{\infty} e^{-sx} dF(x), \ \mathrm{Re} s \geq 0, \ F^{\sim}(0) = 1$$

is called the **Laplace–Stieltjes transform** of the CDF F. Since $\left|e^{-sX}\right| \leq 1$ if $\mathrm{Re} s \geq 0$, then the function $F^{\sim}(s)$ is well defined. If f is a PDF, then the function

$$f^*(s) = \int_{0}^{\infty} e^{-sx} f(x) dx, \ \mathrm{Re} s \geq 0,$$

is called the **Laplace transform** of the function f. These notations will be used even if the functions F and f do not possess the necessary properties of distribution and PDFs but $F^{\sim}(s)$ and $f^*(s)$ are well defined. If f is a PDF related to the CDF F, then the equality

$$F^{\sim}(s) = f^*(s) = sF^*(s). \tag{1.3}$$

holds.

Proof It is clear that

$$F^{\sim}(s) = \int_0^\infty e^{-sx} dF(x) = \int_0^\infty e^{-sx} f(x) dx = f^*(s),$$

and integrating by part we have

$$F^{\sim}(s) = \int_0^\infty e^{-sx} dF(x) = \int_0^\infty s e^{-sx} F(x) dx = s F^*(s).$$

Since the preceding equation is true between the two introduced transforms, it is enough to consider the Laplace–Stieltjes transform only and to enumerate its main properties.

(a) $F^{\sim}(s)$, Re$s \geq 0$ is a continuous function and $0 \leq |F^{\sim}(s)| \leq 1$, Re$s \geq 0$.
(b) $F^{\sim}_{aX+b}(s) = e^{-bs} F^{\sim}(as)$
(c) For all positive integers k

$$(-1)^k F^{\sim(k)}(s) = \int_0^\infty x^k e^{-sx} dF(x), \quad \text{Re}s > 0.$$

If the kth moment $\mu_k = \mathbf{E}\left(X^k\right)$ exists, then $\mu_k = (-1)^k F^{\sim(k)}(0)$.

(d) If the nonnegative random variables X and Y are independent, then

$$F^{\sim}_{X+Y}(s) = F^{\sim}_X(s) F^{\sim}_Y(s).$$

(e) For all continuity points of the CDF F the inversion formula

$$F(x) = \lim_{a \to \infty} \sum_{n \leq ax} (-1)^n (F^{\sim}(a))^{(n)} \frac{a^n}{n!}.$$

is true.

Covariance and Correlation Let X and Y be two random variables with finite variances σ_X^2 and σ_Y^2, respectively. The **covariance** between the pair of random variables (X, Y) is defined as

$$\mathbf{Cov}\,(X, Y) = \mathbf{E}\left((X - \mathbf{E}\,(X))(Y - \mathbf{E}\,(Y))\right).$$

The covariance can be rewritten in the simple computational form

$$\mathbf{Cov}\,(X, Y) = \mathbf{E}\,(XY) - \mathbf{E}\,(X)\mathbf{E}\,(Y).$$

If the variances σ_X^2 and σ_Y^2 satisfy the conditions $\mathbf{D}(X) > 0$, $\mathbf{D}(Y) > 0$, then the quantity

$$\mathbf{Corr}(X, Y) = \mathbf{Cov}\left(\frac{X - \mathbf{E}(X)}{\mathbf{D}(X)}, \frac{Y - \mathbf{E}(Y)}{\mathbf{D}(Y)}\right) = \frac{\mathbf{Cov}(X, Y)}{\mathbf{D}(X)\mathbf{D}(Y)}$$

is called the **correlation** between the pair of random variables (X, Y).

Correlation can be used as a measure of the dependence between random variables. It is always true that

$$-1 \leq \mathbf{Corr}(X, Y) \leq 1,$$

provided that the variances of random variables X and Y are finite and nonzero.

Proof Since by the Cauchy-Schwartz inequality for all random variable U and V with finite second moments

$$(\mathbf{E}(UV))^2 \leq \mathbf{E}(U^2)\mathbf{E}(V^2),$$

therefore

$$(\mathbf{Cov}(X, Y))^2 \leq \mathbf{E}((X - \mathbf{E}(X))^2)\mathbf{E}((Y - \mathbf{E}(Y))^2) = \mathbf{D}^2(X)\mathbf{D}^2(Y),$$

from which the inequality $|\mathbf{Corr}(X, Y)| \leq 1$ immediately follows.

It can also be proved that the equality $|\mathbf{Corr}(X, Y)| = 1$ holds if and only if a linear relation exists between random variables X and Y with probability 1, that is, there are two constants a and b for which $\mathbf{P}(Y = aX + b) = 1$.

Both covariance and correlation play essential roles in multivariate statistical analysis. Let $X = (X_1, ..., X_n)^T$ be a column vector whose n elements $X_1, ..., X_n$ are random variables. Here it should be noted that in probability theory and statistics usually column vectors are applied, but in queuing theory row vectors are used if Markov processes are considered. We define

$$\mathbf{E}(X) = (\mathbf{E}(X_1), ..., \mathbf{E}(X_n))^T,$$

provided that the expected values of components exist. The upper index T denotes the transpose of vectors or matrices. Similarly, if a matrix $W = (W_{ij}) \in \mathbb{R}^{k \times m}$ is given whose elements W_{ij} are random variables of finite expected values, then we define

$$\mathbf{E}(W) = (\mathbf{E}(W_{ij})), \quad 1 \leq i \leq k, \ 1 \leq j \leq m).$$

If the variances of components of a random vector $X = (X_1, ..., X_k)^T$ are finite, then the matrix

$$R = \mathbf{E}\left((X - \mathbf{E}(X))(X - \mathbf{E}(X))^T\right) \tag{1.4}$$

is called a **covariance matrix** of X. It can be seen that the (i, j) entries of the matrix $\mathbf{R}$ are $R_{ij} = \mathbf{Cov}(X_i, X_j)$, which are the covariances between the random variables X_i and X_j.

The covariance matrix can be defined in cases where the components of X are complex valued random variables replacing in definition (1.4) $(X - \mathbf{E}(X))^T$ by $(X - \mathbf{E}(X))^{*T}$ the complex composed of transpose.

An important property of a covariance matrix $\mathbf{R}$ is that it is nonnegative definite, i.e., for all real or complex k-dimensional column vectors $\mathbf{z} = (z_1, ..., z_k)^T$ the inequality

$$\mathbf{z}^T \mathbf{R} \mathbf{z} \geq 0$$

holds.

The matrix $\mathbf{r} = (r_{i,j})$ with components $r_{i,j} = \mathbf{Corr}(X_i, X_j)$, $1 \leq i \leq k$, $1 \leq j \leq m$ is called a **correlation matrix** of random vector X.

Conditional Expectation and Its Properties The notion of conditional expectation is defined with the help of results of set and measure theories. We present the general concept and important properties and illustrate the important special cases.

Let $(\Omega, \mathscr{A}, \mathbf{P})$ be a fixed probability space, and let X be a random variable whose expected value exists. Let $\mathscr{C}$ be an arbitrary sub-σ-algebra of $\mathscr{A}$. We wish to define the conditional expectation $Z = \mathbf{E}(X|\mathscr{C})$ of X given $\mathscr{C}$ as a $\mathscr{C}$-measurable random variable for which the random variable Z satisfies the condition $\mathbf{E}\left(\mathbf{E}(X|\mathscr{C})\mathscr{I}_{\{C\}}\right) = \mathbf{E}\left(X\mathscr{I}_{\{C\}}\right)$ for all $C \in \mathscr{C}$. It is known from the measure theory (see, e.g., p. 200 in [1]) that there exists a unique random variable Z with probability 1 that satisfies the required conditions.

Definition 1.21 Random variable Z is called the **conditional expectation** of X given σ-algebra $\mathscr{C}$ if the following conditions hold:

(a) Z is a $\mathscr{C}$-measurable random variable.
(b) $\mathbf{E}\left(\mathbf{E}(X|\mathscr{C})\mathscr{I}_{\{C\}}\right) = \mathbf{E}\left(X\mathscr{I}_{\{C\}}\right)$ for all $C \in \mathscr{C}$.

Definition 1.22 Let $A \in \mathscr{A}$ be an event. The random variable $\mathbf{P}(A|\mathscr{C}) = \mathbf{E}\left(\mathscr{I}_{\{A\}}|\mathscr{C}\right)$ is called the **conditional expectation** of event A given σ-algebra $\mathscr{C}$.

Important Properties of Conditional Expectation Let $\mathscr{C}$, $\mathscr{C}_1$, and $\mathscr{C}_2$ be sub-σ-algebras of $\mathscr{A}$, and let X, X_1, and X_2 be random variables with finite expected values. Then the following relations hold with probability 1:

1. $\mathbf{E}(\mathbf{E}(X|\mathscr{C})) = \mathbf{E}(X)$.
2. $\mathbf{E}(cX|\mathscr{C}) = c\mathbf{E}(X|\mathscr{C})$ for all constant c.
3. If $\mathscr{C}_0 = \{\varnothing, \Omega\}$ is the trivial σ-algebra, then $\mathbf{E}(X|\mathscr{C}_0) = \mathbf{E}(X)$.
4. If $\mathscr{C}_1 \subset \mathscr{C}_2$, then $\mathbf{E}(\mathbf{E}(X|\mathscr{C}_1)|\mathscr{C}_2) = \mathbf{E}(\mathbf{E}(X|\mathscr{C}_2)|\mathscr{C}_1) = \mathbf{E}(X|\mathscr{C}_1)$.

5. If random variable X does not depend on the σ-algebra $\mathscr{C}$, i.e., if for all Borel sets $D \in \mathscr{B}$ and for all events $A \in \mathscr{C}$ the equality $\mathbf{P}(X \in D, A) = \mathbf{P}(X \in D)\mathbf{P}(A)$ holds, then $\mathbf{E}(X|\mathscr{C}) = \mathbf{E}(X)$.
6. $\mathbf{E}(X_1 + X_2|\mathscr{C}) = \mathbf{E}(X_1|\mathscr{C}) + \mathbf{E}(X_2|\mathscr{C})$.
7. If the random variable X_1 is $\mathscr{C}$-measurable, then $\mathbf{E}(X_1 X_2|\mathscr{C}) = X_1 \mathbf{E}(X_2|\mathscr{C})$.

Definition 1.23 Let Y be a random variable, and denote by $\mathscr{A}_Y$ the σ-algebra generated by random variable Y, i.e., let $\mathscr{A}_Y$ be the minimal sub-σ-algebra of $\mathscr{A}$ for which Y is $\mathscr{A}_Y$-measurable. The random variable $\mathbf{E}(X|Y) = \mathbf{E}(X|\mathscr{C}_Y)$ is called the **conditional expectation** of X given random variable Y.

Main Properties of Conditional Expectation First, consider the case where random variable Y is discrete and takes values in the set $\mathscr{Y} = \{y_1, ..., y_n\}$ and $\mathbf{P}(Y = y_i) > 0$, $1 \leq i \leq n$. We then define the events $C_i = \{Y = y_i\}$, $1 \leq i \leq n$. It is clear that the collection of events $\{C_1, ..., C_n\}$ forms a complete system of events, i.e., they are mutually exclusive, $\mathbf{P}(C_i) > 0$, $1 \leq i \leq n$ and $\mathbf{P}(C_1) + ... + \mathbf{P}(C_n) = 1$. The σ-algebra $\mathscr{C}_Y = \sigma(C_1, ..., C_n) \subset \mathscr{A}$, which is generated by random variable Y, is the set of events consisting of all subsets of $\{C_1, ..., C_n\}$. Note that here we can write "algebra" instead of "σ-algebra" because the set $\{C_1, ..., C_n\}$ is finite. Since the events C_i have positive probability, the conditional probabilities

$$\mathbf{E}(X|C_i) = \frac{\mathbf{E}\left(X \mathscr{I}_{\{C_i\}}\right)}{\mathbf{P}(C_i)}$$

are well defined.

Theorem 1.8 *The conditional expectation $\mathbf{E}(X|\mathscr{C}_Y)$ satisfies the relation*

$$\mathbf{E}(X|\mathscr{C}_Y) = \mathbf{E}(X|\mathscr{C}_Y)(\omega) = \sum_{k=1}^{n} \mathbf{E}(X|C_k)\mathscr{I}_{\{C_k\}} \text{ with probability } 1. \qquad (1.5)$$

Note that Eq. (1.5) can also be rewritten in the form

$$\mathbf{E}(X|Y) = \mathbf{E}(X|Y)(\omega) = \sum_{k=1}^{n} \mathbf{E}(X|Y = y_k)\mathscr{I}_{\{Y=y_k\}}. \qquad (1.6)$$

Proof Since the relation

$$\{\mathbf{E}(X|\mathscr{C}_Y) \leq x\} = \cup\{C_i : \mathbf{E}(X|C_i) \leq x\} \in \mathscr{C}_Y$$

holds for all $x \in \mathbb{R}$, then $\mathbf{E}(X|\mathscr{C}_Y)$ is a $\mathscr{C}_Y$ measurable random variable. On the other hand, if $C \in \mathscr{C}_Y$, $C \neq \{\varnothing\}$, then $C = \cup\{C_i : i \in K\}$ stands with an appropriately chosen set of indices $K \subset \{1, ..., n\}$, and we obtain

$$\mathbf{E}\left(\mathbf{E}\left(X|\mathscr{C}_Y\right)\mathscr{I}_{\{C\}}\right) = \mathbf{E}\left(\sum_{k \in K} \mathbf{E}\left(X|C_k\right)\mathscr{I}_{\{C_k\}}\right)$$

$$= \sum_{k \in K} \mathbf{E}\left(X|C_k\right)\mathbf{P}\left(C_k\right) = \sum_{k \in K} \mathbf{E}\left(X\mathscr{I}_{\{C_k\}}\right) = \mathbf{E}\left(X\mathscr{I}_{\{C\}}\right).$$

If $C = \{\varnothing\}$, then $\mathbf{E}\left(\mathbf{E}\left(X|\mathscr{C}_Y\right)\mathscr{I}_{\{C\}}\right) = \mathbf{E}\left(X\mathscr{I}_{\{C\}}\right) = 0$. Thus we have proved that random variable (1.5) satisfies all required properties of conditional expectation.

Remark 1.11 From the expression (1.6) the following relation can be obtained:

$$\mathbf{E}\left(X\right) = \mathbf{E}\left(\mathbf{E}\left(X|Y\right)\right) = \int_{-\infty}^{\infty} \mathbf{E}\left(X|Y = y\right) \mathrm{d}F_Y(y). \tag{1.7}$$

This relation remains valid if, instead of the finite set $\mathscr{Y} = \{y_1, ..., y_n\}$, we choose the countable infinite set $\mathscr{Y} = \{y_i, \ i \in I\}$ for which $\mathbf{P}\left(Y = y_i\right) > 0, i \in I$.

Remark 1.12 Denote the function g by the relation

$$g(y) = \begin{cases} \mathbf{E}\left(X|Y = y_k\right), & \text{if } y = y_k \text{ for an index } k, \\ 0, & \text{otherwise.} \end{cases} \tag{1.8}$$

Then, using formula (1.6), the conditional expectation of X given Y can be obtained with the help of function g as follows:

$$\mathbf{E}\left(X|Y\right) = g(Y) \tag{1.9}$$

with probability 1.

Continuous Random Variables (X, Y) Consider a pair of random variables (X, Y) having joint density $f_{X,Y}(x, y)$ and marginal densities $f_X(x)$ and $f_Y(y)$, respectively. Then the conditional density $f_{X|Y}(x|y)$ exists, and according to (1.1), can be defined as

$$f_{X|Y}(x|y) = \begin{cases} \dfrac{f_{XY}(x, y)}{f_Y(y)}, & \text{if } f_Y(y) > 0, \\ 0, & \text{otherwise .} \end{cases}$$

Define $g(y) = \mathbf{E}\left(X|Y = y\right) = \int_{-\infty}^{\infty} x f_{X|Y}(x|y)\mathrm{d}x$. Then the conditional expectation of X given Y can be determined with probability 1 as follows:

$$\mathbf{E}\left(X|Y\right) = g(Y),$$

and so we can define

$$\mathbf{E}\,(X|Y=y) = g(y).$$

Proof It is clear that $g(Y)$ is a $\mathscr{C}_Y$ measurable random variable; therefore, it is enough to prove that the equality

$$\mathbf{E}\left(\mathbf{E}\,(X|Y)\mathscr{I}_{\{Y\in D\}}\right) = \mathbf{E}\left(X\mathscr{I}_{\{Y\in D\}}\right)$$

holds for all Borel set D of a real number line. It is not difficult to see that

$$\mathbf{E}\left(\mathbf{E}\,(X|Y)\mathscr{I}_{\{Y\in D\}}\right) = \mathbf{E}\left(g(Y)\mathscr{I}_{\{Y\in D\}}\right) = \int\limits_{D}\int\limits_{-\infty}^{\infty} x\,\frac{f_{XY}(x,\,y)}{f_Y(y)}\,f_Y(y)\mathrm{d}x\mathrm{d}y$$

$$= \int\limits_{D}\int\limits_{-\infty}^{\infty} x f_{XY}(x,\,y)\mathrm{d}x\mathrm{d}y$$

and on the other hand,

$$\mathbf{E}\left(X\mathscr{I}_{\{Y\in D\}}\right) = \int\limits_{D}\int\limits_{-\infty}^{\infty} x f_{XY}(x,\,y)\mathrm{d}x\mathrm{d}y.$$

Remark 1.13 In the case where a pair of random variables have joint normal distribution, the conditional expectation $\mathbf{E}\,(X|Y)$ is a linear function of random variable Y with probability 1, that is, the regression function g is a linear function and the relation

$$\mathbf{E}\,(X|Y) = \mathbf{E}\,(X) + \frac{\mathbf{Cov}\,(X,\,Y)}{\mathbf{D}\,(X)}(X - \mathbf{E}\,(X))$$

holds.

General Case By the definition of conditional expectation, $\mathbf{E}\,(X|Y)$ is $\mathscr{C}_Y$ measurable; therefore, there is a Borel-measurable function g such that $\mathbf{E}\,(X|Y)$ can be given with probability 1 in the form (see p.17 in [1])

$$\mathbf{E}\,(X|Y) = g(Y). \tag{1.10}$$

Due to this relation we can define the conditional expectation $\mathbf{E}\,(X|Y=y)$ as the function

$$\mathbf{E}\,(X|Y=y) = g(y),$$

which is called a **regression function**. It is clear that the regression function is not necessarily unique and it is determined on a Borel set of the real line D satisfying the condition $\mathbf{P}(Y \in D) = 1$.

Remark 1.14 Let X and Y be two random variables. Assume that X has finite variance. Consider the quadratic distance $\mathbf{E}\left([X - h(Y)]^2\right)$ for the set $\mathcal{H}_Y$ of all Borel-measurable function h, for which $h(Y)$ has finite variance. Then the assertion

$$\min\left\{\mathbf{E}\left([X - h(Y)]^2\right) : h \in \mathcal{H}_Y\right\} = \mathbf{E}\left([X - g(Y)]^2\right)$$

holds. This relation implies that the best approximation of X by Borel measurable functions of Y in a quadratic mean is the regression $\mathbf{E}(X|Y) = g(Y)$.

Formula of Total Expected Value A useful formula can be given to compute the expected value of the random variable X if the regression function $\mathbf{E}(X|Y = y)$ can be determined.

Making use of relation 1 given as general property of conditional expectation and Eq. (1.10), it is clear that

$$\mathbf{E}(X) = \mathbf{E}(\mathbf{E}(X|Y)) = \mathbf{E}(g(Y))$$

$$= \int_{-\infty}^{\infty} g(y)\mathrm{d}F_Y(y) = \int_{-\infty}^{\infty} \mathbf{E}(X|Y = y)\mathrm{d}F_Y(y).$$

From this relation we have the so-called **formula of total expected value**. If random variable Y has discrete or continuous distributions, then we have the formulas

$$\mathbf{E}(X) = \sum_{i \in I} \mathbf{E}(X|Y = y_i)\mathbf{P}(Y = y_i)$$

and

$$\mathbf{E}(X) = \int_{-\infty}^{\infty} \mathbf{E}(X|Y = y)f_Y(y)\mathrm{d}y.$$

1.2　Frequently Used Discrete and Continuous Distributions

In this part we consider some frequently used distributions and give their definitions and important characteristics. In addition to the formal description of the distributions, we will give appropriate mathematical models that lead to a given distribution. If the distribution function of a random variable is given as a function $F_X(x; a_1, ..., a_n)$ depending on a positive integer n and constants $a_1, ..., a_n$, then $a_1, ..., a_n$ are called the parameters of distribution function F_X.

1.2.1 Discrete Distributions

Bernoulli Distribution $Be(p)$, $0 \leq p \leq 1$. The PDF of random variable X with values $\{0, 1\}$ is called Bernoulli distribution if

$$p_k = \mathbf{P}\,(X = k) = \begin{cases} p, & \text{if } k = 1, \\ 1 - p, & \text{if } k = 0. \end{cases}$$

Expected value and variance: $\mathbf{E}\,(X) = p$, $\mathbf{D}^2\,(X) = p(1 - p)$;

Generating function: $G(z) = 1 - p + pz$;

Characteristic function: $\varphi(t) = 1 - p + pe^{\Im t}$.

Example Let X be the number of heads appearing in one toss of a coin, where

$$p = \mathbf{P}\,(\text{head appearing in a toss}).$$

Then X has a $Be(p)$ distribution.

Binomial Distribution $B(n, p)$. The distribution of a discrete random variable X with values $\{0, 1, ..., n\}$ is called binomial with parameters n and p, $0 < p < 1$, if its PDF is

$$p_k = \mathbf{P}\,(X = k) = \binom{n}{k} p^k (1 - p)^{n-k}, \quad k = 0, 1, ..., n.$$

Expected value and variance: $\mathbf{E}\,(X) = np$, $\mathbf{D}^2\,(X) = np(1 - p)$;

Generating function: $G(z) = (pz + (1 - p))^n$;

Characteristic function: $\varphi(t) = (1 + p(e^{\Im t} - 1))^n$.

Example Consider an experiment in which we observe that an event A with probability $p = \mathbf{P}\,(A)$, $0 < p < 1$, occurs (success) or not (failure). Repeating the experiment n times independently, define random variable X by the frequency of event A. Then the random variable has $B(n, p)$ PDF.

Note that if the $Be(n, p)$ random variables $X_1, ..., X_n$ are independent, then the random variable $X = X_1 + ... + X_n$ has a $B(n, p)$ distribution.

Polynomial Distribution (Also Referred to as Multinomial) The PDF of a random vector $X = (X_1, ..., X_k)^T$ taking values in the set $\{(n_1, ..., n_k) : n_i \geq 0, \ n_1 + ... + n_k = n\}$ is called polynomial with the parameters n and $p_1, ..., p_k$ ($p_i > 0$, $p_1 + ... + p_k = 1$) if X has a PDF

$$p_{n_1,...,n_k} = \mathbf{P}\,(X_1 = n_1, ..., X_k = n_k) = \frac{n!}{n_1!...n_k!} p_1^{n_1}...p_k^{n_k}.$$

Note that each coordinate variable X_i of random vector X has $B(p_i, n)$ binomial distribution whose expected value and variance are np_i and $np_i(1 - p_i)$.

Expected value $\qquad\qquad\qquad\qquad \mathbf{E}(X) = (np_1, ..., np_n)^T;$

Covariance matrix $\qquad\quad \mathbf{R} = (R_{ij})_{1 \le i,j \le k}$, where $R_{ij} = \begin{cases} np_i(1 - p_i), & \text{if } i = j, \\ np_i p_j, & \text{if } i \ne j; \end{cases}$

Characteristic function: $\qquad\qquad \varphi(t_1, ..., t_k) = (p_1 e^{\Im t_1} + ... + p_k e^{\Im t_k})^n.$

Example Let $A_1, ..., A_k$ be k disjoint events for which $p_i = \mathbf{P}(A_i) > 0, \ p_1 + ... + p_k = 1$. Consider an experiment with possible outcomes $A_1, ..., A_k$ and repeat it n times independently. Denote by X_i the frequency of event A_i in the series of n observations. Then the distribution of X is polynomial with the parameters n and $p_1, ..., p_k$.

Geometric Distribution The PDF of random variable X taking values in $\{1, 2, ...\}$ is called a geometric distribution with the parameter $p, \ 0 < p < 1$, if its PDF is

$$p_k = \mathbf{P}(X = k) = (1 - p)^{k-1}p, \ k = 1, 2, ...$$

Expected value and variance: $\qquad \mathbf{E}(X) = \frac{1}{p}, \quad \mathbf{D}^2(X) = \frac{1-p}{p^2};$

Generating function: $\qquad\qquad\qquad G(z) = \frac{pz}{1-(1-p)z};$

Characteristic function: $\qquad\qquad\quad \varphi(t) = \frac{p}{1-(1-p)e^{\Im t}}.$

Theorem 1.9 *If X has a geometric distribution, then X has a so-called **memoryless property**, that is, for all nonnegative integer numbers i, j the following relation holds:*

$$\mathbf{P}(X \ge i + j | X \ge i) = \mathbf{P}(X \ge j).$$

Proof It is easy to verify that for $k \ge 1$

$$\mathbf{P}(X \ge k) = \sum_{\ell=k}^{\infty} \mathbf{P}(X = \ell) = \sum_{\ell=k}^{\infty}(1 - p)^{\ell-1}p$$

$$= (1 - p)^{k-1}p \sum_{\ell=0}^{\infty}(1 - p)^{\ell} = (1 - p)^{k-1};$$

therefore,

$$\mathbf{P}(X \ge i + j | X \ge i) = \frac{\mathbf{P}(X \ge i + j, X \ge i)}{\mathbf{P}(X \ge i)}$$

$$= \frac{\mathbf{P}(X \ge i + j)}{\mathbf{P}(X \ge i)}$$

$$= \frac{(1-p)^{i+j-1}}{(1-p)^{i-1}} = (1-p)^j, \quad j = 0, 1, \dots.$$

□

Note that a geometric distribution is sometimes defined on the set $\{0, 1, 2, \dots\}$ instead of $\{1, 2, \dots, \}$; in this case, the PDF is determined by

$$p_k = (1-p)^k p, \quad k = 0, 1, 2, \dots.$$

Example Consider a sequence of experiments and observe whether an event A, $p = \mathbf{P}(A) > 0$, occurs (success) or does not (failure) in each step. If the event occurs in the kth step first, then define the random variable as $X = k$. In other words, let X be the number of Bernoulli trials of the first success. Then random variable X has a geometric distribution with the parameter p.

Negative Binomial Distribution The distribution of random variable X taking values in $\{0, 1, \dots\}$ is called a negative binomial distribution with the parameter p, $0 < p < 1$, if

$$p_k = \mathbf{P}(X = k + r) = \binom{r+k-1}{k}(1-p)^k p^r, \quad k = 0, 1, \dots.$$

Expected value and variance: $\quad \mathbf{E}(X) = r\frac{1}{p}, \quad \mathbf{D}^2(X) = r\frac{1-p}{p^2};$

Generating function: $\quad G(z) = \left(\frac{pz}{1-(1-p)z}\right)^r;$

Characteristic function: $\quad \varphi(t) = p^r \left(1 - (1-p)e^{3t}\right)^{-r}.$

Example Let p, $0 < p < 1$ and the positive integer r be two given constants. Suppose that we are given a coin that has a probability p of coming up heads. Toss the coin repeatedly until the rth head appears and define by X the number of tosses. Then random variable X has a negative binomial distribution with parameters (p, r).

Note that from this example it immediately follows that X has a geometric distribution with the parameter p when $r = 1$.

Poisson Distribution The PDF of a random variable X is called a Poisson distribution with parameter λ ($\lambda > 0$) if X takes values in $\{0, 1, \dots\}$ and

$$p_k = \mathbf{P}(X = k) = \frac{\lambda^k}{k!}e^{-\lambda}, \quad k = 0, 1, \dots.$$

Expected value and variance: $\quad \mathbf{E}(X) = \lambda, \quad \mathbf{D}^2(X) = \lambda;$

Generating function: $\quad G(z) = e^{\lambda(z-1)};$

Characteristic function: $\quad \varphi(t) = e^{\lambda(e^{3t}-1)}.$

The following theorem establishes that a binomial distribution can be approximated with a Poisson distribution with the parameter λ when the parameters (p, n) of binomial distribution satisfy the condition $np \to \lambda, n \to \infty$.

Theorem 1.10 *Consider a binomial distribution with the parameter (p, n). Assume that for a fixed constant λ, $\lambda > 0$, the convergence $np \to \lambda$, $n \to \infty$ holds; then the limit of probabilities satisfies the relation*

$$\binom{n}{k} p^k (1 - p)^{n-k} \to \frac{\lambda^k}{k!} e^{-\lambda}, \quad k = 0, 1, \dots.$$

Proof For any fixed $k \geq 0$ integer number we have

$$\binom{n}{k} p^k (1 - p)^{n-k} = \frac{(np)((n-1)p) \dots ((n-k+1)p)}{k!} e^{(n-k) \log(1-p)}.$$

Since $np \to \lambda$, $n \to \infty$, therefore $p \to 0$, and we obtain

$$\frac{(np)((n-1)p) \dots ((n-k+1)p)}{1 \cdot 2 \cdot \dots \cdot k} \to \frac{\lambda^k}{k!}, \quad np \to \lambda.$$

On the other hand, if $p \to 0$, then we get the asymptotic relation $\log(1 - p) = -p + o(p)$. Consequently,

$$(n - k) \log(1 - p) = -(n - k)(p + o(p)) \to -\lambda, \quad np \to \lambda, n \to \infty;$$

therefore, using the last two asymptotic relations, the assertion of the theorem immediately follows. □

1.2.2 Continuous Distributions

Uniform Distribution Let a, b $(a < b)$ be two real numbers. The distribution of random variable X is called uniform on the interval (a, b) if its PDF is given by

$$f(x) = \begin{cases} \frac{1}{b-a}, & \text{if } x \in (a, b), \\ 0, & \text{if } x \notin (a, b). \end{cases}$$

Expected value and variance: $\mathbf{E}(X) = \frac{a+b}{2}, \quad \mathbf{D}^2(X) = \frac{(b-a)^2}{12}$;

Characteristic function: $\varphi(t) = \frac{1}{b-a} \frac{e^{\Im t b} - e^{\Im t a}}{\Im t}$.

Note that if X has uniform distribution on the interval (a, b), then the random variable $Y = \frac{X-a}{b-a}$ is distributed uniformly on the interval $(0, 1)$.

Exponential Distribution Exp(λ), $\lambda > 0$ The distribution of a random variable X is called exponential with the parameter λ, $\lambda > 0$, if its PDF

$$f(x) = \begin{cases} \lambda e^{-\lambda x}, & \text{if } x > 0, \\ 0, & \text{if } x \le 0. \end{cases}$$

Expected value and variance: $\mathbf{E}(X) = \frac{1}{\lambda}$, $\quad \mathbf{D}^2(X) = \frac{1}{\lambda^2}$;

Characteristic function: $\varphi(t) = \frac{\lambda}{\lambda - \Im t}$.

The Laplace and Laplace–Stieltjes transform of the density and distribution function of an Exp(λ) distribution are determined as

$$\mathbf{E}\left(e^{-sX}\right) = f^*(s) = F^\sim(s) = \frac{\lambda}{s + \lambda}.$$

The exponential distribution, similar to the geometric distribution, has the memoryless property.

Theorem 1.11 *For arbitrary constants $t, s > 0$ the relation*

$$\mathbf{P}(X > t + s \mid X > t) = \mathbf{P}(X > s)$$

holds.

Proof It is clear that

$$\mathbf{P}(X > t + s \mid X > t) = \frac{\mathbf{P}(X > t + s, X > t)}{\mathbf{P}(X > t)}$$

$$= \frac{\mathbf{P}(X > t + s)}{\mathbf{P}(X > t)} = \frac{e^{-\lambda(t+s)}}{e^{-\lambda t}} = e^{-\lambda s}.$$

$\square$

Hyperexponential Distribution Let the PDF of random variable X be a mixture of exponential distributions with the parameters $\lambda_1, \ldots, \lambda_n$ and with weights $a_1, \ldots, a_n$ ($a_k > 0$, $a_1 + \ldots + a_n = 1$). Then the PDF

$$f(x) = \begin{cases} \sum_{k=1}^{n} a_k \lambda_k e^{-\lambda_k x}, & \text{if } x > 0, \\ 0, & \text{if } x \le 0, \end{cases}$$

of random variable X is called hyperexponential.

Expected value and variance: $\mathbf{E}(X) = \sum_{k=1}^{n} \frac{a_k}{\lambda_k}$, $\mathbf{D}^2(X) = 2 \sum_{k=1}^{n} \frac{a_k}{\lambda_k^2} - \left(\sum_{k=1}^{n} \frac{a_k}{\lambda_k} \right)^2$;

Characteristic function: $\varphi(t) = \sum_{k=1}^{n} a_k \frac{\lambda_k}{\lambda_k - \Im t}$.

Denote by $\Gamma(x) = \int_0^{\infty} y^{x-1} e^{-y} dy$, $x > -1$ the well-known **gamma function** Γ in analysis, which is necessary for the definition of gamma distribution.

Gamma Distribution Gamma(α, λ), $\alpha, \lambda > 0$ The distribution of a random variable X is called **gamma distribution** with the parameters $\alpha, \lambda > 0$, if its PDF is

$$f(x) = \begin{cases} \frac{\lambda^{\alpha}}{\Gamma(\alpha)} x^{\alpha-1} e^{-\lambda x}, & \text{if } x > 0, \\ 0, & \text{if } x \le 0. \end{cases}$$

Expected value and variance: $\mathbf{E}(X) = \frac{\alpha}{\lambda}$, $\mathbf{D}^2(X) = \frac{\alpha}{\lambda^2}$;

Characteristic function: $\varphi(t) = \left(\frac{\lambda}{\lambda - \Im t} \right)^{\alpha}$.

Remark 1.15 A gamma distribution with the parameters $\alpha = n$, $\lambda = n\mu$ is called an **Erlang distribution.**

Remark 1.16 If the independent identically distributed random variables $X_1, X_2, \ldots$ have an exponential distribution with the parameter λ, then the distribution of the sum $Z = X_1 + \ldots + X_n$ is gamma distribution with the parameter (n, λ). This relation is easy to see because the characteristic function of an exponential distribution with the parameter λ is $(1 - \Im t / \lambda)^{-1}$; then the characteristic function of its nth convolution power is $(1 - \Im t / \lambda)^{-n}$, which equals the characteristic function of a Gamma(n, λ) distribution.

Beta Distribution Beta(a, b), $a, b > 0$ The distribution of random variable X is called a beta distribution if its PDF is

$$f(x) = \begin{cases} \frac{\Gamma(a+b)}{\Gamma(a)\Gamma(b)} x^{a-1} (1-x)^{b-1}, & \text{if } x \in (0, 1), \\ 0, & \text{if } x \notin (0, 1). \end{cases}$$

Expected value and variance: $\mathbf{E}(X) = \frac{a}{a+b}$, $\mathbf{D}^2(X) = \frac{ab}{(a+b)^2(a+b+1)}$;

Characteristic function in the form of power series: $\varphi(t) = \frac{\Gamma(\alpha+\beta)}{\Gamma(\alpha)} \sum_{k=0}^{\infty} \frac{(\Im t)^k}{k!} \frac{\Gamma(\alpha+k)}{\Gamma(\alpha+\beta+k)}$.

Gaussian (Also Called Normal) Distribution $N(\mu, \lambda)$, $-\infty < \mu < \infty$, $0 < \sigma < \infty$ The distribution of random variable X is called Gaussian with the parameters (μ, σ), if it has PDF

$$f(x) = \frac{1}{\sqrt{2\pi}\sigma}e^{-(x-\mu)^2/2\sigma^2}, \quad -\infty < x < \infty.$$

Expected value and variance: $\mu = \mathbf{E}(X)$ and $\sigma^2 = \mathbf{D}^2(X)$;

Characteristic function: $\varphi(t) = \exp\left\{\Im\mu t - \frac{\sigma^2}{2}t^2\right\}.$

The $N(0, 1)$ distribution is usually called a standard Gaussian or standard normal distribution, and its PDF equals to

$$f(x) = \frac{1}{\sqrt{2\pi}}e^{-x^2/2}, \quad -\infty < x < \infty.$$

It is easy to verify that if a random variable has an $N(\mu, \sigma)$ distribution, then the centered and linearly normed random variable $Y = (X - \mu)/\sigma$ has a standard Gaussian distribution.

Multidimensional Gaussian (Normal) Distribution $N(\mu, \mathbf{R})$ Let $\mathbf{Z} = (Z_1, ..., Z_n)$ be an n dimensional random vector whose coordinates $Z_1, ..., Z_n$ are independent and have a standard $N(0, 1)$ Gaussian distribution. Let $\mathbf{V} \in \mathbb{R}^{m \times n}$ be an $(m \times n)$ matrix and $\mu = (\mu_1, ..., \mu_m)^T \in \mathbb{R}^m$ an m-dimensional vector. Then the distribution of the m-dimensional random vector X defined by the equation $\mathbf{X} = \mathbf{VZ} + \mu$ is called an m-dimensional Gaussian distribution.

Expected value and variance matrix:

$$\mathbf{E}(\mathbf{X}) = \mu_{\mathbf{X}} = \mu \text{ and } \mathbf{D}^2(\mathbf{X}) = \mathbf{R_X} = \mathbf{E}\left((\mathbf{X} - \mu)(\mathbf{X} - \lambda)^T\right) = \mathbf{VV}^T;$$

Characteristic function:

$$\varphi(\mathbf{t}) = \exp\left\{\Im\mathbf{t}^T\mu - \frac{1}{2}\mathbf{t}^T\mathbf{R_X}\mathbf{t}\right\}, \text{ where } \mathbf{t} = (t_1, ..., t_m)^T \in \mathbb{R}^m.$$

If $\mathbf{V}$ is a nonsingular quadratic matrix ($m = n$ and $\det \mathbf{V} \neq 0$), then the random vector $\mathbf{X}$ has a density in the form

$$f_{\mathbf{X}}(\mathbf{x}) = \frac{1}{(2\pi)^{n/2}(\det \mathbf{R_X})^{1/2}}\exp\left\{-\frac{1}{2}(\mathbf{x} - \mu)^T\mathbf{R_X}^{-1}(\mathbf{x} - \mu)\right\}, \quad \mathbf{x} = (x_1, ..., x_n)^T \in \mathbb{R}^n$$

Example If the random vector $\mathbf{X} = (X_1, X_2)^T$ has a two-dimensional Gaussian distribution with expected value $\mu = (\mu_1, \mu_2)^T$ and covariance matrix

$$\mathbf{R_X} = \begin{bmatrix} a & b \\ b & c \end{bmatrix},$$

then its PDF has the form

$$f_{\mathbf{X}}(\mathbf{x}) = \frac{\sqrt{ac-b^2}}{2\pi} \exp\left\{ -\frac{1}{2}[a(x_1-\mu_1)^2 + 2b(x_1-\mu_1)(x_2-\mu_2) + c(x_2-\mu_2)^2] \right\},$$

where a, b, c, μ_1, μ_2 are constants satisfying the conditions $a > 0$, $c > 0$, and $b^2 < ac$.

Note that the marginal distributions of random variables X_1 and X_2 are $N(\mu_1, \sigma_1)$ and $N(\mu_2, \sigma_2)$ Gaussian, respectively, where

$$\sigma_1 = \sqrt{\frac{a}{ac-b^2}}, \quad \sigma_2 = \sqrt{\frac{c}{ac-b^2}} \quad \text{and} \quad b = \mathbf{Cov}\,(X_1, X_2).$$

Distribution Functions Associated with Gaussian Distributions Let $Z, Z_1, Z_2, \ldots$ be independent random variables whose distributions are standard Gaussian, i.e., with the parameters $(0, 1)$. There are many distributions, for example, the χ^2 and the logarithmically normal distributions defined subsequently (further examples are the frequently used t, F, and Wishart distributions in statistics [3]), that can be given as distributions of appropriately chosen functions of random variables $Z, Z_1, Z_2, \ldots$.

χ^2**-Distribution** The distribution of the random variable $X = Z_1^2 + \ldots + Z_n^2$ is called a χ^2-distribution with parameter n. The PDF

$$f_n(x) = \begin{cases} \frac{1}{2^{n/2}\Gamma(n/2a)} x^{n/2-1} e^{-x/2}, & \text{if } x > 0, \\ 0, & \text{if } x \leq 0. \end{cases}$$

Expected value and variance: $\mathbf{E}\,(X) = n$, $\quad \mathbf{D}^2\,(X) = 2n$;
Characteristic function: $\quad \varphi(t) = (1 - 2\Im t)^{-n/2}$.

Logarithmic Gaussian (Normal) Distribution If random variable Z has an $N(\mu, \sigma)$ Gaussian distribution, then the distribution of random variable $X = e^Z$ is called a logarithmic Gaussian (normal) distribution. The PDF is

$$f(x) = \begin{cases} \frac{1}{\sqrt{2\pi}\sigma x} \exp\left\{ \frac{(\log x - \mu)^2}{2\sigma^2} \right\}, & \text{if } x > 0, \\ 0, & \text{if } x \leq 0. \end{cases}$$

Expected value and variance: $\quad \mathbf{E}\,(X) = e^{\sigma^2/2+\mu}$, $\quad \mathbf{D}^2\,(X) = e^{\sigma^2/2+\mu}\left(e^{e^2} - 1\right)$.

For this and the next distributions the characteristic function does not have a closed form expression.

Weibull Distribution   The Weibull distribution is a generalization of the exponential distribution for which the behavior of the tail distribution is modified by a positive constant k, as follows

$$F(x) = \begin{cases} 1 - e^{-(x/\lambda)^k}, & \text{if } x > 0, \\ 0, & \text{if } x \le 0; \end{cases}$$

$$f(x) = \begin{cases} \left(\frac{k}{\lambda}\right)\left(\frac{x}{\lambda}\right)^{k-1} e^{-(x/\lambda)^k}, & \text{if } x > 0, \\ 0, & \text{if } x \le 0. \end{cases}$$

Expected value and variance:

$$\mathbf{E}(X) = \lambda \Gamma(1 + 1/k), \quad \mathbf{D}^2(X) = \lambda^2 \left(\Gamma(1 + 2/k) - \Gamma^2(1 + 1/k) \right).$$

Pareto Distribution   Let c and λ be positive numbers. The density function and the PDF of a Pareto distribution are defined as follows:

$$F(x) = \begin{cases} 1 - \left(\frac{x}{c}\right)^{-\lambda}, & \text{if } x > c, \\ 0, & \text{if } x \le c; \end{cases}$$

$$f(x) = \begin{cases} \left(\frac{\lambda}{c}\right)\left(\frac{x}{c}\right)^{-\lambda-1}, & \text{if } x > c, \\ 0, & \text{if } x \le c. \end{cases}$$

Since the PDF of the Pareto distribution is a simple power function in consequence of this property, it tends to zero with polynomial order as x goes to infinity and the nth moment exists if and only if $n < \lambda$.

Expected value (if $\lambda > 1$) and variance (if $\lambda > 2$):

$$\mathbf{E}(X) = \frac{c\lambda}{\lambda - 1}, \quad \mathbf{D}(X) = \frac{c^2 \lambda}{(\lambda - 1)^2 (\lambda - 2)}.$$

Finally, we note that the set of nonnegative continuous distributions are often characterized by the normalized measure of variance the **coefficient of variation** for which the following inequalities hold:

Exponential distribution:	$CV = 1$,
Hyper exponential distribution:	$CV \ge 1$,
Erlang distribution:	$CV \le 1$.

1.3   Limit Theorems

1.3.1   Convergence Notions

There are many convergence notions in the theory of analysis, for example, pointwise convergence, uniform convergence, and convergences defined by various metrics. In the theory of probability, several kinds of convergences are also used that are related to the sequences of random variables or to their sequence of distribution functions. The following notion is the so-called weak convergence of distribution functions.

Definition 1.24 The sequence of distribution functions F_n, $n = 1, 2, \ldots$ **weakly converges** to the distribution function F (abbreviated $F_n \overset{w}{\to} F$, $n \to \infty$) if the convergence $F_n(x) \to F(x)$, $n \to \infty$, holds in all continuity points of F.

If the distribution function F is continuous, then the convergence $F_n \overset{w}{\to} F$, $n \to \infty$ holds if and only if $F_n(x) \to F(x)$, $n \to \infty$ for all $x \in \mathbb{R}$. The weak convergence of the sequence F_n, $n = 1, 2, \ldots$ is equivalent to the condition that the convergence

$$\int\limits_{-\infty}^{\infty} g(x) \mathrm{d}F_n(x) \to \int\limits_{-\infty}^{\infty} g(x) \mathrm{d}F(x)$$

is true for all bounded and continuous functions g.

In addition, the weak convergence of distribution functions can be given with the help of an appropriate metric in the space $\mathbb{F} = \{F\}$ of all distribution functions. Let G and H be two distribution functions (i.e., $G, H \in \mathbb{F}$) and define the **Levy metric** (see [7]) as follows:

$$L(G, H) = \inf\{\varepsilon : G(x) \le H(x + \varepsilon) + \varepsilon, \ H(x) \le G(x + \varepsilon) + \varepsilon, \ \text{for all } x \in \mathbb{R}\}.$$

Then it can be proved that the weak convergence $F_n \overset{w}{\to} F$, $n \to \infty$, of the distribution functions F, F_n, $n = 1, 2, \ldots$, holds if and only if $\lim\limits_{n \to \infty} L(F_n, F) = 0$.

The most frequently used convergence notions in probability theory for a sequence of random variables are the convergence in distribution, convergence in probability, convergence with probability 1, or almost surely (a.s.), and convergence in mean square (convergence in L_2), which will be introduced subsequently. In cases of the last three convergences, it is assumed that the random variables are defined on a common probability space $(\Omega, \mathscr{A}, \mathbf{P})$.

Definition 1.25 The sequence of random variables $X_1, X_2, \ldots$ **converges in distribution** to a random variable X (abbreviated $X_n \overset{d}{\to} X$, $n \to \infty$) if their distribution functions satisfy the weak convergence

$$F_{X_n} \overset{w}{\to} F_X, \ n = 1, 2, ..$$

Definition 1.26 The sequence of random variables $X_1, X_2, \dots$ **converges in probability** to a random variable X ($X_n \overset{P}{\to} X, \ n \to \infty$) if the convergence

$$\lim_{n \to \infty} \mathbf{P}\left(|X_n - X| > \varepsilon\right) = 0$$

holds for all positive constant ε.

Definition 1.27 The random variables $X_1, X_2, \dots$ **converge with probability** 1 (or **almost surely**) to a random variable X (abbreviated $X_n \overset{a.s.}{\to} X, \ n \to \infty$) if the condition

$$\mathbf{P}\left(\lim_{n \to \infty} X_n = X\right) = 1$$

holds.

The limit $\lim_{n \to \infty} X_n = X$ exists with probability 1 if there are defined random variables with probability 1 $X' = \limsup_{n \to \infty} X_n$ and $X''(\omega) = \liminf_{n \to \infty} X_n$ for which the relation

$$\mathbf{P}\left(X'(\omega) = X''(\omega)\right) = \mathbf{P}\left(X'(\omega) = X(\omega)\right) = \mathbf{P}\left(X''(\omega) = X(\omega)\right) = 1$$

is true. This means that there is an event $A \in \mathscr{A}, \mathbf{P}(A) = 0$, such that the equality

$$X'(\omega) = X''(\omega) = X(\omega), \ \omega \in \Omega \setminus A$$

holds.

Theorem 1.12 ([6]) *The convergence* $\lim_{n \to \infty} X_n = X$ *with probability one is true if and only if for all* $\varepsilon > 0$

$$\mathbf{P}\left(\sup_{k \geq n}|X_k - X| > \varepsilon\right) \to 0, \ n \to \infty.$$

Definition 1.28 Let $X_n, \ n \geq 1$ and X be random variables with finite variance. The sequence $X_1, X_2, \dots$ **converges in mean square** to random variable X (abbreviated $X_n \overset{L_2}{\to} X, \ n \to \infty$) if

$$\mathbf{E}\left(|X_n - X|^2\right) \to 0, \ n \to \infty.$$

This type of convergence is often called as L_2 convergence of random variables.

The enumerated convergence notions are not equivalent to each other, but we can mention several connections between them. The convergence in distribution follows from all the others. The convergence in probability follows from the convergence with probability 1 and from the convergence in mean square. It can be proved that if the sequence $X_1, X_2, \dots$ is convergent in probability to the random variable X, then there exists a subsequence $X_{n_1}, X_{n_2}, \dots$ such that it converges with probability 1 to the random variable X.

1.3.2   The Laws of Large Numbers

The intuitive introduction of probability implicitly uses the limit behavior of the average

$$\overline{S}_n = \frac{X_1 + \dots + X_n}{n}, \; n = 1, 2, \dots$$

of independent identically distributed random variables $X_1, X_2, \dots$ The main question is: under what condition does the sequence $\overline{S}_n$ converges to a constant μ in probability (weak law of large numbers) or with probability 1 (strong law of large numbers) as n goes to infinity?

Consider an experiment in which we observe that an event A occurs or not. Repeating the experiment n times independently, define the frequency of event A by $S_n(A)$ and the relative frequency by $\overline{S}_n(A)$.

Theorem 1.13 (Bernoulli) *The relative frequency of an event A tends in probability to the probability of the event $p = \mathbf{P}(A)$, that is, for all $\varepsilon > 0$ the relation*

$$\lim_{n \to \infty} \mathbf{P}\left(\left|\overline{S}_n(A) - p\right| > \varepsilon\right) = 0$$

holds.

If we introduce the notation

$$X_i = \begin{cases} 1, & \text{if the } i\text{-th outcome in } A, \\ 0, & \text{otherwise,} \end{cases}$$

then the assertion of the last theorem can be formulated as follows

$$\overline{S}_n = \frac{X_1 + \dots + X_n}{n} \xrightarrow{p} p, \; n \to \infty,$$

which is a simple consequence of the Chebishev inequality because the X_i are independent and identically distributed and $\mathbf{E}(X_i) = p = \mathbf{P}(A)$, $\mathbf{D}^2(X_i) =$

$p(1 - p)$, $i = 1, 2, ...$ This result can be generalized without any difficulties as follows.

Theorem 1.14 *Let $X_1, X_2, ...$ be independent and identically distributed random variables with common expected value μ and finite variance σ^2. Then the convergence in probability*

$$\overline{S}_n = \frac{X_1 + ... + X_n}{n} \overset{p}{\to} \mu, \, n \to \infty.$$

is true.

Proof Example 1.3, which is given after the proof of Chebishev inequality, shows that for all $\varepsilon > 0$ the inequality

$$\mathbf{P}\left(\left|\tfrac{X_1 + ... + X_n}{n} - \mu\right| \geq \varepsilon\right) \leq \frac{\sigma^2}{n\varepsilon^2}$$

is valid. From this, the convergence in probability $\overline{S}_n \overset{p}{\to} \mu$, $n \to \infty$ follows. It is not difficult to see that the convergence in L_2 is also true, i.e., $\overline{S}_n \overset{L_2}{\to} \mu$, $n \to \infty$.

$\square$

It should be noted that the inequality $\mathbf{P}\left(\left|\tfrac{X_1 + ... + X_n}{n} - \mu\right| \geq \varepsilon\right) \leq \frac{\sigma^2}{n\varepsilon^2}$, which guarantees the convergence in probability, gives an upper bound for the probability $\mathbf{P}\left(\left|\tfrac{X_1 + ... + X_n}{n} - \mu\right| \geq \varepsilon\right)$ also.

The Kolmogorov strong law of large number gives a necessary and sufficient condition for the convergence with probability 1 (see, e.g., [1, 5]).

Theorem 1.15 (Kolmogorov) *If the sequence of random variables $X_1, X_2, ...$ is independent and identically distributed, then the convergence*

$$\frac{X_1 + ... + X_n}{n} \overset{a.s.}{\to} \mu, \, n \to \infty$$

holds for a constant μ if and only if the random variables X_i have finite expected value and $\mathbf{E}(X_i) = \mu$.

Corollary 1.1 *If $\overline{S}_n(A)$ defines the relative frequency of an event A occurring in n independent experiments, then the Bernoulli law of large numbers*

$$\overline{S}_n(A) \overset{p}{\to} p = \mathbf{P}(A), \, n \to \infty$$

is valid. By the Kolmogorov strong law of large numbers, this convergence is true with probability 1 also, that is,

$$\overline{S}_n(A) \overset{a.s.}{\to} p = \mathbf{P}(A), \, n \to \infty.$$

1.3.3   Central Limit Theorem, Lindeberg–Feller Theorem

The basic problem of central limit theorems is as follows. Let $X_1, X_2, \ldots$ be independent and identically distributed random variables with a common distribution function $F_X(x)$. The question is, under what conditions does a sequence of constants μ_n and σ_n, $\sigma_n \neq 0$, $n = 1, 2, \ldots$ exist such that the sequence of centered and linearly normed sums

$$\overline{S}_n = \frac{X_1 + \ldots + X_n - \mu_n}{\sigma_n}, \; n = 1, 2, \ldots \tag{1.11}$$

converges in distributions

$$F_{\overline{S}_n} \xrightarrow{w} F, \; n \to \infty$$

and have a nondegenerate limit distribution function F? A distribution function $F(x)$ is nondegenerate if there is no point $x_0 \in \mathbb{R}$ satisfying the condition $F(x_0) - F(x_0-) = 1$, that is, the distribution does not concentrate at one point.

Theorem 1.16 *If the random variables $X_1, X_2, \ldots$ are independent and identically distributed with finite expected value $\mu = \mathbf{E}(X_1)$ and variance $\sigma^2 = \mathbf{D}^2(X_1) > 0$, then*

$$\mathbf{P}\left(\tfrac{X_1 + \ldots + X_n - n\mu}{\sqrt{n}\sigma} \leq x\right) \to \Phi(x) = \int\limits_{-\infty}^{x} \frac{1}{\sqrt{2\pi}} e^{-u^2/2} du, \; n \to \infty$$

holds for all $x \in \mathbb{R}$, where the function $\Phi(x)$ denotes the distribution function of standard normal random variables.

If the random variables $X_1, X_2, \ldots$ are independent but not necessarily identically distributed, then a general, so-called Lindeberg–Feller theorem is valid (see, e.g., [1, 5]).

Theorem 1.17 *Let $X_1, X_2, \ldots$ be independent random variables whose variances are finite. Denote*

$$\mu_n = \mathbf{E}(X_1) + \ldots + \mathbf{E}(X_n), \quad \sigma_n = \sqrt{\mathbf{D}^2(X_1) + \ldots + \mathbf{D}^2(X_n)}, \quad n = 1, 2, \ldots$$

The limit

$$\mathbf{P}\left(\tfrac{X_1 + \ldots + X_n - \mu_n}{\sigma_n} \leq x\right) \to \Phi(x), \; n \to \infty$$

is true for all $x \in \mathbb{R}$ if and only if the Lindeberg–Feller condition holds:

$$\lim_{n\to\infty} \max_{1\le j\le n} \frac{1}{\sigma_j^2} \mathbf{E}\left(X_j^2 \mathscr{I}_{\{|X_j|>\varepsilon\sigma_n\}}\right) = 0, \ x \in \mathbb{R}, \ \varepsilon > 0,$$

where $\mathscr{I}_{\{\}}$ denotes the indicator variable.

1.3.4 Infinitely Divisible Distributions and Convergence to the Poisson Distribution

There are many practical problems for which model (1.11) and results related to it are not satisfactory. The reason is that the class of possible limit distributions is insufficiently large; for instance, it does not consist of discrete distributions. An example of this is a Poisson distribution, which is an often-used distribution in queuing theory. The general notions and results of present section may be found, e.g., in [1, 2, 5].

As a generalization of model (1.11), consider the sequence of series of random variables (sometimes called a sequence of random variables of triangular arrays)

$$\left\{X_{n,1}, ..., X_{n,k_n}\right\}, \ n = 1, 2, ..., \ k_n \to \infty$$

satisfying the following conditions for all fixed positive integers n:

1. The random variables $X_{n,1}, ..., X_{n,k_n}$ are independent.
2. The random variables $X_{n,1}, ..., X_{n,k_n}$ are **infinitesimal** (in other words, **asymptotically negligible**) if the limit for all $\varepsilon > 0$

$$\lim_{n\to\infty} \max_{1\le j\le k_n} \mathbf{P}\left(\left|X_{n,j}\right| > \varepsilon\right) = 0$$

holds.

Considering the sums of series of random variables

$$S_n = X_{n,1} + ... + X_{n,k_n}, \ n = 1, 2, ...$$

the class of possible limit distributions (the so-called infinitely divisible distributions) is already a sufficiently large class containing, for example, a Poisson distribution.

Definition 1.29 A random variable X is called **infinitely divisible** if it can be given in the form

$$X \stackrel{d}{=} X_{n,1} + ... + X_{n,n}$$

for every $n = 1, 2, \ldots$, where the random variables $X_{n,1}, \ldots, X_{n,n}$ are independent and identically distributed.

Infinitely divisible distributions (to which, for example, the normal and Poisson distributions belong) can be given with the help of their characteristic functions.

Theorem 1.18 *If random variable X is infinitely divisible, then its characteristic function has the form* (***Lévy-Khinchin canonical form***)

$$\log f(t) = \Im t \mu - \frac{\sigma^2}{2} t^2 + \int\limits_{-\infty}^{0} \left(e^{\Im t x} - 1 - \frac{\Im t x}{1 + x^2} \right) dL(x)$$

$$+ \int\limits_{0}^{\infty} \left(e^{\Im t x} - 1 - \frac{\Im t x}{1 + x^2} \right) dR(x),$$

where the functions L and R satisfy the following conditions:

(a) μ *and* σ *(* $\sigma \geq 0$ *) are real constants.*

(b) $L(x)$, $x \in (-\infty, 0)$ *and* $R(x)$, $x \in (0, \infty)$ *are monotonically increasing functions on the intervals* $(-\infty, 0)$ *and* $(0, \infty)$, *respectively.*

(c) $L(-\infty) = R(\infty) = 0$ *and the inequality condition*

$$\int\limits_{-\infty}^{0} x^2 dL(x) + \int\limits_{0}^{\infty} x^2 dR(x) < \infty$$

holds.

If an infinitely divisible distribution has finite variance, then its characteristic function can be given in a more simple form (**Kolmogorov formula**):

$$\log f(t) = \Im t \mu + \int\limits_{-\infty}^{\infty} \left(e^{\Im t x} - 1 - \Im t x \right) \frac{1}{x^2} dK(x),$$

where μ is a constant and $K(x)$ ($K(-\infty) = 0$) is a monotonically nondecreasing function.

As special cases of the Kolmogorov formula, we get the normal and Poisson distributions.

(a) An infinitely divisible distribution is normal with parameters (μ, σ) if the function $K(x)$ is defined as

$$K(x) = \begin{cases} 0, & \text{if } x \leq 0, \\ \sigma^2, & \text{if } x > 0. \end{cases}$$

Then the characteristic function is

$$\log f(t) = \Im t\mu - \frac{\sigma^2}{2}t^2.$$

(b) An infinitely divisible distribution is Poisson with the parameter λ ($\lambda > 0$) if $\mu = \lambda$ and the function $K(x)$ is defined as

$$K(x) = \begin{cases} 0, & \text{if } x \le 1, \\ \lambda, & \text{if } x > 1. \end{cases}$$

In this case the characteristic function can be given as follows

$$\log f(t) = \Im t\mu + \int\limits_{-\infty}^{\infty} \left(e^{\Im tx} - 1 - \Im tx\right)\frac{1}{x^2}dK(x) = \lambda(e^{\Im t} - 1).$$

The following theorem gives an answer to the question of the conditions under which the limit distribution of sums of independent infinitesimal random variables is Poisson. This result will be used later when considering sums of independent arrival processes of queues.

Theorem 1.19 (Gnedenko, Marczinkiewicz) *Let* $\left\{X_{1,n}, ..., X_{k_n,n}\right\}$, $n = 1, 2, ...$ *be a sequence of series of independent infinitesimal random variables. The sequence of distributions of sums*

$$X_n = X_{n1} + ... + X_{n,k_n}, \quad n \ge 1$$

converges weakly to a Poisson distribution with the parameter λ ($\lambda > 0$) *as* $n \to \infty$ *if and only if the following conditions hold for all* ε ($0 < \varepsilon < 1$):

(A) $\displaystyle\sum_{j=1}^{k_n} \int\limits_{\mathbb{R}_\varepsilon} dF_{nj}(x) \to 0,$

(B) $\displaystyle\sum_{j=1}^{k_n} \int\limits_{|x-1|<\varepsilon} dF_{nj}(x) \to \lambda,$

(C) $\displaystyle\sum_{j=1}^{k_n} \int\limits_{|x|<\varepsilon} dF_{nj}(x) \to 0,$

(D) $\displaystyle\sum_{j=1}^{k_n} \left[\int\limits_{|x|<\varepsilon} x^2 dF_{nj}(x) - \left(\int\limits_{|x|<\varepsilon} x dF_{nj}(x) \right)^2 \right] \to 0,$

where $F_{nj}(x) = \mathbf{P}\left(X_{nj} \le x\right)$ *and* $\mathbb{R}_\varepsilon = \mathbb{R} \setminus (\{|x| < \varepsilon\} \cup \{|x - 1| < \varepsilon\}).$

Note that the conditions (A) and (B) guarantee the convergence of Poisson part to the appropriate Poisson distribution of the limit, (C) means that there is no centralization, and from (D) it follows that the limit distribution does not contain a Gaussian part.

1.4 Exercises

Exercise 1.1 Let X be a nonnegative random variable with CDF F_X. Given $0 \leq t \leq X$ [$\mathbf{P}(X > t) \neq 0$], find the CDF of residual life time X.

Exercise 1.2 Let X and Y be independent random variables with a Poisson distribution of parameters λ and μ, respectively. Verify that

(a) The sum $X + Y$ has a Poisson distribution with the parameter $\lambda + \mu$.
(b) For any nonnegative integers $m \leq n$ the conditional distribution $\mathbf{P}(X = m \mid X + Y = n)$ is binomial with the parameter $(n, \frac{\lambda}{\lambda+\mu})$, i.e.

$$\mathbf{P}(X = m \mid X + Y = n) = \binom{m}{n} \left(\frac{\lambda}{\lambda + \mu}\right)^m \left(1 - \frac{\lambda}{\lambda + \mu}\right)^{n-m}.$$

Exercise 1.3 Let X and Y be independent random variables having a uniform distribution on the interval $(0, 1)$ and an exponential distribution with the parameter 1, respectively. Find the probability (concrete number) that $X < Y$.

Exercise 1.4 Divide the interval $(0, 1)$ into three parts with two independently and randomly chosen points U_1 and U_2 of the interval $(0, 1)$. Find the probability of the event A that the three parts can determine a triangle.

Exercise 1.5 Show that for a nonnegative random variable X with a finite n-th ($n \geq 1$) moment it is true that $\mathbf{E}(X^n) = \int\limits_0^\infty \mathbf{P}(x \leq X)nx^{n-1}dx$.

Exercise 1.6 Let X and Y be independent random variables with a uniform distribution on the interval $(0, 1)$. Find the quantities

(a) $\mathbf{E}(|X - Y|), \mathbf{D}^2(|X - Y|)$,
(b) $\mathbf{P}(|X - Y|) > \frac{1}{2}$.

Exercise 1.7 Let X and Y be independent random variables having an exponential distribution with the parameters λ and μ, respectively.

(a) Determine the density function of the random variable $Z = X + Y$.
(b) Find the density function of the random variable $W = \min(X, Y)$.

Exercise 1.8 Let $X_1, \ldots, X_n$ be independent random variables having an exponential distribution with the parameter λ.
Find the expected values of the random variables $V_n = \max(X_1, \ldots, X_n)$, and $W_n = \min(X_1, \ldots, X_n)$.

Exercise 1.9 Let X and Y be independent random variables with density functions $f_X(x)$ and $f_Y(x)$, respectively. Determine the conditional expected value $E(X \mid X < Y)$.

Exercise 1.10 Determine the conditional expectation $E(X \mid Y = y)$ and $E(X \mid Y)$ if the joint PDF of the random variables X and Y has the form

(a) $f_{X,Y}(x, y) = \begin{cases} 2, & \text{if } 0 < x, y \text{ and } x + y < 1, \\ 0, & \text{otherwise;} \end{cases}$

(b) $f_{X,Y}(x, y) = \begin{cases} 3(x + y), & \text{if } 0 < x, y \text{ and } x + y < 1, \\ 0, & \text{otherwise.} \end{cases}$

Exercise 1.11 Let $X_1, X_2, \ldots$ be independent random variables with an exponential distribution of the parameter λ. Let N be a geometrically distributed random variable with the parameter p $[p_k = P(N = k) = p(1 - p)^k, \ k = 1, 2, \ldots]$, which does not depend on random variables $(X_1, X_2, \ldots)$. Prove that the sum $Y = X_1 + \ldots + X_N$ has an exponential distribution with the parameter $p\lambda$.

Exercise 1.12 Consider the distribution function of the sum Y_{40} of independent random variables $X_1, \ldots, X_{40}$ having an exponential distribution with the parameter 1. Give an estimate for the probability $p = P\left(\dfrac{|Y_{40} - E(Y_{40})|}{D(Y_{40})} > 0.05 \right)$ calculated with the help of the central limit theorem. We can numerically calculate this probability because the random variable Y_{40} has a gamma distribution with the parameter $(40, 1)$. Using this fact, what result can we obtain for the considered probability? (On the numerical calculation of the gamma distribution see, for example, NIST: National Institute of Standards and Technology. Digital library of mathematical functions. http://dlmf.nist.gov)

1.5 Solutions

Solution 1.1 Suppose $0 \le t \le z$, then

$$P(X \le z \mid X > t) = \frac{P(X \le z, X > t)}{P(X > t)} = \frac{P(t < X \le z)}{P(X > t)}$$

$$= \frac{P(X \le z) - P(X \le t)}{1 - P(X \le t)} = \frac{F_X(z) - F_X(t)}{1 - F_X(t)}.$$

Solution 1.2

(a) Since the random variables X and Y are independent, therefore the generating function of the random variable $X + Y$ has the form

$$G_{X+Y}(z) = G_X(z)G_Y(z) = e^{\lambda(z-1)}e^{\mu(z-1)} = e^{(\lambda+\mu)(z-1)},$$

which justifies that $X + Y$ has a Poisson distribution with parameter $\lambda + \mu$.

(b) With a simple calculation we have

$$\mathbf{P}\,(X = m \mid X + Y = n) = \frac{\mathbf{P}\,(X = m, \, X + Y = n)}{\mathbf{P}\,(X + Y = n)} = \frac{\mathbf{P}\,(X = m)\mathbf{P}\,(Y = n - m)}{\mathbf{P}\,(X + Y = n)}$$

$$= \frac{\frac{\lambda^m}{m!}e^{-\lambda}\frac{\mu^{n-m}}{(n-m)!}e^{-\mu}}{\frac{(\lambda+\mu)^n e^{-(\lambda+\mu)}}{n!}} = \binom{n}{m}\frac{\lambda^m \mu^{n-m}}{(\lambda + \mu)^n} = \binom{n}{m}\left(\frac{\lambda}{\lambda + \mu}\right)^m \left(\frac{\mu}{\lambda + \mu}\right)^{n-m}.$$

Solution 1.3 It is clear that the density function of X is $f_X(x) = \mathscr{I}_{\{0<x<1\}}$, and Y has the density function $f_Y(y) = e^{-y}\mathscr{I}_{\{y>0\}}$, therefore

$$\mathbf{P}\,(X < Y) = \int\limits_{-\infty}^{\infty} \int\limits_{-\infty}^{\infty} \mathscr{I}_{\{x<y\}} f_X(x) f_Y(y)\mathrm{d}x\mathrm{d}y = \int\limits_{0}^{\infty} \int\limits_{0}^{1} \mathscr{I}_{\{x<y\}} e^{-y}\mathrm{d}x\mathrm{d}y$$

$$= \int\limits_{0}^{\infty} \int\limits_{0}^{\min(1,y)} e^{-y}\mathrm{d}x\mathrm{d}y = \int\limits_{0}^{\infty} \min(1, y)e^{-y}\mathrm{d}y = \int\limits_{0}^{1} ye^{-y}\mathrm{d}y + \int\limits_{1}^{\infty} e^{-y}\mathrm{d}y$$

$$= \left([-ye^{-y}]_0^1 + \int\limits_{0}^{1} e^{-y}\mathrm{d}y\right) + [-e^{-y}]_1^{\infty} = 1 - e^{-1} = 0.63.$$

Solution 1.4 The random variables U_1 and U_2 are independent and uniformly distributed on the interval $(0, 1)$. The length of the three parts are:

$$U_1, U_2 - U_1, 1 - U_2, \text{ if } U_1 \leq U_2,$$
$$U_2, U_1 - U_2, 1 - U_1, \text{ if } U_1 > U_2.$$

The three parts determine a triangle if and only if the triangle inequality is satisfied, then using the formula of the total probability we obtain

$$\mathbf{P}\,(A) = \mathbf{P}\,(A \mid U_1 \leq U_2)\mathbf{P}\,(U_1 \leq U_2) + \mathbf{P}\,(A \mid U_1 > U_2)\mathbf{P}\,(U_1 > U_2)$$

$$= 2\mathbf{P}\,(A \mid U_1 \leq U_2)\mathbf{P}\,(U_1 \leq U_2)$$

$$= \int\limits_{0}^{1} \int\limits_{0}^{1} \mathscr{I}_{\{x\leq(y-x)+(1-y), y-x\leq x+(1-y), (1-y)\leq x+(y-x)\}}\mathrm{d}x\mathrm{d}y$$

$$= \int\limits_{0}^{1} \int\limits_{0}^{1} \mathscr{I}_{\{x\leq\frac{1}{2}, y\leq x+\frac{1}{2}, \frac{1}{2}\leq y\}}\mathrm{d}x\mathrm{d}y = \int\limits_{0}^{1/2} \int\limits_{1/2}^{1/2+x} \mathrm{d}y\mathrm{d}x = \int\limits_{0}^{1/2} x\mathrm{d}x = \frac{1}{4}.$$

Solution 1.5 Denote by $F_X(x)$ the CDF of the random variable X. Since $E(X^n) <$ ∞, then for $a \to \infty$, $E\left(X^n \mathscr{I}_{\{X>a\}}\right) = \int_a^\infty x^n dF_X(x) = -\int_a^\infty x^n d(1 - F_X(x)) \to 0$ and consequently $-\int_a^\infty x^n d(1 - F_X(x)) \geq a^n(1 - F(a)) \to 0$. Integrating by part, we have

$$E\left(X^n \mathscr{I}_{\{X \leq a\}}\right) = -\int_0^a x^n d(1 - F_X(x)) = -\left[x^n(1 - F_X(x))\right]_0^a + \int_0^a (1 - F_X(x)) d(x^n)$$

$$= -a^n(1 - F(a)) + \int_0^a (1 - F_X(x))nx^{n-1}dx,$$

from this

$$E(X^n) = \lim_{a \to \infty} \left[-a^n(1 - F(a)) + \int_0^a (1 - F_X(x))nx^{n-1}dx \right]$$

$$= \int_0^\infty (1 - F_X(x))nx^{n-1}dx = \int_0^\infty P(x \leq X)nx^{n-1}dx$$

follows.

Solution 1.6

(a) Since $f_X(u) = f_Y(u) \equiv \mathscr{I}_{\{0 < u < 1\}}$, then

$$E(|X - Y|) = \int_0^1 \int_0^1 |x - y| f_X(x) f_Y(y) dxdy = \int_0^1 \int_0^1 |x - y| dxdy$$

$$= 2 \int_0^1 \int_0^1 |x - y| \mathscr{I}_{\{x \leq y\}} dxdy$$

$$= 2 \int_0^1 \int_0^y (y - x) dxdy = 2 \int_0^1 \left[y^2 - \frac{1}{2}y^2 \right] dy = \frac{1}{3},$$

and

$$\mathbf{D}^2\left(|X-Y|\right) = \mathbf{E}\left(|X-Y|^2\right) - \left(\mathbf{E}\left(|X-Y|\right)\right)^2 = \mathbf{E}\left(X^2 - 2EXY + EY^2\right) - \frac{1}{9}$$

$$= 2\int_0^1 x^2 dx - 2\left(\int_0^1 x dx\right)^2 - \frac{1}{9} = 2\cdot\frac{1}{3} - 2\frac{1}{2^2} - \frac{1}{9} = \frac{1}{18}.$$

(b) It is easy to see

$$\mathbf{P}\left(|X-Y| > \tfrac{1}{2}\right) = \int_0^1\int_0^1 \mathscr{I}_{\left\{|x-y|>\frac{1}{2}\right\}} dx dy = 2\int_0^1\int_0^1 \mathscr{I}_{\left\{|x-y|>\frac{1}{2},x\geq y\right\}} dx dy$$

$$= 2\int_0^{1/2}\int_y^1 dx dy = 2\int_0^{1/2}\left(\frac{1}{2} - y\right) dy$$

$$= 2\int_0^{1/2} y dy = 2\left[\frac{1}{2}\frac{1}{2^2}\right] = \frac{1}{4}.$$

Solution 1.7

(a) Applying the convolution formula for the sum of independent random variables, we have

$$f_Z(z) = \int_{-\infty}^{\infty} f_X(x) f_Y(z-x) dx = \int_{-\infty}^{\infty} \lambda e^{-\lambda x} \mu e^{-\mu(z-x)} \mathscr{I}_{\{x>0, z-x>0\}} dx$$

$$= \lambda\mu e^{-\mu z}\int_0^z e^{-(\lambda-\mu)x} dx = \begin{cases} \lambda^2 z e^{-\lambda z}, & \text{if } \lambda = \mu \\ \frac{\lambda\mu}{\lambda-\mu}\left[e^{-\mu z} - e^{-\lambda z}\right], & \text{if } \lambda \neq \mu \end{cases}.$$

As a result we get a gamma distribution with the parameter $(\lambda, 2)$ if $\lambda = \mu$.

(b) It is clear that

$$F_Z(z) = \mathbf{P}\left(Z \leq z\right) = \mathbf{P}\left(\min(X, Y) \leq z\right) = 1 - \mathbf{P}\left(\min(X, Y) > z\right) =$$

$$= 1 - \mathbf{P}\left(X > z\right)\mathbf{P}\left(Y > z\right) = 1 - (1 - F_X(z)(1 - F_Y(z))$$

$$= F_X(z) + F_Y(z) - F_X(z) F_Y(z),$$

from which

$$f_Z(z) = f_X(z) + f_Y(z) - f_X(z) F_Y(z) - F_X(z) f_Y(z).$$

Using the exponential distributions with the parameters λ and μ, we have

$$f_Z(z) = \lambda e^{-\lambda z}\left(1 - \left[1 - e^{-\mu z}\right]\right) + \mu e^{-\mu z}\left(1 - \left[1 - e^{-\lambda z}\right]\right)$$

$$= \lambda e^{-\lambda z - \mu z} + \mu e^{-\lambda z - \mu z} = (\lambda + \mu)e^{-(\lambda+\mu)z}.$$

Solution 1.8 Clearly, for $x \geq 0$

$$F_{V_n}(x) = \mathbf{P}(V_n \leq x) = \mathbf{P}(X_1 \leq x, \ldots, X_n \leq x) = (\mathbf{P}(X_1 \leq x))^n = (1 - e^{-\lambda x})^n,$$

$$F_{W_n}(x) = \mathbf{P}(W_n \leq x) = 1 - \mathbf{P}(W_n > x) = 1 - \mathbf{P}(\min(X_1, \ldots, X_n) > x)$$

$$= 1 - \mathbf{P}(X_1 > x, \ldots, X_n > x) = 1 - (\mathbf{P}(X_1 > x))^n$$

$$= 1 - (e^{-\lambda x})^n = 1 - e^{-\lambda n x}.$$

Using $\int_{-\infty}^{\infty} x \, dF(x) = \int_{-\infty}^{\infty} (1 - F(x)) dx$ (see Exercise 1.5) and introducing a new variable $y = 1 - e^{-\lambda x}$ in the integral (i.e., $x = -\frac{1}{\lambda}\log(1 - y)$), we get the expected value of V_n as follows

$$\mathbf{E}(V_n) = \int_0^{\infty} (1 - F_V(x)) dx = \int_0^{\infty} (1 - (1 - e^{-\lambda x})^n) dx$$

$$= \frac{1}{\lambda} \int_0^1 (1 - y^n) \frac{1}{1 - y} dy = \frac{1}{\lambda} \int_0^1 (1 + y + \ldots + y^{n-1}) dy = \frac{1}{\lambda} \sum_{i=1}^n \frac{1}{i}.$$

From the formula $F_{W_n}(x) = 1 - e^{-\lambda n x}$ it obviously follows that W_n has an exponential distribution with parameter λn and so $\mathbf{E}(W_n) = \frac{1}{\lambda n}$.

Solution 1.9 By definition, the conditional distribution function of random variable X has the following expression given $X < Y$

$$\mathbf{P}(X \leq x \mid X < Y) = \frac{\mathbf{P}(X \leq x, X < Y)}{\mathbf{P}(X < Y)}$$

$$= \frac{1}{\mathbf{P}(X < Y)} \int_{-\infty}^{\infty} \int_{-\infty}^{\infty} \mathscr{I}_{\{u \leq x, u < y\}} f_X(u) f_Y(y) dx dy$$

$$= \frac{1}{\mathbf{P}(X < Y)} \int_{-\infty}^{x} f_X(u) \left(\int_u^{\infty} f_Y(y) dy\right) dx = \frac{1}{\mathbf{P}(X < Y)} \int_{-\infty}^{x} f_X(u) (1 - F_Y(u)) du.$$

This means that the conditional density function is $\dfrac{1}{\mathbf{P}(X < Y)} f_X(u)(1 - F_Y(u))$,

where

$$\mathbf{P}(X < Y) = \int_{-\infty}^{\infty} \int_{-\infty}^{\infty} \mathscr{I}_{\{x<y\}} f_X(x) f_Y(y) \mathrm{d}x \mathrm{d}y = \int_{-\infty}^{\infty} f_X(x) \left(\int_{x}^{\infty} f_Y(y) \mathrm{d}y \right) \mathrm{d}x$$

$$= \int_{-\infty}^{\infty} f_X(x)(1 - F_Y(x)) \, \mathrm{d}x = \mathbf{E}((1 - F_Y(X))).$$

Consequently,

$$\mathbf{E}(X \mid X < Y) = \frac{\int_{-\infty}^{\infty} x f_X(x)(1 - F_Y(x)) \, \mathrm{d}x}{\int_{-\infty}^{\infty} f_X(x)(1 - F_Y(x)) \, \mathrm{d}x} = \frac{\mathbf{E}(X(1 - F_Y(X)))}{\mathbf{E}(1 - F_Y(X))}.$$

Solution 1.10

(a) Since $f_Y(y) = \int_{0}^{1} f_{X,Y}(x, y)\mathrm{d}x = \int_{0}^{1-y} 2\mathrm{d}x = 2(1 - y), \quad 0 < y < 1$, thus

the conditional density function is $f_{X|Y}(x|y) = \dfrac{f_{X,Y}(x,y)}{f_Y(y)} = \dfrac{2}{2(1-y)} = \dfrac{1}{1-y}$, if $0 < x, y$ and $x + y < 1$, so

$$\mathbf{E}(X \mid Y = y) = \int_{0}^{1-y} x f_{X|Y}(x|y)\mathrm{d}x = \int_{0}^{1-y} x \frac{1}{1-y}\mathrm{d}x = \frac{1}{2}\frac{(1-y)^2}{1-y} = \frac{1}{2}(1-y)$$

and

$$\mathbf{E}(X \mid Y) = \frac{1}{2}(1 - Y).$$

(b) Analogously, we get $f_Y(y) = \int_{0}^{1} f_{X,Y}(x, y)\mathrm{d}x = \int_{0}^{1-y} 3(x + y)\mathrm{d}x =$

$3\left[\frac{1}{2}(1 - y)^2 + y(1 - y)\right] = \frac{3}{2}(1 - y^2)$, for $0 < y < 1$ and $f_{X|Y}(x|y) = \dfrac{3(x+y)}{\frac{3}{2}(1-y^2)} = 2\dfrac{x+y}{1-y^2}$, for $0 < y < 1, x + y < 1$. From this, it follows that

$$\mathbf{E}(X \mid Y = y) = \int_{0}^{1-y} x f_{X|Y}(x|y)\mathrm{d}x = \int_{0}^{1-y} x \frac{2(x + y)}{1 - y^2}\mathrm{d}x$$

$$= \frac{1}{3}\frac{2 - y - y^2}{1 + y}, 0 < y < 1,$$

$$\mathbf{E}(X \mid Y) = \frac{1}{3}\frac{2 - Y - Y^2}{1 + Y}.$$

Solution 1.11

First Solution Since the sum of random variables $Y_n = X_1 + \ldots + X_n$ has a gamma distribution with the parameter (n, λ) thus

$$f_{Y_n}(x) = \frac{\lambda^n}{(n-1)!}x^{n-1}e^{-\lambda x}, \; x > 0$$

and

$$F_Y(y) = \mathbf{P}(X_1 + \ldots + X_N \le y) = \sum_{n=1}^{\infty} \mathbf{P}(X_1 + \ldots + X_N \le y \mid N = n)\mathbf{P}(N = n)$$

$$= \sum_{n=1}^{\infty} \mathbf{P}(Y_n \le y)\mathbf{P}(N = n) = \sum_{n=1}^{\infty} \int_{-\infty}^{y} \frac{\lambda^n}{(n-1)!}x^{n-1}e^{-\lambda x}\left[p(1-p)^{n-1}\right]dx$$

$$= p\lambda \int_{-\infty}^{y} \left(\sum_{n=0}^{\infty} \frac{[(1-p)\lambda x]^n}{n!}\right)e^{-\lambda x}dx = p\lambda \int_{-\infty}^{y} \left(e^{(1-p)\lambda x}e^{-\lambda x}\right)dx$$

$$= p\lambda \int_{-\infty}^{y} e^{-p\lambda x}dx.$$

From this

$$f_Y(y) = p\lambda e^{-p\lambda x}, \; y > 0,$$

therefore the random variable Y really has an exponential distribution with the parameter $p\lambda$.

Second Solution Clearly, the generating function of the random variable N is $G_N(z) = \frac{pz}{1-(1-p)z}$, and the random variable X_1 with exponential distribution of the parameter λ has Laplace transform $F_{X_1}(s) = \frac{\lambda}{s+\lambda}$, then applying Theorem A.4 we have

$$\mathbf{E}\left(exp\left\{\sum_{k=1}^{N} X_k\right\}\right) = G_N(F_{X_1}(s)) = \frac{p\frac{\lambda}{s+\lambda}}{1-(1-p)\frac{\lambda}{s+\lambda}} = \frac{p\lambda}{s+p\lambda},$$

which is the Laplace transform of the exponential distribution of the parameter $p\lambda$.

Solution 1.12 By the use of the central limit theorem the random variable $(Y_{40} - \mathbf{E}(Y_{40}))/DY_{40}$ has approximately $N(0, 1)$ normal distribution, thus

$$\mathbf{P}\left(\frac{|Y_{40}-\mathbf{E}(Y_{40})|}{\mathbf{D}(Y_{40})} > 0.05\right) = \mathbf{P}\left(\frac{|Y_{40}-40|}{\sqrt{40}} > 0.05\right) \approx$$

$$\approx 1 - (\Phi(0.05) - \Phi(-0.05)) = 0.0612,$$

where $\Phi(x) = \frac{1}{2\pi}\int\limits_{-\infty}^{x} e^{-u^2/2}du$ denotes the standard normal distribution function. Compute numerically the probability

$$\mathbf{P}\left(\frac{|Y_{40}-40|}{\sqrt{40}} > 0.05\right) = 1 - \mathbf{P}(39.6837 \leq Y_{40} \leq 40.3163)$$

$$= 1 - (0.5409 - 0.5010) = 0.0601.$$

It can be seen that the difference between the estimated and numerically computed values is only 0.0011.

References

1. Chow, Y., Teicher, H.: Probability Theory. Springer, New York (1978)
2. Gnedenko, B.V., Kolmogorov, A.N.: Independent Random Variables. Addison-Wesley, Cambridge (1954)
3. Johnson, N.L., Kotz, S.: Distributions in statistics: continuous multivariate distributions. In: Applied Probability and Statistics. Wiley, Hoboken (1972)
4. Kallenberg, O.: Foundations of Modern Probability. Springer, New York (2002)
5. Petrov, V.V.: Sums of Independent Random Variables, vol. 82. Springer, Berlin (2012)
6. Shiryaev, A.N.: Probability. Springer, Berlin (1994)
7. Zolotarev, V.M.: Modern Theory of Summation of Random Variables. VSP, Utrecht (1997)

Chapter 2
Introduction to Stochastic Processes

2.1 Stochastic Processes

When considering technical, economic, ecological, or other problems, in several cases the quantities $\{X_t,\ t \in \mathscr{T}\}$ being examined can be regarded as a collection of random variables. This collection describes the changes (usually in time and in space) of considered quantities. If the set $\mathscr{T}$ is a subset of the set of real numbers, then the set $\{t \in \mathscr{T}\}$ can be interpreted as time and we can say that the random quantities X_t vary in time. In this case the collection of random variables $\{X_t,\ t \in \mathscr{T}\}$ is called a **stochastic process**. In mathematical modeling of randomly varying quantities in time, one might rely on the highly developed theory of stochastic processes.

Definition 2.1 Let $\mathscr{T} \subset \mathbb{R}$. A **stochastic process** X is defined as a collection $X = \{X_t,\ t \in \mathscr{T}\}$ of indexed random variables X_t, which are given on the same probability space $(\Omega, \mathscr{A}, \mathbf{P}())$.

Depending on the notational complexity of the parameter, we occasionally interchange the notation X_t with $X(t)$.

It is clear that $X_t = X_t(\omega)$ is a function of two variables. For fixed $t \in \mathscr{T}$, X_t is a random variable, and for fixed $\omega \in \Omega$, X_t is a function of the variable $t \in \mathscr{T}$, which is called a **sample path** of the stochastic process.

Depending on the set $\mathscr{T}$, X is called **discrete-time** stochastic process, if the index set $\mathscr{T}$ consists of consecutive integers, for example, $\mathscr{T} = \{0, 1, \ldots\}$ or $\mathscr{T} = \{\ldots, -1, 0, 1, \ldots\}$. Further, X is called **continuous-time** stochastic process, if $\mathscr{T}$ equals an interval of the real line, for example, $\mathscr{T} = [a, b]$, $\mathscr{T} = [0, \infty)$ or $\mathscr{T} = (-\infty, \infty)$.

Note that in the case of discrete time, X is a sequence $\{X_n,\ n \in \mathscr{T}\}$ of random variables, while it determines a random function in the continuous-time case. It should be noted that similar to the notion of real valued stochastic processes, we

© Springer Nature Switzerland AG 2019
L. Lakatos et al., *Introduction to Queueing Systems with Telecommunication Applications*, https://doi.org/10.1007/978-3-030-15142-3_2

may define complex or vector valued stochastic processes also if X_t take values in a complex plane or in higher-dimensional Euclidean space.

2.2 Finite Dimensional Distributions of Stochastic Processes

A stochastic process $\{X_t,\ t \in \mathscr{T}\}$ can be characterized in a statistical sense by its finite-dimensional distributions.

Definition 2.2 The **finite-dimensional distributions** of a stochastic process $\{X_t,\ t \in \mathscr{T}\}$ are defined by the family of all joint distribution functions

$$F_{t_1,\dots,t_n}(x_1,\dots,x_n) = \mathbf{P}\left(X_{t_1} < x_1, \dots, X_{t_n} < x_n\right),$$

where $n = 1, 2, \dots$ and $t_1, \dots, t_n \in \mathscr{T}$.

The family of introduced distribution functions

$$\mathscr{F} = \left\{F_{t_1,\dots,t_n},\ t_1,\dots,t_n \in \mathscr{T},\ n = 1, 2, \dots\right\}$$

satisfies the following, specified consistency conditions:

(a) For all positive integers n, m and indices $t_1, \dots, t_{n+m} \in \mathscr{T}$

$$\lim_{x_{n+1} \to \infty} \dots \lim_{x_{n+m} \to \infty} F_{t_1,\dots,t_n,t_{n+1},\dots,t_{n+m}}(x_1, \dots, x_n, x_{n+1}, \dots, x_{n+m})$$

$$= F_{t_1,\dots,t_n}(x_1, \dots, x_n), \qquad x_1, \dots, x_n \in R\ .$$

(b) For all permutations $(i_1, \dots, i_n)$ of the numbers $\{1, 2, \dots, n\}$

$$F_{s_1,\dots,s_n}(x_{i_1}, \dots, x_{i_n}) = F_{t_1,\dots,t_n}(x_1, \dots, x_n), \qquad x_1, \dots, x_n \in R\ ,$$

where $s_j = t_{i_j},\ j = 1, \dots, n$.

Definition 2.3 If the family $\mathscr{F}$ of joint distribution functions defined previously satisfies conditions (a) and (b), then we say that $\mathscr{F}$ satisfies the **consistency conditions**.

The following theorem is a basic one in probability theory and ensures the existence of a stochastic process (in general of a collection of random variables) with given finite-dimensional distribution functions satisfying the consistency conditions.

Theorem 2.1 (Kolmogorov Consistency Theorem) *Suppose a family of distribution functions* $\mathscr{F} = \left\{F_{t_1,\dots,t_n},\ t_1, \dots, t_n \in \mathscr{T},\ n = 1, 2, \dots\right\}$ *satisfies the consistency conditions (a) and (b). Then there exists a probability space* $(\Omega, \mathscr{A}, \mathbf{P})$, *and on that a stochastic process* $\{X_t,\ t \in \mathscr{T}\}$, *whose finite-dimensional distributions are identical to* $\mathscr{F}$.

For our considerations, it usually suffices to provide the finite-dimensional distribution functions of the stochastic processes, in which case the process is defined in the **weak sense** and it is irrelevant on which probability space it is given. In some instances the behavior of the random path is significant (e.g., continuity in time), which is related to a given probability space $(\Omega, \mathscr{A}, P)$ where the process $\{X_t,\ t \in \mathscr{T}\}$ is defined. In this case the process is given in a **strict sense**.

2.3 Stationary Processes

The class of stochastic processes that show a stationary statistical property in time plays a significant role in practice. Among these processes the most important ones are the stationary processes in the strict and the weak senses. The main notions are given here for the one-dimensional processes, but the notion for the high-dimensional processes can be introduced similarly.

Definition 2.4 A process $\{X_t,\ t \in \mathscr{T}\}$ is called **stationary in strict sense** if the joint distribution functions of random variables

$$(X_{t_1}, ..., X_{t_n}) \quad \text{and} \quad (X_{t_1+t}, ..., X_{t_n+t})$$

are identical for all t, positive integer n, and $t_1, ..., t_n \in \mathscr{T}$ satisfying the conditions $t_i + t \in \mathscr{T}$, $i = 1, ..., n$.

Note that this definition remains valid in the case of vector-valued stochastic processes. Consider a stochastic process X with finite second moment, that is, $\mathbf{E}\left(X_t^2\right) < \infty$, for all $t \in \mathscr{T}$. Denote the expected value and covariance functions by

$$\mu_X(t) = \mathbf{E}\left(X_t\right),\ t \in \mathscr{T},$$
$$R_X(s, t) = \mathbf{Cov}\left(X_s, X_t\right)$$
$$= \mathbf{E}\left((X_t - \mu_X(t))(X_s - \mu_X(s))\right),\ s, t \in \mathscr{T}.$$

Definition 2.5 A process $\{X_t,\ t \in \mathscr{T}\}$ is called **stationary in the weak sense** if X_t has finite second moment for all $t \in \mathscr{T}$ and the expected value and covariance function satisfy the following relation:

$$\mu_X(t) = \mu_X,\ t \in \mathscr{T},$$
$$R_X(s, t) = R_X(t - s),\ s, t \in \mathscr{T},$$

The function R_X is called the **covariance function**.

It is clear that if a stochastic process with finite second moment is stationary in a strict sense, then it is stationary in the weak sense also, because the expected value and covariance function depend also on the two-dimensional joint distribution, which is time-invariant if the time shifts. Besides the covariance function $R_X(t)$, the **correlation function** $r_X(t)$ is also used, which is defined as follows:

$$r_X(t) = \frac{1}{R_X(0)} R_X(t) = \frac{1}{\sigma_X^2} R_X(t).$$

2.4   Gaussian Process

In practice, we often encounter stochastic processes whose finite-dimensional distributions are Gaussian. These stochastic processes are called **Gaussian**. In queuing theory Gaussian processes often appear when asymptotic methods are applied.

Note that the expected values and covariances determine the finite-dimensional distributions of the Gaussian process (see multidimensional Gaussian distribution in Sect. 1.2.2); therefore, a Gaussian process is stationary in both strict and weak senses if and only if the expected value and covariance function is invariant to translations in time. For this reason, notions of the strict and the weak stationarity for Gaussian processes are equivalent. We also mention here that the discrete-time Gaussian process consists of independent Gaussian random variables if these random variables are uncorrelated.

2.5   Stochastic Process with Independent and Stationary Increments

In several practical modeling problems, stochastic processes have independent and stationary increments. These processes play a significant role in both theory and practice. Among such processes the Brownian motion and the Poisson processes are defined below.

Definition 2.6 If for any integer $n \geq 1$ and parameters $t_0, ..., t_n \in \mathscr{T}$, $t_0 < ... < t_n$, the increments

$$X_{t_1} - X_{t_0}, ..., X_{t_n} - X_{t_{n-1}}$$

of a stochastic process $X = \{X_t, \ t \in \mathscr{T}\}$ are independent random variables, then X is called a stochastic process with **independent increments**. The process X has **stationary increments** if the distribution of $X_{t+h} - X_t$, $t, t + h \in \mathscr{T}$ does not depend on t.

2.6 Brownian Motion

As a special but important case of stochastic processes with independent and stationary increments, we mention here the **Brownian motion** (also called **Wiener process**), which gives the mathematical model of diffusion. A process $X = \{X_t,\ t \in [0, \infty)\}$ is called a Brownian motion if the increments of the process are independent and for any positive integer n and $0 \leq t_0 < \ldots < t_n$ the joint density function of random variables $X_{t_0}, \ldots, X_{t_n}$ can be given in the form

$$f(x_0, \ldots, x_n; t_0, \ldots, t_n) = (2\pi)^{-n/2} \left[t_0(t_1 - t_0) \ldots (t_n - t_{n-1}) \right]^{-1/2}$$

$$\cdot \exp \left\{ -\frac{1}{2} \left(\frac{x_0^2}{t_0} + \frac{(x_1 - x_0)^2}{t_1 - t_0} + \ldots + \frac{(x_n - x_{n-1})^2}{t_n - t_{n-1}} \right) \right\}.$$

It can be seen from this formula that the Brownian motion is a Gaussian process and the increments

$$X_{t_j} - X_{t_{j-1}}, \quad j = 1, \ldots, n,$$

are independent Gaussian random variables with expected values 0 and variances $t_j - t_{j-1}$. The expected value function and the covariance function are determined as

$$\mu_X(t) = 0, \quad R_X(s, t) = \min(t, s), \quad t, s \geq 0.$$

2.7 Poisson Process

2.7.1 Definition of Poisson Process

Besides the Wiener process defined above, we discuss in this chapter another important process with independent increments in probability theory, the Poisson process. This process plays a fundamental role not only in the field of queueing theory but also in many areas of theoretical and applied sciences, and we will deal with this process later as a Markov arrival process, birth-and-death process, and renewal process. Its significance in probability theory and practice is that it can be used to model different event occurrences in time and space in, for example, queuing systems, physics, insurance, and population biology. There are several introductions and equivalent definitions of the Poisson process in the literature according to its different characterizations. First we present the notion in the simple (classical) form and after that in a more general context.

In queueing theory, a frequently used model for the description of the arrival process of costumers is as follows. Assume that costumers arrive at the system one after another at $(0 <)\tau_1 < \tau_2 < ...;\ \tau_n \to \infty$ as $n \to \infty$. The differences in occurrence times, called **interarrival times**, are denoted by

$$X_1 = \tau_1,\ X_2 = \tau_2 - \tau_1, ...,\ X_n = \tau_n - \tau_{n-1}, ...$$

Define the process $\{N(t),\ t \geq 0\}$ with $N(0) = 0$ and

$$N(t) = \max\{n :\ \tau_n \leq t\} = \max\{n :\ X_1 + ... + X_n \leq t\},\ t > 0.$$

This process counts the number of customers arriving at the system in the time interval $(0, t]$ and it is called the **counting process** for the sequence $\tau_1 < \tau_2 < ....$ Obviously, the process takes nonnegative integer values only, is nondecreasing, and $N(t) - N(s)$ equals the number of occurrences in the time interval $(s, t]$ for all $0 < s < t$.

In the special case, when $X_1, X_2, ...$ is a sequence of independent and identically distributed random variables with exponential distribution $\text{Exp}(\lambda)$, the increments $N(t) - N(s)$ have a Poisson distribution with the parameter $\lambda(t - s)$. In addition, the counting process $N(t)$ possesses an essential property, that is, it evolves in time **without aftereffects**. This means that the past and current occurrences have no effects on subsequent occurrences. This feature leads to the property of independent increments.

Definition 2.7 We say that the process $N(t)$ is a **Poisson process** with **rate** λ if

1. $N(0) = 0$,
2. $N(t),\ t \geq 0$ is a process with independent increments,
3. The distribution of increments is Poisson with the parameter $\lambda(t - s)$ for all $0 < s < t$.

By definition, the distributions of the increments $N(t + h) - N(t), t \geq 0, h > 0$, do not depend on the moment t; therefore, it is a process with stationary increments and is called a **homogeneous** Poisson process at rate λ. Next, we introduce the Poisson process in a more general setting, and as a special case we have the homogeneous case. After that we will deal with the different characterizations of Poisson processes, which in some cases can serve as a definition of the process. At the end of this chapter, we will introduce the notion of the high-dimensional Poisson process (sometimes called a spatial Poisson process) and give its basic properties.

Definition 2.8 Let $\{\Lambda(t),\ t \geq 0\}$ be a nonnegative, monotonically nondecreasing, continuous from-right real-valued function for which $\Lambda(0) = 0$. We say that a stochastic process $\{N(t),\ t \geq 0\}$ taking nonnegative integers is a **Poisson process** if

1. $N(0) = 0$,
2. $N(t)$ is a process with independent increments,
3. The CDFs of the increments $N(t) - N(s)$ are Poisson with the parameter $\Lambda(t) - \Lambda(s)$ for all $0 \leq s \leq t$, that is,

$$\mathbf{P}\left(N(t) - N(s) = k\right) = \frac{(\Lambda(t) - \Lambda(s))^k}{k!} e^{-(\Lambda(t) - \Lambda(s))}, \ k = 0, 1, \dots.$$

Since for any fixed $t > 0$ the distribution of $N(t) = N(t) - N(0)$ is Poisson with mean $\Lambda(t)$, that is the reason that $N(t)$ is called a Poisson process. We can state that the process $N(t)$ is a monotonically nondecreasing jumping process whose increments $N(t) - N(s)$, $0 \le s < t$, take nonnegative integers only and the increments have Poisson distributions with the parameter $(\Lambda(t) - \Lambda(s))$. Thus the random variables $N(t)$, $t \ge 0$ have Poisson distributions with the parameter $\Lambda(t)$; therefore, the expected value of $N(t)$ is $\mathbf{E}\left(N(t)\right) = \Lambda(t)$, $t \ge 0$, which is called **a mean value function**.

We also note that using the property of independent increments, the joint distribution of the random variables $N(t_1), \dots, N(t_n)$ can be derived for all positive integers n and all $0 < t_1 < \dots < t_n$ without difficulty because for any integers $0 \le k_1 \le \dots \le k_n$ we get

$$\mathbf{P}\left(N(t_1) = k_1, \dots, N(t_n) = k_n\right)$$

$$= \mathbf{P}\left(N(t_1) = k_1, N(t_2) - N(t_1) = k_2 - k_1, \dots, N(t_n) - N(t_{n-1}) = k_n - k_{n-1}\right)$$

$$= \frac{(\Lambda(t_1))^{k_1}}{k_1!} e^{-\Lambda(t_1)} \prod_{i=2}^{n} \frac{(\Lambda(t_i) - \Lambda(t_{i-1}))^{k_i - k_{i-1}}}{(k_i - k_{i-1})!} e^{-(\Lambda(t_i) - \Lambda(t_{i-1}))}.$$

In accordance with Definition 2.8, the mean value function $\Lambda(t) = \mathbf{E}\left(N(t)\right)$ is monotonically nondecreasing; thus the set of discontinuity points $\{s_n\}$ of $\Lambda(t)$ is finite or countably infinite. It can happen that the set of discontinuity points $\{s_n\}$ has more than one convergence point, and in this case we cannot give the points of $\{s_n\}$ as an ordered sequence $s_1 < s_2 < \dots$. Define the jumps of the function $\Lambda(t)$ at discontinuity points s_n as follows:

$$\lambda_n = \Lambda(s_n + 0) - \Lambda(s_n - 0) = \Lambda(s_n) - \Lambda(s_n - 0).$$

By definition, the increments of a Poisson process are independent; thus it is easy to check that the following decomposition exists

$$N(t) = N_r(t) + N_s(t),$$

where $N_r(t)$ and $N_s(t)$ are independent Poisson processes with mean value functions

$$\Lambda_r(t) = \Lambda(t) - \sum_{s_n \le t} \lambda_n \quad \text{and} \quad \Lambda_s(t) = \sum_{s_n \le t} \lambda_n.$$

The **regular** part $N_r(t)$ of $N(t)$ has jumps equal to 1 only, whose mean value function $\Lambda_r(t)$ is continuous. Thus we can state that the process $N_r(t)$ is continuous in probability, that is, for any points $t, 0 \leq t < \infty$, the relation

$$\lim_{s \to 0} \mathbf{P}\left(N_r(t+s) - N_r(t) > 0\right) = \lim_{s \to 0} \mathbf{P}\left(N_r(t+s) - N_r(t) \geq 1\right) = 0$$

is true. The second part $N_s(t)$ of $N(t)$ is called a **singular** Poisson process because it can have jumps only in discrete points $\{s_n\}$. Then

$$\mathbf{P}\left(N_s(s_n) - N_s(s_n - 0)\right) = k) = \frac{\lambda_n^k}{k!} \, e^{-\lambda_n}, k = 0, 1, 2, \dots .$$

Definition 2.9 If the mean value function $\Lambda(t)$ of a Poisson process $\{N(t), \ t \geq 0\}$ is differentiable with the derivative $\lambda(s), \ s \geq 0$ satisfying $\Lambda(t) = \int_0^t \lambda(s) \mathrm{d}s$, then the function $\lambda(s)$ is called a **rate** (or **intensity**) **function** of the process.

In accordance with our first definition 2.7, we say that the Poisson process $N(t)$ is **homogeneous** with the rate λ if the rate function is a constant $\lambda(t) = \lambda, \ t \geq 0$. In this case, $\Lambda(t) = \lambda t, \ t \geq 0$ is satisfied; consequently, the distributions of all increments $N(t) - N(s), 0 \leq s < t$ are Poisson with the parameter $\lambda(t - s)$ and $\mathbf{E}\left(N(t) - N(s)\right) = \lambda(t - s)$. This shows that the average number of occurrences is proportional to the length of the corresponding interval and the constant of proportionality is λ. These circumstances justify the name of the rate λ.

If the rate can vary with time, that is, the rate function does not equal a constant, the Poisson process is called **inhomogeneous**.

2.7.2   Construction of Homogeneous Poisson Process

The construction of Poisson processes plays an essential role both from a theoretical and a practical points of view. In particular, it is essential in simulation methods. The Poisson process $N(t)$ and the sequence of the random jumping points $0 < \tau_1 < \tau_2 < \dots$ of the process uniquely determine each other. This fact provides an opportunity to give another definition of the Poisson process on the real number line.

Theorem 2.2 *If the process $N(t), t \geq 0$ is a Poisson process at rate λ, then its interarrival times $X_n = \tau_n - \tau_{n-1}, \ n = 1, 2, \dots$ constitute a sequence of independent random variables with common exponential distribution of parameter λ.*

Verifying this statement and taking the subsequent Theorems on Construction 1 and II into account, we conclude that the conditions of Definition 2.7 are equivalent to that of Theorems 2.3 and 2.4. This means that Constructions I and II are interpretable as definitions of Poisson processes at rate λ. Furthermore, we note that the following chain of conclusions is valid: Definition 2.7 $\Longrightarrow$ Theorem 2.2 $\Longrightarrow$ Theorem 2.3 $\Longrightarrow$ Theorem 2.4 $\Longrightarrow$ Definition 2.7.

Proof (Theorem 2.2) It is clear that the inequality $\tau_n > t$ holds if and only if $N(\tau_{n-1} + t) - N(\tau_{n-1}) = 0$ and the distribution of the random variable $N(\tau_{n-1} + t) - N(\tau_{n-1})$ does not depend on the value of random variable τ_{n-1} because Poisson process is a process without aftereffects, consequently,

$$\mathbf{P}(\tau_n > t) = \mathbf{P}(N(\tau_{n-1} + t) - N(\tau_{n-1}) = 0) =$$
$$= \mathbf{P}(N(t) - N(0) = 0) = e^{-\lambda t},$$

thus we obtain the CDF of τ_n as $\mathbf{P}(\tau_n \leq t) = 1 - e^{-\lambda t}$.

Now we prove that the following two constructions of Poisson processes are valid (see, for example, pp. 117–118 in [6]).

Theorem 2.3 (Construction I) *Let* $X_1, X_2, \ldots$ *be independent and identically distributed random variables whose common CDF is exponential with parameter* 1. *Define*

$$M(t) = \sum_{m=1}^{\infty} \mathscr{I}_{\{X_1 + \ldots + X_m \leq t\}}, \ t \geq 0. \tag{2.1}$$

Then the process $M(t)$ *is homogeneous Poisson process with rate equal to* 1.

Theorem 2.4 (Construction II) *Let* $U_1, U_2, \ldots$ *be a sequence of independent and identically distributed random variables having common uniform distribution on the interval* $(0, T)$, *and let* N *be a random variable independent of* U_i *with a Poisson distribution with the parameter* λT. *Define*

$$N(t) = \sum_{m=1}^{N} \mathscr{I}_{\{U_m \leq t\}}, \ 0 \leq t \leq T. \tag{2.2}$$

Then $N(t)$ *is a homogeneous Poisson process on the interval* $[0, T]$ *at rate* λ.

We begin with the proof of Construction II. Then, using this result, we verify Construction I.

Proof (Construction II) Let K be a positive integer and $t_1, \ldots, t_K$ positive constants such that $t_0 = 0 < t_1 < t_2 < \ldots < t_K = T$. Since, by Eq. (2.2), $N(T) = N$ and $N(t) = \sum_{m=1}^{N} \mathscr{I}_{\{U_m \leq t\}}$, the increments of $N(t)$ on the intervals $(t_{k-1}, t_k]$, $k = 1, \ldots, K$, can be given in the form

$$N(t_k) - N(t_{k-1}) = \sum_{n=1}^{N} \mathscr{I}_{\{t_{k-1} < U_n \leq t_k\}}, \ k = 1, \ldots, K.$$

Determine the joint characteristic function of the increments $N(t_k) - N(t_{k-1})$. Let $s_k \in \mathbb{R}$, $k = 1, ..., K$, be arbitrary; then

$$\varphi(s_1, ..., s_K) = \mathbf{E}\left(\exp\left\{\sum_{k=1}^{K} \Im s_k (N(t_k) - N(t_{k-1}))\right\}\right)$$

$$= \mathbf{P}(N = 0) + \sum_{n=1}^{\infty} \mathbf{E}\left(\exp\left\{\sum_{k=1}^{K} \Im s_k (N(t_k) - N(t_{k-1}))\right\} \middle| N = n\right) \mathbf{P}(N = n)$$

$$= e^{-\lambda T} + \sum_{n=1}^{\infty} \mathbf{E}\left(\exp\left\{\sum_{k=1}^{K} \Im s_k \sum_{\ell=1}^{n} \mathscr{I}_{\{t_{k-1} < U_\ell \leq t_k\}}\right\}\right) \mathbf{P}(N = n)$$

$$= e^{-\lambda T} + \sum_{n=1}^{\infty} \prod_{\ell=1}^{n} \mathbf{E}\left(\exp\left\{\sum_{k=1}^{K} \Im s_k \mathscr{I}_{\{t_{k-1} < U_\ell \leq t_k\}}\right\}\right) \frac{(\lambda T)^n}{n!} e^{-\lambda T}$$

$$= e^{-\lambda T} \sum_{n=0}^{\infty} \left(\sum_{k=1}^{K} \frac{t_k - t_{k-1}}{T} e^{\Im s_k}\right)^n \frac{(\lambda T)^n}{n!} = e^{-\lambda T} \exp\left\{\sum_{k=1}^{K} e^{\Im s_k} \lambda(t_k - t_{k-1})\right\},$$

and using the relation $T = t_K - t_0 = \sum_{k=1}^{K} (t_k - t_{k-1})$ we get

$$\varphi(s_1, ..., s_K) = \prod_{k=1}^{K} \exp\left\{\lambda(t_k - t_{k-1})(e^{\Im s_k} - 1)\right\}.$$

Since the characteristic function $\varphi(s_1, ..., s_K)$ derived here is equal to the joint characteristic function of independent random variables having a Poisson distribution with the parameters $\lambda(t_k - t_{k-1})$, $k = 1, ..., K$, the proof is complete. $\square$

For the proof of Construction I it is sufficient to justify that the conditions of Construction II hold. We need the following well-known lemma of probability theory.

Lemma 2.1 *Let T be a positive constant and k be a positive integer. Let $U_1, ..., U_k$ be independent and identically distributed random variables having a common uniform distribution on the interval $(0, T)$. Define by $U_{1k} \leq ... \leq U_{kk}$ the ordered random variables $U_1, ..., U_k$. Then the joint PDF of random variables $U_{1k}, ..., U_{kk}$ is*

$$f_{U_{1k}, ..., U_{kk}}(t_1, ..., t_k) = \begin{cases} \dfrac{k!}{T^k}, & \text{if } 0 < t_1 \leq t_2 \leq ... \leq t_k < T, \\ 0, & \text{otherwise.} \end{cases}$$

Proof Since $U_{1k} \leq ... \leq U_{kk}$, it is enough to determine the joint PDF of random variables $U_{1k}, ..., U_{kk}$ on the set

$$\mathscr{K} = \{(t_1, ..., t_k) : \ 0 \le t_1 \le ... \le t_k < T\}.$$

Under the assumptions of the lemma, the random variables $U_1, ..., U_k$ are independent and uniformly distributed on the interval $(0, T)$; thus for every permutation $i_1, ..., i_k$ of the numbers $1, 2, ..., k$ (the number of all permutations is equal to $k!$)

$$\mathbf{P}\left(U_{i_1} \le ... \le U_{i_k}, U_{i_1} \le t_1, ..., U_{i_k} \le t_k\right)$$
$$= \mathbf{P}\left(U_1 \le ... \le U_k, U_1 \le t_1, ..., U_k \le t_k\right),$$

then

$$F_{U_{1k},...,U_{kk}}(t_1, ..., t_k) = \mathbf{P}\left(U_{1k} \le t_1, ..., U_{kk} \le t_k\right)$$
$$= k!\mathbf{P}\left(U_1 \le ... \le U_k, U_1 \le t_1, ..., U_k \le t_k\right)$$
$$= k! \int_0^{t_1} ... \int_0^{t_k} \frac{1}{T^k} \mathscr{I}_{\{u_1 \le ... \le u_k\}} du_k ... du_1$$
$$= \frac{k!}{T^k} \int_0^{t_1} \int_{u_1}^{t_2} ... \int_{u_{k-1}}^{t_k} du_k ... du_1.$$

From this we immediately have

$$f_{U_{1k},...,U_{kk}}(t_1, ..., t_k) = \frac{k!}{T^k}, \quad (t_1, ..., t_k) \in \mathscr{K},$$

which completes the proof. $\qquad\qquad\qquad\qquad\qquad\qquad\qquad\qquad\qquad\qquad\qquad$ □

Proof (Construction I) We verify that for any $T > 0$ the process $M(t)$, $0 \le t \le T$ satisfies the conditions of Construction II; therefore, it is a homogeneous Poisson process with rate λ.

In Construction II, $N(T) = N$, where the distribution of random variable N is Poisson with the parameter λT. From Eq. (2.2) it follows that the process $N(t)$, $0 \le t \le T$, can be rewritten in the form

$$N(t) = \sum_{m=1}^N \mathscr{I}_{\{U_m \le t\}} = \sum_{n=1}^N \mathscr{I}_{\{U_{nN} \le t\}},$$

where the random variables $U_1, U_2, ...$ are independent and uniformly distributed on the interval $(0, T)$ and for every $k \ge 1$ and under the condition $N(T) = k$ the random variables $U_{1k} \le U_{2k} \le ... \le U_{kk}$ are the ordered random variables $U_1, ... U_k$. Note that we used these properties only to determine the joint characteristic function of the increments. Define

$$\tau_n = X_1 + \ldots + X_n, \quad n = 1, 2, \ldots,$$

where, by assumptions, $X_1, X_2, \ldots$ are independent and identically distributed random variables with a common exponential CDF of parameter λ. Then using the relation (2.1), for any $0 \leq t \leq T$,

$$M(t) = \sum_{n=1}^{\infty} \mathscr{I}_{\{\tau_n \leq t\}} = \begin{cases} \displaystyle\sum_{n=1}^{M(T)} \mathscr{I}_{\{\tau_n \leq t\}}, & \text{if } T \geq \tau_1, \\ 0, & \text{if } T < \tau_1. \end{cases}$$

By the previous note it is enough to prove that

(a) The random variable $M(T)$ has a Poisson CDF with the parameter λT;
(b) For every positive integer k and under the condition $M(T) = k$, the joint CDF of the random variables $\tau_1, \ldots, \tau_k$ are identical with the CDF of the random variables $U_{1k}, \ldots, U_{kk}$.

(a) First we prove that for any positive t the CDF of the random variable $M(t)$ is Poisson with the parameter (λT). Since the common CDF of independent and identically distributed random variables X_i is exponential with the parameter λ, the random variable τ_n has a gamma(n, λ) distribution whose PDF (see the description of gamma distribution in Sect. 1.2.2) is

$$f_{\tau_n}(x) = \begin{cases} \dfrac{\lambda^n}{\Gamma(n)} x^{n-1} e^{-\lambda x}, & \text{if } x > 0, \\ 0, & \text{if } x \leq 0. \end{cases}$$

From the exponential distribution of the first arrival we have

$$\mathbf{P}(M(t) = 0) = \mathbf{P}(X_1 > t) = e^{-\lambda t}.$$

Using the theorem of the total expected value, for every positive integer k we obtain

$$\mathbf{P}(M(t) = k) = \mathbf{P}(X_1 + \ldots + X_k \leq t < X_1 + \ldots + X_{k+1})$$

$$= \mathbf{P}(\tau_k \leq t < \tau_k + X_{k+1})$$

$$= \int_0^t \mathbf{P}(\tau_k \leq t < \tau_k + X_{k+1} \mid \tau_k = z) \frac{\lambda^k}{\Gamma(k)} z^{k-1} e^{-\lambda z} dz$$

$$= \int_0^t \mathbf{P}(t - z < X_{k+1}) \frac{\lambda^k}{\Gamma(k)} z^{k-1} e^{-\lambda z} dz$$

$$= \int_0^t e^{-\lambda(t-z)} \frac{\lambda^k}{\Gamma(k)} z^{k-1} e^{-\lambda z} dz$$

$$= \frac{\lambda^k}{\Gamma(k)} e^{-\lambda t} \int_0^t z^{k-1} dz = \frac{(\lambda t)^k}{\Gamma(k)k} e^{-\lambda t} = \frac{(\lambda t)^k}{k!} e^{-\lambda t},$$

thus the random variable $M(t)$, $t \geq 0$ has a Poisson distribution with the parameter λt.

(b) Let T be a fixed positive number and let $U_1, ...U_k$ be independent random variables uniformly distributed on the interval $(0, 1)$. Denote by $U_{1k} \leq ... \leq U_{kk}$ the ordered random variables $U_1, ...U_k$. Now we verify that for every positive integer k the joint CDF of random variables $\tau_1, ..., \tau_k$ under the condition $M(T) = k$ is identical with the joint CDF of the ordered random variables $U_{1k}, ..., U_{kk}$ (see Theorem 2.3 of Ch. 4. in [3]).

For any positive numbers $t_1, ..., t_k$, the joint conditional CDF of random variables $\tau_1, ..., \tau_k$ given $M(t) = k$ can be written in the form

$$\mathbf{P}\left(\tau_1 \leq t_1, ..., \tau_k \leq t_k \mid M(T) = k\right) = \frac{\mathbf{P}\left(\tau_1 \leq t_1, ..., \tau_k \leq t_k, M(T) = k\right)}{\mathbf{P}\left(M(T) = k\right)}.$$

By the result proved in part (a), the denominator has the form

$$\mathbf{P}\left(M(T) = k\right) = \frac{(\lambda T)^k}{k!} e^{-\lambda T}, \quad k = 0, 1, ...,$$

while the numerator can be written as follows:

$$\mathbf{P}\left(\tau_1 \leq t_1, ..., \tau_k \leq t_k, M(T) = k\right)$$

$$= \mathbf{P}\left(X_1 \leq t_1, X_1 + X_2 \leq t_2, ..., X_1 + ... + X_k \leq t_k, X_1 + ... + X_{k+1} > T\right)$$

$$= \int_0^{t_1} \int_0^{t_2 - u_1} \int_0^{t_3 - (u_1 + u_2)} \cdots \int_0^{t_k - (u_1 + ... + u_{k-1})} \int_{T - (u_1 + ... + u_k)}^{\infty} \prod_{i=1}^{k+1} \left(\lambda e^{-\lambda u_i}\right) du_{k+1} ... du_1$$

$$= \lambda^k \int_0^{t_1} \int_0^{t_2 - u_1} \int_0^{t_3 - (u_1 + u_2)} \cdots \int_0^{t_k - (u_1 + ... + u_{k-1})} e^{-\lambda(u_1 + ... + u_k)} e^{-\lambda(T - u_1 + ... + u_k)} du_{k+1} ... du_1$$

$$= \lambda^k e^{-\lambda T} \int_0^{t_1} \int_0^{t_2 - u_1} \int_0^{t_3 - (u_1 + u_2)} \cdots \int_0^{t_k - (u_1 + ... + u_{k-1})} du_k ... du_1.$$

Setting $v_1 = u_1$, $v_2 = u_1 + u_2$, ..., $v_k = u_1 + ... + u_k$, the last integral takes the form

$$\frac{(\lambda T)^k}{k!} e^{-\lambda T} \frac{k!}{T^k} \int_0^{t_1} \int_{v_1}^{t_2} \int_{v_2}^{t_3} ... \int_{v_{k-1}}^{t_k} dv_k...dv_1,$$

thus

$$\mathbf{P}(\tau_1 \leq t_1, ..., \tau_k \leq t_k \,|M(T) = k) = \frac{k!}{T^k} \int_0^{t_1} \int_{v_1}^{t_2} \int_{v_2}^{t_3} ... \int_{v_{k-1}}^{t_k} dv_k...dv_1.$$

From this we get that the joint conditional PDF of random variables $\tau_1, ..., \tau_k$ given $M(T) = k$ equals the constant value $\frac{k!}{T^k}$, which, by the preceding lemma, is identical with the joint PDF of random variables $U_{1k}, ..., U_{kk}$. Using the proof of Construction II, we obtain that Construction I has the result of a homogeneous Poisson process at rate λ on the interval $(0, T]$, and at the same time on the whole interval $(0, \infty)$, because T was chosen arbitrarily.   □

2.7.3 Basic Properties of a Homogeneous Poisson Process

Let $N(t)$, $t \geq 0$ be a homogeneous Poisson process with rate λ. We enumerate below the main properties of $N(t)$.

(a) For any $t \geq 0$ the CDF of $N(t)$ is Poisson with the parameter λt, that is,

$$\mathbf{P}(N(t) = k) = \frac{(\lambda t)^k}{k!} e^{-\lambda t}, \ k = 0, 1, ...$$

(b) The increments of $N(t) - N(s)$, $0 \leq s < t$, are independent and have a Poisson distribution with the parameter $\lambda(t - s)$.

(c) The sum of two independent homogeneous Poisson processes $N_1(t; \lambda_1)$ and $N_2(t; \lambda_2)$ at rates λ_1 and λ_2, respectively, is a homogeneous Poisson process with rate $(\lambda_1 + \lambda_2)$.

(d) Given $0 < t < T < \infty$, a positive integer N_0 and an integer k satisfy the inequality $0 \leq k \leq N_0$. The conditional CDF of the random variable $N(t)$ given $N(T) = N_0$ is binomial with the parameters $(N_0, 1/T)$.

Proof

$$\mathbf{P}\left(N(t) = k \mid N(T) = N_0\right) = \frac{\mathbf{P}\left(N(t) = k, N(T) = N_0\right)}{\mathbf{P}\left(N(T) = N_0\right)}$$

$$= \frac{\mathbf{P}\left(N(t) = k, N(T) - N(t) = N_0 - k\right)}{\mathbf{P}\left(N(T) = N_0\right)}$$

$$= \frac{(\lambda t)^k}{k!} e^{-\lambda t} \frac{(\lambda(T - t))^{N_0 - k}}{(N_0 - k)!} e^{-\lambda(T-t)} \left(\frac{(\lambda T)^{N_0}}{N_0!} e^{-\lambda T}\right)^{-1}$$

$$= \binom{N_0}{k} \left(\frac{t}{T}\right)^k \left(1 - \frac{t}{T}\right)^{N_0 - k}.$$

$\square$

(e) The following asymptotic relations are valid as $h \to 0+$

$$\mathbf{P}\left(N(h) = 0\right) = 1 - \lambda h + o(h),$$

$$\mathbf{P}\left(N(h) = 1\right) = \lambda h + o(h),$$

$$\mathbf{P}\left(N(h) \geq 2\right) = o(h). \qquad \text{(orderliness)}$$

Lemma 2.2 *For every nonnegative integer m the inequality*

$$\left| e^x - \sum_{k=0}^m \frac{x^k}{k!} \right| < \frac{|x|^{m+1}}{(m+1)!} e^{|x|} = o(|x|^m), \quad x \to 0$$

holds.

Proof The assertion of the lemma follows from the nth order Taylor approximation to e^x with the Lagrange form of the remainder term (see, Sect 7.7. of [1]), but one can obtain it by simple computations. Using the Taylor expansion

$$e^x = \sum_{k=0}^\infty \frac{x^k}{k!}$$

of the function e^x, which implies that

$$\left| e^x - \sum_{k=0}^m \frac{x^k}{k!} \right| = \left| \sum_{k=m+1}^\infty \frac{x^k}{k!} \right| \leq \frac{|x|^{m+1}}{(m+1)!} \sum_{k=0}^\infty \frac{(m+1)!}{(m+1+k)!} |x|^k$$

$$< \frac{|x|^{m+1}}{(m+1)!} \sum_{k=0}^\infty \frac{|x|^k}{k!} = \frac{|x|^{m+1}}{(m+1)!} e^{|x|} = o(|x|^m), \quad x \to 0.$$

$\square$

Proof of Property (e) From the preceding lemma we have as $h \to 0+$

$$\mathbf{P}\left(N(h) = 0\right) = e^{-\lambda h} = 1 - \lambda h + o(h),$$

$$\mathbf{P}\left(N(h) = 1\right) = \frac{\lambda h}{1!} e^{-\lambda h} = \lambda h (1 - \lambda h + o(h)) = \lambda h + o(h),$$

$$\mathbf{P}\left(N(h) \geq 2\right) = 1 - \left(e^{-\lambda h} + \frac{(\lambda h)^1}{1!} e^{-\lambda h} \right) = o(h).$$

$\square$

(f) Given that exactly one event of a homogeneous Poisson process $[N(t),\ t \geq 0]$ has occurred during the interval $(0, t]$, the time of occurrence of this event is uniformly distributed over $(0, t]$.

Proof of Property (f) Denote by λ the rate of the process $N(t)$. Immediate application of the conditional probability gives for all $0 < x < t$

$$\mathbf{P}\left(X_1 \leq x | N(t) = 1\right) = \frac{\mathbf{P}\left(X_1 \leq x,\ N(t) = 1\right)}{\mathbf{P}\left(N(t) = 1\right)}$$

$$= \frac{\mathbf{P}\left(N(x) = 1,\ N(t) - N(x) = 0\right)}{\mathbf{P}\left(N(t) = 1\right)}$$

$$= \frac{\mathbf{P}\left(N(x) = 1\right)\mathbf{P}\left(N(t - x) = 0\right)}{\mathbf{P}\left(N(t) = 1\right)}$$

$$= \left(\frac{(\lambda x)^1}{1!} e^{-\lambda x} \frac{[\lambda(t - x)]^0}{0!} e^{-\lambda(t-x)} \right) \left(\frac{(\lambda t)^1}{1!} e^{-\lambda t} \right)^{-1} = \frac{x}{t}.$$

$\square$

(g) **Strong Markov property.** [4] Let $\{N(t),\ t \geq 0\}$ be a homogeneous Poisson process with the rate λ, and assume that $N(t)$ is $\mathscr{A}_t$- measurable for all $t \geq 0$, where $\mathscr{A}_t \subset \mathscr{A},\ t \geq 0$, is a monotonically increasing family of σ- algebras. Let τ be a random variable such that the condition $\{\tau \leq t\} \in \mathscr{A}_t$ holds for all $t \geq 0$. This type of random variable is called a **Markov point** with respect to the family of σ- algebra $\mathscr{A}_t,\ t \geq 0$. For example, the constant $\tau = t$ and the so-called **first hitting time**, $\tau_k = \sup\{s : N(s) < k\}$, where k is a positive integer, are Markov points. Denote

$$N_\tau(t) = N(t + \tau) - N(\tau),\ t \geq 0.$$

Then the process $N_\tau(t),\ t \geq 0$, is a homogeneous Poisson process with the rate λ, which does not depend on the Markov point τ or on the process $\{N(t),\ 0 \leq t \leq \tau\}$.

(h) **Random deletion (filtering) of a Poisson process** Let $N(t)$, $t \geq 0$ be a homogeneous Poisson process with rate $\lambda > 0$. Let us suppose that we delete points in the process $N(t)$ independently with probability $(1 - p)$, where $0 < p < 1$ is a fixed number. Then the new process $M(t)$, $t \geq 0$, determined by the undeleted points of $N(t)$ constitutes a homogeneous Poisson process with rate $p\lambda$.

Proof of Property (h) Let us represent the Poisson process $N(t)$ in the form

$$N(t) = \sum_{k=1}^{\infty} \mathscr{I}_{\{\tau_k \leq t\}}, \quad t \geq 0,$$

where $\tau_k = X_1 + \ldots + X_k$, $k = 1, 2, \ldots$ and $X_1, X_2, \ldots$ are independent exponentially distributed random variables with the parameter λ. The random deletion in the process $N(t)$ can be realized with the help of a sequence of independent and identically distributed random variables $I_1, I_2, \ldots$, which do not depend on the process $N(t)$, $t \geq 0$ and have a distribution $\mathbf{P}(I_k = 1) = p$, $\mathbf{P}(I_k = 0) = 1 - p$. The deletion of a point τ_k in the process $N(t)$ happens only in the case $I_k = 0$. Let $T_0 = 0$ and denote by $0 < T_1 < T_2 < \ldots$ the sequence of remaining points. Thus the new process can be given in the form

$$M(t) = \sum_{k=1}^{\infty} \mathscr{I}_{\{T_k \leq t\}} = \sum_{k=1}^{\infty} \mathscr{I}_{\{\tau_k \leq t, \, I_k = 1\}}, \quad t \geq 0.$$

Using the property of the process $N(t)$ and the random sequence I_k, $k \geq 1$, it is clear that the sequence of random variables $Y_k = T_k - T_{k-1}$, $k = 1, 2, \ldots$, are independent and identically distributed; therefore, it is enough to prove that they have an exponential distribution with parameter $p\lambda$, i.e., $\mathbf{P}(Y_k < y) = 1 - e^{-p\lambda y}$.

The sequence of the remaining points T_k can be given in the form $T_k = \tau_{n_k}$, $k = 1, 2, \ldots$, where the random variables n_k are defined as follows:

$$n_1 = \min\{j : \; j \geq 1, \; I_j = 1\},$$

$$n_k = \min\{j : \; j > n_{k-1}, \; I_j = 1\}, \; k \geq 2.$$

Let us compute the distribution of the random variable

$$Y_1 = T_1 = X_1 + \ldots + X_{n_1}.$$

By the use of the formula of total probability, we obtain

$$\mathbf{P}\left(Y_1 < y\right) = \mathbf{P}\left(X_1 + \ldots + X_{n_1} < y\right)$$

$$= \sum_{k=1}^{\infty} \mathbf{P}\left(X_1 + \ldots + X_{n_1} < y | n_1 = k\right)\mathbf{P}\left(n_1 = k\right)$$

$$= \sum_{k=1}^{\infty} \mathbf{P}\left(X_1 + \ldots + X_k < y\right)\mathbf{P}\left(n_1 = k\right).$$

The sum $X_1 + \ldots + X_k$ of independent exponentially distributed random variables X_i has a gamma distribution with the density function

$$f(y; k, \lambda) = \frac{\lambda^k}{(k-1)!}y^{k-1}e^{-\lambda y}, \ \ y > 0,$$

whereas, on the other hand, the random variable n_1 has a geometric distribution with the parameter p, i.e.,

$$\mathbf{P}\left(n_1 = k\right) = (1-p)^{k-1}p;$$

therefore, we get

$$\sum_{k=1}^{\infty} \mathbf{P}\left(X_1 + \ldots + X_k < y\right)\mathbf{P}\left(n_1 = k\right) = \sum_{k=1}^{\infty} \int_0^y \frac{\lambda^k}{(k-1)!}x^{k-1}e^{-\lambda x}(1-p)^{k-1}p\,dx$$

$$= \lambda p \int_0^y \left(\sum_{k=0}^{\infty} \frac{[(1-p)\lambda x]^k}{k!}\right)e^{-\lambda x}dx = \lambda p \int_0^y e^{(1-p)\lambda x}e^{-\lambda x}dx$$

$$= \lambda p \int_0^y e^{-p\lambda x}dx = 1 - e^{-p\lambda y}.$$

(i) **Modeling an inhomogeneous Poisson process.** Let $\{\Lambda(t), \ t \geq 0\}$ be a non-negative, monotonically nondecreasing, continuous-from-left function such that $\Lambda(0) = 0$. Let $N(t), \ t \geq 0$, be a homogeneous Poisson process with rate 1. Then the process defined by the equation

$$N_\Lambda(t) = N(\Lambda(t)), \ t \geq 0,$$

is a Poisson process with mean value function $\Lambda(t), \ t \geq 0$.

Proof of Property (i) Obviously, $N(\Lambda(0)) = N(0) = 0$, and the increments of the process $N_\Lambda(t)$ are independent and the CDF of the increments are

Poissonian, because for any $0 \leq s \leq t$ the CDF of the increment $N_\Lambda(t) - N_\Lambda(s)$ is Poisson with the parameter $\Lambda(t) - \Lambda(s)$,

$$\mathbf{P}(N_\Lambda(t) - N_\Lambda(s) = k) = \mathbf{P}(N(\Lambda(t)) - N(\Lambda(s)))$$

$$= \frac{(\Lambda(t) - \Lambda(s))^k}{k!} e^{-(\Lambda(t) - \Lambda(s))}, \ k = 0, 1, ...$$

2.7.4 Higher-Dimensional Poisson Process

The Poisson process can be defined, in higher dimensions, as a model of random points in space. To do this, we first concentrate on the process on the real number line, from the aspect of a possible generalization.

Let $\{N(t), \ t \geq 0\}$ be a Poisson process on a probability space $(\Omega, \mathscr{A}, P)$. Assume that it has a rate function $\lambda(t), \ t \geq 0$; thus the mean value function has the form

$$\Lambda(t) = \int_0^t \lambda(s) \mathrm{d}s, \ t \geq 0,$$

where the function $\lambda(t)$ is nonnegative and locally integrable function. Denote by $\tau_1, \tau_2, ...$ the sequence of the random jumping points of $N(t)$. Since the mean value function is continuous, the jumps of $N(t)$ are exactly 1; moreover, the process $N(t)$ and the random points $\Pi = \{\tau_1, \tau_2, ..\}$ determine uniquely each other. If we can characterize the countable set Π of random points $\{\tau_1, \tau_2, ..\}$, then at the same time we can give a new definition of the Poisson process $N(t)$.

Denote by $\mathscr{B}_+ = \mathscr{B}(\mathbb{R}_+)$ the Borel σ-algebra of the half line $\mathbb{R}_+ = [0, \infty)$, i.e., the minimal σ-algebra that consists of all open intervals of $\mathbb{R}_+$. Let $B_i = (a_i, b_i], \ i = 1, ..., n$, be nonoverlapping intervals of $\mathbb{R}_+$; then obviously $B_i \in \mathscr{B}_+$. Introduce the random variables

$$\Pi(B_i) = \#\{\Pi \cap B_i\} = \#\{\tau_j : \tau_j \in B_i\}, \ i = 1, ..., n,$$

where $\#\{\cdot\}$ means the number of elements of a set; then

$$\Pi(B_i) = N(b_i) - N(a_i).$$

By the use of the properties of Poisson processes, the following statements hold:

(i) The random variables $\Pi(B_i)$ are independent because the increments of the process $N(t)$ are independent.

(ii) The CDF of $\Pi(B_i)$ is Poisson with parameter $\Lambda(B_i)$, i.e.,

$$\mathbf{P}\left(\Pi(B_i) = k\right) = \frac{(\Lambda(B_i))^k}{k!} e^{-\Lambda(B_i)},$$

where $\Lambda(B_i) = \int_{B_i} \lambda(s) \mathrm{d}s, \ 1 \le i \le n$.

Observe that by the definition of random variables $\Pi(B_i)$, it is unimportant whether or not the set of random points $\Pi = \{\tau_i\}$ is ordered and $\Pi(B_i)$ is determined by the number of points τ_i only, which is included in the interval $(a_i, b_i]$. This circumstance is important because we want to define the Poisson processes on higher-dimensional spaces, which do not constitute an ordered set, contrary to the one-dimensional case.

More generally, let $B_i \in \mathscr{B}(\mathbb{R}_+), \ 1 \le i \le n$, be disjoint Borel sets and denote $\Pi(B_i) = \#\{\Pi \cap B_i\}$. It can be checked that $\Pi(B_i)$ are random variables defined by the random points $\Pi = \{\tau_1, \tau_2, ..\}$ and they satisfy properties (1) and (2). On this basis, the Poisson process can be defined in higher-dimensional Euclidean spaces and, in general, in metric spaces also (see Chap. 2 of [4] and Chap. 3 of [5]).

Before discussing the higher-dimensional case, we note that a homogeneous Poisson point process Π of the rate λ allows a simple representation on the whole real line with a collection of random points $... - \tau_{-2} < -\tau_{-1} < 0 < \tau_1 < \tau_2 < ...$, where the random variables $... \tau_{-2} - \tau_{-1}, \tau_{-1}, \tau_1, \tau_2 - \tau_1, ...$ are independent and identically distributed with an exponential distribution of the parameter λ. It is clear that if we restrict ourselves to an interval $(a, b]$, a homogeneous Poisson process $N(t), \ t \ge a, \ N(a) = 0$ at rate λ can be defined as $N(t) = \#\{\Pi \cap (a, t]\}, \ a < t \le b$.

Consider the d-dimensional Euclidean space $S = \mathbb{R}^d$ and denote by $\mathscr{B}(S)$ the Borel σ- algebra of the subsets of S. We will define the Poisson process Π as a random set function satisfying properties (i) and (ii).

Let $\Pi : \Omega \to \mathscr{S}$ be a random point set in $\mathscr{S}$, where $\mathscr{S}$ denotes the set of all subsets of S consisting of countable points. Then the quantities $\Pi(A) = \#\{\Pi \cap A\}$ define random variables for all $A \in \mathscr{B}(S)$. Introduce the set function $\Lambda(A) = E(\Pi(A)), \ A \in \mathscr{B}(S)$. Evidently, for any disjoint sets $A_1, A_2, ... \in \mathscr{B}(S)$,

$$\Pi(A) = \sum_{i=1}^{\infty} \Pi(A_i) \quad \text{and} \quad \Lambda(A) = \sum_{i=1}^{\infty} \Lambda(A_i),$$

where $A = \cup_{i=1}^{\infty} A_i$. The last relation means that the set function $\Lambda(A), \ A \in \mathscr{B}(S)$ satisfies the conditions of a measure, i.e., it is a nonnegative, σ-additive set function on the measurable space $(S, \mathscr{B}(S))$, which justifies to call Λ as **mean measure**.

Definition 2.10 We say that Π is a Poisson process with mean measure Λ on the space S if $\Pi \in \mathscr{S}$ is a random countable set of points in S and the following conditions are satisfied:

(1) The random variables $\Pi(A_i) = \#\{\Pi \cap A_i\}$ are independent for all disjoint sets $A_1, ..., A_n \in \mathscr{B}(S)$.
(2) For any $A \in \mathscr{B}(S)$, the CDF of random variables $\Pi(A)$ are Poisson with the parameter $\Lambda(A)$, where $0 \le \Lambda(A) \le \infty$.

The function $\Lambda(A)$, $A \in \mathscr{B}(S)$ is called the **mean measure** of a Poisson process (see, [4], p. 14).

Properties:

1. Since the random variable $\Pi(A)$ has a Poisson distribution with the parameter $\Lambda(A)$, then $\mathbf{E}(\Pi(A)) = \Lambda(A)$ and $\mathbf{D}^2(\Pi(A)) = \Lambda(A)$.
2. If $\Lambda(A)$ is finite, then the random variable $\Pi(A)$ is finite with probability 1, and if $\Lambda(A) = \infty$, then the number of elements of the random point set $\Pi \cap A$ is countable infinite with probability 1.

Like the one-dimensional case, when the Poisson process has a rate function, it is an important class of Poisson processes for which there exists a nonnegative locally integrable function λ, called **intensity function**, with the property

$$\Lambda(B) = \int_B \lambda(s)\mathrm{d}s, \ B \in \mathscr{B}(S)$$

(here the integral is defined with respect to the Lebesgue measure $\mathrm{d}s$). Then the mean measure Λ is **nonatomic**, that is, there is no point $s_0 \in \mathscr{B}(S)$ such that $\Lambda(\{s_0\}) > 0$.

3. By the use of properties 1 and 3, it is easy to obtain the relation

$$\mathbf{D}^2(\Pi(A)) = \sum_{i=1}^{\infty} \mathbf{D}^2(\Pi(A_i)) = \sum_{i=1}^{\infty} \Lambda(A_i) = \Lambda(A).$$

4. For any $B, C \in \mathscr{B}(S)$,

$$\mathbf{Cov}(\Pi(B), \Pi(C)) = \Lambda(B \cap C).$$

Proof Since $\Pi(B) = \Pi(B \cap C) + \Pi(B \backslash C)$ and $\Pi(C) = \Pi(B \cap C) + \Pi(C \backslash B)$, where the sets $A \cap C$, $A \backslash C$ and $C \backslash A$ are disjoint, the $\Pi(A \cap C)$, $\Pi(A \backslash C)$, and $\Pi(C \backslash A)$ are independent random variables, and thus

$$\mathbf{Cov}(\Pi(A), \Pi(C)) = \mathbf{Cov}(\Pi(A \cap C), \Pi(A \cap C))$$

$$= \mathbf{D}^2(\Pi(A \cap C)) = \Lambda(A \cap C).$$

5. For any (not necessarily disjoint) sets $A_1, ..., A_n \in \mathscr{B}(S)$ the joint distribution of random variables $\Pi(A_1), ..., \Pi(A_n)$ is uniquely determined by the mean measure Λ.

Proof Denote the set of the 2^n pairwise disjoint sets by

$$\mathscr{C} = \{C = B_1 \cap ... \cap B_n, \text{ where } B_i \text{ means the set either } A_i, \text{ or } \bar{A}_i\},$$

then the random variables $\Pi(C)$ are independent and have a Poisson distribution with the parameter $\Lambda(C)$. Consequently, the random variables $\Pi(A_1), \ldots, \Pi(A_n)$ can be given as a sum from a 2^n number of independent random variables $\Pi(C)$, $C \in \mathscr{C}$, having a Poisson distribution with the parameter $\Lambda(C)$; therefore, the joint distribution of random variables $\Pi(A_i)$ is uniquely determined by $\Pi(C)$, $C \in \mathscr{C}$ and also the mean measure Λ.   □

Remark 2.1 Let $S = \mathbb{R}^d$ and assume

$$\Lambda(A) = \int_A \lambda(x)\mathrm{d}x, \ A \in \mathscr{B}(S),$$

where $\lambda(x)$ is a nonnegative and locally integrable function and $\mathrm{d}x = \mathrm{d}x_1 \ldots \mathrm{d}x_n$. If $|A|$ denotes the n-dimensional (Lebesgue) measure of a set A and the function $\lambda(x)$ is continuous at a point $x_0 \in S$, then

$$\Lambda(A) \sim \lambda(x_0)\,|A|$$

if the set A is included in a small neighborhood of the point x_0.

The Poisson process Π is called **homogeneous** if $\lambda(x) = \lambda$ for a positive constant λ. In this case for any $A \in \mathscr{B}(S)$ the inequality $\Lambda(A) = \lambda\,|A|$ holds.

The following three theorems state general assertions on the Poisson processes defined in higher-dimensional spaces (see, Chap. 2 of [4] and p. 190 of [5]).

Theorem 2.5 (Existence Theorem) *If the mean measure Λ is σ-finite on the space S, i.e., it can be expressed in the form*

$$\Lambda = \sum_{i=1}^{\infty} \Lambda_i, \ \text{where} \ \Lambda_i(S) < \infty,$$

then there exists a Poisson process Π on the space S and has mean measure Λ.

Theorem 2.6 (Superposition Theorem) *If $\Pi_i, i = 1, 2, \ldots$, is a sequence of independent Poisson processes with mean measure $\Lambda_1, \Lambda_2, \ldots$ on the space S, then the superposition $\Pi = \cup_{i=1}^{\infty}\Pi_i$ is a Poisson process with mean measure $\Lambda = \sum_{i=1}^{\infty} \Lambda_i$.*

Theorem 2.7 (Restriction Theorem) *Let Π be a Poisson process on the space S with mean measure Λ. Then for any $S_0 \in \mathscr{B}(S)$ the process*

$$\Pi_0 = \Pi \cap S_0$$

can be defined as a Poisson process on S with mean measure

$$\Lambda_0(A) = \Lambda(A \cap S_0).$$

The process Π_0 can be interpreted as a Poisson process on the space S_0 with mean measure Λ_0, where Λ_0 is called the restriction of mean measure Λ to S_0.

2.7.5 Campbell's and Displacement Theorems

Let $N(s)$, $s \geq 0$, $N(0) = 0$ be a homogeneous Poisson process with rate $\lambda > 0$. Denote by $0 < \tau_1 < \tau_2 < \ldots$ the sequence of its jump times and let $R(t) = \sum_{\tau_k \leq t} f(\tau_k)$, $t \geq 0$ be the sum of f over the jump times of the process $N(s)$, $0 \leq s \leq t$, where $f(x)$ is a nonnegative measurable deterministic function.

The formulas (2.3) and (2.4) of the following theorem determine the expected value and variation of the random variable $R(t)$, which are known as Campbell's formulas [2]. Moreover, the Laplace transform $\mathbf{E}\left(e^{-sR(t)}\right)$ can also be given in an explicit form.

Theorem 2.8 (Campbell) *If the integral $\int_0^t f(x)dx$ exists, then*

$$\mathbf{E}\left(R(t)\right) = \lambda \int_0^t f(x)dx, \quad \mathbf{D}^2\left(R(t)\right) = \lambda \int_0^t f^2(x)dx. \quad (2.3)$$

The Laplace transform of the random variable $R(t)$ has the following form

$$\mathbf{E}\left(e^{-sR(t)}\right) = \exp\left\{\lambda \int_0^t \left(e^{-sf(x)} - 1\right) dx\right\}. \quad (2.4)$$

Proof The proof will be based on the construction of Theorem 2.4, which allows us to represent the process $N(s)$, $0 \leq s \leq t$ by its jump times τ_k, $k = 1, 2, \ldots$ in the following way: let N be a random variable of Poisson distribution with the parameter λt; moreover, let $U_1, U_2, \ldots$ be a sequence of independent and uniformly distributed random variables on the interval $(0, t)$, that is independent of the random variable N. Denote by $U_{N,1} \leq U_{N,2} \leq \ldots \leq U_{N,N}$ the ordered statistics of random variables $U_1, U_2, \ldots, U_N$. Then the process of random sums

$$N(s) = \sum_{k=1}^N \mathscr{I}_{\{U_{N,k} \leq s\}}, \quad 0 \leq s \leq t$$

constitutes a homogeneous Poisson process with rate λ and the random variables $\tau_k = U_{N,k}$, $k = 1, 2, \ldots$ are the jump times of the process $N(t)$. The preceding sum is defined to be zero if $N = 0$. It is clear from the definition of random variables τ_k that $R(t)$ can be rewritten in the form

$$R(t) = \sum_{k=1}^{N} f(\tau_k) = \sum_{i=1}^{N} f(U_{N,i}) = \sum_{i=1}^{N} f(U_i). \tag{2.5}$$

Then using the Wald's Lemma (Exercise 2.1), we obtain

$$\mathbf{E}\left(R(t)\right) = \mathbf{E}\left(N\right)\mathbf{E}\left(f(U_1)\right) = \lambda t \int_0^t \frac{1}{t} f(x)\mathrm{d}x = \lambda \int_0^t f(x)\mathrm{d}x.$$

Moreover, $\mathbf{E}\left(f(U_1)\right) = \int_0^t \frac{1}{t} f(x)\mathrm{d}x$, $\mathbf{E}\left(f^2(U_1)\right) = \int_0^t \frac{1}{t} f^2(x)\mathrm{d}x$ and $\mathbf{E}\left(N\right) = \lambda t$, $\mathbf{E}\left(N^2\right) = \lambda t + (\lambda t)^2$, then

$$\mathbf{D}^2\left(R(t)\right) = \mathbf{E}\left(N\right)\mathbf{D}^2\left(f(U_1)\right) + \mathbf{D}^2\left(N\right)(\mathbf{E}\left(f(U_1)\right))^2$$

$$= (\lambda t)[\mathbf{E}\left(f^2(U_1)\right) - (\mathbf{E}\left(f(U_1)\right))^2] + (\lambda t)(\mathbf{E}\left(f(U_1)\right))^2$$

$$= (\lambda t)\mathbf{E}\left(f^2(U_1)\right) = \lambda t \int_0^t \frac{1}{t} f^2(x)\mathrm{d}x = \lambda \int_0^t f^2(x)\mathrm{d}x.$$

To prove Eq. (2.4) we observe that the independence of random variables $U_1, U_2, \ldots$ implies

$$\mathbf{E}\left(e^{-sR(t)} \mid N = k\right) = \mathbf{E}\left(\exp\{-s(f(U_1) + \ldots + f(U_k))\}\right)$$

$$= [\mathbf{E}\left(\exp\{-sf(U_1)\}\right)]^k, \ k \geq 0$$

and it is obvious that

$$\mathbf{E}\left(\exp\{-sf(U_1)\}\right) = \frac{1}{t} \int_0^t e^{-sf(x)}\mathrm{d}x,$$

then we can write

$$\mathbf{E}\left(e^{-sR(t)}\right) = \sum_{k=0}^{\infty} \mathbf{E}\left(e^{-sR(t)} \mid N = k\right)\mathbf{P}\left(N = k\right)$$

$$= \sum_{k=0}^{\infty} (\mathbf{E}\left(\exp\{-sf(U_1)\}\right))^k \frac{(\lambda t)^k}{k!} e^{-\lambda t}$$

$$= \exp\{\mathbf{E}\left(\exp\{-sf(U_1)\}\right)\lambda t - \lambda t\} = \exp\left\{\lambda \int_0^t (e^{-sf(x)} - 1)\mathrm{d}x\right\},$$

which completes the proof.

The assertions of Theorem 2.8 remain valid in a more general context (see, [4, 5]). Let Poisson process Π be defined on the d-dimensional Euclidean space $S = \mathbb{R}^d$ with mean measure Λ that may also consist of atoms; $f : S \to \mathbb{R} = [0, \infty)$ is a measurable function for which the integrals $\int_S f(x)\Lambda(dx)$ and $\int_S f^2(x)\Lambda(dx)$ exist. Then

$$\mathbf{E}\,(R) = \int_S f(x)\Lambda(dx), \qquad \mathbf{D}^2\,(R) = \int_S f^2(x)\Lambda(dx).$$

and

$$\mathbf{E}\left(e^{-sR}\right) = \exp\left\{-\int_S (1 - e^{-sf(x)})\Lambda(dx)\right\}.$$

Related to Poisson processes, we touch upon an important version of the so-called displacement theorem (2.9) and some consequences of that.

Theorem 2.9 (Displacement Theorem) *Let Π be a Poisson point process on the real line $\mathbb{R}$ with rate function $\lambda(t)$. Suppose that at time $t \in \mathbb{R}$ for each $\tau \in \Pi$, a random variable $W_\tau(t)$ is assigned with values in the real line $\mathbb{R}'$ such that the points of Π are randomly and independently displaced to some new location $\tau' = W_\tau(t) \in \mathbb{R}'$. Furthermore the random variables $W_\tau(t)$ take a value in a Borel set $B \in \mathscr{B}'(\mathbb{R}')$ according to a probability kernel given $\tau = x \in \mathbb{R}$*

$$P_t(x, B) = \mathbf{P}\,(W_\tau(t) \in B \mid \tau = x),$$

where the probability kernel $P_t(\cdot, B)$ is supposed to be a measurable function on $\mathbb{R}$ and $P_t(x, \cdot)$ to be a probability measure on $\mathbb{R}'$. Then the set of random points $\Pi' = \{\tau'\}$ constitutes a Poisson point process on $\mathbb{R}'$ with mean measure

$$\Theta(B) = \int_{-\infty}^{t} P_t(x, B)\lambda(x)dx,$$

provided that the integral exists. Moreover, if the probability measure $P_t(x, \cdot)$ has a density function $p_t(x, y)$, i.e., it can be represented in the form $P_t(x, B) = \int_B p_t(x, y)dy$, $B \in \mathscr{B}'$, then the rate function $\vartheta(y)$ of the Poisson process Π' exists and

$$\vartheta(y) = \int_{-\infty}^{t} p_t(x, y)\lambda(x)dx. \tag{2.6}$$

2.8   Exercises

Exercise 2.1 Let $X_1, X_2, \ldots$ be independent identically distributed random variables with finite absolute moment $\mathbf{E}\left(|X_1|\right) < \infty$. Let N be a random variable taking nonnegative integer numbers and independent of the random variables $(X_i, i = 1, 2, \ldots)$. Define $S_0 = 0$, and $S_n = X_1 + \ldots + X_n$, $n \geq 1$. Prove that

(a) $\mathbf{E}\left(S_N\right) = \mathbf{E}\left(X_1\right)\mathbf{E}\left(N\right)$,
(b) $\mathbf{D}^2\left(S_N\right) = \mathbf{E}\left(N\right)\mathbf{D}^2\left(X_1\right) + \mathbf{D}^2\left(N\right)(\mathbf{E}\left(X_1\right))^2$, provided that $\mathbf{E}\left(N^2\right) < \infty$, $\mathbf{E}\left(X_1^2\right) < \infty$.

Which are known as Wald identities or Wald lemma.

Exercise 2.2 Let $X_0, X_1, \ldots$ be independent random variables with joint distribution $\mathbf{P}\left(X_i = 1\right) = \mathbf{P}\left(X_i = -1\right) = \frac{1}{2}$.

Define $Z_0 = 0$, $Z_k = Z_{k-1} + X_k$, $k = 0, 1, \ldots$ Determine the expectation and covariance function of the process $(Z_k, k = 1, 2, \ldots)$ (random walk on the integer numbers).

Let a and b be real numbers, $|b| < 1$. Denote $W_0 = aX_0$, $W_k = bW_{k-1} + X_k$, $k = 1, 2, \ldots$ [here the process $(W_k, k = 0, 1, \ldots$ constitutes a first degree autoregressive process with the initial value aX_0 and with the innovation process $(X_k, k = 1, 2, \ldots)$]. If we fix the value b, for which value of a will the process W_k be stationary in weak sense?

Determine the expectation and covariance function of the process W_k.

Exercise 2.3 Let a and b real numbers and let U be a random variable uniformly distributed on the interval $(0, 2\pi)$. Denote by $X_t = a\cos(bt + U)$, $-\infty < t < \infty$. Prove that the random cosine process $(X_t, -\infty < t < \infty)$ is stationary.

Exercise 2.4 Let $N(t), T \geq 0$ be a homogeneous Poisson process with intensity λ.

(a) Determine the covariance and correlation function of $N(t)$.
(b) Determine the conditional expectation $\mathbf{E}\left(N(t + s) \mid N(t)\right)$.

2.9   Solutions

Solution 2.1

(a) The random variable N does not depend on the independent and identically distributed random variables $X_1, X_2, \ldots$, then by the use of the properties of conditional expectation we have

$$\mathbf{E}\left(S_N\right) = \mathbf{E}\left(\mathbf{E}\left(\sum_{i=1}^{N} X_i \mid N\right)\right)$$

$$= \mathbf{E}\left(\sum_{i=1}^{N} \mathbf{E}\left(X_i\right)\right) = \mathbf{E}\left(N\mathbf{E}\left(X_1\right)\right) = \mathbf{E}\left(N\right)\mathbf{E}\left(X_1\right)$$

(b) Clearly, $\mathbf{D}^2 (S_N)$ $=$ $\mathbf{E}(S_N^2)$ $-$ $(\mathbf{E}(S_N))^2$ and
$\mathbf{E}((X_i - \mathbf{E}(X_1))((X_j - \mathbf{E}(X_1)))) = 0$ if $i \neq j$. Analogously to the case
of (a), we obtain

$$\mathbf{E}(S_N^2) = \mathbf{E}\left(\mathbf{E}\left(\left\{\sum_{i=1}^{N}(X_i - \mathbf{E}(X_1)) + N\mathbf{E}(X_1)\right\}^2 \mid N\right)\right)$$

$$= \mathbf{E}\left(N\mathbf{D}^2(X_1) + N^2(\mathbf{E}(X_1))^2\right) = \mathbf{E}(N)\mathbf{D}^2(X_1) + \mathbf{E}(N^2)(\mathbf{E}(X_1))^2,$$

thus finally we get

$$\mathbf{D}^2(S_N) = (\mathbf{E}(N)\mathbf{D}^2(X_1) + \mathbf{E}(N^2)(\mathbf{E}(X_1))^2) - (\mathbf{E}(N))^2(\mathbf{E}(X_1))^2$$

$$= \mathbf{E}(N)\mathbf{D}^2(X_1) + \mathbf{D}^2(N)(\mathbf{E}(X_1))^2$$

Note that the preceding identities remain valid if the event $\{N = n\}$ depends only
on $(X_1, \ldots, X_n)$ for all $n = 1, 2, \ldots$

Solution 2.2

(a) It is clear that $\mathbf{E}(X_k) = \frac{1}{2} \cdot 1 + \frac{1}{2} \cdot (-1) = 0$, $\sigma_X^2 = \mathbf{D}^2\{X_k\} = \mathbf{E}(X_k^2) - (\mathbf{E}(X_k))^2 = \frac{1}{2} \cdot 1 + \frac{1}{2} \cdot (-1)^2 = 1$, moreover, $\mathbf{Cov}(X_i, X_i) = \sigma_X^2$ and by the
independence of the random variables X_i, we have $\mathbf{Cov}(X_i, X_j) = 0$, if $i \neq j$.
Since $Z_k = Z_{k-1} + X_k = \ldots = X_k + \ldots + X_1$, $k = 1, 2, \ldots$, then

$$\mathbf{E}(Z_k) = \mathbf{E}(X_k) + \ldots + \mathbf{E}(Z_1) = 0,$$

$$\mathbf{D}^2(Z_k) = \mathbf{D}^2(X_k) + \ldots + \mathbf{D}^2(X_1) = k\sigma_X^2.$$

(b) With a simple calculation, for $k = 1, 2, \ldots$, we get

$$W_k = bW_{k-1} + X_k = X_k + bX_{k-1} + b^2 W_{k-2} = X_k + bX_{k-1} + \ldots + b^{k-1}X_1 + b^k aX_0,$$

then the expectation function $\mathbf{E}(W_k) = \mathbf{E}(X_k + bX_{k-1} + \ldots + b^{k-1}X_1 + b^k aX_0) = 0$, the deviation of W_k is the following

$$\mathbf{D}^2(W_k) = \mathbf{D}^2(X_k + bX_{k-1} + \ldots + b^{k-1}X_1 + b^k aX_0)$$

$$= \mathbf{D}^2(X_k) + b^2\mathbf{D}^2(X_{k-1}) + \ldots + b^{2(k-1)}\mathbf{D}^2(X_1) + b^{2k}a\mathbf{D}^2(X_0)$$

$$= \sigma_X^2(1 + b^2 + \ldots + b^{2(k-1)}) + b^{2k}a\sigma_X^2 = \frac{1 - b^{2k}}{1 - b^2}\sigma_X^2 + b^{2k}a\sigma_X^2.$$

Since the random variables X_k are independent with expected value 0, the
covariance function is

$$R_W(k+m, k) = \mathbf{Cov}\,(W_{k+m}, W_k) = \mathbf{E}\,(W_{k+m}W_k)$$

$$= \mathbf{E}\,\big([X_{k+m} + bX_{k+m-1} + \ldots + b^{m-1}X_k + b^m W_k]W_k\big)$$

$$= b^m \mathbf{E}\,\big(W_k^2\big) = b^m \mathbf{D}^2\,(W_k) = b^m \left(\frac{1-b^{2k}}{1-b^2} + b^{2k}a\right)\sigma_X^2.$$

From this, it can be seen that if we choose $a = \frac{1}{1-b^2}$, then the process $(W_k, k = 0, 1, \ldots)$ will be stationary with expectation function $\mathbf{E}\,(W_k) = 0$, $k = 0, 1, \ldots$ and covariance function $R_W(k+m, k) = R_W(k, k+m) = \frac{b^m}{1-b^2}\sigma_X^2$, $k, m = 0, 1, \ldots$.

Solution 2.3 Since the expectation function of the process $(X_t, -\infty < t < \infty)$ takes the form $\mu_t = \mathbf{E}\,(X_t) = \int\limits_0^{2\pi} a\cos(bt + x)\frac{1}{2\pi}dx = \frac{a}{2\pi}\int\limits_0^{bt+2\pi} \cos x\,dx = 0$, therefore it does not depend on the parameter t. The covariance function is

$$R_X(t, s) = \mathbf{Cov}\,(X_t, X_s) = \mathbf{E}\,(X_t X_s) = \int\limits_0^{2\pi} a^2 \cos(bt + x)\cos(bs + x)\frac{1}{2\pi}dx$$

$$= \frac{a^2}{2\pi}\int\limits_0^{2\pi}\frac{1}{2}\left[\cos(b(t+s)+2x) + \cos(b(t-s))\right]dx = \frac{a^2}{4\pi}\cos(b(t-s))$$

$$= \frac{a^2}{4\pi}\cos(b\,|t-s|),$$

which means that $(X_t, -\infty < t < \infty)$ is a stationary process.

Solution 2.4

(a) Since $N(t)$ has a Poisson distribution with the parameter λt, then $\mathbf{E}\,(N(t)) = \lambda t$ and $\mathbf{D}^2\{N(t)\} = \lambda t$, $t \geq 0$. By the use of the property of independent increments we can get the covariance and correlation functions for $t, s \geq 0$:

$$\mathbf{Cov}\,(N(t+s), N(t)) = \mathbf{Cov}\,([N(t+s)-N(t)]+N(t), N(t)) = \mathbf{Cov}\,(N(t),$$

$$N(t)) = \lambda t,$$

$$\mathbf{Corr}\,(N(t+s), N(t)) = \frac{\mathbf{Cov}\,(N(t), N(t))}{\mathbf{D}\,(N(t+s))\mathbf{D}\,(N(t))} = \frac{\lambda t}{\sqrt{\lambda(t+s)}\sqrt{\lambda t}} = \frac{1}{\sqrt{1+\frac{s}{t}}}.$$

(b) Repeating the use of the property of independent increments of the process $N(t)$, we have

$$\mathbf{E}\,(N(t+s) \mid N(t)) = \mathbf{E}\,([N(t+s) - N(t)] + N(t) \mid N(t))$$

$$= \mathbf{E}\,([N(t+s) - N(t)] \mid N(t)) + \mathbf{E}\,(N(t) \mid N(t))$$

$$= \mathbf{E}\,(N(t+s) - N(t)) + N(t) = \lambda s + N(t).$$

References

1. Apostol, T.: Calculus I. Wiley, Hoboken (1967)
2. Campbell, N.: Discontinuities in light emission. In: Proceedings of the Cambridge Philosophical Society, vol. 15, pp. 310–328. Cambridge University, Cambridge (1910)
3. Karlin, S., Taylor, H.M.: A First Course in Stochastic Processes. Academic, New York (1975)
4. Kingman, J.F.C.: Poisson Processes. Clerendon, Oxford (1993)
5. Serfozo, R.: Basics of Applied Stochastic Processes. Springer, Berlin (2009)
6. Snyder, D.L.: Random Point Processes. Wiley, New York (1975)

Chapter 3
Markov Chains

In the early twentieth century, Markov (1856–1922) introduced in [18] a new class of models called Markov chains, applying sequences of dependent random variables that enable one to capture dependencies over time. Since that time, Markov chains have developed significantly, which is reflected in the achievements of Kolmogorov, Feller, Doob, Dynkin, and many others. The significance of the extensive theory of the Markov chains and the continuous-time variant called Markov processes is that it can be successfully applied to the modeling behavior of many problems in, for example, physics, biology, and economics, where the outcome of one experiment can affect the outcome of subsequent experiments. The terminology is not consistent in the literature, and many authors use the same name (Markov chain) for both discrete and continuous cases. We also apply this terminology.

Heuristically, the property that characterizes the Markov chains can be expressed by the so-called memoryless notion (Markov property) as follows: a Markov chain is a stochastic process for which future behavior, given the past and the present, depends only on the present and not on the past.

This chapter presents a brief introduction to the theory of discrete-time Markov chains (DTMCs) and to the continuous-time variant, continuous-time Markov chains (CTMCs), that will be applied to the modeling and analysis of queueing systems. Note that DTMCs and CTMCs taking values in a set of countable elements have many similar properties; however, in contrast to the discrete-time processes, the characteristics of a sample path essentially differ in continuous cases.

We limit ourselves here to the definition of Markov processes and to their basic properties with countable state space in discrete-time $\mathscr{T} = \{0, 1, \ldots\}$ and continuous-time $\mathscr{T} = [0, \infty)$. In connection with the classic results discussed in this chapter, we refer mainly to the classical works [7] and [8].

Consider a discrete-time or continuous-time stochastic process $X = (X_t, t \in \mathscr{T})$ given on a probability space $(\Omega, \mathscr{A}, P)$ and taking values in a countable set, called the **state space**, $\mathscr{X} = \{x_0, x_1, \ldots\}$. The state space $\mathscr{X}$ is called **finite** if it consists of a finite number of elements. The sample path of a discrete-time process with discrete

© Springer Nature Switzerland AG 2019
L. Lakatos et al., *Introduction to Queueing Systems with Telecommunication Applications*, https://doi.org/10.1007/978-3-030-15142-3_3

sample space is defined in the space of sequences $\mathscr{S} = \{x_{k_0}, x_{k_1}, \ldots\}$, $x_{k_i} \in \mathscr{X}$, while it is an element of the space of all functions $\mathscr{S} = \{x_t : x_t \in \mathscr{X}, t \geq 0\}$ in continuous-time cases.

We say that the process is **in the state** $x \in \mathscr{X}$ at the time $t \in \mathscr{T}$ if $X_t = x$. The process starts from a state $x_0 \in \mathscr{X}$ determined by the distribution of the random variable X_0, which is the **initial distribution** of the process. If there exists a state $x_0 \in \mathscr{X}$ for which $\mathbf{P}(X_0 = x_0) = 1$, then the state x_0 is called the **initial state**. The state of the process can change from time to time, and these changes in state are known as **transitions**. The probabilities of these state changes are called **transition probabilities**, which with the initial distribution determine the statistical behavior of the process.

If we denote by $\mathscr{B}_X$ the σ algebra of all subsets of the state space $\mathscr{X}$, then the pair $(\mathscr{X}, \mathscr{B}_X)$ is a measurable space and the connection $\{X_t \in A\} \in \mathscr{A}$ holds for all $t \in \mathscr{T}$ and $\{A \in \mathscr{B}_X\} \in \mathscr{A}$.

Definition 3.1 A stochastic process $(X_t, t \in \mathscr{T})$ with the discrete state space $\mathscr{X}$ is called a **Markov chain** if for every nonnegative integer n and for all $t_0 < \ldots < t_n < t_{n+1}$, $t_i \in \mathscr{T}$, $x_0, \ldots, x_{n+1} \in \mathscr{X}$

$$\mathbf{P}\big(X_{t_{n+1}} = x_{n+1} \mid X_{t_0} = x_0, \ldots, X_{t_n} = x_n\big) = \mathbf{P}\big(X_{t_{n+1}} = x_{n+1} \mid X_{t_n} = x_n\big), \tag{3.1}$$

provided that this conditional probability exists. Let $x, y \in \mathscr{X}$, $s \leq t$, $s, t \in \mathscr{T}$; then the function

$$p_{x,y}(s, t) = \mathbf{P}(X_t = y \mid X_s = x)$$

is called a **transition probability function** of a Markov chain. If the equation $p_{x,y}(s, t) = p_{x,y}(t - s)$ holds for all $x, y \in \mathscr{X}$, $s \leq t$, $s, t \in \mathscr{T}$, then the Markov chain is called **(time) homogeneous**; otherwise it is known as **inhomogeneous**.

In both discrete- and continuous-time cases, this definition expresses the aforementioned memoryless property of a Markov chain, and it ensures that the transition probabilities depend only on the present state X_s, not on how the present state was reached. We start with the discussion of DTMCs.

3.1   Discrete-Time Markov Chains

Given a Markov chain $X = (X_t, t \in \mathscr{T})$, $\mathscr{T} = \{0, 1, \ldots\}$ on a probability space $(\Omega, \mathscr{A}, P)$ taking values in a finite or countably infinite set of elements $\mathscr{X}$. It is conventional to denote the finite state space by the set $\mathscr{X} = \{0, 1, \ldots, K\}$ ($0 < K < \infty$) and the countably infinite one by the set $\mathscr{X} = \{0, 1, \ldots\}$. This notation is quite reasonable for queueing systems, and in general, it does not lead to a separate problem if the elements of $\mathscr{X}$ serve to distinguish the states only; otherwise, the state space is chosen based on practical requirements. Assume that the events $\{X_t = i\}$, $i \in \mathcal{X}$, are disjoint for all $t \in \mathcal{T}$.

In the discrete-time case we can give an alternative definition of a Markov chain instead of Eq. (3.1).

Definition 3.2 A discrete-time stochastic process X with state space $\mathcal{X}$ is called a Markov chain if for every $n = 0, 1, \ldots$ and for all states $i_0, \ldots, i_{n+1} \in \mathcal{X}$

$$p_{i_n, i_{n+1}}(n, n+1) = \mathbf{P}(X_{n+1} = i_{n+1} \mid X_0 = i_0, \ldots, X_n = i_n)$$

$$= \mathbf{P}(X_{n+1} = i_{n+1} \mid X_n = i_n), \qquad (3.2)$$

provided that a conditional probability exists. The probability

$$p_{i,j}(n, n+1) = \mathbf{P}(X_{n+1} = j \mid X_n = i), \ i, j \in \mathcal{X}, n = 0, 1, \ldots$$

is called a **one-step transition probability**, which is the probability of a transition from a state i to a state j in a single step from time n to time $n + 1$.

The relation (3.2) is simpler in our case than that of Eq. (3.1), but it can be easily checked that they are equivalent to each other. Here we can define, from a practical point of view, the transition probability $p_{i,j}(s, t) = 0$, when the probability of the event $\{X_s = i\}$ equals 0 at the time point s because if $\mathbf{P}\{X_s = i\} = 0$ holds, then the sample path arrives at the state i with probability 0 at time s; therefore, the quantity $p_{i,j}(s, t)$ can be defined freely in this case.

Definition 3.3 We say that a stochastic process X with state space $\mathcal{X}$ is **Markov chain of m-order** (or a **Markov chain with memory** m), if for every $n = 1, 2, \ldots$ and for arbitrary states $i_k \in \mathcal{X}$, $k = 0, \ldots, n + m$,

$$\mathbf{P}(X_{n+m} = i_{n+m} \mid X_0 = i_0, \ldots, X_{n+m-1} = i_{n+m-1})$$
$$= \mathbf{P}(X_{n+m} = i_{n+m} \mid X_n = i_n, \ldots, X_{n+m-1} = i_{n+m-1}),$$

provided that the conditional probability exists.

It is not difficult to verify that an m-order Markov chain can be represented as a first-order one if we introduce a new m-dimensional process as follows. Define the vector-valued process $Y = (Y_0, Y_1, \ldots)$,

$$Y_n = (X_n, \ldots, X_{n+m-1}), \ n = 0, 1, \ldots$$

with state space

$$\mathcal{X}' = \{(k_1, \ldots, k_m) : \ k_1, \ldots, k_m \in \mathcal{X}\}.$$

Then the process Y is a first-order Markov chain because

$$\mathbf{P}(Y_{n+1} = (i_{n+1}, \ldots, i_{n+m}) \mid Y_0 = (i_0, \ldots, i_{m-1}), \ldots, Y_n = (i_n, \ldots, i_{n+m-1}))$$
$$= \mathbf{P}(X_{n+m} = i_{n+m}, \ldots, X_{n+1} = i_{n+1} \mid X_0 = i_0, \ldots, X_{n+m-1} = i_{n+m-1})$$
$$= \mathbf{P}(X_{n+m} = i_{n+m}, \ldots, X_{n+1} = i_{n+1} \mid X_n = i_n, \ldots, X_{n+m-1} = i_{n+m-1})$$
$$= \mathbf{P}(Y_{n+1} = (i_{n+1}, \ldots, i_{n+m}) \mid Y_n = (i_n, \ldots, i_{n+m-1})).$$

This is why we consider only first-order Markov chains and why, later on, we will write only *Markov chain* instead of *Markov chain of first order*.

In the theory of DTMCs, the initial distribution

$$P = (p_i, \ i \in \mathscr{X}), \ \text{where} \ p_i = \mathbf{P}(X_0 = i),$$

and the transition probabilities [see Eq. (3.2)]

$$p_{ij}(n, n + 1), \ i, j \in \mathscr{X}, \ n = 0, 1, \ldots,$$

play a fundamental role because the statistical behavior of a Markov chain is completely determined by them (Theorem 3.1).

The states i and j, which play a role in Definition 3.2, can be identical, which means that the process can remain in the same state at the next time point. We say that a Markov chain X is (**time**)**homogeneous** if the transition probabilities do not depend on the time shifting, that is,

$$p_{ij} = \mathbf{P}(X_{n+1} = j \mid X_n = i) = \mathbf{P}(X_1 = j \mid X_0 = i), \ i, j \in \mathscr{X}, n = 0, 1, \ldots.$$

If a Markov chain is not homogeneous, then it is called **inhomogeneous**.

3.1.1 Homogeneous Markov Chains

From a practical point of view, the class of homogeneous Markov chains plays a significant role; therefore, in this chapter we will mainly investigate the properties of this class of processes. However, many results for homogeneous cases remain valid in inhomogeneous case, too.

By definition, for a homogeneous Markov chain the one-step transition probability (or simply transition probability) $p_{i,j}$, $i, j \in \mathscr{X}$, equals the probability that, starting from the initial state $X_0 = i$ at time 0, the process will be in the state j at the next time point 1, and this probability does not change if we take the transition probability in arbitrary time $n = 1, 2, \ldots,$

$$p_{ij} = \mathbf{P}\{X_1 = j \mid X_0 = i\} = \mathbf{P}\{X_{n+1} = j \mid X_n = i\}.$$

The transition probabilities satisfy the equation

$$\sum_{j \in \mathscr{X}} p_{ij} = 1.$$

This equation expresses the obvious fact that starting in a state i at the next time point the process takes certainly some state $j \in \mathscr{X}$. The following theorem states that the initial distribution and the transition probabilities determine the finite-dimensional distribution of a homogeneous Markov chain, and as a consequence we obtain that a Markov chain can be given in a statistical sense with the state space, the initial distribution, and the transition probabilities.

Theorem 3.1 *The finite-dimensional distributions of a Markov chain X are uniquely determined by the initial distribution and the transition probabilities and*

$$\mathbf{P}(X_0 = i_0, \ldots, X_n = i_n) = p_{i_{n-1}i_n} p_{i_{n-2}i_{n-1}} \cdot \ldots \cdot p_{i_0 i_1} p_{i_0}. \tag{3.3}$$

Proof Let n be a positive integer, and let $i_0, \ldots, i_n \in \mathscr{X}$. First, assume that $\mathbf{P}(X_0 = i_0, \ldots, X_n = i_n) > 0$. By the definition of conditional probability,

$\mathbf{P}(X_0 = i_0, \ldots, X_n = i_n)$
$= \mathbf{P}(X_n = i_n \mid X_0 = i_0, \ldots, X_{n-1} = i_{n-1})\mathbf{P}(X_0 = i_0, \ldots, X_{n-1} = i_{n-1}) = \ldots$
$= \mathbf{P}(X_n = i_n \mid X_0 = i_0, \ldots, X_{n-1} = i_{n-1})$
$\cdot \mathbf{P}(X_{n-1} = i_{n-1} \mid X_0 = i_0, \ldots, X_{n-2} = i_{n-2}) \cdot \ldots \cdot \mathbf{P}(X_1 = i_1 \mid X_0 = i_0)\mathbf{P}(X_0 = i_0).$

Using the Markov property we can rewrite this formula in the form

$\mathbf{P}(X_0 = i_0, \ldots, X_n = i_n)$
$= \mathbf{P}(X_n = i_n \mid X_{n-1} = i_{n-1}) \cdot \ldots \cdot \mathbf{P}(X_1 = i_1 \mid X_0 = i_0)\mathbf{P}(X_0 = i_0)$
$= p_{i_{n-1}i_n} p_{i_{n-2}i_{n-1}} \cdot \ldots \cdot p_{i_0 i_1} p_{i_0}.$

If $\mathbf{P}(X_0 = i_0, \ldots, X_n = i_n) = 0$, then either $\mathbf{P}(X_0 = i_0) = p_{i_0} = 0$ or there exists an index m, $0 \le m \le n - 1$, for which

$$\mathbf{P}(X_0 = i_0, \ldots, X_m = i_m) > 0, \quad \text{and} \quad \mathbf{P}(X_0 = i_0, \ldots, X_{m+1} = i_{m+1}) = 0.$$

Consequently,

$\mathbf{P}(X_0 = i_0, \ldots, X_{m+1} = i_{m+1})$
$= \mathbf{P}(X_{m+1} = i_{m+1} \mid X_0 = i_0, \ldots, X_m = i_m)\mathbf{P}(X_0 = i_0, \ldots, X_m = i_m)$
$= p_{i_m i_{m+1}}\mathbf{P}(X_0 = i_0, \ldots, X_m = i_m),$

and therefore $p_{i_m i_{m+1}} = 0$. This means that the product $p_{i_{n-1}i_n} p_{i_{n-2}i_{n-1}} \cdot \ldots \cdot p_{i_0 i_1} p_{i_0}$ equals 0 in both cases, and so assertion (3.3) of the theorem is true. $\qquad \square$

Remark 3.1 Obviously, from relation (3.1) it immediately follows that for any $A_i \subset \mathscr{X}$, $0 \le i \le n$ the probability $\mathbf{P}(X_0 \in A_0, \ldots, X_n \in A_n)$ can be given in the form

$$\mathbf{P}(X_0 \in A_0, \dots, X_n \in A_n) = \sum_{i_0 \in A_0} \cdots \sum_{i_n \in A_n} \mathbf{P}(X_0 = i_0, \dots, X_n = i_n),$$

where the probabilities are determined by relation (3.3), that is, with the help of the initial distribution and the transition probabilities.

The following remark clarifies an essential property of the homogeneous Markov chain, and on that basis limit theorems can be proved. This property relates the behavior of Markov chains to renewal and regenerative processes, which we will discuss later on in Sects. 4.1 and 4.2.

Remark 3.2 From the memoryless property of a Markov chain X it follows that we can divide the time access into disjoint parts where the process behavior is mutually independent and follows the same probabilistic rules. We define the limits of these independent parts by the time instants when the process stays in state $i_0 \in \mathcal{X}$.

Formally, we define the sequence of random time points $0 \le \tau_1 < \tau_2 < \cdots$ by the condition $X_{\tau_n} = i_0$, $n = 1, 2, \dots$ and $X_s \ne i_0$ if $s \notin \{\tau_1, \tau_2, \dots\}$. In this way $0 \le \tau_1 < \tau_2 < \cdots$ are the times of the first, second, etc. visits to the state i_0 and i_0 is not visited between τ_n and τ_{n+1}, $n = 1, 2, \dots$. We define Y_n and $Z_{n,k}$ by $Y_n = \tau_{n+1} - \tau_n$ and $Z_{n,k} = X_{\tau_n+k}$, $0 \le k < Y_n$. Y_n is the time between the nth and the $n + 1$th visits to i_0, and $Z_{n,k}$ is the state of the process at k steps after the nth visit to i_0, having that the next visit to i_0 is after $\tau_n + k$. Using the memoryless property of the Markov chain X we obtain that the random variables $(Y_n, Z_{n,k}, 0 \le k < Y_n)$, $n = 1, 2, \dots$, are independent and their stochastic behaviors are identical. This fact ensures that the process is regenerative (Sect. 4.2).

In many cases, the study of Markov chains will be made simpler by the use of transition probability matrices.

Definition 3.4 The matrices associated with the transition probabilities of a Markov chain $\mathcal{X}$ with finite or countably infinite elements are

$$\boldsymbol{\Pi} = \begin{bmatrix} p_{00} & p_{01} & \cdots & p_{0N} \\ p_{10} & p_{11} & \cdots & p_{1N} \\ \vdots & \vdots & \ddots & \vdots \\ p_{N0} & p_{N1} & \cdots & p_{NN} \end{bmatrix}, \quad \text{and} \quad \boldsymbol{\Pi} = \begin{bmatrix} p_{00} & p_{01} & p_{12} & \cdots \\ p_{10} & p_{11} & p_{12} & \cdots \\ p_{20} & p_{21} & p_{22} & \cdots \\ \vdots & \vdots & \vdots & \ddots \end{bmatrix}$$

These matrices are called (one-step) **transition probability matrices**.

A matrix with nonnegative entries $\mathbf{A} = \left[a_{ij}\right]_{i,j \in \mathcal{X}}$ is called a **stochastic matrix** if for every row the sum of row elements equals 1. Then all transition probability matrices are stochastic ones:

(a) The elements of $\boldsymbol{\Pi}$ are obviously nonnegative,

$$p_{ij} \ge 0, \ i, j \in \mathcal{X}.$$

(b) For every i the sum of ith row elements of $\boldsymbol{\Pi}$ equals 1,

$$\sum_{j \in \mathscr{X}} p_{ij} = 1, \ i \in \mathscr{X}.$$

The first of the following three examples shows that a sequence of independent and identically distributed discrete random variables is a homogeneous Markov chain. The second one shows that the sequence of sums of these random variables also constitutes a homogeneous Markov chain. If in the second case the random variables are independent, but not identically distributed, then the defined sequences will be an inhomogeneous Markov chain. The third example describes the stochastic behavior of a random walk on the real number line; and in this case it is reasonable to choose the state space to be the set of all integer numbers, that is, $\mathscr{X} = \{0, \pm 1, \pm 2, \ldots\}$.

Let $Z_0, Z_1, \ldots$ be a sequence of independent and identically distributed random variables with common CDF

$$\mathbf{P}(Z_m = k) = p_k, \ p_k \geq 0, \ k = 0, 1, \ldots, \ m = 0, 1, \ldots$$

Example 3.1 Define the discrete-time stochastic process X with the relation $X_n = Z_n$, $n = 0, 1, \ldots$ Then X is a homogeneous Markov chain with initial distribution $\mathbf{P}(X_0 = k) = p_k$, $k = 0, 1, \ldots$ and transition probability matrix

$$\boldsymbol{\Pi} = \begin{bmatrix} p_0 & p_1 & p_2 & \cdots \\ p_0 & p_1 & p_2 & \cdots \\ p_0 & p_1 & p_2 & \cdots \\ \vdots & \vdots & \vdots & \ddots \end{bmatrix}.$$

Example 3.2 Consider the process $X_n = Z_1 + \ldots + Z_n$, $n = 0, 1, \ldots$, with the initial distribution $\mathbf{P}(X_0 = 0) = 1$, i.e., the initial state is 0. The one-step transition probabilities are

$$p_{ij}(n, n+1) = \mathbf{P}(X_{n+1} = j \mid X_n = i)$$

$$= \mathbf{P}(Z_1 + \ldots + Z_{n+1} = j \mid Z_1 + \ldots + Z_n = i)$$

$$= \mathbf{P}(Z_{n+1} = j - i) = \begin{cases} p_{j-i}, & \text{if } j \geq i, \\ 0, & \text{if } j < i. \end{cases}$$

This means that the process X is a homogeneous Markov chain with the transition probability matrix

$$\boldsymbol{\Pi} = \begin{bmatrix} p_0 & p_1 & p_2 & p_3 & p_4 & \cdots \\ 0 & p_0 & p_1 & p_2 & p_3 & \cdots \\ 0 & 0 & p_0 & p_1 & p_2 & \cdots \\ 0 & 0 & 0 & p_0 & p_1 & \cdots \\ \vdots & \ddots & \ddots & \ddots & \ddots & \ddots \end{bmatrix}.$$

Example 3.3 Now let

$$\mathbf{P}(Z_i = +1) = p, \ \mathbf{P}(Z_i = -1) = 1 - p \quad (0 < p < 1), \ i = 1, 2, ..$$

be the common distribution function of a sequence of independent random variables $Z_0, Z_1, \ldots$, and define the process $X_n = Z_1 + \ldots + Z_n$, $n = 1, 2, \ldots$. Let $\mathbf{P}(X_0 = 0) = 1$ be the initial distribution of the process X. Then the process X is a homogeneous Markov chain with initial state $X_0 = 0$ and transition probability matrix

$$p_{ij}(n, n + 1) = \mathbf{P}(X_{n+1} = j \mid X_n = i)$$

$$= \mathbf{P}(Z_1 + \ldots + Z_{n+1} = j \mid Z_1 + \ldots + Z_n = i)$$

$$= \mathbf{P}(Z_{n+1} = j - i) = \begin{cases} p, & \text{if } j = i + 1, \\ 1 - p, & \text{if } j = i - 1, \\ 0, & \text{if } |i - j| \neq 1. \end{cases}$$

The process X describes the random walk on the number line starting from the origin and moves at every step one unit to the right with probability p and to the left with probability $(1 - p)$, with these moves being independent of each other. The case $p = 1/2$ corresponds to the symmetric random walk.

Figure 3.1 demonstrates the transitions of the random walk, while Fig. 3.2 shows the transitions of a Markov chain with a finite state space.

Fig. 3.1 Random walk

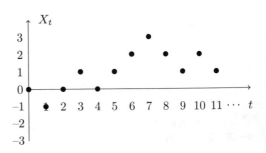

Fig. 3.2 Markov chain with finite state space

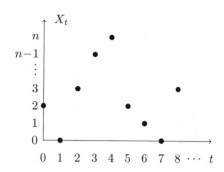

3.1.2 The m-Step Transition Probabilities

Let X be a DTMC with discrete state space $\mathscr{X}$. Denote by

$$p_{ij}(s, t) = \mathbf{P}(X_t = j \mid X_s = i)$$

the transition probabilities of X and by

$$\Pi(s, t) = \left[p_{ij}(s, t) \right]_{i, j \in \mathscr{X}}, \ i, j \in \mathscr{X} \ \text{and} \ 0 \leq s \leq t < \infty$$

the transition probability matrices. We set for $s = t$,

$$p_{ij}(s, s) = \begin{cases} 1, & \text{if } i = j, \\ 0, & \text{if } i \neq j. \end{cases}$$

If the Markov chain X is homogeneous, then the transition probability $p_{ij}(s, t)$ depends only on the difference $t - s$. Thus, using the notation $t = s + m$, we have

$$p_{ij}(s, s + m) = p_{ij}(m), \ s, m = 0, 1, \ldots, \ i, j \in \mathscr{X}.$$

Definition 3.5 The quantities $p_{ij}(m)$, $m = 0, 1, \ldots$, $i, j \in \mathscr{X}$ are called the **m-step transition probabilities** of the Markov chain X, and the matrix $\Pi(m) = \left[p_{ij}(m) \right]_{i, j \in \mathscr{X}}$ associated with them is called an **m-step transition probability matrix**.

Theorem 3.2 (Chapman–Kolmogorov Equation) *For every nonnegative integer number r, s, the $(r + s)$-step transition probabilities of the homogeneous Markov chain satisfy the equation*

$$p_{ij}(r + s) = \sum_{k \in \mathscr{X}} p_{ik}(r) p_{kj}(s). \tag{3.4}$$

Proof Assume the initial state of the process is i, that is, the process starts from the state i at the time point 0. First we note that the relation

$$p_{ik}(r) = \mathbf{P}(X_r = k \mid X_0 = i) = \frac{\mathbf{P}(X_0 = i, X_r = k)}{\mathbf{P}(X_0 = i)} = 0$$

holds for some state k if and only if $\mathbf{P}(X_0 = i, X_r = k) = 0$. On the other hand, since $\{X_r = k\}$, $k \in \mathscr{X}$ form a complete system of events, $\sum_{k \in \mathscr{X}} \mathbf{P}(X_r = k) = 1$, and, in accordance with the definitions of the $(r+s)$-step transition probability and the conditional probability, we obtain

$$
\begin{aligned}
p_{ij}(r+s) &= \mathbf{P}(X_{r+s} = j \mid X_0 = i) \\
&= \frac{\mathbf{P}(X_{r+s} = j, \ X_0 = i)}{\mathbf{P}(X_0 = i)} = \sum_{k \in \mathscr{X}} \frac{\mathbf{P}(X_{r+s} = j, X_0 = i, X_r = k)}{\mathbf{P}(X_0 = i)} \\
&= \sum_{k \in \mathscr{X}} \mathscr{I}_{\{p_{ik} \neq 0\}} \frac{\mathbf{P}(X_0 = i, X_r = k)}{\mathbf{P}(X_0 = i)} \frac{\mathbf{P}(X_{r+s} = j, X_0 = i, X_r = k)}{\mathbf{P}(X_0 = i, X_r = k)} \\
&= \sum_{k \in \mathscr{X}} \mathscr{I}_{\{p_{ik} \neq 0\}} \mathbf{P}(X_r = k \mid X_0 = i)\mathbf{P}(X_{r+s} = j \mid X_r = k, X_0 = i) \\
&= \sum_{k \in \mathscr{X}} \mathscr{I}_{\{p_{ik} \neq 0\}} p_{ik}(0, r) p_{kj}(r, r+s) = \sum_{k \in \mathscr{X}} p_{ik}(r) p_{kj}(s).
\end{aligned}
$$

$\square$

If we use the matrix notation $\boldsymbol{\Pi}(s, t) = \left[p_{ij}(s, t)\right]_{i,j \in \mathscr{X}}$, then the Chapman–Kolmogorov equation can be rewritten in the matrix form

$$\boldsymbol{\Pi}(s, t) = \boldsymbol{\Pi}(s, r)\boldsymbol{\Pi}(r, t),$$

where s, r, t, and n are integer numbers satisfying the inequality $0 \leq s \leq r \leq t$, $n \geq 1$. Successively repeating this relation we have

$$\boldsymbol{\Pi}(0, n) = \boldsymbol{\Pi}(0, 1)\boldsymbol{\Pi}(1, n) = \ldots = \boldsymbol{\Pi}(0, 1)\boldsymbol{\Pi}(1, 2) \cdot \ldots \cdot \boldsymbol{\Pi}(n-1, n).$$

Consequently, the m-step transition probability matrix of a homogeneous Markov chain can be given in the form

$$\boldsymbol{\Pi}(m) = \boldsymbol{\Pi}^m,$$

where $\boldsymbol{\Pi} = \boldsymbol{\Pi}(0, 1)$ is the (one-step) transition probability matrix of the Markov chain.

3.1.3 Classification of States of Homogeneous Markov Chains

The behavior of a Markov chain and its asymptotic properties essentially depend on the transition probabilities, which reflect the connections among the different states.

Denote by $P_i(t) = \mathbf{P}(X_t = i), i \in \mathcal{X}$ the distribution of the Markov chain X at the time $t \geq 0$. One of the most important questions in the theory of Markov chains concerns the conditions under which a limit distribution exists for all initial states $X_0 = k \in \mathcal{X}$,

$$\lim_{t \to \infty} P(t) = \pi = (\pi_i, i \in \mathcal{X})$$

of the time-dependent distribution $P(t) = (P_i(t), i \in \mathcal{X})$, where $\pi_i \geq 0$, $\sum_{i \in \mathcal{X}} \pi_i = 1$. In the answer to this question, the arithmetic properties of the transition probabilities play important role.

To demonstrate this fact, consider the case where the sample space $\mathcal{X}$ can be divided into two disjoint (nonempty) sets $\mathcal{X}_1$ and $\mathcal{X}_2$ such that

$$p_{ij} = p_{ji} = 0, \quad \text{for all } i \in \mathcal{X}_1 \text{ and } j \in \mathcal{X}_2.$$

Obviously, if $X_0 = i_0 \in \mathcal{X}_1$ is the initial state, then the relation $X_t \in \mathcal{X}_1$ is valid for all $t \geq 0$, and in the opposite case, $X_t \in \mathcal{X}_2$, for all $t \geq 0$ holds if the initial state i_0 satisfies the condition $i_0 \in \mathcal{X}_2$. This means that in this case we can in fact consider two Markov chains $(\mathcal{X}_k, (P_i(0), i \in \mathcal{X}_k), \Pi_k)$ $k = 1, 2$, that can be investigated independently of each other.

Definition 3.6 The state $j \in \mathcal{X}$ is **accessible** from state $i \in \mathcal{X}$ (denoted by $i \to j$) if there exists a positive integer m such that $p_{ij}(m) > 0$. If the states $i, j \in \mathcal{X}$ are mutually accessible from each other, then we say that they **communicate** (denoted by $i \longleftrightarrow j$).

$p_{ii}(0) = 1$, $i \in \mathcal{X}$ represents the assumption that "every state is accessible in 0 steps from itself." If the state $j \in \mathcal{X}$ is not accessible from the state $i \in \mathcal{X}$ (denoted by $i \nrightarrow j$), then $p_{ij}(m) = 0$, $m \geq 1$. It is easy to check that $i \longleftrightarrow j$ is an **equivalence relation**: it is reflexive, transitive, and symmetric. Furthermore, if the states i and j do not communicate, then either $p_{ij}(m) = 0$, $m \geq 1$, or $p_{ji}(m) = 0$, $m \geq 1$. If a state i satisfies the condition $p_{ii} = p_{ii}(1) = 1$, then state i is called **absorbing**. This means that if the process visits an absorbing state at time t, then it remains there forever and no more state transitions occur.

If the state space $\mathcal{X}$ does not consist of the states i and j such that $i \to j$, but $j \nrightarrow i$, then $\mathcal{X}$ can be given as a union of finite or countable disjoint sets

$$\mathcal{X} = \mathcal{X}_1 \cup \mathcal{X}_2 \cup \ldots,$$

where for every k the states of $\mathcal{X}_k$ communicate, while for every $k, n,\ k \neq n$, the states of $\mathcal{X}_k$ cannot be accessible from the states of $\mathcal{X}_n$.

Definition 3.7 A set of states is called **irreducible** if all pairs of its elements communicate.

In the theory of Markov chains, irreducible classes play an important role because they can be independently analyzed.

Definition 3.8 A Markov chain is called **irreducible** if all pairs of its states communicate.

Clearly, if a Markov chain is irreducible, then it consists of only one irreducible class of states, that is, for every $i, j \in \mathcal{X}$ there exists an integer $m \geq 1$ (depending on i and j) such that $p_{ij}(m) > 0$.

Definition 3.9 For every i denote by $d(i)$ the greatest common divisor of integer numbers $m \geq 1$ for which $p_{ii}(m) > 0$. If $p_{ii}(m) = 0$ for every m, then we set $d(i) = 0$. Then the number $d(i)$ is called the **period** of the Markov chain. If $d(i) = 1$ for every state, then the Markov chain is called **aperiodic**.

Example 3.4 (Periodic Markov Chain) Consider the random walk on the number line demonstrated earlier in Example 3.3. Starting from an arbitrary state i we can return to state i with positive probabilities in steps $2, 4, \ldots$ only. It is clear that in this case, $p_{ii}(2k) > 0$ and $p_{ii}(2(k-1)+1) = 0$ for every $i \in \mathcal{X}$ and $k = 1, 2, \ldots$; therefore, $d(i) = 2$. At the same time, the Markov chain is obviously irreducible.

Theorem 3.3 *Let X be a homogeneous Markov chain with state space $\mathcal{X}$, and let $\mathcal{X}' \subset \mathcal{X}$ be a nonempty irreducible class. Then for every $i, j \in \mathcal{X}'$, the periods of i and j are the same, i.e., $d(i) = d(j)$.*

Proof Let $i, j \in \mathcal{X}', i \neq j$, be two arbitrary states. Since $\mathcal{X}'$ is an irreducible class, there exist $t, s \geq 1$ integers such that the inequalities $p_{ij}(t) > 0$ and $p_{ji}(s) > 0$ hold. From this, by the Chapman–Kolmogorov equation, we obtain

$$p_{ii}(t + s) \geq p_{ij}(t)p_{ji}(s) > 0 \quad \text{and} \quad p_{jj}(t + s) \geq p_{ji}(s)p_{ij}(t) > 0;$$

therefore, the numbers $d(i)$ and $d(j)$ differ from 0. Choose arbitrarily an integer $m \geq 1$ such that $p_{ii}(m) > 0$. Repeatedly applying the Chapman–Kolmogorov equation, we have for any $k \geq 1$

$$p_{jj}(t + s + km) \geq p_{ji}(s)p_{ii}(km)p_{ij}(t) \geq p_{ji}(t)\left(p_{ii}(m)\right)^k p_{ij}(s) > 0.$$

Thus by the definition of the period of the state j, $d(j)$ is a divisor of both $(t+s+m)$ and $(t+s+2m)$, and hence it is also a divisor of their difference $(t+s+2m) - (t+s+m) = m$. From this it immediately follows that $d(j)$ is a divisor of every m for which $p_{ii}(m) > 0$, and thus it is a divisor of $d(i)$; therefore, $d(j) \leq d(i)$. Changing the role of i and j we get the reverse inequality $d(j) \geq d(i)$, and consequently $d(j) = d(i)$. $\qquad\square$

Notice that from this theorem it follows that the states of an irreducible class have a common period $d(\mathcal{X}')$ called the **period of the class**. As a consequence, we have the following assertion.

Corollary 3.1 *If the Markov chain X is homogeneous and irreducible with state space $\mathcal{X}$, then every state has the same period $d = d(\mathcal{X}) > 0$ and is periodic or aperiodic depending on $d > 1$ or $d = 1$, respectively.*

The main property of the numbers for which the probabilities of returning to a state i in k steps are positive, i.e., $p_{ii}(k) > 0$, is given by the following assertion.

Theorem 3.4 *Let X be a homogeneous irreducible Markov chain with state space $\mathcal{X}$. Then for every state $i \in \mathcal{X}$ there exists an integer M_i such that $p_{ii}(d(i)m) > 0$ if $m \geq M_i$.*

Proof By the previous theorem, $d(i) \geq 1$. Let $m_1, \ldots, m_L$ be different positive integer numbers such that, on one hand, $p_{ii}(m_k) > 0$, $1 \leq k \leq L$, and on the other hand, $d(i)$ can be given as the greatest common divisor of integers $m_1, \ldots, m_L$. Now, we will use the well-known assertion from the number theory that there exists an integer M_i such that for every integer $m \geq M_i$, the equation $md(i) = r_1 m_1 + \ldots + r_L m_L$ has a solution with nonnegative integers $r_1, \ldots, r_L$. Applying this fact and the Chapman–Kolmogorov equation we obtain

$$p_{ii}(md(i)) \geq (p_{ii}(m_1))^{r_1} \cdot \ldots \cdot (p_{ii}(m_L))^{r_L} > 0,$$

and consequently, the assertion of the theorem is true. □

Consider now the homogeneous irreducible Markov chain with period $d(\mathcal{X}) > 1$. We show that for the transitions among the states there exists a cyclic property, demonstrated in Example 3.5, of the random walk on the number line: if the walk starts from state 0, then the process can take only even integers in even steps and only odd integers in odd steps. The cyclic property in this case means that after even numbers follow odd numbers and after odd numbers follow even numbers as states. This division of states is generalized subsequently for Markov chains with arbitrary period d.

Let $i_0 \in \mathcal{X}$ be an arbitrary fixed state, and define the sets

$$\mathcal{X}_k = \{j \in \mathcal{X} : p_{i_0 j}(k + md) > 0, \text{ for some } m \geq 0\}, k = 0, 1, \ldots, d - 1.$$

That is, $\mathcal{X}_k$ is the set of states that are available from i_0 in $k + md$ $(m = 0, 1, \ldots,)$ steps.

Theorem 3.5 *The sets $\mathcal{X}_1, \ldots, \mathcal{X}_{d-1}$ are disjoint, $\mathcal{X} = \mathcal{X}_0 \cup \ldots \cup \mathcal{X}_{d-1}$ and the Markov chain allows for only the following cyclic transitions among the sets $\mathcal{X}_k$:*

$$\mathcal{X}_0 \to \mathcal{X}_1 \to \ldots \to \mathcal{X}_{d-1} \to \mathcal{X}_0 \tag{3.5}$$

Proof First we prove that the sets $\mathcal{X}_0, \ldots, \mathcal{X}_{d-1}$ are disjoint and their union is $\mathcal{X}$. In contrast, assume that there exist integers k_1, k_2, m_1, m_2 such that $0 \leq k_1 < k_2 \leq d-1, m_1, m_2 \geq 1, p_{i_0 j}(k_1 + m_1 d) > 0$, and $p_{i_0 j}(k_2 + m_2 d) > 0$. Since the Markov chain is irreducible, there exists an integer $K \geq 1$ such that $p_{j i_0}(K) > 0$. Using the Chapman–Kolmogorov equation we have

$$p_{i_0 i_0}(k_1 + m_1 d + K) \geq p_{i_0 j}(k_1 + m_1 d) p_{j i_0}(K) > 0,$$

$$p_{i_0 i_0}(k_2 + m_2 d + K) \geq p_{i_0 j}(k_2 + m_2 d) p_{j i_0}(K) > 0.$$

By the definition of the period d, d is a divisor of both $(k_1 + m_1 d + K)$ and $(k_2 + m_2 d + K)$, thus it is also a divisor of their difference, that is, $(k_2 - k_1) + (m_2 - m_1)d$. Consequently, d is a divisor of the difference $(k_2 - k_1)$, which is a contradiction, because $0 < k_2 - k_1 \leq d - 1$. The irreducibility condition ensures that if all states $i \in \mathcal{X}$ are accessible from the state i_0, then $\mathcal{X} = \mathcal{X}_0 \cup \ldots \cup \mathcal{X}_{d-1}$.

We now verify that for every k, $0 \leq k \leq d - 1$, $i \in \mathcal{X}_k$ and $j \in \mathcal{X}$ such that $p_{ij} > 0$, the relation $j \in \mathcal{X}_K, 0 \leq K < d$, is true, where

$$K = \begin{cases} k+1, & \text{if } 0 \leq k < d-1, \\ 0, & \text{if } k = d-1. \end{cases}$$

This property guarantees the transitions between the states in (3.5).

Since $i \in \mathcal{X}_k$, then, by the definition of the sets $\mathcal{X}_k$, there exists an integer $m \geq 0$ such that $p_{i_0 i}(k + md) > 0$. From this, by the use of Chapman–Kolmogorov equality, we have

$$p_{i_0 j}(k + 1 + md) \geq p_{i_0 i}(k + md) p_{ij} > 0.$$

In view of the fact that

$$k + 1 + md = \begin{cases} K + m d, & \text{if } 0 \leq k < d-1, \\ 0 + (m+1) d, & \text{if } k = d-1, \end{cases}$$

from the definition of $\mathcal{X}_K$ follows the relation $j \in \mathcal{X}_K$. $\square$

As a consequence of Theorem 3.5, we have the next important corollary, which allows us to consider an aperiodic Markov chain instead of a periodic one.

Corollary 3.2 *Theorem 3.5 states that starting from a state of $\mathcal{X}_k$, $k = 0, 1, \ldots, d - 1$, after exactly d steps the process returns to a state $\mathcal{X}_k$. If we define the quantities*

$$p_{ij}^{(k)} = \mathbf{P}(X_d = j \mid X_0 = i), \ i, j \in \mathcal{X}_k,$$

then $\sum_{j \in \mathscr{X}_k} p_{ij}^{(k)} = 1$, $i \in \mathscr{X}_k$ *follows. This means that the matrices* $\mathbf{P}^{(k)} = \left[p_{ij}^{(k)} \right]_{i,j \in \mathscr{X}_k}$ *are stochastic; they can be interpreted as one-step transition probability matrices, and consequently the processes*

$$Y^{(k)} = (Y_0, Y_1, \ldots), \ k = 0, 1, \ldots, d-1,$$

with the state space $\mathscr{X}_k$ *and transition probability matrix* $\mathbf{P}^{(k)}$, *are homogeneous and irreducible Markov chains, and so, instead of the original chain, d homogeneous irreducible Markov chains can be considered independently.*

If the states of the Markov chain are numbered according to the $\mathscr{X}_k$, $k = 0, 1, \ldots, d - 1$, sets, then the transition probability matrix has the following structure:

$\mathscr{X}_0$	$\mathbf{0}$	$\boldsymbol{P}_{0 \to 1}$			
$\mathscr{X}_1$		$\mathbf{0}$	$\boldsymbol{P}_{1 \to 2}$		
			$\ddots$	$\ddots$	
$\mathscr{X}_{d-2}$				$\mathbf{0}$	$\boldsymbol{P}_{d-2 \to d-1}$
$\mathscr{X}_{d-1}$	$\boldsymbol{P}_{d-1 \to 0}$				$\mathbf{0}$

3.1.4 Recurrent Markov Chains

We consider the question of what conditions ensure the existence of limit theorems for homogeneous aperiodic Markov chains, that is, under what conditions does there exist the limit distribution $\pi = (\pi_i, i \in \mathscr{X})$, $(\pi_i \geq 0, \sum_{i \in \mathscr{X}} \pi_i = 1)$, such that, independently of initial distribution $(p_i, i \in \mathscr{X})$, the limit is

$$\lim_{n \to \infty} P_i(n) = \lim_{n \to \infty} \mathbf{P}(X_n = i) = \pi_i, \ i \in \mathscr{X}.$$

To provide an answer to this question, it is necessary to consider some quantities such as the probability and the expected value of returning to a given state of a Markov chain or arriving at a state j from another state i. Let $i, j \in \mathscr{X}$ be two arbitrary states, and introduce the following notations:

$T_{ij} = \inf\{n : n > 1, \ X_n = j \mid X_0 = i\}$

$f_{ij}(0) = 0,$

$f_{ij}(1) = \mathbf{P}(X_1 = j \mid X_0 = i),$

$f_{ij}(n) = \mathbf{P}(X_1 \neq j, X_2 \neq j, \ldots, X_{n-1} \neq j, X_n = j \mid X_0 = i), \ n = 2, 3, \ldots$

If $i \neq j$, then the quantities $f_{ij}(n) = \mathbf{P}\{T_{ij} = n\}$ mean the **first hit** (or **first passage**) **probabilities** for state j from i, which is the probability that starting from the state i at time point 0, the process will be first in the state j during n steps (or in time n). If $i = j$, then the quantity $f_{ii}(n)$ means the **first return probability** in the state i in n steps.

Denote $f_{ij} = \sum_{k=1}^{\infty} f_{ij}(k)$, $i, j \in \mathscr{X}$. Obviously, the quantity f_{ij} means the probability that the Markov chain starts from a state i at time 0 and at some time arrives at the state j, that is, $f_{ij} = \mathbf{P}\{T_{ij} < \infty\}$.

Definition 3.10 A state i is called **recurrent** if the process returns in the state i with probability 1, that is, $f_{ii} = \mathbf{P}\{T_{ii} < \infty\} = 1$. If $f_{ii} < 1$, then the state i is called **transient**.

From the definition it follows that when i is a transient state, then a process with positive probability will never return to the state i. The following theorem describes the connection between the return probabilities and the m-step transition probabilities of a Markov chain in the form of a so-called discrete renewal equation.

Theorem 3.6 *For every* $i, j \in \mathscr{X}$, $n = 1, 2, \ldots$,

$$p_{ij}(n) = \sum_{k=1}^{n} f_{ij}(k) p_{jj}(n - k). \tag{3.6}$$

Proof By the definition $p_{jj}(0) = 1$, in the case $n = 1$ we have $p_{ij}(1) = f_{ij}(1) p_{jj}(0) = f_{ij}(1)$. Now let $n \geq 2$. Using conditional probability and the Markov property we get

$$\mathbf{P}(X_n = j, X_1 = j \mid X_0 = i) = \mathbf{P}(X_n = j \mid X_1 = j, X_0 = i)\mathbf{P}(X_1 = j \mid X_0 = i)$$
$$= p_{jj}(n - 1) p_{ij}(1) = f_{ij}(1) p_{jj}(n - 1).$$

Similarly, we obtain

$$\mathbf{P}(X_n = j, X_k = j, X_m \neq j, 1 \leq m \leq k - 1 \mid X_0 = i)$$
$$= f_{ij}(k) p_{jj}(n - k), \quad n = 1, 2, \ldots$$

On the basis of the last two equations, it follows that

$$p_{ij}(n) = \mathbf{P}(X_n = j, X_1 = j \mid X_0 = i)$$

$$+ \sum_{k=2}^{n} \mathbf{P}(X_n = j, X_k = j, X_m \neq j, 1 \leq m \leq k - 1 \mid X_0 = i)$$

$$= f_{ij}(1) p_{jj}(n - 1) + \sum_{k=2}^{n} f_{ij}(k) p_{jj}(n - k), \quad n = 1, 2, \ldots \qquad \square$$

The notion of the recurrence of a state is defined by the return probabilities, but the following theorem makes it possible to provide a condition for it with the use of n-step transition probabilities $p_{ii}(n)$ and to classify the Markov chains.

Theorem 3.7

(a) The state $i \in \mathcal{X}$ is recurrent if and only if

$$\sum_{n=1}^{\infty} p_{ii}(n) = \infty.$$

(b) If i and j are communicating states and i is recurrent, then j is also recurrent.
(c) If a state $j \in \mathcal{X}$ is transient, then for arbitrary $i \in \mathcal{X}$,

$$\sum_{n=1}^{\infty} p_{ij}(n) < \infty \quad \text{and consequently} \quad \lim_{n \to \infty} p_{ij}(n) = 0.$$

Proof

(a) By the definition $p_{ii}(0) = 1$ and using the relation (3.6) of the preceding theorem we obtain

$$\sum_{n=1}^{\infty} p_{ii}(n) = \sum_{n=1}^{\infty} \sum_{k=1}^{n} f_{ii}(k) p_{ii}(n-k) = \sum_{k=1}^{\infty} \sum_{n=k}^{\infty} f_{ii}(k) p_{ii}(n-k)$$

$$= \sum_{k=1}^{\infty} f_{ii}(k) \left(p_{ii}(0) + \sum_{n=1}^{\infty} p_{ii}(n) \right).$$

From this equation, if the sum $\sum_{n=1}^{\infty} p_{ii}(n)$ is finite, then we get

$$f_{ii} = \left(1 + \sum_{n=1}^{\infty} p_{ii}(n) \right)^{-1} \sum_{n=1}^{\infty} p_{ii}(n) < 1,$$

consequently, i is not a recurrent state.

If $\sum_{n=1}^{\infty} p_{ii}(n) = \infty$, then obviously $\lim_{N \to \infty} \sum_{n=1}^{N} p_{ii}(n) = \infty$. Since for all positive integers N the relation

$$\sum_{n=1}^{N} p_{ii}(n) = \sum_{n=1}^{N} \sum_{k=1}^{n} f_{ii}(k) p_{ii}(n-k)$$

$$= \sum_{k=1}^{N} \sum_{n=k}^{N} f_{ii}(k) p_{ii}(n-k) \le \sum_{k=1}^{N} f_{ii}(k) \sum_{n=0}^{N} p_{ii}(n)$$

$$\le \left(1 + \sum_{n=1}^{N} p_{ii}(n)\right) \sum_{k=1}^{N} f_{ii}(k)$$

holds, from the limit $\sum_{n=1}^{N} p_{ii}(n) \to \infty$

$$1 \ge f_{ii} = \sum_{k=1}^{\infty} f_{ii}(k) \ge \sum_{k=1}^{N} f_{ii}(k) \ge \left(1 + \sum_{k=1}^{N} p_{ii}(k)\right)^{-1} \sum_{k=1}^{N} p_{ii}(k) \to 1, \ N \to \infty$$

follows. Consequently, $f_{ii} = 1$, and thus the state i is recurrent.

(b) Since the states i and j communicate, there exist integers $n, m \ge 1$ such that $p_{ij}(m) > 0$ and $p_{ji}(n) > 0$. By the Chapman–Kolmogorov equation for every integer $k \ge 1$,

$$p_{ii}(m + k + n) \ge p_{ij}(m) p_{jj}(k) p_{ji}(n),$$

$$p_{jj}(m + k + n) \ge p_{ji}(n) p_{ii}(k) p_{ij}(m).$$

From this

$$\sum_{k=1}^{\infty} p_{ii}(k) \ge \sum_{k=1}^{\infty} p_{ii}(m + n + k) \ge p_{ij}(m) p_{ji}(n) \sum_{k=1}^{\infty} p_{jj}(k),$$

$$\sum_{k=1}^{\infty} p_{jj}(k) \ge \sum_{k=1}^{\infty} p_{jj}(m + n + k) \ge p_{ij}(m) p_{ji}(n) \sum_{k=1}^{\infty} p_{ii}(k).$$

Both series $\sum_{k=1}^{\infty} p_{ii}(k)$ and $\sum_{k=1}^{\infty} p_{jj}(k)$ are simultaneously convergent or divergent because $p_{ij}(m) > 0$ and $p_{ji}(n) > 0$; thus, by the assertion (a) of the theorem, the states i and j are recurrent or transient at the same time.

(c) Applying the discrete renewal Eq. (3.6) and result (a), assertion (c) immediately follows.   □

Definition 3.11 A Markov chain is called **recurrent** or **transient** if every state is recurrent or transient.

Remark 3.3 Using the n-step transition probabilities $p_{ii}(n)$, a simple formula can be given for the expected value of the number of returns to a state $i \in \mathcal{X}$. Let $X_0 = i$ be the initial state of the Markov chain. The expected value of the return number is expressed as

$$\mathbf{E}\left(\sum_{k=1}^{\infty} \mathscr{I}_{\{X_k=i\}} | X_0 = i\right) = \sum_{k=1}^{\infty} \mathbf{E}\left(\mathscr{I}_{\{X_k=i\}} | X_0 = i\right)$$

$$= \sum_{k=1}^{\infty} \mathbf{P}(X_k = i \mid X_0 = i) = \sum_{k=1}^{\infty} p_{ii}(k).$$

The assertion of Theorem 3.7 can be interpreted in another way: a state $i \in \mathscr{X}$ is recurrent if and only if the expected value of the number of returns equals infinity.

Example 3.5 (Recurrent Markov Chain) Consider the random walk process $X = (X_n, \ n = 0, 1, \ldots)$ described in Example 3.3. The process, starting from the origin, at all steps moves one unit to the right with probability p and to the left with probability $(1 - p)$, independently of each other. We have proved earlier that the process X is a homogeneous, irreducible, and periodic Markov chain with period 2. Here we discuss the conditions under which the Markov chain will be recurrent.

By the condition $X_0 = 0$, it is clear that $p_{00}(2k + 1) = 0, \ k = 0, 1, \ldots$. The process can return in $2k$ steps to the state 0 only if it moves, in some way, k times to the left and k times right, the probability of which is

$$p_{00}(2k) = \binom{2k}{k} p^k (1 - p)^k = \frac{(2k)!}{k!k!} [p(1 - p)]^k.$$

Using the well-known Stirling's formula, which gives an asymptotic relation for $k!$ as $k \to \infty$ as follows (see p. 616 of [1]):

$$\sqrt{2\pi} k^{k+1/2} e^{-k} < k! < \sqrt{2\pi} k^{k+1/2} e^{-k}(1 + \frac{1}{4k});$$

then

$$k! \approx \left(\frac{k}{e}\right)^k \sqrt{2\pi k};$$

and thus we have

$$p_{00}(2k) \approx \left(\frac{2k}{e}\right)^{2k} \sqrt{2\pi(2k)} \left(\left(\frac{k}{e}\right)^k \sqrt{2\pi k}\right)^{-2} [p(1 - p)]^k = \frac{[4p(1 - p)]^k}{\sqrt{\pi k}}.$$

By the inequality between arithmetic and geometrical means, the numerator has an upper bound

$$4 [p(1 - p)] \le 4 \left[\frac{p + (1 - p)}{2}\right]^2 = 1,$$

where the equality holds if and only if $p = 1 - p$, that is, $p = 1/2$. In all other cases the product is less than 1; consequently, the sum of return probabilities $p_{00}(2k)$ is divergent if and only if $p = 1/2$ (symmetric random walk); otherwise, it is convergent. As a consequence of Theorem 3.7, we obtain that the state 0, and together with it all states of the Markov chain, is recurrent if and only if $p = 1/2$.

Note that a similar result is valid if we consider the random walk with integer coordinates in a plane. It can be verified that only in the case of symmetric random walk will the state $(0, 0)$ be recurrent, when the probabilities of the movements left-right-up-down are $1/4 - 1/4$. In addition, if a random walk is defined in a similar way in higher-dimensional (≥ 3) spaces, then the Markov chain will no longer be recurrent.

3.2   Fundamental Limit Theorem of Homogeneous Markov Chains

3.2.1   Positive Recurrent and Null Recurrent Markov Chains

Let X be a homogeneous Markov chain with a finite ($N < \infty$) or countably infinite ($N = \infty$) state space $\mathcal{X} = \{0, 1, \ldots, N\}$ and (one-step) transition probability matrix $\boldsymbol{\Pi} = \left[p_{ij}\right]_{i, j \in \mathcal{X}}$. Let $P = (p_i = \mathbf{P}(X_0 = i), \ i \in \mathcal{X})$ be the initial distribution. Denote by $P(n) = (P_i(n) = \mathbf{P}(X_n = i), \ i \in \mathcal{X}), \ n = 0, 1, \ldots$, the time-dependent distribution of the Markov chain; then $P(0) = P$.

The main question to be investigated here concerns the conditions under which there exists a limit distribution of m-step transition probabilities

$$\lim_{m \to \infty} p_{ij}(m) = \pi_j, \quad \text{where } \pi_j \geq 0 \text{ and } \sum_{i \in \mathcal{X}} \pi_i = 1$$

and how it can be determined. The answer is closely related to the behavior of the recurrent states i of a Markov chain. Note that the condition of recurrence $f_{ii} = \sum_{k=1}^{\infty} f_{ii}(k) = 1$ does not ensure the existence of a limit distribution. The main characteristics are the expected values of the return times $\mu_i = \mathbf{E}(T_{ii}) = \sum_{k=1}^{\infty} k f_{ii}(k)$, and the recurrent states will be classified according to whether or not the μ_i are finite because the condition $\mu_i < \infty$, $i \in \mathcal{X}$, guarantees the existence of a limit distribution.

Definition 3.12 A recurrent state $i \in \mathcal{X}$ is called **positive recurrent** (or **nonnull recurrent**) if the return time has a finite expected value μ_i; otherwise, if $\mu_i = \infty$, then it is called **null recurrent**.

Theorem 3.8 *Let X be a homogeneous, irreducible, aperiodic, and recurrent Markov chain. Then for all states $i, j \in \mathcal{X}$,*

$$\lim_{m \to \infty} p_{ij}(m) = \frac{1}{\mu_j}.$$

Note that this theorem not only gives the limit of the m-step transition probabilities with the help of the expected value of the return times, but it interprets the notion of positive and null recurrence. By definition, a recurrent state j is positive recurrent if $1/\mu_j > 0$ and null recurrent if $1/\mu_j = 0$ (here and subsequently, we write $1/\infty = 0$). The assertion given in the theorem is closely related to the discrete renewal Eq. (3.6), and using it we can prove a limit theorem, as the following lemma shows (see [4] and Ch. XIII. of [5]).

Lemma 3.1 (Erdős, Feller, Pollard) *Let $(q_i, \ i \geq 0)$ be an arbitrary distribution on the natural numbers, i.e., $q_i \geq 0$, $\sum_{i=0}^{\infty} q_i = 1$. Assume that the distribution $(q_i, \ i \geq 0)$ is not latticed, that is the greatest common divisor of the indices with the probabilities $q_i > 0$ equals 1. If the sequence $\{v_n, \ n \geq 0\}$, satisfies the discrete renewal equation*

$$v_0 = 1, \quad v_n = \sum_{k=1}^{n} q_k v_{n-k}, \ n \geq 1,$$

then

$$\lim_{n \to \infty} v_n = \frac{1}{\mu},$$

where $\mu = \sum_{k=1}^{\infty} k q_k$ and $\frac{1}{\mu} = 0$, if $\mu = \infty$.

The proof of Theorem 3.8 uses the following result from analysis.

Lemma 3.2 *Assume that the sequence $(q_1, q_2, \ldots)$ of nonnegative real numbers satisfies the condition $\sum_{i=0}^{\infty} q_i = 1$. If the sequence of real numbers $(w_n, \ n \geq 0)$ is convergent, $\lim_{n \to \infty} w_n = w$, then*

$$\lim_{n \to \infty} \sum_{k=0}^{n} q_{n-k} w_n = w.$$

Proof It is clear that the elements of $\{w_n\}$ are bounded; then there exists a number W such that $|w_n| \leq W$, $n \geq 0$. From the conditions $\lim_{n \to \infty} w_n = w$ and $\sum_{i=0}^{\infty} q_i = 1$

it follows for any $\varepsilon > 0$ that there exist integers $N(\varepsilon)$ and $K(\varepsilon)$ such that

$$|w_n - w| < \varepsilon, \; n > N(\varepsilon) \quad \text{and} \quad \sum_{k=K(\varepsilon)}^{\infty} q_k < \varepsilon.$$

It is easy to check that for every $n > n(\varepsilon) = 2\max(N(\varepsilon), K(\varepsilon))$,

$$|w_n - w| \leq \left| \sum_{k=0}^{n} q_k w_{n-k} - \sum_{k=0}^{n} q_k w \right| + \sum_{k=n+1}^{\infty} q_k |w|$$

$$\leq \sum_{k=0}^{n(\varepsilon)} q_k |w_{n-k} - w| + \sum_{k=n(\varepsilon)+1}^{n} q_k |w_{n-k} - w| + \sum_{k=n+1}^{\infty} q_k |w|$$

$$\leq \sum_{k=0}^{n(\varepsilon)} q_k \varepsilon + \sum_{k=n(\varepsilon)+1}^{n} q_k (W + |w|) + \sum_{k=n+1}^{\infty} q_k |w|$$

$$\leq \varepsilon + \varepsilon(W + |w|) + \varepsilon|w| = \varepsilon(1 + W + 2|w|).$$

Since $\varepsilon > 0$ can be chosen freely, we get the convergence $w_n \to w$, $n \to \infty$. □

Proof (Theorem 3.8)

(a) We prove first the assertion for the case $i = j$. By the discrete renewal equation

$$p_{ii}(0) = 1, \quad p_{ii}(n) = \sum_{k=1}^{n} f_{ii}(k) p_{ii}(n - k), \quad n = 1, 2, \ldots,$$

where the state i is recurrent, $f_{ii} = \sum_{k=1}^{\infty} f_{ii}(k) = 1$ ($f_{ii} \geq 0$). Using the assertion of Lemma 3.1 we have

$$\lim_{n \to \infty} p_{ii}(n) = \frac{1}{\mu_i}.$$

(b) Now let $i \neq j$ and apply Lemma 3.2. Since the Markov chain is irreducible and recurrent, $f_{ij} = \sum_{k=1}^{\infty} f_{ij}(k) = 1$ ($f_{ij} \geq 0$). Then, as $n \to \infty$,

$$\lim_{n \to \infty} p_{ij}(n) = \lim_{n \to \infty} \sum_{k=1}^{n} f_{ij}(k) p_{jj}(n - k) = \sum_{k=1}^{\infty} f_{ij}(k) \frac{1}{\mu_j} = \frac{1}{\mu_j}.$$

□

Similar results can be easily proven for periodic cases. Let X be a homogeneous, irreducible, and recurrent Markov chain with period $d > 1$. Then the state space $\mathscr{X}$

can be decomposed into disjoint subsets $\mathscr{X}_0, \ldots, \mathscr{X}_{d-1}$ [see Eq. (3.5)] such that the Markov chain allows only for cyclic transitions between the states of the sets $\mathscr{X}_i$: $\mathscr{X}_0 \to \mathscr{X}_1 \to \ldots \to \mathscr{X}_{d-1} \to \mathscr{X}_0$. Let $0 \le k,\ m \le d - 1$ be arbitrarily fixed integers; then, starting from a state $i \in \mathscr{X}_k$, the process arrives at a state of $j \in \mathscr{X}_m$ in exactly

$$\ell = \begin{cases} m - k, & \text{if } k < m, \\ m - k + d, & \text{if } m \le k, \end{cases}$$

steps. From this follows $p_{ij}(s) = 0$ if $s - \ell$ is not divisible by d.

Theorem 3.9 *Let X be a homogeneous, irreducible, and recurrent Markov chain with period $d > 1$ and $i \in \mathscr{X}_k$, $j \in \mathscr{X}_m$ arbitrarily fixed states. Then*

$$\lim_{n \to \infty} p_{ij}(\ell + nd) = \frac{d}{\mu_j},$$

where $\mu_j = \sum\limits_{k=1}^{\infty} k\, f_{jj}(k) = \sum\limits_{r=1}^{\infty} rd\, f_{jj}(rd)$.

Proof First assume $k = m$, and consider the transition probabilities $p_{ij}(nd)$ for $i, j \in \mathscr{X}_k$. This is equivalent to the case (see Conclusion 3.2 according to the cyclic transitions of a Markov chain) where we investigate the Markov chain $\overline{X}$ with the state space $\mathscr{X}_k$ and it has (one-step) transition probability matrix $\overline{\Pi} = [\overline{p}_{ij}]_{i,j \in \mathscr{X}_k}$, $\overline{p}_{ij} = p_{ij}(d)$, $i, j \in \mathscr{X}_k$. Obviously, the Markov chain $\overline{X}$ that originated from X is a homogeneous, irreducible, recurrent, and aperiodic Markov chain. Using the limit theorem 3.1, we obtain

$$\lim_{n \to \infty} \overline{p}_{ii}(n) = \lim_{n \to \infty} p_{ii}(nd) = \frac{1}{\sum\limits_{k=1}^{\infty} k f_{ii}(kd)} = \frac{d}{\sum\limits_{k=1}^{\infty} kd f_{ii}(kd)} = \frac{d}{\sum\limits_{k=1}^{\infty} k f_{ii}(k)} = \frac{d}{\mu_i},$$

where $f_{ii}(r) = 0$, if $r \neq d, 2d, \ldots$

Assume now that $k \neq m$. Then $f_{ij}(k) = 0$ and $p_{ij}(k) = 0$ if $k \neq \ell + nd$, $n \ge 0$; moreover, the Markov chain $\overline{X}$ is recurrent because

$$f_{ij} = \sum_{s=1}^{\infty} f_{ij}(s) = \sum_{r=1}^{\infty} f_{ij}(\ell + rd) = 1;$$

then

$$p_{ij}(\ell+nd) = \sum_{k=1}^{\ell+nd} f_{ij}(k) p_{jj}(\ell+nd-k) = \sum_{r=1}^{\ell+nd} f_{ij}(\ell+rd) p_{jj}(rd) \to \frac{d}{\mu_j}, \quad n \to \infty.$$

$\square$

Theorem 3.10 *If the homogeneous Markov chain X is irreducible and has a positive recurrent state $i \in \mathscr{X}$, then all its states are positive recurrent.*

Proof Let $j \in \mathscr{X}$ be arbitrary. Since the Markov chain is irreducible, there exist integers $s, t > 0$ such that $p_{ij}(s) > 0$, $p_{ji}(t) > 0$. Denote by d the period of the Markov chain. It is clear that $d > 0$ because $p_{ii}(s + t) \geq p_{ij}(s)p_{ji}(t) > 0$. Moreover,

$$p_{ii}(s + nd + t) \geq p_{ij}(s)p_{jj}(nd)p_{ji}(t),$$

$$p_{jj}(s + nd + t) \geq p_{ji}(t)p_{ii}(nd)p_{ij}(s).$$

Applying Theorem 3.9 and taking the limit as $n \to \infty$ we have

$$\frac{1}{\mu_i} \geq p_{ij}(s)\frac{1}{\mu_j}p_{ji}(t), \quad \frac{1}{\mu_j} \geq p_{ij}(s)\frac{1}{\mu_i}p_{ji}(t);$$

thus

$$\frac{1}{\mu_i} \geq p_{ij}(s)p_{ji}(t)\frac{1}{\mu_j} \geq [p_{ij}(s)p_{ji}(t)]^2\frac{1}{\mu_i}.$$

From the last inequality it immediately follows that when the state i is recurrent, at same time j is also recurrent.   □

Summing up the results derived previously, we can state the following theorem.

Theorem 3.11 *Let X be a homogeneous irreducible Markov chain, then*

1. *All states are aperiodic or all states are periodic with the same period.*
2. *All states are transient or all states are recurrent, and in the latter case*

   – All are positive recurrent or all are null recurrent.

3.2.2   Stationary Distribution of Markov Chains

Retaining the notations introduced previously, $P(n) = (P_i(n) = \mathbf{P}(X_n = i),$ $i \in \mathscr{X})$ denotes the distribution of a Markov chain depending on the time $n \geq 0$. Then $P(0) = (P_i(0) = p_i, i \in \mathscr{X})$ is the initial distribution.

Definition 3.13 Let $\pi = (\pi_i, i \in \mathscr{X})$ be a distribution, i.e., $\pi_i \geq 0$ and $\sum_{i \in \mathscr{X}} \pi_i = 1$. π is called a **stationary distribution** of the Markov chain X if by choosing $P(0) = \pi$ as the initial distribution, the distribution of the process does not depend on the time, that is,

$$P(n) = \pi, \ n \geq 0.$$

A stationary distribution is also called an **equilibrium distribution** of a chain.

With Markov chains, the main problem is the existence and determination of stationary distributions. Theorem 3.8 deals with the convergence of the sequence of n-step transition probabilities $P(n)$ as $n \to \infty$, and if it converges, then the limit gives the stationary distribution of the chain. The proofs of these results are not too difficult but consist of many technical steps [7, 8], and so we omit them here.

Theorem 3.12 *Let X be a homogeneous, irreducible, recurrent, and aperiodic Markov chain. Then the following assertions hold:*

(a) The limit

$$\pi_i = \lim_{n \to \infty} P_i(n) = \frac{1}{\mu_i}, \ i \in \mathcal{X},$$

exists and does not depend on the initial distribution.
(b) If all states are recurrent null states, then the stationary distribution does not exist and $\pi_i = 0$ for all $i \in \mathcal{X}$.
(c) If all states are positive recurrent, then the stationary distribution $\pi = (\pi_i, \ i \in \mathcal{X})$ does exist and $\pi_i = 1/\mu_i > 0$ for all $i \in \mathcal{X}$ and $P(n) \to \pi$, as $n \to \infty$. The stationary distribution is unique and it satisfies the system of linear equations

$$\sum_{i \in \mathcal{X}} \pi_i = 1, \tag{3.7}$$

$$\pi_i = \sum_{j \in \mathcal{X}} \pi_j p_{ji}, \ i \in \mathcal{X}. \tag{3.8}$$

Remark 3.4 Since the Markov chain is irreducible, it is enough to require in part (b) the existence of a positive recurrent state because from the existence of a single positive recurrent state and the fact that the Markov chain is irreducible it follows that all states are positive recurrent.

Equation (3.8) of Theorem (3.12) can be rewritten in the more concise form $\pi = \pi \boldsymbol{\Pi}$, where $\boldsymbol{\Pi}$ is the one-step transition probability matrix of the chain.

The initial distribution does not play a role in Eqs. (3.7) and (3.8); therefore, when the stationary distribution π exists, it does not depend on the initial distribution, only on the transition probability matrix $\boldsymbol{\Pi}$.

Remark 3.5 If there is a solution of Eqs. (3.7) and (3.8) for an irreducible Markov chain, then the solution is unique.

Given that the stationary distribution π exists, it can be easily proven that π satisfies the system of linear Eq. (3.8), and at the same time, these circumstances lead to an iterative method of solution (see Eq. (3.9) below). This iterative procedure to determine the stationary distribution can be applied to chains with finite state spaces.

The time-dependent distribution $P(n) = (P_0(n), P_1(n), \ldots)$ satisfies the equation for all $n = 0, 1, \ldots,$

$$P(n) = P(n-1)\boldsymbol{\Pi}. \tag{3.9}$$

Repeating this equation n times, we have

$$P(n) = P(0)\boldsymbol{\Pi}^n, \quad n = 0, 1, \ldots$$

Since it is assumed that the stationary distribution π exists, we can write

$$\pi = \lim_{n \to \infty} P(n);$$

thus from the equation

$$\lim_{n \to \infty} P(n) = \lim_{n \to \infty} P(n-1)\boldsymbol{\Pi}$$

it follows

$$\pi = \pi\boldsymbol{\Pi}.$$

Definition 3.14 A state i of an irreducible homogeneous Markov chain X is called **ergodic** if the state i is aperiodic and positive recurrent, i.e., $d(i) = 1$, $\mu_i < \infty$. If all states of the chain are ergodic, then the Markov chain is called ergodic.

Here we define the ergodic property only of Markov chains. This property can be defined for much more complex stochastic processes as well.

By Theorem 3.12, a homogeneous, aperiodic, positive recurrent Markov chain is always ergodic. Since an irreducible Markov chain with finite state space is positive recurrent, the following statement is also true.

Theorem 3.13 *A homogeneous, irreducible, aperiodic Markov chain with finite state space is ergodic.*

In practical applications, the equilibrium distributions of Markov chains play an essential role. In what follows, we give two theorems without proofs whose conditions ensure the existence of the stationary distribution of a homogeneous, irreducible, aperiodic Markov chain X with state space $\mathscr{X} = \{0, 1, \ldots\}$. The third theorem gives an upper bound for the convergence rate to the stationary distribution of the iterative procedure (3.9).

Theorem 3.14 (Klimov [16]) *If there exists a function $g(i)$, $i \in \mathscr{X}$, a state $i_0 \in \mathscr{X}$, and a positive constant ε such that the relations*

$$\mathbf{E}(g(X_{n+1}) \mid X_n = i) \leq g(i) - \varepsilon, \quad i \geq i_0, \; n \geq 0,$$

$$\mathbf{E}(g(X_{n+1}) \mid X_n = i) < \infty, \quad i \geq 0, \; n \geq 0$$

hold, then the chain X is ergodic.

$g(i)$ is commonly referred to as Lyapunov function.

Theorem 3.15 (Foster [6]) *Assume that there exist constants $a, b > 0$ and $\ell \geq 0$ such that the inequalities*

$$\mathbf{E}(X_{n+1} \mid X_n = i) \leq a, \ \ i \leq \ell,$$

$$\mathbf{E}(X_{n+1} \mid X_n = i) \leq i - b, \ \ i > \ell$$

are valid. Then the Markov chain X is ergodic.

3.2.3   Rate of Convergence to the Stationary Distribution

We have discussed in the preceding section that an irreducible aperiodic DTMC X with finite state space $\mathscr{X} = \{0, 1, \ldots, N\}$ always has the stationary limit distribution. Besides this, the analysis of the convergence rate of the n-step transition probability $p_{ij}(n)$ of the chain is also an important problem from both a theoretical and practical (numerical computational) point of view. In this section, we will focus our attention to a brief discussion of this problem (related to the convergence rate in a more general setting see [20]).

Preserving the notations used previously, the following upper bound is valid for the n-step transition probability of a Markov chain:

Theorem 3.16 (Bernstein [3]) *Assume that there exist a state $i_0 \in \mathscr{X}$ and a constant $\lambda > 0$ such that for all $i \in \mathscr{X}$ the inequality $p_{ii_0} \geq \lambda$ holds. Then, independently of i,*

$$\lim_{n \to \infty} p_{ij}(n) = \lim_{n \to \infty} \mathbf{P}(X_n = j \mid X_0 = i) = \pi_j, \ i, j \in \mathscr{X},$$

where $\pi = (\pi_i, i \in \mathscr{X})$ denotes the stationary distribution of the Markov chain; moreover,

$$\sum_{i \in \mathscr{X}} |p_{ij}(n) - \pi_j| \leq 2(1 - \lambda)^n, \ n \geq 1.$$

We observe that this theorem does not provide a generally good rate of convergence and it does not use the basic structure of the one-step transition matrix Π. Another way to analyze the convergence rate of $p_{ij}(n) \to \pi_j$ as $n \to \infty$ is the use of the fundamental Perron–Frobenius theorem which characterizes the square matrices with nonnegative entries. In general, the Perron–Frobenius theorem (see [22, Theorem 1.1]) constitutes an effective theoretical framework related to different theoretical and practical problems.

One can prove (see Exercise 3.10) that for any one-step transition matrix $\Pi = \begin{pmatrix} 1-a & a \\ b & 1-b \end{pmatrix}$, $0 < a, b < 1$ and $\Pi_\infty = \begin{pmatrix} b/(a+b) & a/(a+b) \\ b/(a+b) & a/(a+b) \end{pmatrix}$ we have the formula (for the n-step transition matrix)

$$\Pi^n = \Pi_\infty + (1-a-b)^n \begin{pmatrix} a/(a+b) & -a/(a+b) \\ -b/(a+b) & b/(a+b) \end{pmatrix}, \quad n \geq 1, \tag{3.10}$$

which gives the exact rate of convergence to the stationary distribution (here, obviously, $|1 - a - b| < 1$). We will show hereinafter that under the adequate general condition the geometric rate of convergence follows from the Perron–Frobenius theory.

The following version of Perron–Frobenius theorem (see [22, pp. 3, 8]) is true for arbitrary primitive $N \times N$ stochastic matrix M (a square nonnegative matrix M is called primitive if there exists a positive integer m such that $M^m > 0$).

Theorem 3.17 (Perron–Frobenius) *Suppose M is an $N \times N$ primitive stochastic matrix. Then there exists an eigenvalue $\lambda_1 = 1$ such that:*

(a) strictly positive left and right eigenvectors can be associated with λ_1, which are unique to constant multiples;
(b) $\lambda_1 > |\lambda|$ for any eigenvalue $\lambda \neq \lambda_1$;
(d) λ_1 is a simple root of the characteristic equation $\det(M - \lambda I) = 0$.

In the following we will restrict ourselves to the one-step transition matrix Π of an irreducible aperiodic MC X. In this case Π is primitive then Theorem 3.17 is applicable and we have an opportunity to estimate the convergence rate of n-step transition probabilities to the stationary distribution.

Let us consider the case when the (left) eigenvectors $v_1, \ldots, v_N$ associated with the eigenvalues $\lambda_1, \ldots, \lambda_N$ are linearly independent (with respect to the general case see p. 8 in [22]). Obviously, the eigenvalues of Π can be given in the form $\lambda_1 = 1 > |\lambda_2| \geq \ldots \geq |\lambda_N|$. The strictly positive (left) eigenvector v_1 is unique to constant multiples, and it may be chosen to satisfy the condition $v_1 e^T = 1$, where actually $e = (1, \ldots, 1)$ is the right eigenvector of the eigenvalue $\lambda_1 = 1$. It means that with this choice the vector $v_1 = \pi$ is identical to stationary distribution of the MC X because the equations $v_1 \Pi = v_1$, $v_1 e^T = 1$ hold. Denote by V the $N \times N$ matrix with the (row) vectors $v_1, \ldots, v_N$. In this case we have $V\Pi = \Lambda V$, where $\Lambda = \text{diag}(\lambda_1, \ldots, \lambda_N)$ denotes the diagonal matrix with the diagonal elements $\lambda_1, \ldots, \lambda_N$. By the assumption that the eigenvectors $v_1, \ldots, v_N$ are linearly independent, then the matrix V has the inverse V^{-1} which satisfies the equation

$$\Pi V^{-1} = V^{-1} \Lambda.$$

This means that the column vectors of V^{-1} denoted by $u_1, \ldots, u_N$, are the right (column) eigenvectors of the matrix Π Then we have $u_1 = e^T$ and yields the spectral representation of the matrix Π^n, $n \geq 1$ as follows

$$\Pi^n = \left(V^{-1}\Lambda V\right)^n = V^{-1}\Lambda^n V = \sum_{k=1}^{N} \lambda_k^n u_k v_k = e^T v_1 + \lambda_2^n \sum_{k=2}^{N} \left(\frac{\lambda_k}{\lambda_2}\right)^n u_k v_k.$$

From this we immediately obtain the limiting matrix

$$\Pi^n \to \Pi_\infty = \pi^T e = \begin{bmatrix} \pi_1 & \dots & \pi_N \\ \cdot & \dots & \cdot \\ \pi_1 & \dots & \pi_N \end{bmatrix}$$

and the convergence rate of n-step transition probabilities to the stationary distribution

$$|p_{ij}(n) - \pi_j| = O\left(|\lambda_2^n|\right), \quad n \to \infty.$$

Note that the quantity $\lambda_2 = 1 - a - b$ in (3.10) is actually the second eigenvalue besides the first one $\lambda_1 = 1$ related to the matrix $\Pi = \begin{pmatrix} 1-a & a \\ b & 1-b \end{pmatrix}$, because the characteristic equation takes the form

$$\det\left(\Pi - \lambda I\right) = \det\begin{pmatrix} a - \lambda & 1 - a \\ 1 - b & b - \lambda \end{pmatrix} = (1 - \lambda)(1 - a - b) = 0.$$

3.2.4 Ergodic Theorems for Markov Chains

Let X be a homogeneous irreducible and positive recurrent Markov chain with state space $\mathscr{X} = \{0, 1, \dots\}$ and i a fixed state. Compute the time and the relative frequencies when the process stays in the state i on the time interval $[0, T]$ as follows:

$$S_i(T) = \sum_{n=0}^{T} \mathscr{I}_{\{X_n = i\}},$$

$$\overline{S}_i(T) = \frac{1}{T} \sum_{n=0}^{T} \mathscr{I}_{\{X_n = i\}} = \frac{1}{T} S_i(T).$$

Let us consider when and in what sense there exists a limit of the relative frequencies $\overline{S}_i(T)$ as $T \to \infty$ and, if it exists, how it can be determined. This problem has, in particular, practical importance when applying simulation methods. To clarify the stochastic background of the problem, we introduce the following notations.

Assume that a process starts at time 0 from the state i. Let $0 = T_0^{(i)} < T_1^{(i)} < T_2^{(i)} < \dots$ be the sequence of the consecutive random time points when a Markov chain arrives at the state i, that is, $T_k^{(i)}$, $k = 1, 2, \ldots$, are the return time points to the state i of the chain. This means that

$$X(T_n^{(i)}) = i, \ n = 0, 1, \ldots \ \text{ and } \ X(k) \neq i, \ \text{ if } \ k \neq T_0^{(i)}, T_1^{(i)}, \ldots.$$

Denote by

$$\tau_k^{(i)} = T_k^{(i)} - T_{k-1}^{(i)}, \ k = 1, 2, \ldots,$$

the time length between the return time points. Since the Markov chain has the memoryless property, these random variables are independent; moreover, from the homogeneity of the Markov chain it follows that $\tau_n^{(i)}, n \geq 1$, are also identically distributed. The common distribution of these random variables $\tau_n^{(i)}$ is the distribution of the return times from the state i to i, namely, $(f_{ii}(n), n \geq 1)$.

Heuristically, it is clear that when the return time has a finite expected value μ_i, then during the time T the process returns to the state i on average T/μ_i times. This means that the quantity $\overline{S}_i(T)$ fluctuates around the value $1/\mu_i$ and has the same limit as $T \to \infty$. This result can be given in exact mathematical form on the basis of the law of the large numbers as follows.

Theorem 3.18 *If X is an ergodic Markov chain, then, with probability 1,*

$$\lim_{T \to \infty} \overline{S}_i(T) = \frac{1}{\mu_i}, \quad i \in \mathcal{X}. \tag{3.11}$$

If the Markov property is satisfied, then not only are the return times independent and identically distributed, but the stochastic behaviors of the processes on the return periods are identical as well. This fact allows us also to prove the central limit theorem for the parameters of an ergodic Markov chain.

Theorem 3.19 *Let X be an ergodic Markov chain and $g(i)$, $i \in \mathcal{X}$, be a real valued function such that $\sum_{i \in \mathcal{X}} \pi_i |g(i)| < \infty$. Then the convergence*

$$\lim_{T \to \infty} \frac{1}{T} \sum_{n=1}^{T} g(X_n) = \sum_{i \in \mathcal{X}} \pi_i g(i)$$

is true with probability 1, where $\pi_i, i \in \mathcal{X}$, denotes the stationary distribution of the Markov chain, which exists under the given condition.

3.2.5   Estimation of Transition Probabilities

In modeling ergodic Markov chains an important question is to estimate the transition probabilities by the observation of the chain. The relative frequencies give corresponding estimates of the probabilities because by Theorem 3.18 they tend to them with probability 1 under the given conditions. Note that from the heuristic approach discussed previously it follows under quite general conditions that not only can the law of large numbers be derived for the relative frequencies, but the central limit theorems can as well.

Consider now the estimate of transition probabilities with the maximum likelihood method. Let X be an ergodic Markov chain with finite state space $\mathscr{X} = \{0, 1, \ldots, N\}$ and with the (one-step) transition probability matrix $\boldsymbol{\Pi} = (p_{ij})_{i,j \in \mathscr{X}}$. Assume that we have an observation of n elements $X_1 = i_1, \ldots, X_n = i_n$ starting from the initial state $X_0 = i_0$, and we will estimate the entries of the matrix Π. By the Markov property, the conditional likelihood function can be given in the form

$$\mathbf{P}(X_1 = i_1, \ldots, X_n = i_n \mid X_0 = i_0) = p_{i_0 i_1} \cdots p_{i_{n-1} i_n}.$$

Denote by n_{ij}, $i, j \in \mathscr{X}$, the number of one-step transitions from the state i to j in the sample path $i_0, i_1, \ldots, i_n$, and let $0^0 = 1$, $0/0 = 0$. Then the conditional likelihood function given the $X_0 = i_0$ initial state is

$$L(i_1, \ldots, i_n; \boldsymbol{\Pi} \mid i_0) = \prod_{i=0}^{N} \left(\prod_{j=0}^{N} p_{ij}^{n_{ij}} \right). \tag{3.12}$$

Applying the maximum likelihood method, maximize the expression in p_{ij} under the conditions

$$p_{ij} \geq 0, \ i, j \in \mathscr{X}, \ \sum_{j \in \mathscr{X}} p_{ij} = 1, \ i \in \mathscr{X}.$$

It is clear that there are no relations between the products playing a role in the parenthesis of Eq. (3.12) for different i; therefore, the maximization problem can be solved by means of $N + 1$ different, but similar, optimization problems:

$$\max \left\{ \prod_{j=0}^{N} p_{ij}^{n_{ij}} : p_{ij} \geq 0, \ \sum_{j \in \mathscr{X}} p_{ij} = 1 \right\}, \ i = 0, 1, \ldots, N.$$

Obviously it is enough to solve it only for one state i since the others can be derived analogously to that one.

Let $i \in \mathscr{X}$ be a fixed state, and denote $n_i = \sum_{j \in \mathscr{X}} n_{ij}$. Apply the Lagrange multiplier method; then for every $m = 0, \ldots, N$,

$$\frac{\partial}{\partial p_{im}} \left(\prod_{j=0}^{N} p_{ij}^{n_{ij}} + \lambda(p_{i0} + p_{11} + \ldots + p_{iN} - 1) \right) = \frac{n_{im}}{p_{im}} \prod_{j=0}^{N} p_{ij}^{n_{ij}} + \lambda = 0;$$

consequently, for a constant λ_i we have

$$\frac{n_{im}}{p_{im}} = \lambda \prod_{j=0}^{N} p_{ij}^{-n_{ij}} = \lambda_i, \ m = 0, \dots, N.$$

From this it follows that the equations

$$n_{im} = \lambda_i \, p_{im}, \quad m = 0, \dots, N$$

hold; then

$$\sum_{m=0}^{N} n_{im} = n_i = \lambda_i \sum_{m=0}^{N} p_{im} = \lambda_i.$$

These relations lead to the conditional maximum likelihood estimates for the transition probabilities p_{im} as follows:

$$\widehat{p}_{im} = \frac{n_{im}}{\lambda_i} = \frac{n_{im}}{n_i}, \ 0 \le i, m \le N.$$

It can be verified that these estimates $\widehat{p}_{im}$ converge to p_{im} with probability 1 as $n \to \infty$.

3.3   Continuous-Time Markov Chains

Like the case of the DMTCs, we assume that the state space $\mathscr{X}$ is a finite $\{0, 1, \dots, N\}$ or countably infinite set $\{0, 1, \dots\}$ and assume that the time parameter varies in $\mathscr{T} = [0, \infty)$. According to the general definition (3.1), a process $X = (X_t, \ t \ge 0)$ is said to be CTMC with state space $\mathscr{X}$ if for every positive integer n and $0 \le t_0 < t_1 < \dots < t_n, i_0, \dots, i_n \in \mathscr{X}$, the equation

$$\mathbf{P}\big(X_{t_n} = i_n \mid X_{t_{n-1}} = i_{n-1}, \dots, X_{t_0} = i_0\big)$$
$$= \mathbf{P}\big(X_{t_n} = i_n \mid X_{t_{n-1}} = i_{n-1}\big) = p_{i_{n-1}, i_n}(t_{n-1}, t_n).$$

holds, provided that a conditional probability exists. The Markov chain X is (time) homogeneous, if the transition probability function $p_{ij}(s, t)$ satisfies the condition $p_{ij}(s, t) = p_{ij}(t - s)$ for all $i, j \in \mathscr{X}, 0 \le s \le t$. Denote by $\boldsymbol{\Pi}(s, t) = \big[p_{ij}(s, t), i, j \in \mathscr{X}\big]$ the transition probability matrix.

In the case of a CTMC the time index $t \in [a, b]$ can take uncountably many values for arbitrary $0 \le a < b < \infty$; therefore, the collection of random variables $X_t, \ t \in [a, b]$, is also uncountable. If we consider the questions in accordance with the sample paths of the chain, then these circumstances can lead to measurability problems (discussed later). However, the Markov processes that will be investigated later are the so-called stepwise processes, and they ensure the necessary measurability property.

We will deal mainly with the part of the theory that is relevant to queueing theory, and we touch upon only some questions in general cases showing the root of the measurability problems. A discussion of jumping processes, which is more general than the investigation of stepwise Markov chains, can be found in [8, Chap. III].

If the Markov chain $\{X_t, \ t \geq 0\}$, is homogeneous, then the transition probability functions $p_{ij}(s, t)$ can be given in a simpler form:

$$p_{ij}(s, s + t) = p_{ij}(t), \ i, j \in \mathcal{X}, \ s, t \geq 0,$$

and thus the matrix form of transition probabilities is

$$\Pi(s, s + t) = \Pi(t), \ s, t \geq 0.$$

As was done previously, denote by

$$P(t) = (P_0(t), P_1(t), \ldots), \ t \geq 0,$$

the time-dependent distribution of the chain, where $P_i(t) = \mathbf{P}(X_t = i), \ i \in \mathcal{X}$; then $P(0)$ means the initial distribution, while if there exists a state $k \in \mathcal{X}$ such that $\mathbf{P}(X_0 = k) = 1$, then k is the initial state.

3.3.1 Characterization of Homogeneous Continuous-Time Markov Chains

We now deal with the main properties of the homogeneous CTMCs. Similar to the discrete-time case, the transition probabilities satisfy the following conditions.

(a) $p_{ij}(s) \geq 0, \ s \geq 0, \ p_{ij}(0) = \delta_{ij}, \ i, j \in \mathcal{X}$, where δ_{ij} is the Kronecker δ-function (which equals 1 if $i = j$ and 0 if $i \neq j$)

(b) $\sum_{j \in \mathcal{X}} p_{ij}(s) = 1, \ s \geq 0, \ i \in \mathcal{X}$,

(c) $p_{ij}(s + t) = \sum_{k \in \mathcal{X}} p_{ik}(s) p_{kj}(t), \ s, t \geq 0, \ i, j \in \mathcal{X}$.

An additional condition is needed for our considerations.

(d) The transition probabilities of the Markov chain X satisfy the conditions

$$\lim_{h \to 0+} p_{ij}(h) = p_{ij}(0) = \delta_{ij}, \ i, j \in \mathcal{X} \tag{3.13}$$

Remark 3.6 Condition (b) expresses that $\boldsymbol{\Pi}(s)$, $s \geq 0$, is a stochastic matrix. We will not consider the so-called **killed Markov chains**, where the lifetime $[0, \tau]$ of the chain is random (where the process is defined) and with probability 1 is finite, i.e., $\mathbf{P}\{\tau < \infty\} = 1$. It should be noted that the condition (b) ensures that the chain is defined on the whole interval $[0, \infty)$ because the process will be certainly in some state $i \in \mathscr{X}$ for any time $s \geq 0$.

Condition (c) is the Chapman–Kolmogorov equation related to the continuous-time case. It can be given in matrix form as follows:

$$\boldsymbol{\Pi}(s + t) = \boldsymbol{\Pi}(s)\boldsymbol{\Pi}(t), \ s, t \geq 0.$$

Similar to the discrete-time case, the time-dependent distribution of the chain satisfies the equation

$$P(s + t) = P(s)\boldsymbol{\Pi}(t), \ s, t \geq 0,$$

and thus for all $t > 0$,

$$P(t) = P(0)\boldsymbol{\Pi}(t).$$

The last relation means that the initial distribution and the transition probabilities uniquely determine the distribution of the chain at all time points $t \geq 0$.

Instead of (d) it is enough to assume that the condition

$$\lim_{h \to 0+} p_{ii}(h) = 1, \ i \in \mathscr{X},$$

holds, because for every $i, j \in \mathscr{X}, i \neq j$, the relation

$$0 \leq p_{ij}(h) \leq \sum_{j \neq i} p_{ij}(h) = 1 - p_{ii}(h) \to 0, \ h \to 0+,$$

is true.

Under conditions (a)–(d), the following relations are valid.

Theorem 3.20 *The transition probabilities $p_{ij}(t)$, $0 \leq t < \infty$, $i \neq j$, are uniformly continuous.*

Proof Using conditions (a)–(d) we obtain

$$\left| p_{ij}(t + h) - p_{ij}(t) \right| = \left| \sum_{k \in \mathscr{X}} p_{ik}(h)p_{kj}(t) - \sum_{k \in \mathscr{X}} \delta_{ik}p_{kj}(t) \right|$$

$$\leq \sum_{k \in \mathscr{X}} \left| p_{ik}(h) - \delta_{ik} \right| p_{kj}(t)$$

$$\leq 1 - p_{ii}(h) + \sum_{k \neq i} p_{ik}(h) = 2(1 - p_{ii}(h)) \to 0, \quad h \to 0+.$$

$\square$

Theorem 3.21 ([8, p. 200]) *For all $i, j \in \mathscr{X}, i \neq j$, the finite limit*

$$q_{ij} = \lim_{h \to 0+} \frac{p_{ij}(h)}{h}$$

exists.

For every $i \in \mathscr{X}$ there exists a finite or infinite limit

$$q_i = \lim_{h \to 0+} \frac{1 - p_{ii}(h)}{h} = -p'_{ii}(0).$$

The quantities q_{ij} and q_i are the most important characteristics of a homogeneous continuous-time Markov chain. Subsequently we will use the notation $q_{ii} = -q_i$, $i \in \mathscr{X}$, also and interpret the meaning of these quantities.

Definition 3.15 The quantity q_{ij} is called the **transition rate** of intensity from the state i to the state j, while q_i is called the **transition rate** from the state i.

We classify the states in accordance with whether or not the rate q_i is finite. If $q_i < \infty$, then i is called **stable state**, while if case $q_i = +\infty$, then we say that i is **instantaneous** state. Note that there exists a Markov chain with property $q_i = +\infty$ [8, pp. 207–210].

Definition 3.16 A stable noninstantaneous state i is called **regular** if

$$\sum_{i \neq j} q_{ij} = -q_{ii} = q_i,$$

and a Markov chain is **locally regular** if all states are regular.

Corollary 3.3 *As a consequence of Theorem 3.21, we obtain that locally regular Markov chains satisfy the following asymptotic properties as $h \to 0+$:*

$$\mathbf{P}(X_{t+h} \neq i \mid X_t = i) = q_i h + o(h),$$

$$\mathbf{P}(X_{t+h} = i \mid X_t = i) = 1 - q_i h + o(h),$$

$$\mathbf{P}(X_{t+h} = j \mid X_t = i) = q_{ij} h + o(h), \quad j \neq i.$$

From Theorem 3.21 it also follows that Markov chains with a finite state space are locally regular because all $q_{ij}, i \neq j$, are finite and, consequently, all q_i are also finite.

The condition

$$q = \sup_{i \in \mathscr{X}} q_i < \infty \qquad (3.14)$$

will play an important role in our subsequent investigations. We introduce the notation

$$\boldsymbol{Q} = \left[q_{ij}\right]_{i,j \in \mathscr{X}} = \left[p'_{ij}(0)\right]_{i,j \in \mathscr{X}} = \boldsymbol{\Pi}'(0)$$

for locally regular Markov chains. Recall that

$$\lim_{t \to 0+} \boldsymbol{\Pi}(t) = \boldsymbol{\Pi}(0) = I, \qquad (3.15)$$

where I is the identity matrix with suitable dimension.

Definition 3.17 The matrix $\boldsymbol{Q}$ is called a **rate matrix** or **infinitesimal matrix** of a continuous-time Markov chain.

The following assertions hold for all locally regular Markov chains under the initial condition (3.15) [8, pp. 204–206].

Theorem 3.22 *The transition probabilities of a locally regular Markov chain satisfy the Kolmogorov backward differential equation:*

$$\boldsymbol{\Pi}'(t) = \boldsymbol{Q}\,\boldsymbol{\Pi}(t),\ t \geq 0 \qquad (I)$$

If the condition (3.14) is fulfilled, then the Kolmogorov forward differential equation

$$\boldsymbol{\Pi}'(t) = \boldsymbol{\Pi}(t)\,\boldsymbol{Q},\ t \geq 0 \qquad (II)$$

is valid. Under the condition (3.14) the differential Eqs. (I) and (II), referred to as first- and second-system Kolmogorov equations, have unique solutions.

Theorem 3.23 *Under the condition (3.14), the unique solution of both the first- and second-system Kolmogorov equations is*

$$\boldsymbol{\Pi}(t) = e^{\boldsymbol{Q}t} = \sum_{i=1}^{\infty} \frac{t^i}{i!}\boldsymbol{Q}^i(t),$$

3.3.2   Stationary Distribution of CTMCs

The results characterizing the stationary behavior of a CTMC are very similar to the ones characterizing the stationary behavior of a DTMC provided in Sects. 3.2.2 and 3.2.4.

Theorem 3.24 *Let $X(t)$ be a homogeneous, irreducible, recurrent CTMC. Then the following assertions hold:*

(a) The limit

$$\pi_i = \lim_{t \to \infty} P_i(t), \ i \in \mathscr{X},$$

exists and does not depend on the initial distribution.

(b) If all states are recurrent null states, then the stationary distribution does not exist and $\pi_i = 0$ for all $i \in \mathscr{X}$.

(c) If all states are positive recurrent, then the stationary distribution $\pi = (\pi_i, \ i \in \mathscr{X})$ does exist and π_i for all $i \in \mathscr{X}$ and $P(t) \to \pi$, as $t \to \infty$. The stationary distribution is unique and it satisfies the system of linear equations

$$\sum_{i \in \mathscr{X}} \pi_i = 1, \tag{3.16}$$

$$0 = \sum_{j \in \mathscr{X}} \pi_j q_{ji}, \ i \in \mathscr{X}. \tag{3.17}$$

Instead of providing the counterpart results of Sects. 3.2.2 and 3.2.4, we establish the relation between the stationary behavior of DTMC and CTMC.

Theorem 3.25 *For a homogeneous, irreducible, recurrent CTMC with rate matrix Q satisfying condition (3.14), the unique solution of the system of linear equations*

$$\sum_{i \in \mathscr{X}} \pi_i = 1, \quad 0 = \sum_{j \in \mathscr{X}} \pi_j q_{ji}, \ i \in \mathscr{X}.$$

is identical with the unique solution of the system of linear equations

$$\sum_{i \in \mathscr{X}} \pi_i' = 1, \quad \pi_i' = \sum_{j \in \mathscr{X}} \pi_j' p_{ji}, \ i \in \mathscr{X},$$

where

$$p_{ij} = \begin{cases} \frac{q_{ij}}{q}, & \text{if } i \neq j, \\ \frac{q_{ij}}{q} + 1, & \text{if } i = j. \end{cases} \tag{3.18}$$

That is, a CTMC with rate matrix Q and a DTMC with transition probability matrix Π defined by (3.18) has the same stationary distribution.

3.3.3 Stepwise Markov Chains

The results of Theorem 3.22 are related to the analytical properties of transition probabilities and do not deal with the stochastic behavior of sample paths. In this part we investigate the so-called stepwise Markov chains and their sample paths. We introduce the embedded Markov chain and consider the transition probabilities and holding times. In the remaining part of this chapter we assume that the Markov chain is locally regular and condition (3.14) holds.

Definition 3.18 A Markov chain X is a **jump process** if for any $t \geq 0$ there exists a random time $\Delta = \Delta(t, \omega) > 0$ such that

$$X_s = X_t, \quad \text{if } s \in [t, t + \Delta).$$

In the definition, Δ can be the remaining time the process stays at state X_t, and the definition requires that this time be positive.

Definition 3.19 We say that the Markov chain has a **jump** at time $t_0 > 0$ if there exists a monotonically increasing sequence $t_1, t_2, \ldots$ such that $t_n \to t_0$, $n \to \infty$ and at the same time $X_{t_n} \neq X_{t_0}$, $n = 1, 2, \ldots$. A Markov chain is called a **stepwise** process if it is a jump process and the number of jumps is finite for all sample paths on all finite intervals $[0, t]$.

It should be noted that a stepwise process is continuous from the right and has a limit from the left at all jumping points.

Denote by $(\tau_0 =) 0 < \tau_1 < \tau_2 < \ldots$ the sequence of consecutive jumping points; then all finite time intervals consist, at most, of finite jumping points. Between two jumping points the state of the process does not change, and this time is called **holding time**.

Definition 3.20 A stepwise Markov chain is called **regular** if the sequence of holding times $\zeta_k = \tau_{k+1} - \tau_k$, $k = 0, 1, \ldots$ satisfies the condition

$$\mathbf{P}\left(\sum_{k=0}^{\infty} \zeta_k = \infty\right) = 1.$$

By the definition of stepwise process, we have

$$X_s \equiv X_{\tau_i}, \quad s \in [\tau_i, \tau_{i+1}), \quad i = 0, 1, \ldots$$

Denote by $Y_k = X_{\tau_k}$, $k = 0, 1, \ldots$, the states at time points where the transitions change, and define for $i \neq j$

$$\pi_{ij} = \begin{cases} \frac{q_{ij}}{q_i}, & \text{if } q_i > 0, \\ 0, & \text{if } q_i = 0. \end{cases} \tag{3.19}$$

In addition, let

$$\pi_{ii} = 1 - \sum_{j \neq i} \pi_{ij} . \tag{3.20}$$

By the Markov property, the process $(Y_k, \ k \geq 0)$, is a discrete-time homogeneous Markov chain with the state space $\mathscr{X} = \{0, 1, \ldots\}$ and the transition probabilities

$$\mathbf{P}(Y_{n+1} = j \mid Y_n = i) = \pi_{ij}, \ ij \in \mathscr{X}, \ n \geq 0.$$

The process $(Y_k, \ k \geq 0)$ is called an **embedded Markov chain** of the continuous-time stepwise Markov chain X.

Note that the condition $q_i = 0$ corresponds to the case where i is an absorbing state, and in other cases the holding times for arbitrary state i have an exponential distribution with parameter q_i whose density function is $q_i \, e^{-q_i x}$, $x > 0$.

3.3.4 Construction of Stepwise Markov Chains

The construction derived here gives a method for simulating stepwise Markov chains at the same time. Thus, we construct a CTMC $\{X_t, \ t \geq 0\}$, with initial distribution $P(0) = (P_0(0), P_1(0), \ldots)$ and transition probability matrix $\boldsymbol{\Pi}(t) = \left[p_{ij}(t) \right]$, $t \geq 0$, satisfying condition (3.14).

Using notations (3.19) and (3.20), define the random time intervals with length $S_0, S_1, \ldots$, nonnegative random variables $K_0, K_1, ..$ taking integer numbers and the random jumping points $\tau_m = S_0 + \ldots + S_{m-1}$, $m = 1, 2, \ldots$, by the following procedure.

(a) Generate a random variable K_0 with distribution $P(0)$ [i.e., $\mathbf{P}(K_0 = k) = P_k(0)$, $k \in \mathscr{X}$] and a random variable S_0 distributed exponentially with parameter q_{K_0} conditionally dependent on K_0. Define $X_t = K_0$ if $0 \leq t < S_0$.
(b) In the mth steps $(m = 1, 2, \ldots)$ generate a random variable K_m with distribution $P^{(m)} = (\pi_{K_{m-1},j}, \ j \in \mathscr{X})$, and a random variable S_m distributed exponentially with parameter q_{K_m}. Define $X_t = K_m$ if $\tau_m \leq t < \tau_{m+1}$, $m = 0, 1, \ldots$

Then the stochastic process $\{X_t, t \geq 0\}$ is a stepwise Markov chain with initial distribution $P(0)$ and transition probability matrix $\boldsymbol{\Pi}(t)$, $t \geq 0$.

3.3.5 Poisson Process as Continuous-Time Markov Chain

Theorem 3.26 Let $(N_t, \ t \geq 0)$ be a homogeneous Poisson process with intensity rate λ, $N_0 = 0$. Then the process N_t is a homogeneous Markov chain.

Proof Choose arbitrarily a positive integer n, integers $0 \leq i_1 \leq \ldots \leq i_{n+1}$, and real numbers $t_0 = 0 < t_1 < \ldots < t_{n+1}$. It can be seen that

$$\mathbf{P}\big(N_{t_{n+1}} = i_{n+1} \mid N_{t_n} = i_n, \ldots, N_{t_1} = i_1\big)$$

$$= \frac{\mathbf{P}\big(N_{t_{n+1}} = i_{n+1}, N_{t_n} = i_n, \ldots, N_{t_1} = i_1\big)}{\mathbf{P}\big(N_{t_n} = i_n, \ldots, N_{t_1} = i_1\big)}$$

$$= \frac{\mathbf{P}\big(N_{t_{n+1}} - N_{t_n} = i_{n+1} - i_n, \ldots, N_{t_2} - N_{t_1} = i_2 - i_1, N_{t_1} = i_1\big)}{\mathbf{P}\big(N_{t_n} - N_{t_{n-1}} = i_n - i_{n-1}, \ldots, N_{t_2} - N_{t_1} = i_2 - i_1, N_{t_1} = i_1\big)}.$$

Since the increments of the Poisson process are independent, the last fraction can be written in the form

$$\frac{\mathbf{P}\big(N_{t_{n+1}} - N_{t_n} = i_{n+1} - i_n\big) \cdot \ldots \cdot \mathbf{P}\big(N_{t_2} - N_{t_1} = i_2 - i_1\big)\mathbf{P}\big(N_{t_1} = i_1\big)}{\mathbf{P}\big(N_{t_n} - N_{t_{n-1}} = i_n - i_{n-1}\big) \cdot \ldots \cdot \mathbf{P}\big(N_{t_2} - N_{t_1} = i_2 - i_1\big)\mathbf{P}\big(N_{t_1} = i_1\big)}$$

$$= \mathbf{P}\big(N_{t_{n+1}} - N_{t_n} = i_{n+1} - i_n\big).$$

From the independence of the increments $N_{t_{n+1}} - N_{t_n}$ and $N_{t_n} = N_{t_n} - N_0$ it follows that the events $\big\{N_{t_{n+1}} - N_{t_n} = i_{n+1} - i_n\big\}$ and $\big\{N_{t_n} = i_n\big\}$ are also independent, and thus

$$\mathbf{P}\big(N_{t_{n+1}} - N_{t_n} = i_{n+1} - i_n\big) = \mathbf{P}\big(N_{t_{n+1}} - N_{t_n} = i_{n+1} - i_n \mid N_{t_n} = i_n\big)$$

$$= \mathbf{P}\big(N_{t_{n+1}} = i_{n+1} \mid N_{t_n} = i_n\big),$$

and finally we have

$$\mathbf{P}\big(N_{t_{n+1}} = i_{n+1} \mid N_{t_n} = i_n, \ldots, N_{t_1} = i_1\big) = \mathbf{P}\big(N_{t_{n+1}} = i_{n+1} \mid N_{t_n} = i_n\big).$$

$\square$

It is easy to determine the rate matrix of a homogeneous Poisson process with intensity λ. Clearly, the transition probability of the process is

$$p_{ij}(h) = \mathbf{P}(N_{t+h} = j \mid N_t = i) = \mathbf{P}(N_h = j - i) = \frac{(\lambda h)^{j-i}}{(j-i)!}e^{-\lambda h}, \quad j \geq i,$$

and

$$p_{ij}(h) \equiv 0, \quad j < i.$$

If $j < i$, then obviously $q_{ij} \equiv 0$. Let now $i \leq j$; then

$$q_{ij} = \lim_{h \to 0+} \frac{p_{ij}(h)}{h} = \lim_{h \to 0+} \frac{1}{h}\frac{(\lambda h)^{j-i}}{(j-i)!}e^{-\lambda h} = \begin{cases} \lambda, & \text{if } j = i + 1, \\ 0, & \text{if } j > i + 1. \end{cases}$$

Finally, let $i = j$. By the use of the L'Hospital's rule

$$q_i = \lim_{t \to 0+} \frac{1 - p_{ii}(h)}{h} = \lim_{t \to 0+} \frac{1 - e^{-\lambda h}}{h} = \lambda.$$

Thus, summing up the obtained results, we have the rate matrix

$$Q = \begin{bmatrix} -\lambda & \lambda & 0 & 0 & \cdots \\ 0 & -\lambda & \lambda & 0 & \\ 0 & 0 & -\lambda & \lambda & \ddots \\ \vdots & & & \ddots & \ddots & \ddots \end{bmatrix}. \tag{3.21}$$

The Poisson process is regular because for all $i \in \mathscr{X}$

$$\sum_{j \neq i} q_{ij} = \lambda = q_i < \infty.$$

3.3.6 Reversible DTMC

Let $(X_n, n = 0, 1, \ldots)$ be an irreducible DTMC with transition probabilities $\Pi = (p_{ij}, \ i, j \in \mathscr{X})$ and stationary distribution $\pi = (\pi_i, \ i \in \mathscr{X})$. Let T be a positive integer and define by $\overline{X}_n = X_{T-n}, \ n = 0, 1, \ldots, T$ in reversed time the process $X_n, \ n = 0, 1, \ldots, T$. In the following, we can obtain with a simple calculation that the reversed process satisfies the Markov property and at same time, we get the one-step transition function of that. For arbitrary states $i, j, i_2, \ldots, i_T \in \mathscr{X}$ yields

$$\mathbf{P}\left(\overline{X}_T = j \mid \overline{X}_{T-1} = i, \overline{X}_{T-2} = i_2, \ldots, \overline{X}_0 = i_0\right)$$
$$= \mathbf{P}\left(X_0 = j \mid X_1 = i, X_2 = i_2, \ldots, X_T = i_T\right)$$
$$= \frac{\mathbf{P}\left(X_0 = j, X_1 = i, X_2 = i_2, \ldots, X_T = i_T\right)}{\mathbf{P}\left(X_1 = i, X_2 = i_2 \ldots, X_T = i_T\right)}$$
$$= \frac{\mathbf{P}\left(X_2 = i_2 \ldots, X_T = i_T \mid X_0 = j, X_1 = i\right)\mathbf{P}\left(X_0 = j, X_1 = i\right)}{\mathbf{P}\left(X_2 = i_2 \ldots, X_T = i_T \mid X_1 = i\right)\mathbf{P}\left(X_1 = i\right)}$$
$$= \frac{\mathbf{P}\left(X_0 = j \mid X_1 = i\right)\mathbf{P}\left(X_1 = i\right)}{\pi_i}$$
$$= \mathbf{P}\left(X_0 = j \mid X_1 = i\right) = \mathbf{P}\left(\overline{X}_T = j \mid \overline{X}_{T-1} = i\right)$$

This relation ensures that the reversed process $\overline{X}_n$ satisfies the Markov property and, moreover, we get the one-step transition probabilities $\overline{p}_{ij}$ of $\overline{X}_n$ as follows

$$\bar{p}_{ij} = \mathbf{P}\left(\overline{X}_T = j \mid \overline{X}_{T-1} = i\right) = \mathbf{P}\left(X_0 = j \mid X_1 = i\right) = \frac{\mathbf{P}\left(X_0 = j, X_1 = i\right)}{\mathbf{P}\left(X_1 = i\right)}$$

$$= \frac{\mathbf{P}\left(X_1 = i \mid X_0 = j\right)\mathbf{P}\left(X_0 = j\right)}{\mathbf{P}\left(X_1 = i\right)} = p_{ji}\frac{\pi_j}{\pi_i}.$$

Definition 3.21 A discrete-time Markov chain is called **reversible** if for every state i, j the equation

$$\pi_i p_{ij} = \pi_j p_{ji}, \ i, j \in \mathcal{X}$$

holds, where π_i is the equilibrium probability of the states i.

It is important to note that a necessary condition for reversibility of a MCs is that $p_{ij} > 0$ if and only if $p_{ji} > 0$ for all states $i, j \in \mathcal{X}$ because in the equation of the definition the stationary probabilities $\pi_i, \ i \in \mathcal{X}$ are positive.

The equation of the definition is usually called **local** (or **detailed**) **balance condition** because of its similarity to the (global) balance Eq. (3.8) or, more precisely, to its form

$$\sum_{j \in \mathcal{X}} \pi_i p_{ij} = \sum_{j \in \mathcal{X}} \pi_j p_{ji}, \ i, j \in \mathcal{X}.$$

The notation of reversibility of Markov chains originates from the fact that if the initial distribution of the chain equals the stationary one, then the forward and reverse conditional transition probabilities are identical, that is,

$$\mathbf{P}\left(X_n = i \mid X_{n+1} = j\right) = \mathbf{P}\left(X_{n+1} = i \mid X_n = j\right).$$

The following Kolmogorov criterion characterizes the reversibility of a MC X_n without knowing the stationary distribution and based only on the one-step transition probabilities (see [23, Theorem 97, p. 64]).

Theorem 3.27 *The following statements are equivalent.*

(1) The Markov chain X_n is reversible.
(2) (Kolmogorov Criterion) For each $i_0, \ldots, i_k \geq 0$ with $i_k = i_0$,

$$\prod_{m=1}^{k} p_{i_{m-1}i_m} = \prod_{m=1}^{k} p_{i_m i_{m-1}}.$$

(3) For each sequence of transitions $i_0, \ldots, i_k \in \mathcal{X}$ the product $\prod_{m=1}^{k} \frac{p_{i_{m-1}i_m}}{p_{i_m i_{m-1}}}$ depends on $i_0, \ldots, i_k$ and k only through i_0 and i_k.

Corollaries of Theorem 3.27

1. The theorem says that the Kolmogorov criterion is equivalent to the reversibility condition and thus to the detailed balance conditions.
2. The Kolmogorov criterion permits to give a sufficient condition for reversibility for a class of MCs having special importance in analysis of queueing models. Suppose the transition probabilities p_{ij}, $i, j \in \mathcal{X}$ of a stationary MC satisfy the conditions:

   (a) $p_{ij} > 0$ if and only if $p_{ji} > 0$.
   (b) For all pairs of states i and j, $i \neq j$ there is a unique sequence of distinct states $i, i_1, \ldots, i_{N-1}, j$ such that

$$p_{ii_1} p_{i_1 i_2} \cdots p_{i_{N-1} j} = p_{j, i_{N-1}} p_{i_{N-1} i_{N-2}} \cdots p_{i_1 i}.$$

   In this case, as a direct consequence of the Kolmogorov criterion, the property (b) is satisfied, hence the MC is reversible.

3. If the MC X_n is reversible, then $\pi_i p_{ij} = \pi_j p_{ji}$, $i, j \in \mathcal{X}$ and, moreover, due to the property (3) of Theorem 3.27 for a fixed state i_0, $j \neq i_0$ and any sequence of transitions depending on j: $i_0, \ldots, i_{k_j} = j \in \mathcal{X}$, $\mathbf{P}\left(i_0 \to \ldots \to i_{k_j}\right) > 0$ we obtain that

$$r_j = \prod_{m=1}^{k_j} \frac{p_{i_{m-1} i_m}}{p_{i_m i_{m-1}}} = \prod_{m=1}^{k_j} \frac{1}{p_{i_{m-1} i_m}} p_{i_{m-1} i_m} \frac{\pi_{i_m}}{\pi_{i_{m-1}}} = \frac{\pi_j}{\pi_{i_0}}, \quad j \in \mathcal{X} \setminus \{i_0\}.$$

From this equation we get

$$\sum_{j \in \mathcal{X}} r_j = 1 = \sum_{j \in \mathcal{X}} \frac{\pi_j}{\pi_{i_0}} = \frac{1 - \pi_{i_0}}{\pi_{i_0}}$$

and the stationary distribution is

$$\pi_{i_0} = \frac{1}{1 + \sum_{j \in \mathcal{X}} r_j}, \quad \pi_j = \frac{r_j}{1 + \sum_{i \in \mathcal{X}} r_i}, \quad j \in \mathcal{X}.$$

3.3.7 Reversible CTMC

In the case of CTMCs, a definition of reversibility can be applied analogously to the discrete-time case, the only difference is that transition rates are used instead of transition probabilities (detailed descriptions of reversible CTMCs are given, e.g., in [23, Section 4.11]).

Definition 3.22 A CTMC $(X(t), t \geq 0)$ is called **reversible** if for all pairs $i, j \in \mathcal{X}$ of states the equation

$$\pi_i q_{ij} = \pi_j q_{ji}$$

holds, where π_i is the equilibrium probability of the state $i \in \mathcal{X}$ and the transition rates are denoted by q_{ij}, $i, j \in \mathcal{X}$.

The reversibility property and the local balance equations are often valid for Markov chains describing the processes in queueing networks (see Sect. 10.1); in consequence, the equilibrium probabilities can be computed in a simple, so-called **product form**.

3.3.8   Markov Reward Model

In practice, some functional of the CTMC $(X(t), t \geq 0)$ is often the main interest. If the functional of interest is a simple function of the CTMC behavior, then we refer to Markov reward model, where *rate reward* is accumulated during a sojourn in a state and *impulse reward* is accumulated at state transitions. The rate and impulse reward are time independent and they depend only on the associated state and the state transition.

Let $Y(t)$ denote the reward accumulated by the CTMC $X(\tau)$ up to time t. $Y(0) = 0$ and for the rate and impulse rate accumulation we have

$$Y(t + \Delta) = Y(t) + r_i \Delta \quad \text{if } X(\tau) = i, \forall \tau \in (t, t + \Delta),$$

$$Y(t^+) = Y(t-) + d_{ij} \quad \text{if } X(t^-) = i, X(t^+) = j.$$

That is, while $X(t)$ stays in state i reward accumulates at rate r_i and when the CTMC experiences a state transition from state i to j the reward gain is d_{ij}. As a result

$$Y(t) = \int_{\tau=0}^{t} r_{X(s)} + \mathscr{I}_{\{X(s^-)=i, X(s^+)=j\}} d_{ij} \, ds. \tag{3.22}$$

The r_i rates and the d_{ij} values are nonnegative and consequently, $Y(t)$ tends to infinity as t increases, in general. The two main reward measures considered for Markov reward models are the *accumulated reward*, i.e., the distribution of $Y(t)$ and the *completion time*, $C(w)$, which is the time needed to reach reward level w ($w > 0$), that is

$$C(w) = \min(t \mid Y(t) = w). \tag{3.23}$$

Since $Y(t)$ is monotone increasing, the distribution of the completion time and the accumulated reward are connected by

$$F_{Comp}(t, w) = \mathbf{P}(C(w) \leq t) = \mathbf{P}(Y(t) \geq w) = 1 - \mathbf{P}(Y(t) < w).$$

Unfortunately, the distribution of any of them is harder to compute, while we have a simpler expression for the mean of $Y(t)$.

Let $\tau_i(t)$ be the time spent in state i and $N_{ij}(t)$ the number of state transitions from i to j done by the CTMC $X(t)$ in the interval $(0, t)$. With these notations

$$Y(t) = \sum_{i \in S} r_i \, \tau_i(t) + \sum_{i \in S} \sum_{j \in S, j \neq i} N_{ij}(t) \, d_{ij}.$$

For $\mathbf{E}(\tau_i(t))$ and $\mathbf{E}(N_{ij}(t))$ we have

$$\mathbf{E}(\tau_i(t)) = \mathbf{E}\left(\int_0^t \mathscr{I}_{\{X(s)=i\}} \, ds \right) = \int_{s=0}^t \mathbf{E}\left(\mathscr{I}_{\{X(s)=i\}} \right) ds$$

$$= \int_{s=0}^t \mathbf{P}(X(s) = i) \, ds = \int_0^t P_i(s) \, ds \, ,$$

and

$$\mathbf{E}(N_{ij}(t)) = \mathbf{E}\left(\lim_{\Delta \to 0} \sum_{k=0}^{\lfloor t/\Delta \rfloor - 1} \mathscr{I}_{\{X(k\Delta)=i, X((k+1)\Delta)=j\}} \right)$$

$$= \lim_{\Delta \to 0} \sum_{k=0}^{\lfloor t/\Delta \rfloor - 1} \mathbf{E}(I(X(k\Delta) = i, X((k+1)\Delta) = j))$$

$$= \lim_{\Delta \to 0} \sum_{k=0}^{\lfloor t/\Delta \rfloor - 1} \mathbf{P}(X(k\Delta) = i, X((k+1)\Delta) = j)$$

$$= \lim_{\Delta \to 0} \sum_{k=0}^{\lfloor t/\Delta \rfloor - 1} \mathbf{P}(X(k\Delta) = i)\mathbf{P}(X((k+1)\Delta) = j \mid X(k\Delta) = i)$$

$$= \lim_{\Delta \to 0} \sum_{k=0}^{\lfloor t/\Delta \rfloor - 1} p_i(k\Delta)(q_{ij}\Delta + \sigma(\Delta))$$

$$= q_{ij} \lim_{\Delta \to 0} \sum_{k=0}^{\lfloor t/\Delta \rfloor - 1} p_i(k\Delta) \, \Delta + \lim_{\Delta \to 0} \sum_{k=0}^{\lfloor t/\Delta \rfloor - 1} \sigma(\Delta)$$

$$= q_{ij} \int_{s=0}^t p_i(s) ds + \lim_{\Delta \to 0} \frac{t}{\Delta} \sigma(\Delta) = q_{ij} \int_{s=0}^t P_i(s) ds \, ,$$

where $\sigma(\Delta)$ is a small error term satisfying $\lim_{\Delta \to 0} \sigma(\Delta)/\Delta = 0$. Finally, for the mean accumulated reward we have

$$\mathbf{E}(Y(t)) = \mathbf{E}\left(\sum_{i \in S} r_i \, \tau_i(t) + \sum_{i \in S} \sum_{j \in S, j \neq i} \sum_{k=1}^{N_{ij}(t)} d_{ij}\right)$$

$$= \sum_{i \in S} r_i \, \mathbf{E}(\tau_i(t)) + \sum_{i \in S} \sum_{j \in S, j \neq i} \mathbf{E}(N_{ij}(t)) \, d_{ij}$$

$$= \sum_{i \in S}\left(r_i + \sum_{j \in S, j \neq i} q_{ij} \, d_{ij}\right) \int_0^t P_i(s) \, \mathrm{d}s \ ,$$

which depends only on the accumulated state probabilities $\displaystyle\int_0^t P_i(s) \, \mathrm{d}s$.

3.4   Birth–Death and Birth–Death-Immigration Processes

The birth-and-death process (BDPs) and its generalization, the birth–death-immigration process constitute an essential subclass of continuous-time Markov chains, which, in many cases, constitutes a basis of the evolution of various stochastic systems in the theory of queueing systems (see, Chap. 7), population biology, ecology, and others. We note here that the problems at issue are parts of a broader mathematical field i.e., the theory of branching processes (see, e.g., [9, 10]); however, we are only focusing on problems connected to BDP and BDP with immigrations, thus we will not touch upon the general theory. In the meantime, we note that the given approach gives an opportunity to present the separate cases in a structural form and put them in the Tables 3.1 and 3.2.

Table 3.1 Subclasses of the general BDPs considered in this section

Model	Processes	Birth rate	Death rate	Parameters
(A)	Time-homogeneous BDP, $r \geq 0$	$a_k(t) \equiv a_k$	$b_k(t) \equiv b_k$	$a_k, b_k \geq 0$
(B)	Time-inhomogeneous Kendall BDP with immigrations, $r \geq 0$	$ka(t) + \kappa(t)$	$kb(t)$	$a(t), b(t) \geq 0,$ $\kappa(t) \geq 0$
(B1)	Poisson process, $r = 0$	$\kappa(t)$	0	$\kappa(t) \geq 0$
(B2)	Pure birth (Yule-Furry) process, $r = 0$	$ka(t)$	0	$a(t) \geq 0$
(B3)	Pure death process, $r \geq 1$	0	$kb(t)$	$b(t) \geq 0$
(B4)	Kendall BDP without immigration, $r \geq 1$	$ka(t)$	$kb(t)$	$a(t), \, b(t) \geq 0$
(B5)	Kendall BDP withl immigration, $r = 0$	$ka(t) + \kappa(t)$	$kb(t)$	$a(t), \, b(t) \geq 0,$ $\kappa(t) \geq 0$

Table 3.2 Generating functions and PDFs for inhomogeneous Kendall BDPs

BDP models	Parameters	Generating function	Distribution
(B) BDP with immigrations, $r \geq 1$	$a(s)$, $b(s)$ and $\kappa(s)$ are not identically 0	$G(z,t) = [G_{V_1}(z,t)]^r \, G_W(z,t)$	
(B1) Poisson process, $r = 0$	$a(s) \equiv 0$, $b(s) \equiv 0$, $\kappa(s)$ is not identically 0	$G_W(z,t) = \exp\{(z-1)\lambda(t)\}$, where $\lambda(t) = \int_0^t \kappa(s)ds$	Poisson PDF with parameter $\lambda(t)$
(B2) Pure birth process, $r = 0$	$b(s) \equiv 0$, $\kappa(s) \equiv 0$, $s \geq 0$,	$G_{V_1}(z,t) = \frac{\alpha(t)z}{1-(1-\alpha(t))z}$, where $\alpha(t) = \exp\left\{-\int_0^t a(s)ds\right\}$	Geometric PDF with parameter $\alpha(t)$
(B3) Pure death process, $r = 1$	$a(s)$, $\kappa(s) \equiv 0$	$G_{V_1}(z,t) = (1-\alpha(t)) + \alpha(t)z$, where $\alpha(t) = \exp\left\{-\int_0^t b(s)ds\right\}$	Bernoulli PDF with parameter $\alpha(t)$
(B4) BDP without immigrations, $r = 1$	$a(s)$, $b(s)$ are not identically 0, $\kappa(s) \equiv 0$	$G_{V_1}(z,t) = \frac{\alpha(t)+B(t)+(1-\alpha(t)-\beta(t))z}{1-\beta(t)z}$, where $\alpha(t) = \frac{A(t)+B(t)-1}{A(t)+B(t)}$ and $\beta(t) = \frac{B(t)}{A(t)+B(t)}$	Modified geometric PDF parametrized by $\alpha(t)$ and $\beta(t)$
(B5) BDP with immigrations, $r = 0$	$a(s)$, $b(s)$ and $\kappa(s)$ are not identically 0	$G_W(z,t) = \exp\left\{\int_0^t (G_{V_1}(z,s,t) - 1)\kappa(s)ds\right\}$	There is no general explicit formula for PDFs

A BDP represents the number of individuals (population size) in a system. The individuals reproduce (split into two fresh individuals) or die and the birth and death rates can be time-homogeneous or inhomogeneous and the rates can depend on the population size, but they, at any time $t \geq 0$, do not depend on the age of individuals being in the system at moment t (Markov property). In the case of the birth-death-immigration process, immigrants (or additionally arriving individuals at the system) enter the system from outside according to a Poisson point process $0 < \tau_1 < \tau_2 < \ldots$ independently of the population size of the system.

Note that the notions of individuals, birth, death, etc. can be interpreted very differently in practical application of queueing systems. The left side of Fig. 3.3 shows typical evolution trees related to two individuals being in the system at time $t = 0$ and related to individuals (immigrants) after entering the system at time points $0 < \tau_1 < \tau_2 < \ldots$. On the right side of the Fig. 3.3, parallel to the trees, the population size process $X(t)$ is represented depending on time t.

The formalized mathematical model of BDPs is defined on the basis of the CTMC. Denote by $P_j(t) = \mathbf{P}(X(t) = j)$, $j \geq 0$, $t \geq 0$ the time-dependent distribution and let $X(0) = r$, $r \geq 0$ be the initial state of the process, i.e.,

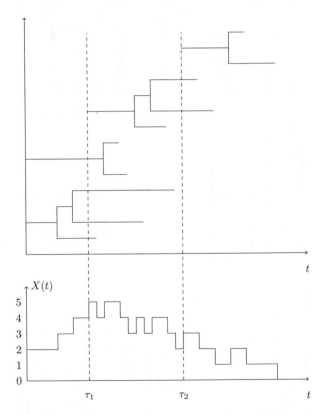

Fig. 3.3 Typical evolution trees and the resulting population size process (the initial value $r = 2$)

$P_r(0) = \mathbf{P}(X(0) = r) = 1$. Let us also denote the transition probability functions as follows

$$P_{i,k}(s,t) = \mathbf{P}(X(t) = k | X(s) = i) \text{ and } P_{i,k}(t) = P_{i,k}(0,t), \ 0 \le s \le t < \infty, \ i, k \ge 0.$$

A BDP $X(t)$, in general, is a continuous-time, right-continuous Markov chain with discrete state space $\{0, 1, \ldots\}$ whose transition functions satisfy the following asymptotic relations as $h \to 0+$:

if $|k - i| \ge 2$, then

$$P_{i,k}(t, t+h) = o(h), \tag{3.24}$$

if $|k - i| \le 1$, then

$$P_{i,k}(t, t+h) = \begin{cases} [a_i(t) + \kappa(t)]h + o(h), & \text{if } k = i+1, \ i \ge 0, \\ 1 - [a_i(t) + b_i(t) + \kappa(t)]h + o(h), & \text{if } k = i, \ i > 0, \\ 1 - [a_0(t) + \kappa(t)]h + o(h), & \text{if } k = i = 0, \\ b_i(t)h + o(h), & \text{if } k = i-1, \ i \ge 1, \end{cases} \tag{3.25}$$

where the nonnegative continuous functions $a_i(t)$, $b_i(t)$ and $\kappa(t)$, $t \ge 0$, are called the **birth** and **death rate** and the **rate of immigrations**, respectively. The parameter $b_0(t)$ is defined to be identically zero, because a state can only take nonnegative integers. When the birth and death rate of a BDP satisfies the conditions $a_k(t) = ka(t)$, $b_k(t) = kb(t)$, i.e., the rates $a_k(t)$ and $b_k(t)$ linearly depend on the states k of a system, then a BDP is called **linear** or **Kendall birth–death-immigration process** [14, 15]. The aim of this section is to consider the transition functions $P_{r,k}(s,t)$ of the BDPs which are contained in Table 3.1.

Remark 3.7 Due to the state transition structure of birth–death-immigration processes the underlying Markov chains are reversible, which might also be utilized for their stationary analysis (c.f. Sect. 3.3.7).

3.4.1 Time-Homogeneous Birth–Death Process

The model (A) of Table 3.1 satisfying the conditions (3.25) with $a_i(t) \equiv a_i$, $b_i(t) \equiv b_i$, $t \ge 0$, represents a general time-homogeneous BDP, which is a special homogeneous stepwise CTMC, and whose transition functions satisfy the system of differential equations

$$P'_{i,0}(t) = -a_0 P_{i,0}(t) + b_i P_{i,1}(t),$$

$$P'_{i,k}(t) = -[a_k + b_k]P_{i,k}(t) + a_{k-1}P_{i,k-1}(t) + b_{k+1}P_{i,k+1}(t), \ k \ge 1. \tag{3.26}$$

This circumstance makes it possible to define a homogeneous BDP from the constructive perspective and this approach is more natural for the applications of simulation techniques.

Definition 3.23 The right-continuous stochastic process $\{X(t), \ t \geq 0\}$ is a *time-homogeneous birth–death process* if

1. Its set of states is $I = \{0, 1, 2, \ldots\}$ [that is $X(t) \in I$];
2. The sojourn time in the state $k \in I, k > 0$, is exponentially distributed with the parameter

$$d_k = a_k + b_k, \quad a_k, b_k \geq 0, k > 0,$$

   and it is independent of the trajectory before arriving at the state k;
3. After the state $k \in I, k \geq 1$, the process visits the state $k + 1$ with probability $p_k = \frac{a_k}{d_k}$ and state $k - 1$ with probability $q_k = 1 - p_k = \frac{b_k}{d_k}$;
4. For the state 0 we consider the following two cases:

   - The process stays an exponentially distributed amount of time in state 0 with parameter $d_0 = a_0 > 0$ and after that visits state 1 (with probability $p_0 = 1$).
   - Once the process arrives at state 0 it remains there forever ($q_0 = 1$, $p_0 = 0$).

$P_k(0) = \mathbf{P}(X(0) = k) = \varphi_k, \ k \in I$ denotes the initial distribution of the process.

If $\{X(t), \ t \geq 0\}$ is a birth–death process, then it is an infinite state continuous-time (time-homogeneous) Markov chain. The parameters a_k and b_k are referred to as the **birth rate** and the **death rate** in state k, respectively, and k is referred to as the **population**. The special case when $b_k \equiv 0$ is referred to as the **birth process** and where $a_k \equiv 0$ as the **death process**.

Let $(T_0 =)0 < T_1 < T_2 < \ldots$ denote the time instants of the population changes (birth and death). The discrete-time $\{X_n, \ n \geq 0\}$ process, where $X_n = X(T_n)$ is the population after the nth change in population [nth jump of $X(t)$], is referred to as the Markov chain embedded in the population changes of $\{X(t), \ t \geq 0\}$. The state-transition probability matrix of the embedded Markov chain is

$$\begin{bmatrix} q_0 & p_0 & 0 & 0 & 0 & \cdots \\ q_1 & 0 & p_1 & 0 & 0 & \cdots \\ 0 & q_2 & 0 & p_2 & 0 & \cdots \\ 0 & 0 & q_3 & 0 & p_3 & \cdots \\ \vdots & \vdots & \vdots & \vdots & \vdots & \ddots \end{bmatrix}.$$

Some properties of time-homogeneous birth–death processes

The transition probability, its Laplace transform, and the initial probabilities for $k \geq 0$, $t \geq 0$, and $Re\ s > 0$ are denoted by

$$P_k(t) = \mathbf{P}(X(t) = k), \quad p_k^*(s) = \int_0^\infty e^{-st} P_k(t)\, dt, \quad P_k(0) = \mathbf{P}(X(0) = k) = \varphi_k.$$

In special cases, the following theorems are true [19]

Theorem 3.28 *If $p_0 = 1$, $0 < p_k < 1$, $k \geq 1$, then the following statements hold:*

1. $P_k(t)$ satisfies the following ordinary differential equations:

$$P_0'(t) = -a_0 P_0(t) + b_1 P_1(t),$$
$$P_k'(t) = a_{k-1} P_{k-1}(t) - (a_k + b_k) P_k(t) + b_{k+1} P_{k+1}(t), \quad k \geq 1.$$

2. For φ_k, $k \geq 0$ and $Re\ s > 0$ the following linear system defines $p_k^*(s)$:

$$s p_0^*(s) - \varphi_0 = -a_0 p_0^*(s) + b_1 p_1^*(s),$$
$$s p_k^*(s) - \varphi_k = a_{k-1} p_{k-1}^*(s) - (a_k + b_k) p_k^*(s) + b_{k+1} p_{k+1}^*(s), \quad k \geq 1.$$

3. For $k \geq 0$ the limits

$$\lim_{t \to \infty} P_k(t) = \pi_k$$

exist and are independent of the initial distribution of the process. If

$$\sum_{k=0}^\infty \rho_k < \infty, \tag{3.27}$$

where $\rho_0 = 1$ and $\rho_k = \dfrac{a_0 a_1 \cdots a_{k-1}}{b_1 b_2 \cdots b_k}$, $k \geq 1$, then $\pi_k > 0$, $k \geq 0$ and

$$\pi_0 = \left(\sum_{j=0}^\infty \rho_j \right)^{-1}, \tag{3.28}$$

$$\pi_k = \rho_k \pi_0. \tag{3.29}$$

If

$$\sum_{k=0}^{\infty} \rho_k = \infty \tag{3.30}$$

then

$$\pi_k = 0, \ k \geq 0. \tag{3.31}$$

Theorem 3.29 (Finite Birth–Death Process) *Let the state space of $X(t)$ be $\{0, 1, 2, \ldots, n\}$, $p_0 = 1$, $0 < p_k < 1$, for $1 \leq k \leq n - 1$ and $p_n = 0$; then the following statements hold:*

1. *$P_k(t)$ satisfy the following ordinary differential equations:*

$$P_0'(t) = -a_0 P_0(t) + b_1 P_1(t),$$
$$P_k'(t) = a_{k-1} P_{k-1}(t) - (a_k + b_k) P_k(t) + b_{k+1} P_{k+1}(t), \quad 1 \leq k \leq n - 1,$$
$$P_n'(t) = a_{n-1} P_{n-1}(t) - b_n P_n(t).$$

2. *If the initial distribution of the process is $\varphi_k = \mathbf{P}(X(0) = k)$, $0 \leq k \leq n$, then for $\mathrm{Re}\ s > 0$ the Laplace transforms of the transition probabilities $p_k^*(s)$ satisfy*

$$s p_0^*(s) - \varphi_0 = -a_0 p_0^*(s) + b_1 p_1^*(s),$$
$$s p_k^*(s) - \varphi_k = a_{k-1} p_{k-1}^*(s) - (a_k + b_k) p_k^*(s) + b_{k+1} p_{k+1}^*(s), \ 1 \leq k \leq n - 1,$$
$$s p_n^*(s) - \varphi_n = a_{n-1} p_{n-1}^*(s) - b_n p_n^*(s).$$

3. *For $0 \leq k \leq n$ the*

$$\lim_{t \to \infty} P_k(t) = \pi_k > 0$$

limit exists and is independent of the initial distribution

$$\pi_j = \rho_j \pi_0, \quad \pi_0 = \left(\sum_{j=0}^{n} \rho_j \right)^{-1},$$

where

$$\rho_0 = 1, \quad \rho_j = \frac{a_0 a_1 \cdots a_{j-1}}{b_1 b_2 \cdots b_j}, \ 1 \leq j \leq n.$$

Theorems on birth–death processes with absorption state.

Denote by r_k the probability of absorption into state 0 given that the initial state is $k \geq 0$. Clearly, $r_0 = 1$.

Theorem 3.30 *The following equations hold.*

1. Let $p_0 = 0$, $0 < p_k < 1$, $k \geq 1$; then for $P_k(t)$ we have

$$P_0'(t) = b_1 P_1(t),$$
$$P_1'(t) = -(a_1 + b_1) P_1(t) + b_2 P_2(t),$$
$$P_k'(t) = a_{k-1} P_{k-1}(t) - (a_k + b_k) P_k(t) + b_{k+1} P_{k+1}(t), \quad k \geq 2.$$

2. For Re $s > 0$ and the initial distribution φ_k, $k \geq 0$, we have

$$s p_0^*(s) - \varphi_0 = b_1 p_1^*(s),$$
$$s p_1^*(s) - \varphi_1 = -(a_1 + b_1) p_1^*(s) + b_2 p_2^*(s),$$
$$s p_k^*(s) - \varphi_k = a_{k-1} p_{k-1}^*(s) - (a_k + b_k) p_k^*(s) + b_{k+1} p_{k+1}^*(s), \quad k \geq 2.$$

3. (a) If $\sum_{i=1}^{\infty} \prod_{m=1}^{i} \rho_m = \infty$, then $r_n = 1$, $n \geq 1$.
   (b) If $\sum_{i=1}^{\infty} \prod_{m=1}^{i} \rho_m < \infty$, then

$$r_1 = \frac{\sum_{i=1}^{\infty} \prod_{m=1}^{i} \rho_m}{1 + \sum_{i=1}^{\infty} \prod_{m=1}^{i} \rho_m} \quad \text{and} \quad r_n = \frac{\sum_{i=n}^{\infty} \prod_{m=1}^{i} \rho_m}{1 + \sum_{i=1}^{\infty} \prod_{m=1}^{i} \rho_m}, \quad n \geq 2.$$

Theorem 3.31 *The following equations hold.*

1. Let $X(t) \in \{0, 1, 2, \ldots, n\}$, $p_0 = 0$, $0 < p_k < 1$ if $1 \leq k \leq n - 1$, and $p_n = 0$; then for $P_k(t)$ we have

$$P_0'(t) = b_1 P_1(t),$$
$$P_1'(t) = -(a_1 + b_1) P_1(t) + b_2 P_2(t),$$
$$P_k'(t) = a_{k-1} P_{k-1}(t) - (a_k + b_k) P_k(t) + b_{k+1} P_{k+1}(t), \quad 2 \leq k \leq n - 1,$$
$$P_n'(t) = a_{n-1} P_{n-1}(t) - b_n P_n(t).$$

2. For $p_k^*(s)$, Re $s > 0$, we have $[\varphi_k = \mathbf{P}(X(0) = k), \; 0 \leq k \leq n]$

$$s p_0^*(s) - \varphi_0 = b_1 p_1^*(s),$$
$$s p_1^*(s) - \varphi_1 = -(a_1 + b_1) p_1^*(s) + b_2 p_2^*(s),$$
$$s p_k^*(s) - \varphi_k = a_{k-1} p_{k-1}^*(s) - (a_k + b_k) p_k^*(s) + b_{k+1} p_{k+1}^*(s), \quad 2 \leq k \leq n - 1,$$
$$s p_n^*(s) - \varphi_n = a_{n-1} p_{n-1}^*(s) - b_n p_n^*(s).$$

3. The process $X(t)$ will be absorbed with probability 1.

Remark 3.8 In Theorems 3.28–3.31 the differential equations for $P_j(t)$ are indeed the Kolmogorov (forward) differential equations for the given systems. The equations for $p_j^*(s)$ can be obtained from the related differential equations for $P_j(t)$ using

$$\int_0^\infty e^{-st} P_j'(t)\, dt = s p_j^*(s) - P_j'(0) .$$

In Theorem 3.30, state 0 is an absorbing state and all other states are transient, because state 0 is reachable from all other states. The stationary distribution of such a system is trivial, $\pi_0 = 1$ and $\pi_k = 0$ for $k \geq 1$. Theorem 3.30 allows us to compute the parameters of the busy period of birth–death Markov chains starting from state k ($\varphi_k = 1$), where the busy period is the time to reach state 0 (which commonly represents the idle state of a system, where the server is not working, in contrast to the $i > 0$ states, where the server is commonly busy). Let T_k denote the length of the busy period starting from state k; then

$$\Pi_k(t) = \mathbf{P}\,(T_k \leq t) = \mathbf{P}\,(X(t) = 0 | X(0) = k)$$

that is, assuming $\varphi_k = P_k(0) = 1$ and $\varphi_j = P_j(0) = 0$ for $\forall j \neq k$ Theorem 3.30 defines the distribution of the length of the busy period.

Proof (Theorems 3.28 and 3.29) The limiting (stationary or equilibrium) distribution as $t \to \infty$ can be easily calculated in Theorems 3.28 and 3.29 because we can obtain simple linear equations for the limits (a) $\pi_k = \lim_{t\to\infty} P_k(t)$ when taking into account that (b) $\lim_{t\to\infty} P_k'(t) = 0$. We prove only the statement 3 of Theorem 3.28 because the limiting distribution in Theorems 3.29 can be reached in similar way. The relations (a) and (b) lead to the following system of linear difference equations:

$$-a_0\pi_0 + b_1\pi_1 = 0, \quad a_{k-1}\pi_{k-1} - (a_k + b_k)\pi_k + b_{k+1}\pi_{k+1} = 0, \quad k \geq 1,$$

$$\sum_{k=0}^\infty \pi_k = 1,$$

from which it is not difficult to derive the explicit form of the limiting distribution given in Theorem 3.28 because we obtain

$$0 = a_0\pi_0 - b_1\pi_1 = a_1\pi_1 - b_2\pi_2 = \cdots$$

then

$$\pi_{k+1} = \frac{a_k}{b_{k+1}}, \quad k = 2, 3, \ldots$$

Proof The proof of the statements of Theorem 3.30 may be performed by a probabilistic method. Note that it is enough to consider the embedded Markov chain of the birth–death process because both processes execute the same transitions. By the transition probabilities of the birth–death process considered here, the first possible movements and their probabilities from the state i, $i \geq 0$ are $\mathbf{P}(0 \to 1) = p_0 = 0$ and $\mathbf{P}(i \to i+1) = p_i$, $\mathbf{P}(i \to i-1) = 1 - p_i$, $i > 0$. Applying a first step analysis, we immediately obtain the recursive equations for the sequence of probabilities r_i, $i \geq 0$ ($r_0 = 1$) as follows

$$r_i = (1 - p_i)r_{i+1} + p_i r_{i-1}, \ i \geq 1.$$

From these, with simple calculation we obtain

$$r_i - r_{i+1} = p_i(r_{i-1} - r_{i+1}) = p_i(r_{i-1} - r_i) + p_i(r_i - r_{i+1})$$

leading to

$$r_i - r_{i+1} = \frac{p_i}{1 - p_i}(r_{i-1} - r_i) = \prod_{m=1}^{i} \frac{p_m}{1 - p_m}(1 - r_1) = \prod_{m=1}^{i} \rho_m(1 - r_1), \ i \geq 1,$$
(3.32)

where $\rho_m = \frac{a_m}{b_m}$, $m \geq 1$.

(a) Case $\sum_{i=1}^{\infty} \prod_{m=1}^{i} \rho_m = \infty$. Summing up these last equations from $i = 1$ to $i = n - 1$, we get

$$r_1 - r_n = (1 - r_1) \sum_{i=1}^{n-1} \prod_{m=1}^{i} \rho_m, \ n \geq 2.$$
(3.33)

It is clear that $0 \leq r_1 - r_n \leq 1$ and, as a consequence of the relation $\sum_{i=1}^{\infty} \prod_{m=1}^{i} \rho_m = \infty$, Eq. (3.33) can hold if and only if $r_1 = 1$ and, consequently, if and only if $r_n \equiv 1$, $n \geq 1$.

(b) Case $\sum_{i=1}^{\infty} \prod_{m=1}^{i} \rho_m < \infty$. Summing up Eq. (3.32) from $i = n$ to ∞, we have

$$r_n = (1 - r_1) \sum_{i=n}^{\infty} \prod_{m=1}^{i} \rho_m, \ n \geq 2.$$
(3.34)

Since the sum $\sum_{i=1}^{\infty} \prod_{m=1}^{i} \rho_m$ with positive members is finite, then, obviously, $r_n \to 0$, $n \to \infty$ and, accordingly, Eq. (3.33) lead to

$$r_1 = \frac{\sum_{i=1}^{\infty} \prod_{m=1}^{i} \rho_m}{1 + \sum_{i=1}^{\infty} \prod_{m=1}^{i} \rho_m}$$

as $n \to \infty$. From this and Eq. (3.34) we obtain

$$r_n = \frac{\sum_{i=n}^{\infty} \prod_{m=1}^{i} \rho_m}{1 + \sum_{i=1}^{\infty} \prod_{m=1}^{i} \rho_m}, \quad n \geq 2.$$

Proof (Theorem 3.31) Let $Z(t)$, $t \geq 0$ be a process satisfying the conditions of Theorem 3.30. Let $t_0 = 0$ and denote by $0 < t_1 < t_2 < \ldots$ the time points when the process changes the states. Then the sequence of random variables $Z(t_i)$, $i = 0, 1, 2, \ldots$ constitutes a finite homogeneous and irreducible Markov chain and, consequently, it is positive recurrent and the random variable $N = \min\{k : Z(t_k) = 0\}$ is finite with probability 1. From this it follows that

$$X(t) = \begin{cases} Z(t), & \text{if } 0 \leq t \leq t_N, \\ 0, & \text{if } t > t_N, \end{cases}$$

defines a process satisfying the conditions of Theorem 3.29 and $X(t)$ will be absorbed with probability 1.

3.4.2   Inhomogeneous Kendall BDP with Immigration

The models (B1)–(B5) in Table 3.1 constitute subclasses of the general Kendall BDPs, in which the birth and death rates have special forms depending on time and proportional to the state of the system as follows: $a_k(t) = ka(t)$ and $b_k(t) = kb(t)$, respectively. Such traffic models can arise during major events attracting a large number of participants such as sport matches, or concerts. Time dependent $a(t)$, $b(t)$ functions can describe the interval of the event or the alternation of the number of participants. The service intensity function $b_k(t) = kb(t)$ is a typical description of (unlimited number of) parallel services, which is also referred to as infinite server model. The arrival intensity function $a_k(t) = ka(t)$ is less typical. It describes the behavior when a special event occurs (e.g., a team score at a sport event) and communication demand proportionally grows with the number of ongoing connections. For realistic models the proportional growth has to be bounded at a feasible limit. We will use the method of generating functions, which is an effective tool for the investigation of the system of differential-difference equations (3.36) [14].

The first task is to consider the (time-)inhomogeneous models (B4) and (B5), after that we can obtain the results for general model (B), and the results relating to the models (B1)–(B3) as special cases can be derived.

We mention that the results of this section allow us to analyze the limiting distribution of population size process in special inhomogeneous cases (see, for example, [11]) by applying the concrete form of generating functions and connection (2.6).

We begin with some notations and observations. Let us denote by $V_0(t)$, $t \geq 0$ a population size process having one individual at $t = 0$. By the assumptions, if $r \geq 1$ individuals are in the system at $t = 0$, then they determine r population size processes, which are independent copies of process V_0 and we denote them by $V_1(t), \ldots, V_r(t)$, $t \geq 0$.

Let us introduce a collection $\{W(s,t), t \geq s, W(s,s) = 1 : s \geq 0\}$ of independent copies of process V_0, independently of processes $V_1(t), \ldots, V_r(t)$. The immigrants arrive at moments $0 < \tau_1 < \tau_2 < \ldots$ according to a Poisson process $N(t)$, ≥ 0 of rate $\kappa(t)$ independently of processes $V_1(t), \ldots, V_r(t)$. The evolution of population size processes associated with immigrants entering the system at $\tau_1, \tau_2, \ldots$ can be modeled by the processes $W(\tau_1, t), W(\tau_2, t), \ldots$ which are, evidently, independent, and do not depend on processes $V_1(t), \ldots, V_r(t)$ and they have the same transition probability function denoted by $P_{i,k}(s,t)$, i.e.,

$$\mathbf{P}\left(V_j(t) = k | V_j(s) = i\right) = \mathbf{P}\left(W(\tau_m, t) = k | W(\tau_m(s)) = i\right)$$

$$= P_{i,k}(s,t), \ m \geq 1, \ 0 \leq s \leq t$$

It is clear that the summarized processes

$$V(t) = \sum_{j=1}^{r} V_j(t) \quad \text{and} \quad W(t) = \sum_{\tau_j \leq t} W(\tau_j, t)$$

are also independent and the BDP process with immigration takes the form $X(t) = V(t) + W(t)$. From this, we can obtain as an immediate consequence that the generating function of $X(t)$ has the product form

$$G(z,t) = G_V(z,t)G_W(z,t) = (G_{V_1}(z,t))^r G_W(z,t), \tag{3.35}$$

where the generating functions of random variables $X(t)$, $V(t)$, $V_1(t)$, and $W(t)$ are denoted by

$$G(z,t) = \mathbf{E}\left(z^{X(t)} | X(0) = r\right), \ G_V(z,t) = \mathbf{E}\left(z^{V(t)} | X(0) = r\right)$$

and

$$G_{V_1}(z,t) = \mathbf{E}\left(z^{V(t)} | X(0) = 1\right), \ G_W(z,t) = \mathbf{E}\left(z^{W(t)}\right), \ t \geq 0, \ 0 \leq z \leq 1,$$

respectively. The relation (3.35) means that it is enough to determine the generating functions $G_V(z,t)$ and $G_W(z,t)$ separately. On the one hand, observe that the distribution of $V(t)$ only depends on the birth and death rate and the initial state r, thus $V(t)$ fulfills the conditions of (B4) with $\kappa(t) \equiv 0$. On the other hand, the process $W(t)$ satisfies the conditions of the model (B5) with initial value $r = 0$. Since a Kendall BDP with immigration is a CTMC, then the assumptions (3.25) related to the inhomogeneous Markov processes lead to Kolmogorov's forward differential equations as follows:

$$P'_{i,0}(t) = -\kappa(t)P_{i,0}(t) + ib(t)P_{i,1}(t),$$

$$P'_{i,k}(t) = -[\kappa(t) + ka(t) + kb(t)]P_{i,k}(t) + [\kappa(t) + (k-1)a(t)]P_{i,k-1}(t) +$$

$$+ (k+1)b(t)P_{i,k+1}(t), \quad k \geq 1. \quad (3.36)$$

Now we proceed to determine the generating function of the time-dependent distribution of the BDP process (B). By definition, the generating function $G(z, t)$ of the probability transition function $P_{r,k}(t)$, $k \geq 0$ is

$$G(z,t) = \mathbf{E}\left(e^{X(t)} | X(0) = 1\right) = \sum_{k=0}^{\infty} P_{r,k}(t)z^k, \quad t \geq 0, \ |z| \leq 1, \ G(z,0) = z,$$

(3.37)

which is, obviously, well defined for all $|z| \leq 1$, $t \geq 0$. It can also be seen that the partial derivatives $\frac{\partial}{\partial t}G(z, t)$ and $\frac{\partial}{\partial z}G(z, t)$ exist and the following relations hold

$$\frac{\partial}{\partial t}G(z,t) = \sum_{k=0}^{\infty} \frac{d}{dt}P_{r,k}(t)z^k, \quad (3.38)$$

$$\frac{\partial}{\partial z}G(z,t) = \sum_{k=1}^{\infty} P_{r,k}(t)kz^{k-1}, \quad t \geq 0, \ |z| < 1. \quad (3.39)$$

It is clear from Eqs. (3.36), (3.37), and (3.38) that

$$\sum_{k=1}^{\infty} [\kappa(t) + (a(t) + b(t))\, k] P_{r,k}(t)z^k$$

$$= \sum_{k=1}^{\infty} \kappa(t)P_{r,k}(t)z^k + (a(t) + b(t))\, z \sum_{k=1}^{\infty} P_{r,k}(t)kz^{k-1}$$

$$= \kappa(t)[G(t,z) - P_{r,0}(t)] + (a(t) + b(t))\, z\frac{\partial}{\partial z}G(t,z), \quad (3.40)$$

and

$$\sum_{k=1}^{\infty} [\kappa(t) + (k-1)a(t)]P_{r,k-1}(t)z^k$$

$$= \kappa(t)z \sum_{k=0}^{\infty} P_{r,k}(t)z^k + a(t)z^2 \sum_{k=1}^{\infty} k P_{r,k}(t)z^{k-1}$$

$$= \kappa(t)zG(t,z) + a(t)z^2\frac{\partial}{\partial z}G(t,z), \quad (3.41)$$

and analogously, we have

$$\sum_{k=1}^{\infty}(k+1)b(t)P_{r,k+1}(t)z^k = b(t)\sum_{k=1}^{\infty}P_{r,k}(t)kz^{k-1} - b(t)P_{r,1}(t)$$

$$= b(t)\frac{\partial}{\partial z}G(t,z) - b(t)P_{r,1}(t). \tag{3.42}$$

From Eq. (3.36) we get

$$\frac{\partial}{\partial t}G(z,t) = \frac{d}{dt}P_{r,0}(t) + \sum_{k=1}^{\infty}\frac{d}{dt}P_{r,k}(t)z^k,$$

thus summarizing the left- and right-hand sides of Eqs. (3.38), (3.40), and (3.41), finally, we obtain the following partial differential equation (PDE) with initial condition $G(z,0) = z^r$

$$\frac{\partial}{\partial t}G(z,t) = [-(a(t)+b(t))z + a(t)z^2 + b(t)]\frac{\partial}{\partial z}G(z,t) + (z-1)\kappa(t)G(z,t)$$

$$= (z-1)(a(t)z - b(t))\frac{\partial}{\partial z}G(z,t) + (z-1)\kappa(t)G(z,t). \tag{3.43}$$

Remark 3.9 It is of interest to note that the expected value $EX(t)$ can be simply calculated even in the general case, using the elementary property of generating functions and Eq. (3.43). Since

$$G(1,t) = 1, \ t \geq 0, \ G(z,0) = z^r,$$

$$\mu(0) = \mathbf{E}(X(0)) = r, \quad \mu(t) = \mathbf{E}(X(t)) = \frac{\partial}{\partial z}G(z,t)|_{z=1}$$

then from Eq. (3.42) we obtain the ordinary differential equation (ODE) with initial value $\mu(0) = r$ [17, p. 201]

$$\frac{d}{dt}\mu(t) = \lim_{z\to 1}\frac{\partial}{\partial t}\left(\frac{\partial}{\partial z}G(z,t)\right) = \lim_{z\to 1}\frac{\partial}{\partial z}\left(\frac{\partial}{\partial t}G(z,t)\right)$$

$$= \lim_{z\to 1}\left[\kappa(t)G(z,t) + (z-1)\kappa(t)\frac{\partial}{\partial z}G(z,t) + (a(t)z - b(t))\frac{\partial}{\partial z}G(z,t) + \right.$$

$$\left. +(z-1)\frac{\partial}{\partial z}\left((a(t)z - b(t))\frac{\partial}{\partial z}G(z,t)\right)\right]$$

$$= \kappa(t)G(1,t) + (a(t) - b(t))\mu(t) = \kappa(t) + (a(t) - b(t))\mu(t).$$

The solution of this first order linear ODE with the initial value $\mu(0) = r$ takes the form

$$\mu(t) = \frac{1}{A(t)} \left(r + \int_0^t \kappa(s)A(s)\mathrm{d}s \right), \quad t \geq 0,$$

where

$$A(t) = \exp \left\{ \int_0^t (b(s) - a(s))\mathrm{d}s \right\}.$$

Note that in the homogeneous case, when the conditions $a(t) = a$, $b(t) = b$, and $\kappa(t) = \kappa$ are satisfied, the solution of the differential equation has a simpler form

$$\mu(t) = \mathbf{E}(X(t)) = re^{(a-b)t} + \frac{\kappa}{a-b} \left(e^{(a-b)t} - 1 \right), \quad t \geq 0.$$

Moreover, when the condition $a < b$ is satisfied, then

$$\lim_{t \to \infty} \mu(t) = \frac{\kappa}{a-b}.$$

Let us introduce the following notations

$$B(t) = \int_0^t a(s)A(s)\mathrm{d}s, \quad C(t) = \int_0^t (a(s) + b(s))A(s)\mathrm{d}s, \quad t \geq 0,$$

$$A(s, t) = \exp \left\{ \int_s^t (b(u) - a(u))\mathrm{d}u \right\}, \quad 0 \leq s \leq t.$$

Theorem 3.32 *The generating functions $G_{V_1}(z, t)$ and $G_W(z, t)$ have the forms*

$$G_{V_1}(z, t) = 1 - \frac{1}{\frac{A(t)}{1-z} + B(t)}, \quad t \geq 0 \tag{3.44}$$

$$G_W(z, t) = \exp \left\{ \int_0^s (G_{V_1}(z, s, t) - 1)\kappa(s) \right\}, \quad t \geq 0, \tag{3.45}$$

where

$$G_{V_1}(z, s, t) = 1 - \frac{1}{\frac{A(s,t)}{1-z} + \int_s^t a(u)A(u, t)\mathrm{d}u}. \tag{3.46}$$

Proof We consider first the generating function $G_{V_1}(z, t)$ (model (B4) with $r = 1$, $\kappa(t) \equiv 0$). In this case $G_{V_1}(z, t)$ satisfies the following PDE

$$\frac{\partial}{\partial t}G_{V_1}(z,t) = [a(t)z^2 - (a(t)+b(t))z + b(t)]\frac{\partial}{\partial z}G_{V_1}(z,t), \quad G_{V_1}(z,0) = z.$$
(3.47)

The classical solution of the initial value problem of the first order linear homogeneous PDE (3.47) can be carried out using the method of the so-called characteristic curves or characteristics [21, p. 24]. This method, in general, reduces the solution of the PDE (3.47) to the solution of a system of ordinary differential equations (ODEs). In our case, the characteristic curve is defined by the graph $(z(t), t)$ in the plane (z, t), where the function $z(t)$ is the solution of the Cauchy problem

$$z'(t) = -a(t)z^2(t) + (a(t)+b(t))z(t) - b(t), \quad z(0) = z_0.$$
(3.48)

Let us denote $u(t) = G_{V_1}(z(t), t)$, where $G_{V_1}(z, t)$ is the solution of the PDE (3.47). It is clear that by the help of the multivariate chain rule we have

$$u'(t) = \frac{\partial}{\partial z}G_{V_1}(z(t), t)z'(t) + \frac{\partial}{\partial t}G_{V_1}(z(t), t)$$

$$= \frac{\partial}{\partial z}G_{V_1}(z(t), t)[-a(t)z^2(t) + (a(t)+b(t))z(t) - b(t)] + \frac{\partial}{\partial t}G_{V_1}(z(t), t) = 0.$$

Consequently, $u(t)$ must be a constant function and then the following relation holds

$$u(t) = u(0) = G_{V_1}(z(0), t) = G_{V_1}(z_0, t), \quad t \geq 0.$$

From this, it is clear that the function $G_{V_1}(z(t), t)$ does not vary along any characteristic curve and using the initial condition $G_{V_1}(z, 0) = z$ related to the PDE (3.47), we have to find $z_0 = z_0(z, t)$ so that the corresponding characteristic curve passes through (z, t). Then the resulting $u(t)$ is the solution of PDE (3.47) in the point (z, t).

Returning to prove the statement (3.44) we first solve the ODE (3.48). It is clear that

$$\int_0^t (b(s) - a(s))A(s)ds = \int_0^t (A(s))' ds = [A(s)]_0^t = A(t) - 1,$$

thus

$$C(t) = \int_0^t [a(s) + b(s)]A(s)ds = A(t) - 1 + 2B(t).$$
(3.49)

Obviously, the sum of the coefficients $-a(t)$, $(a(t) + b(t))$, and $-b(t)$ in the differential equation (3.48) takes value 0 for all $t \geq 0$, then the formula concerning a special type of Riccati's ordinary differential equation [13, p. 23] can be applied. Thus the solution of the Cauchy problem (3.48) is

$$z(t) = \frac{c - C(t) - A(t)}{c - C(t) + A(t)}, \quad z(0) = z_0, \tag{3.50}$$

where c is a constant. Using the properties $A(t)$ and $C(t)$, from Eq. (3.50) we obtain

$$z(0) = \frac{c - A(0)}{c + A(0)} = \frac{c - 1}{c + 1} = z_0,$$

therefore the constant c must take the value

$$c = \frac{1 + z_0}{1 - z_0},$$

corresponding to the initial value z_0. By the use of the relations (3.49) and (3.50) we obtain that

$$z(t) = 1 - \frac{2A(t)}{c - C(t) + A(t)} = 1 - \frac{2A(t)}{c + 1 - 2B(t)}$$

$$= 1 - \frac{2A(t)}{\frac{2}{1 - z_0} - 2B(t)} = 1 + \frac{A(t)}{\frac{1}{z_0 - 1} + B(t)},$$

from which, with a simple calculation

$$z_0 = z_0(z, t) = 1 - \frac{1}{\frac{A(t)}{1 - z} + B(t)}.$$

In accordance with the method of characteristic curves discussed previously, the solution of the PDE (3.47) has the form

$$G_{V_1}(z, t) = z_0(z, t) = 1 - \frac{1}{\frac{A(t)}{1 - z} + \int_0^t a(s)A(s)ds}. \tag{3.51}$$

The statement (3.45) of Theorem 3.32 will be proved by probabilistic approach, using the connection (3.51) and the important fact that one can characterize the joint stochastic behavior of the numbers of population size processes having $k \geq 1$ individuals at time t. Denote

$$W_k = \sum_{\tau_i \leq t} \mathscr{I}_{\{W(\tau_i, t) = k\}}, \quad k \geq 1.$$

Then from the Displacement Theorem 2.9 (with the notation $P_t(s, B) = \sum_{k \in B} P_{1k}(s, t)$, $B \in \mathscr{B}$ Borel set) it follows that the random variables W_k are independent and they have Poisson distributions with parameters

$$\mu_k = \int_0^t P_{1,k}(s, t)\kappa(s)ds, \quad k \geq 0.$$

Using this relation, the generating function of the random variable $X_t = \sum_{k \geq 1} k W_k$ can be easily determined. Since $\sum_{k \geq 0} P_{1,k}(s, t) = 1$, then

$$G_W(z, t) = \mathbf{E}\left(z^{X_t}\right) = \mathbf{E}\left(\prod_{k \geq 0} z^{k W_k}\right) = \prod_{k \geq 0} \mathbf{E}\left((z^k)^{W_k}\right)$$

$$= \prod_{k \geq 0} \exp\{\mu_k(z^k - 1)\} = \exp\left\{\sum_{k \geq 0} \mu_k(z^k - 1)\right\}$$

$$= \exp\left\{\int_0^t \left(\sum_{k \geq 0}(z^k - 1)P_{1,k}(s, t)\right)\kappa(s)ds\right\}$$

$$= \exp\left\{\int_0^t \left(G_{V,1}(z, s, t) - 1\right)\kappa(s)ds\right\}. \tag{3.52}$$

Corollaries of Theorem 3.32

Using the formulas (3.44) and (3.45) of Theorem 3.32, we get with simple calculation the following statements for the special inhomogeneous cases.

(B1) Inhomogeneous Poisson process. Assume $a(t) = b(t) = 0$, $t \geq 0$, $\kappa(t)$ is not identically 0 in the interval $[0, t]$ and $r = 0$. It is clear that in this case $G_{V_1}(z, s, t) = G_{V_1}(z, s, s) = z$, $0 \leq s \leq t$, then

$$G_W(z, t) = \exp\left\{(z - 1)\int_0^t \kappa(s)ds\right\},$$

thus $X(t)$, $t \geq 0$ is an inhomogeneous Poisson process with the rate function $\kappa(t)$, $t \geq 0$, and
the transition probability distribution is:

$$P_{0,k}(t) = \frac{\left(-\int_0^t \kappa(s)ds\right)^k}{k!} \exp\left\{-\int_0^t \kappa(s)ds\right\}, \quad k = 0, 1, \ldots$$

Note that in this case the formula for the generating function can be easily proved directly using Eq. (3.43), because the ODE simplifying to

$$\frac{d}{dz}G(z, t) = (z - 1)\kappa(t)G(z, t), \quad G(z, 0) = 1$$

is valid.

(B2) Inhomogeneous pure birth process. If $b(s) \equiv 0$, $\kappa(s) \equiv 0$, $s \geq 0$, $r = 1$, then

$$A(t) = \exp\left\{ -\int_0^t a(s)\mathrm{d}s \right\},$$

$$B(t) = \int_0^t a(s) \exp\left\{ -\int_0^s a(u)\mathrm{d}u \right\} \mathrm{d}s$$

$$= \left[-\exp\left\{ -\int_0^s a(s)\mathrm{d}s \right\} \right]_0^t = 1 - A(t)$$

and

$$G_{V_1}(z,t) = 1 - \frac{1}{\frac{A(t)}{1-z} + 1 - A(t)} = 1 - \frac{1-z}{1 - z + A(t)z} = \frac{A(t)z}{1 - (1 - A(t))z}$$

which is the generating function of the geometric distribution with parameter $A(t)$.
The transition probability distribution is

$$P_{1,k}(t) = A(t)(1 - A(t))^k, \quad k = 1, 2, \ldots$$

(B3) Inhomogeneous pure death process. If $a(s) \equiv 0$, $\kappa(s) \equiv 0$, then we obtain that $A(t) = \exp\left\{ \int_0^t b(s)\mathrm{d}s \right\}$, $B(t) \equiv 0$ and

$$G_{V_1}(z,t) = 1 - \frac{1}{\frac{A(t)}{1-z}} = \left(1 - \frac{1}{A(t)} \right) + \frac{1}{A(t)}z.$$

The transition probability distribution is:

$$P_{1,1}(t) = \exp\left\{ -\int_0^t b(s)\mathrm{d}s \right\}, \quad P_{1,0}(t) = 1 - P_{1,1}(t), \quad P_{1,k}(t) = 0, \ k \geq 2.$$

(B4) Inhomogeneous BDP (without immigrations). If the BDP V_1 is not a pure death process, i.e., the birth rate $a(s)$ is not identically 0 in the time interval $[0, t]$ (in this case, it is necessarily $B(t) > 0$) and $\kappa(s) \equiv 0$, $s \geq 0$, then the generating function (3.44) of the general time-inhomogeneous Kendall BDP without immigrations simplifies to

$$G_{V_1}(z,t) = \frac{\alpha(t) + [1 - \alpha(t) - \beta(t)]z}{1 - \beta(t)z},$$

where

$$\alpha(t) = \frac{A(t) + B(t) - 1}{A(t) + B(t)}, \quad \beta(t) = \frac{B(t)}{A(t) + B(t)}.$$

Consequently, the distribution of the BDP is a modified geometric distribution parametrized by $\alpha(t)$ and $\beta(t)$ [12, p. 144].

Note that $A(0) + B(0) = 1$ and $\frac{d}{dt}(A(t) + B(t)) = b(t)A(t) \geq 0$, then $\alpha(t) \geq 0$. Moreover, due to $A(t), B(t) > 0$, we have $\rho(t) = \frac{B(t)}{A(t)+B(t)} < 1$, so the power series representation of the generating function $G_{V_1}(z, t)$ is

$$G_{V_1}(z, t) = (\alpha(t) + \beta(t) - 1) + (1 - \beta(t)) \sum_{k=1}^{\infty} (\beta(t))^k z^k.$$

The transition probability distribution is:

$$P_{1,0}(t) = \alpha(t) + \beta(t) - 1, \quad P_{1,k}(t) = (1 - \beta(t)) (\beta(t))^k, \quad k = 1, 2, \ldots$$

Formulas for Different Cases of the Time-Homogeneous Kendall BDPs
In these cases, the birth, death, and immigration rates are constants in time, i.e., $a(t) = a$, $b(t) = b$ and $\kappa(t) = \kappa$, $t \geq 0$. Simple calculations show that the parameters $A(t)$, $B(t)$, and $A(s, t)$ can be given in the following form

$$A(t) = \exp\left\{\int_0^t (b - a)ds\right\} = e^{(b-a)t},$$

$$B(t) = \int_0^t ae^{(b-a)s}ds = \begin{cases} \frac{a}{b-a}\left[e^{(b-a)t} - 1\right], & a \neq b, \\ at, & a = b, \end{cases}$$

and

$$A(s, t) = \begin{cases} \exp\left\{\int_s^t (b - a)du\right\} = e^{(b-a)(t-s)}, & \text{if } a, b > 0,\ a \neq b, \\ 1, & \text{if } a, b > 0,\ a = b. \end{cases} \tag{3.53}$$

These formulas, in different cases, allow ready computation of the generating functions and PDFs.

(B1) Poisson process. If $a = b = 0$, $r = 0$, $\kappa > 0$, then

$$G(z, t) = G_{V,0}(z, t) = e^{(z-1)\kappa t}, \quad P_{1,k}(t) = \frac{(\kappa t)^k}{k!}e^{-\kappa t}, \quad k = 0, 1, \ldots$$

(B2) Pure birth process. If $b = 0$, $\kappa = 0$, $r = 1$, $A(t) = e^{-at}$,

$$G_{V_1}(z, t) = \frac{e^{-at}z}{1 - (1 - e^{-at})z}, \quad P_{1,k}(t) = e^{-at}(1 - e^{-at})^{k-1}, \quad k = 1, 2, \ldots$$

$$P_{1,k}(t) = e^{-at}(1 - e^{-at})^{k-1}, \quad k = 1, 2, \ldots$$

(B3) Pure death process. If $a = 0, \kappa = 0$, then

$$G_{V_1}(z, t) = 1 - e^{-bt} + e^{-bt}z,$$

$$P_{1,0}(t) = 1 - e^{-bt}, \quad P_{1,1}(t) = e^{-bt}, \quad P_{1,k}(t) = 0, \ k \ge 2.$$

(B4) Kendall BDP. If $a, b > 0$, $a \ne b$, $\kappa = 0$, then $A(t) = e^{(b-a)t}$, $B(t) = \frac{a}{b-a}\left[e^{(b-a)t} - 1\right]$, $\alpha(t) = (b - a)\frac{1}{be^{(b-a)t}-a}$, $\beta(t) = \frac{a}{b-a}\frac{e^{(b-a)t}-1}{be^{(b-a)t}-a}$ and

$$G_{V_1}(z, t) = (\alpha(t) + \beta(t) - 1) + (1 - \beta(t)) \sum_{k=1}^{\infty} (\beta(t))^k z^k.$$

If $a, b > 0$, $a = b$, $\kappa = 0$, then $A(t) = 1$, $B(t) = at$, $\rho(t) = \frac{at}{1+at}$,

$$G_{V_1}(z, t) = \frac{at}{1 + at} + \frac{1}{at(1 + at)} \sum_{k=1}^{\infty} \left(\frac{at}{1 + at}\right)^k z^k$$

and

$$P_{1,0}(t) = \frac{at}{1 + at}, \quad P_{1,k} = \frac{1}{1 + at}\left(\frac{at}{1 + at}\right)^{k-1}, \quad k \ge 1.$$

(B5) Kendall BDP with immigrations. We mention that in accordance with (3.35) $G(z, t) = [G_{V_1}(z, t)]^r G_W(z, t)$ is true, thus it is enough to determine the generating function $G_W(z, t)$. It is clear that

$$A(s, t) = \begin{cases} \exp\left\{\int_s^t (b - a)du\right\} = e^{(b-a)(t-s)}, & \text{if } a, b > 0, \ a \ne b, \\ 1 & \text{if } a, b > 0, \ a = b. \end{cases}$$

Substituting $A(s, t)$ into Eq. (3.52) we get the following statement. If $a \ne b$, then

$$G_{V_1}(z, s, t) = 1 - \frac{1}{\frac{1}{1-z}e^{(b-a)(t-s)} + \int_s^t ae^{(b-a)(t-u)}du}$$

$$= 1 - \frac{1}{\frac{1}{1-z}e^{(b-a)(t-s)} + \frac{a}{b-a}[e^{(b-a)(t-s)} - 1]}.$$

If $a = b$, then

$$G_{V_1}(z, s, t) = 1 - \frac{1}{\frac{1}{1-z} + \int_s^t a\,du} = 1 - \frac{1}{\frac{1}{1-z} + a(t - s)}.$$

Using the formula (3.52) and integrating the function $\left(G_{V_1}(z, s, t) - 1\right)\kappa$, the generating function $G_W(z, t)$ has the form [15, p. 248]

$$G_W(z, t) = \begin{cases} \left(\frac{a-b}{ae^{(a-b)t}-b}\right)^{\kappa/a} \left[1 - \frac{a(e^{(a-b)t}-1)}{ae^{(a-b)t}-b}\right]^{-\kappa/a}, & \text{if } a, b > 0,\ a \neq b, \\ (1 + at)^{-\kappa/a} \left[1 - \frac{atz}{1+atz}\right]^{-\kappa/a}, & \text{if } a, b > 0,\ a = b. \end{cases}$$

Remark 3.10 It is clear from the above results that there exists a nondegenerate limit distribution only in case (B5) if the condition $a < b$ is satisfied. The generating function of the limit distribution is

$$\lim_{t \to \infty} G_W(z, t) = \left(1 - \frac{a}{b}\right)^{\kappa/a} \left(1 - \frac{a}{b}z\right)^{-\kappa/a},$$

which means that the limit distribution is negative binomial.

3.4.3   Restricted Process

Let $X(t)$ be an irreducible Markov chain with generator Q on state space $\mathscr{S}$ and let $\mathscr{U}$ and $\mathscr{D}$ be a disjunct division of the state space, i.e., $\mathscr{U} \cup \mathscr{D} = \mathscr{S}$ and $\mathscr{U} \cap \mathscr{D} = \emptyset$. Without loss of generality we assume that the states are numbered such that $i < j$ for $\forall i \in \mathscr{U}$ and $\forall j \in \mathscr{D}$ and we decompose the generator matrix into the subset related matrix blocks

$$Q = \begin{array}{|c|c|} \hline Q_U & Q_{UD} \\ \hline Q_{DU} & Q_D \\ \hline \end{array}.$$

Since the Markov chain is irreducible and both subsets are visited only for a finite amount of time, the eigenvalues of Q_U and Q_D have negative real parts, and consequently Q_U and Q_D are nonsingular. In order to study the consecutive visits to $\mathscr{U}$ and $\mathscr{D}$ we introduce the time to reach $\mathscr{U}$,

$$\gamma_U = \min(t : X(t) \in \mathscr{U} | X(0) \in \mathscr{D}),$$

and, with its help, we investigate the first state visited in $\mathscr{U}$. The probability that starting from $i \in \mathscr{D}$ and reaching $\mathscr{U}$ by first visiting $j \in \mathscr{U}$ is $P_{DU\,ij} = \mathbf{P}(X(\gamma_U) = j | X(0) = i)$ and the matrix of size $|\mathscr{D}| \times |\mathscr{U}|$ composed by these elements is P_{DU}.

Theorem 3.33 *Matrix P_{DU} is obtained from*

$$P_{DU} = (-Q_D)^{-1} Q_{DU}. \tag{3.54}$$

Proof Assuming that the process visits $\mathscr{D}$ for time t and at time t it moves from state $k \in \mathscr{D}$ to state $j \in \mathscr{U}$ the $P_{DU\,ij}$ probability can be obtained as

$$P_{DU\,ij} = \sum_{k \in \mathscr{D}} \int_{t=0}^{\infty} (e^{Q_D t})_{ik} dt\, Q_{DU\,kj} = \sum_{k \in \mathscr{D}} (-Q_D)_{ik}^{-1} Q_{DU\,kj},$$

where we used that the eigenvalues of Q_D have negative real part. The matrix form of the last expression gives the theorem.

Definition 3.24 The restricted Markov chain, $X^U(t)$, is defined based on $X(t)$ such that the time advances while $X(t)$ stays in $\mathscr{U}$ and the time is stopped while $X(t)$ stays in $\mathscr{D}$.

Figure 3.4 provides a visual demonstration of the restricted process. By the definition the state space of the right continuous restricted process, $X^U(t)$, is $\mathscr{U}$.

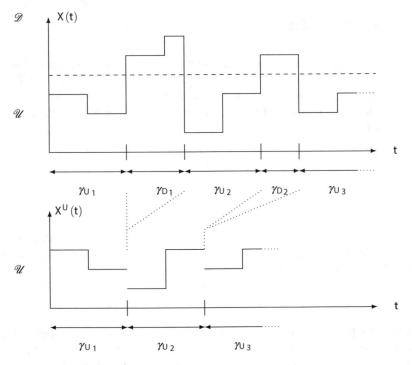

Fig. 3.4 $X(t)$ and its restricted process $X^U(t)$

Theorem 3.34 $X^U(t)$ *is a Markov chain on $\mathscr{U}$ with generator*

$$Q^{(U)} = Q_U + Q_{UD}(-Q_D)^{-1}Q_{DU}.$$

Proof First we show that $(Q_U + Q_{UD}(-Q_D)^{-1}Q_{DU})\mathbb{1} = 0$. From $Q\mathbb{I} = 0$ we have $Q_U\mathbb{1} + Q_{UD}\mathbb{1} = 0$ and $Q_{DU}\mathbb{1} + Q_D\mathbb{1} = 0$. The second equation can also be written as $(-Q_D)^{-1}Q_{DU}\mathbb{1} = \mathbb{1}$, from which

$$(Q_U + Q_{UD}(-Q_D)^{-1}Q_{DU})\mathbb{1} = Q_U\mathbb{1} + Q_{UD}\underbrace{(-Q_D)^{-1}Q_{DU}\mathbb{1}}_{\mathbb{1}} = Q_U\mathbb{1} + Q_{UD}\mathbb{1} = 0.$$

Based on this property it is enough to study the nondiagonal elements of $Q^{(U)}$.

For $i, j \in \mathscr{U}, i \neq j$ we analyze the state transition probability $\mathbf{P}(X^U(\Delta) = j \,|\, X^U(0) = i)$, because the i, j element of generator $Q^{(U)}$ can be obtained as

$$Q^{(U)}_{ij} = \lim_{\Delta \to 0} \frac{1}{\Delta}\mathbf{P}(X^U(\Delta) = j \,|\, X^U(0) = i).$$

or equivalently

$$\mathbf{P}(X^U(\Delta) = j \,|\, X^U(0) = i) = Q^{(U)}_{ij}\Delta + \mathcal{O}(\Delta),$$

where $\lim_{\Delta \to 0} \frac{\mathcal{O}(\Delta)}{\Delta} = 0$. For $\mathbf{P}(X^U(\Delta) = j \,|\, X^U(0) = i)$ we have

$$Q^{(U)}_{ij}\Delta + \mathcal{O}(\Delta) = \Delta Q_{Uij} + \sum_{k \in \mathscr{D}} \Delta Q_{UDik}P_{DUkj} + \mathcal{O}(\Delta)$$

where the first term represents a direct state transition from i to j and the second term represents a state transition to $k \in \mathscr{D}$ multiplied by the probability that the Markov chain returned to $\mathscr{U}$ by first visiting state $j \in \mathscr{U}$. Dividing both sides by Δ and making the $\Delta \to 0$ limit we get

$$Q^{(U)}_{ij} = Q_{Uij} + \sum_{k \in \mathscr{D}} Q_{UDik}P_{DUkj}$$

whose matrix form is

$$Q^{(U)} = Q_U + Q_{UD}P_{DU}. \tag{3.55}$$

Substituting (3.54) into the last expression completes the proof.

3.5 Exercises

Exercise 3.1 Compute the probability that the CTMC with generator matrix
$\begin{pmatrix} -1 & 0.5 & 0.5 \\ 1 & -2 & 1 \\ 1 & 0 & -1 \end{pmatrix}$ stays in state 1 after the second state transition, if the initial
distribution is $(0.5, 0.5, 0)$.

Exercise 3.2 Compute the stationary distribution of the CTMC with generator
matrix $\begin{pmatrix} -3 & 3 & 0 \\ 4 & -4 & 0 \\ 0 & 0 & 0 \end{pmatrix}$, if the initial distribution is $(0.5, 0, 0.5)$.

Exercise 3.3 Z_n and Y_n, $n = 1, 2, \ldots$, are discrete independent random variables.
$\mathbf{P}(Z_n = 0) = 1-p$, $\mathbf{P}(Z_n = 1) = p$ and $\mathbf{P}(Y_n = 0) = 1-q$, $\mathbf{P}(Y_n = 1) = q$. Define
the transition probability matrix of the DTMC X_n if

$$X_{n+1} = (X_n - Y_n)^+ + Z_n,$$

where $(x)^+ = \max(x, 0)$. This equation is commonly referred to as the evolution
equation of the DTMC.

Exercise 3.4 X_n, $n = 1, 2, \ldots$, is a DTMC with transition probability matrix $P = \begin{pmatrix} 3/6 & 1/6 & 2/6 \\ 3/4 & 0 & 1/4 \\ 0 & 1/3 & 2/3 \end{pmatrix}$. Compute $\mathbf{E}(X_0 X_1)$ and $\mathbf{Corr}(X_0, X_1)$ if the initial distribution is
$(0.5, 0, 0.5)$ and the state space is $S = \{0, 1, 2\}$.

Exercise 3.5 The generator of a CTMC is defined by

$$q_{0j} = \begin{cases} \frac{1}{3} & \text{if } j = 1, \\ \frac{1}{3} & \text{if } j = 2, \\ -\frac{2}{3} & \text{if } j = 0, \\ 0 & \text{otherwise;} \end{cases} \qquad q_{ij} = \begin{cases} \frac{1}{3i} & \text{if } j = i+1, \\ \frac{1}{3i} & \text{if } j = i+2, \\ -\frac{2}{3i} - \mu & \text{if } j = i, \\ \mu & \text{if } j = i-1, \\ 0 & \text{otherwise,} \end{cases} \qquad \text{for } i = 1, 2, \ldots.$$

Evaluate the properties of this Markov chain using, e.g., the Foster theorem.

Exercise 3.6 Show examples for

- reducible,
- periodic (and irreducible) and
- transient (and irreducible)

DTMCs. Evaluate $\lim_{n \to \infty} \mathbf{P}(X_n = i)$ for these DTMCs, where i is a state of the
Markov chain.

Exercise 3.7 Two players, A and B, play with dice according to the following rule. They throw the dice and if the number is 1, then A gets 2$ from B, if the number is 2 or 3, then A gets 1$ from B, if the number is greater than 3, then B gets 1$ from A. At the beginning of the game both A and B have 3$. The game lasts until someone could not pay. What is the probability that A wins?

Exercise 3.8 Two players, A and B, play with dice according to the following rule. They throw the dice and if the number is 1, then A gets 2$ from B, if the number is 2 or 3, then A gets 1$ from B, if the number is greater than 3, then B gets 1$ from A. At the beginning of the game both A and B have 3$. If one of them could not pay the required amount he gives all of his money to the other and the game goes on. What is the expected amount of money at A after a very long run? What is the probability that B cannot pay the required amount in the next step of the game after a very long run?

Exercise 3.9 There are two machines at a production site A and B. Their failure times are exponentially distributed with parameter λ_A and λ_B, respectively. Their repair times are also exponentially distributed with parameter μ_A and μ_B, respectively. There is a single repair man associated with the two machines, who can work on one machine at a time. Compute the probability that at least one of the machines works.

Exercise 3.10 Let $X = (X_0, X_1, \ldots)$ be a two-state Markov chain with state space $\mathscr{X} = \{0, 1\}$ and with the probability transition matrix $P = \begin{bmatrix} a & 1-a \\ 1-b & b \end{bmatrix}$, where $0 < a, b < 1$. Prove that $P^n = \frac{1}{2-a-b}\Pi + \frac{(a+b-1)^n}{2-a-b}(I - P)$, where $\Pi = \begin{bmatrix} 1-b & 1-a \\ 1-b & 1-a \end{bmatrix}$ and $I = \begin{bmatrix} 1 & 0 \\ 0 & 1 \end{bmatrix}$.

3.6 Solutions

Solution 3.1 Using the formulae (3.15) and (3.16), let us compute the transition probability of the embedding Markov chain. We get

$$q_0 = 1, \ q_1 = 2, \ q_2 = 1$$

and the transition probability matrix is

$$\Pi = \begin{pmatrix} 0 & 1/2 & 1/2 \\ 1/2 & 0 & 1/2 \\ 1 & 0 & 0 \end{pmatrix}.$$

Since the initial distribution is $p = (1/2, 1/2, 0)$, then the distribution of the embedded Markov chain after the second state transition is

$$p_2 = p\boldsymbol{\Pi}^2 = (5/8, 1/8, 2/8).$$

From this, we have the resulting probability for the state 1 as 5/8.

Solution 3.2 The Markov chain is composed by two irreducible sets of states, $\{1, 2\}$ and $\{3\}$. The probability of being in these irreducible sets are determined by the initial probability vector. The process starts in both sets with probability 0.5.

The stationary solution of the Markov chain on set $\{1, 2\}$ assuming that the process starts in that set is $(0.5, 0.5)$. The overall stationary distribution is $(0.5, 0.5, 0)0.5 + (0, 0, 1)0.5 = (0.25, 0.25, 0.5)$.

Solution 3.3 It can be seen that the state space of the Markov chain $\{X_n, \ n = 1, 2, \ldots\}$ is the set $\{0, 1, \ldots\}$. Using the evolution equation of the DMTC process we get the i, j $(i, j = 0, 1, \ldots)$ element of the transition probability matrix as follows

$$\mathbf{P}(X_{n+1} = j \mid X_n = i)$$
$$= \mathbf{P}\big((X_n - Y_n)^+ + Z_n = j \mid X_n = i\big) = \mathbf{P}\big((i - Y_n)^+ + Z_n = j\big)$$
$$= \mathbf{P}\big((i - Y_n)^+ + Z_n = j \mid Y_n = k, Z_n = m\big)\mathbf{P}(Y_n = k, Z_n = m)$$
$$= \mathbf{P}\big((i - k)^+ + m = j\big)\mathbf{P}(Y_n = k, Z_n = m) = \mathbf{P}(Y_n = k)\mathbf{P}(Z_n = m).$$

Since

$$\mathbf{P}(Y_n = k, Z_n = m) = \mathbf{P}(Y_n = k)\mathbf{P}(Z_n = m) = \begin{cases} (1 - q)(1 - p), & \text{if } k = m = 0, \\ (1 - q)p, & \text{if } k = 0, \ m = 1, \\ q(1 - p), & \text{if } k = 1, \ m = 0, \\ pq, & \text{if } k = m = 1, \end{cases}$$

the transition probability matrix has the form $(i, j = 0, 1, \ldots)$

$$p_{ij} = \begin{cases} (1 - q)(1 - p), & \text{if } j = i, \\ (1 - q)p, & \text{if } j = i + 1, \\ q(1 - p), & \text{if } j = (i - 1)^+, \\ pq, & \text{if } j = (i - 1)^+ + 1, \\ 0, & \text{in other cases.} \end{cases}$$

The transition probability graph with $\bar{p} = 1 - p$ and $\bar{q} = 1 - q$ is

Solution 3.4 Let us denote the transition probability matrix

$$P = (p_{ij}) = \begin{pmatrix} 3/6 & 1/6 & 2/6 \\ 3/4 & 0 & 1/4 \\ 0 & 1/3 & 2/3 \end{pmatrix},$$

and the initial distribution

$$p = (p_0, p_1, p_2) = (1/2, 0, 1/2).$$

Using the Markov property we have

$$\mathbf{E}(X_0 X_1) = \sum_{i=0}^{2} \sum_{j=0}^{2} ij \mathbf{P}(X_0 = i, X_1 = j) = \sum_{i=1}^{2} \sum_{j=1}^{2} ij \mathbf{P}(X_1 = j \mid X_0 = i) \mathbf{P}(X_0 = i)$$

$$= \sum_{j=1}^{2} 2j p_{2j} p_2 = 2 \cdot \frac{1}{3} \cdot \frac{1}{2} + 2 \cdot 2 \cdot \frac{2}{3} \cdot \frac{1}{2} = \frac{5}{3}.$$

It is clear that $\mathbf{Corr}(X_0, X_1) = \frac{\mathbf{E}(X_0 X_1) - \mathbf{E}(X_0)\mathbf{E}(X_1)}{\mathbf{D}(X_0)\mathbf{D}(X_1)}$. Since the distribution of RVs X_1 is

$$q = (q_0, q_1, q_2) = p^T P = (1/4, 1/4, 1/2),$$

then by simple calculations we have

$$\mathbf{E}(X_0) = \sum_{i=0}^{2} i p_i = 1, \quad \mathbf{E}\left(X_0^2\right) = \sum_{i=0}^{2} i^2 p_i = 2, \quad \mathbf{D}(X_0) = \sqrt{\mathbf{E}(X_0^2) - (\mathbf{E}(X_0))^2} = 1,$$

$$\mathbf{E}(X_1) = \sum_{i=0}^{2} i q_i = \frac{5}{4}, \quad \mathbf{E}\left(X_1^2\right) = \sum_{i=0}^{2} i^2 q_i = \frac{9}{4}, \quad \mathbf{D}(X_1) = \sqrt{\mathbf{E}(X_1^2) - (\mathbf{E}(X_1))^2} = \frac{3}{4}$$

and

$$\mathbf{Corr}(X_0, X_1) = \frac{5/3 - 1 \cdot 5/4}{1 \cdot 3/4} = \frac{5}{9}.$$

Solution 3.5 First, let us compute the transition probabilities of the embedding Markov chain. Using the formulae (3.15) and (3.16) we have $q_0 = 2/3$, $q_i = \frac{2+3i\mu}{3i}$, $i = 1, 2, \ldots$, thus

$$\pi_{0j} = \begin{cases} 1/2, & j = 1 \\ 1/2, & j = 2 \\ 0, & j = 0 \\ 0, & \text{otherwise} \end{cases} \qquad \pi_{ij} = \begin{cases} \frac{1}{2+3i\mu}, & j = i+1 \\ \frac{1}{2+3i\mu}, & j = i+2 \\ \frac{3i\mu}{2+3i\mu}, & j = i-1 \\ 0, & j = 1 \\ 0, & \text{otherwise} \end{cases}, \quad i = 1, 2, \dots$$

Denote by $X = (X_0, X_1, \dots)$ the embedding Markov chain of the CTMC, then the Foster theorem (Th. 3.42) says that the Markov chain X is ergodic if there exist constants $a, b > 0$ and $\ell \geq 0$ such that the inequalities

$$\mathbf{E}(X_{n+1} \mid X_n = i) \leq a, \ i \leq \ell,$$

$$\mathbf{E}(X_{n+1} \mid X_n = i) \leq i - b, \ i > \ell$$

hold. Since, for $i = 1, 2, \dots,$

$$\mathbf{E}(X_{n+1} \mid X_n = i) = (i+1)\frac{1}{2+3i\mu} + (i+2)\frac{1}{2+3i\mu} + (i-1)\frac{3i\mu}{2+3i\mu}$$

$$= i - 1 + \frac{5}{2+3i\mu},$$

and

$$\mathbf{E}(X_{n+1} \mid X_n = 0) = 1 \cdot \frac{1}{2} + 2 \cdot \frac{1}{2} = \frac{3}{2},$$

then $\mathbf{E}(X_{n+1} \mid X_n = i) \leq i + 3/2$, $i = 0, 1, \dots$. Choosing $\ell = \lceil 8/3\mu \rceil$, $a = \ell + 3/2$ and $b = 1/2$ we have

$$\mathbf{E}(X_{n+1} \mid X_n = i) \leq a, \ \text{if } i \leq \ell,$$

$$\mathbf{E}(X_{n+1} \mid X_n = i) \leq i - b, \ \text{if } i > \ell.$$

This means that the Foster's conditions of ergodicity hold, i.e., the Markov chain X is ergodic.

Solution 3.6

- Reducible transition probability matrix

$$P = \begin{pmatrix} 1 & 0 & 0 \\ 1/3 & 1/2 & 1/6 \\ 0 & 0 & 1 \end{pmatrix}.$$

In this case the stationary probabilities depend on the initial distribution.

- If the process starts from state 1, then the stationary distribution is $(1, 0, 0)$.
- If the process starts from state 2, then the stationary distribution is $(2/3, 0, 1/3)$.
- If the process starts from state 3, then the stationary distribution is $(0, 0, 1)$.

• Periodic transition probability matrix

$$P = \begin{pmatrix} 0 & 1/2 & 0 & 1/2 \\ 1/2 & 0 & 1/2 & 0 \\ 0 & 1/2 & 0 & 1/2 \\ 1/2 & 0 & 1/2 & 0 \end{pmatrix}.$$

Also in this case the stationary probabilities depend on the initial distribution. If the process starts from state 1 at time $n = 0$, then $\lim_{n \to \infty} \mathbf{P}(X_{2n} = i) = 1/2$ for $i = 1, 3$, $\lim_{n \to \infty} \mathbf{P}(X_{2n} = i) = 0$ for $i = 2, 4$ and $\lim_{n \to \infty} \mathbf{P}(X_{2n+1} = i) = 1/2$ for $i = 2, 4$, $\lim_{n \to \infty} \mathbf{P}(X_{2n+1} = i) = 0$ for $i = 1, 3$. That is, as n tends to infinity the distribution is $(1/2, 0, 1/2, 0)$ in the odd steps and it is $(0, 1/2, 0, 1/2)$ in the even steps.

• Transient transition probability matrix

$$P = \begin{pmatrix} 1/2 & 1/2 & 0 & 0 & 0 & 0 \\ 1/2 & 0 & 1/2 & 0 & 0 & 0 \\ 0 & 1/2 & 0 & 1/2 & 0 & 0 \\ 0 & 0 & 1/2 & 0 & 1/2 & 0 \\ 0 & 0 & 0 & \ddots & 0 & \ddots \end{pmatrix}.$$

In this case $\lim_{n \to \infty} \mathbf{P}(X_n = i) = 0$ for $i = 0, 1, \ldots$.

Solution 3.7 Before giving a solution for Exercise 3.7, we consider the problem in a more general setting (see, for example [24]). Let $K < L < M < N$ be integer numbers and let $X = \{X_n, n = 0, 1, \ldots\}$ be a Markov chain with finite state space $\mathscr{S} = \{K, K + 1, \ldots, N\}$. Let $\mathscr{S}_0 = \{K, K + 1, \ldots, L - 1\}$, $\mathscr{S}_1 = \{L, L + 1, \ldots, M\}$, $\mathscr{S}_2 = \{M + 1, \ldots, N\}$, then they are disjoint and nonempty subsets of $\mathscr{S}$, for which $\mathscr{S}_0 \cup \mathscr{S}_1 \cup \mathscr{S}_2 = \mathscr{S}$. Denote by $P = (p_{ij})$ the transition probability matrix of the Markov chain X.

The problem is to give a system of recurrent equations which describes the probability of the first hit for some state of the set $\mathscr{S}_2$ from a state $i_0 \in \mathscr{S}_1$, which means the probability that the process starts from a state $i_0 \in \mathscr{S}_1$ at the time point 0 and it will be first in a state from the set $\mathscr{S}_2$ without arriving some state from $\mathscr{S}_0$.

Let $X_0 = i_0 \in \mathscr{S}_1$ be an initial state. Let us introduce the set

$$W_{n+1} = \{(i_0, i_1, \ldots, i_n) : i_k \in \mathscr{S}_1, 0 \le k \le n - 1, i_n \in \mathscr{S}_2\}$$

and denote

$$r_n(i_0) = \mathbf{P}((X_0, \ldots, X_n) \in W_{n+1} \mid X_0 = i_0),$$

$$r_n(i) = 1, \text{ if } i \in \mathscr{S}_2, \text{ and } r_n(i) = 0, \text{ if } i \in \mathscr{S}_0$$

and

$$R_n(i_0) = r_1(i_0) + \ldots + r_n(i_0), \ n = 1, 2, \ldots,$$

$$R_n(i) = 1, \text{ if } i \in \mathscr{S}_2, \text{ and } R_n(i) = 0, \text{ if } i \in \mathscr{S}_0.$$

Using the Markov property of the chain X we get the relations

$$r_1(i_0) = \sum_{i_1 \in \mathscr{S}_2} p_{i_0 i_1}$$

and for $n \geq 2$

$$r_n(i_0) = \sum_{i_1 \in \mathscr{S}_1} p_{i_0 i_1} \mathbf{P}((X_0, X_1, \ldots, X_n) \in W_{n+1} \mid X_0 = i_0, X_1 = i_1)$$

$$= \sum_{i_1 \in \mathscr{S}_1} p_{i_0 i_1} \mathbf{P}((X_1, \ldots, X_n) \in W_n \mid X_1 = i_1) = \sum_{i_1 \in \mathscr{S}_1} p_{i_0 i_1} r_{n-1}(i_1).$$

Analogous equations can be derived for the probabilities $R_n(i_0)$, $n \geq 1$ as follows

$$R_n(i_0) = r_1(i_0) + \sum_{k=2}^{n} r_k(i_0) = r_1(i_0) + \sum_{k=2}^{n} \sum_{i_1 \in \mathscr{S}_1} p_{i_0 i_1} r_{k-1}(i_1)$$

$$= r_1(i_0) + \sum_{i_1 \in \mathscr{S}_1} p_{i_0 i_1} \sum_{k=2}^{n} r_{n-1}(i_1) = r_1(i_0) + \sum_{i_1 \in \mathscr{S}_1} p_{i_0 i_1} R_{n-1}(i_1).$$

If $n \to \infty$, then $R_n(i_0) \to R(i)$ and we have the equations

$$R(i_0) = r_1(i_0) + \sum_{i_1 \in \mathscr{S}_1} p_{i_0 i_1} R(i_1), \ R(i_0) = 1, \text{ if } i_0 \in \mathscr{S}_2, \text{ and } R(i_0) = 0, \text{ if } i \in \mathscr{S}_0.$$

Since the sequences $R_n(i), n = 1, 2, \ldots, i \in \mathscr{S}_1$ are monotonically nondecreasing, then the limits $R(i), i \in \mathscr{S}_1$ exist.

Remark Let $i_0 \in \mathscr{S}_1$ and denote

$$T_n = \min \{k : \ X_k \in \mathscr{S}_1, \ 0 \leq k \leq n - 1, \ X_n \in \mathscr{S}_1 \cup \mathscr{S}_2\},$$

$$T_n = n \text{ if } X_k \in \mathscr{S}_1, \ 0 \leq k \leq n$$

and

$$E_n(i_0) = \mathbf{E}(T_n \mid X_0 = i_0).$$

The equations for the expectations $E_n(i)$, $n \geq 0$ are similar to the equations $R_n(i)$, $n \geq 0$ equations are valid.
Denote

$$T(i_0) = \min\{n : X_k \in \mathscr{S}_1, \ 0 \leq k \leq n-1, \ X_n \in \mathscr{S}_1 \cup \mathscr{S}_2 \mid X_0 = i_0\}, \quad i_0 \in \mathscr{S}_1.$$

The RV $T(i_0)$, $i_0 \in \mathscr{S}_1$ is finite with probability 1, therefore after a very long run we have

$$\mathbf{P}(\text{B wins} \mid X_0 = 3) \approx 1 - \mathbf{P}(\text{A wins} \mid X_0 = 3).$$

Now, let us return to Solution 3.7. The possible (generalized) state space of the process X is $S = \{w_B, 0, \ldots, 6, w_A\}$, where $w_B = \{-1\}$, i.e., player B wins and $w_A = \{7, 8\}$, i.e., player A wins. Note that w_A and w_B are absorbing states. Denote by $S_0 = \{w_A\}$, $S_1 = \{0, 1, \ldots, 6\}$ and $S_2 = \{w_A\}$. The probability transition matrix of the MC X is

$$P = \begin{array}{c} \\ w_B \\ 0 \\ 1 \\ 2 \\ 3 \\ 4 \\ 5 \\ 6 \\ w_A \end{array} \begin{array}{c} \begin{array}{ccccccccc} w_B & 0 & 1 & 2 & 3 & 4 & 5 & 6 & w_A \end{array} \\ \left(\begin{array}{ccccccccc} 1 & 0 & 0 & 0 & 0 & 0 & 0 & 0 & 0 \\ 1/2 & 0 & 1/3 & 1/6 & 0 & 0 & 0 & 0 & 0 \\ 0 & 1/2 & 0 & 1/3 & 1/6 & 0 & 0 & 0 & 0 \\ 0 & 0 & 1/2 & 0 & 1/3 & 1/6 & 0 & 0 & 0 \\ 0 & 0 & 0 & 1/2 & 0 & 1/3 & 1/6 & 0 & 0 \\ 0 & 0 & 0 & 0 & 1/2 & 0 & 1/3 & 1/6 & 0 \\ 0 & 0 & 0 & 0 & 0 & 1/2 & 0 & 1/3 & 1/6 \\ 0 & 0 & 0 & 0 & 0 & 0 & 1/2 & 0 & 1/2 \\ 0 & 0 & 0 & 0 & 0 & 0 & 0 & 0 & 1 \end{array} \right) \end{array},$$

then we have a system of equations for the probabilities $R(i)$ as follows ($r_1(3) = 0$ because the initial state is $i_0 = 3$)

$$R(i) = 0, \ i \in S_0, \ R(i) = 1, \ i \in S_2, \tag{3.56}$$

$$R(0) = \frac{1}{3}R(1) + \frac{1}{6}R(2), \quad R(1) = \frac{1}{2}R(0) + \frac{1}{3}R(2) + \frac{1}{6}R(3),$$

$$R(i) = \frac{1}{2}R(i-1) + \frac{1}{3}R(i-1) + \frac{1}{6}R(i+1), \ 2 \leq i \leq 5,$$

$$R(6) = \frac{1}{2}R(5) + \frac{1}{2}.$$

Solving this system of linear equations we get

$$R(5) = 2R(6) - 1, \quad R(4) = \frac{10}{3}R(6) - \frac{7}{3}, \quad R(3) = 5R(6) - 4,$$

$$R(2) = \frac{64}{9}R(6) - \frac{55}{9}, \quad R(1) = \frac{88}{9}R(6) - \frac{79}{9}, \quad R(0) = \frac{355}{27}R(6) - \frac{328}{27}.$$

Thus from the equation $R(0) = \frac{1}{3}R(1) + \frac{1}{6}R(2)$ it follows

$$R(6) = \frac{443}{470} = 0.943.$$

If $X_0 = 3$ is the initial state, then the asymptotic probability that player A wins is (i.e., after a very long run)

$$R(3) = 5R(6) - 4 = 0.713.$$

Comment The system of linear equations for values $\mathbf{P}(\text{A wins} \mid X_0 = i)$, $0 \leq i \leq 6$ can be obtained easier based on intuitively considerations. Denote by $D = \{\text{A wins}\}$ the event that the player A wins if the starting state is $0 \leq i \leq 6$. Then by the Markov property we have

$$\mathbf{P}(D \mid X_0 = i) = \sum_{j \in \mathscr{S}_2} p_{ij} + \sum_{j \in \mathscr{S}_1} p_{ij}\mathbf{P}(D \mid X_1 = j) = \sum_{j \in \mathscr{S}_2} p_{ij} + \sum_{j \in \mathscr{S}_1} p_{ij}\mathbf{P}(D \mid X_0 = j).$$
(3.57)

Denote $R(i) = \mathbf{P}(D \mid X_0 = i)$, $i \in \mathscr{S}_1$, the system of Eqs. (3.56) follows immediately from (3.57).

Solution 3.8 If one of the players cannot pay the required amount of money in a step of the game, then he must give all his money to the other player and the game goes on. Denote X_n, $n = 0, 1, \ldots$ the amount of money the player A has in nth step of game. The state space of the process $X = (X_n, n = 0, 1, \ldots)$ is $\mathscr{X} = \{0, 1, \ldots, 6\}$. Let us introduce a sequence $(Y_1, Y_2, \ldots)$ of independent RVs with identical distribution

$$Y_n = \begin{cases} 1/2, \text{ if } -1, \\ 1/3, \text{ if } 1, \\ 1/6, \text{ if } 2, \end{cases}$$

then the process can be represented with the evolution equation $((x)^+ = \max(x, 0))$

$$X_{n+1} = \begin{cases} (X_n + Y_{n+1})^+, & \text{if } X_n + Y_{n+1} \leq 4, \\ \min(X_n + Y_{n+1}, 6), & \text{if } X_n + Y_{n+1} \geq 5, \end{cases} \quad n = 0, 1, \ldots$$

Contrary to Exercise 3.6, in this case there are no absorbing states and the probability transition matrix is

$$\begin{array}{c} & 0 & 1 & 2 & 3 & 4 & 5 & 6 \\ P = \begin{array}{c} 0 \\ 1 \\ 2 \\ 3 \\ 4 \\ 5 \\ 6 \end{array} & \left(\begin{array}{ccccccc} 1/2 & 1/3 & 1/6 & 0 & 0 & 0 & 0 \\ 1/2 & 0 & 1/3 & 1/6 & 0 & 0 & 0 \\ 0 & 1/2 & 0 & 1/3 & 1/6 & 0 & 0 \\ 0 & 0 & 1/2 & 0 & 1/3 & 1/6 & 0 \\ 0 & 0 & 0 & 1/2 & 0 & 1/3 & 1/6 \\ 0 & 0 & 0 & 0 & 1/2 & 0 & 1/2 \\ 0 & 0 & 0 & 0 & 0 & 1/2 & 1/2 \end{array}\right) \end{array}.$$

The process X is homogeneous, irreducible, and aperiodic MC with finite state space, therefore it is ergodic (see Theorem 3.13) and his stationary distribution, which does not depend on the initial distribution, satisfies the equations

$$\pi P = \pi, \ \pi = (\pi_0, \ldots, \pi_6),$$

$$\pi_0 + \ldots + \pi_6 = 1, \ \pi_i \geq 0,$$

From this we get the system of linear equations

$$\pi_0 = \frac{1}{2}\pi_0 + \frac{1}{2}\pi_1, \ \pi_1 = \frac{1}{3}\pi_0 + \frac{1}{2}\pi_2, \ \pi_2 = \frac{1}{6}\pi_0 + \frac{1}{3}\pi_1 + \frac{1}{2}\pi_3,$$

$$\pi_3 = \frac{1}{6}\pi_1 + \frac{1}{3}\pi_2 + \frac{1}{2}\pi_4, \ \pi_4 = \frac{1}{6}\pi_2 + \frac{1}{3}\pi_3 + \frac{1}{2}\pi_5,$$

$$\pi_5 = \frac{1}{6}\pi_3 + \frac{1}{3}\pi_4 + \frac{1}{2}\pi_6, \ \pi_6 = \frac{1}{6}\pi_4 + \frac{1}{2}\pi_5 + \frac{1}{2}\pi_6, \ \pi_0 + \ldots + \pi_6 = 1.$$

Expressing the probabilities π_i one after the other, we get $\pi_1 = \pi_0$, $\pi_2 = \frac{4}{3}\pi_0$, $\pi_3 = \frac{5}{3}\pi_0$, $\pi_4 = \frac{19}{9}\pi_0$, $\pi_5 = \frac{8}{3}\pi_0$, $\pi_6 = \frac{91}{27}\pi_0$, $1 = \pi_0 + \ldots + \pi_6 = \frac{27}{355}\pi_0$ and from this $\pi_0 = \frac{27}{355} = 0.076$. Finally we can compute the stationary distribution of the MC X

$$\pi = (0.076, \ 0.076, \ 0.101, \ 0.127, \ 0.161, \ 0.203, \ 0.256).$$

Using the stationary distribution, the expected amount of money that A will have after a very long run

$$\lim_{n \to \infty} \mathbf{E}(X_n) = \pi (0, 1, 2, 3, 4, 5, 6)^T = \sum_{i=0}^{6} i\pi_i = 3.904.$$

Denote D the event that B will not be able to pay the required amount in the next step of the game after a very long run. Then

$$\mathbf{P}(D) = \lim_{n \to \infty} \Big(\mathbf{P}(D \mid X_n = 5)\mathbf{P}(X_n = 5) + \mathbf{P}(D \mid X_n = 6)\mathbf{P}(X_n = 6) \Big)$$

$$= \lim_{n \to \infty} \Big(\mathbf{P}(Y_{n+1} = 2)\mathbf{P}(X_n = 5) + \mathbf{P}(Y_{n+1} = 1)\mathbf{P}(X_n = 6) \Big)$$

$$= \frac{1}{6}\pi_5 + \frac{1}{3}\pi_6 = 0.203 \cdot \frac{1}{6} + 0.256 \cdot \frac{1}{3} = 0.119.$$

Solution 3.9 The system has five states as follows:

0—A and B work at same time;
1—A in repair and B works;
2—A works and B in repair;
3—A is waiting for the repair and B in repair ;
4—A in repair and B is waiting for the repair.

Denote by $Z = (Z_t, \; t \geq 0)$ the process with state spaces $\{0, 1, \dots, 4\}$ which describes the state of the system at the time point t, and let $W_0 = 0 < W_1 < W_2 < \dots$ be the consecutive sequence of the transition points of time of the system (i.e., the embedding MC of Z). Denote by P the transition probability matrix of the MC $W = (W_0, W_1, \dots)$.

First Solution Let X, Y, U, and V be independent exponentially distributed random variables with parameters $\lambda_A, \lambda_B, \mu_A$ and μ_B, respectively. First, we compute the probabilities p_{01}, p_{01}, and p_{20}. It is clear that

$$p_{01} = \mathbf{P}(X < Y)$$

$$= \int_0^\infty \int_0^\infty \mathscr{I}_{\{x<y\}}\lambda_A\lambda_B e^{-\lambda_A x}e^{-\lambda_B y}\mathrm{d}x\mathrm{d}y = \int_0^\infty \int_0^y \lambda_A\lambda_B e^{-\lambda_A x}e^{-\lambda_B y}\mathrm{d}x\mathrm{d}y$$

$$= \int_0^\infty \lambda_B[1 - e^{-\lambda_A y}]e^{-\lambda_B y}\mathrm{d}y = 1 - \frac{\lambda_B}{\lambda_A + \lambda_B}\int_0^\infty (\lambda_A + \lambda_B)e^{-(\lambda_A+\lambda_B)y}\mathrm{d}y$$

$$= 1 - \frac{\lambda_B}{\lambda_A + \lambda_B} = \frac{\lambda_A}{\lambda_A + \lambda_B}.$$

Analogously, with the change of parameters, we have

$$p_{10} = \frac{\mu_A}{\mu_A + \lambda_B}, \quad p_{20} = \frac{\mu_B}{\mu_B + \lambda_A}.$$

Denote by $P_{ij}(t)$, $t > 0$ the distribution function of the holding time from a state i to another state j, then $P_{ij}(t) = \mathbf{P}(W_1 - W_0 < t \mid W_0 = i, W_1 = j)$ and

$$P_{01}(t) = \mathbf{P}(X \le t, \ X \le Y), \quad P_{02}(t) = \mathbf{P}(Y \le t, \ X > Y),$$

$$P_{10}(t) = \mathbf{P}(U \le t, \ U \le Y), \quad P_{14}(t) = \mathbf{P}(U \le t, \ U > Y),$$

$$P_{20}(t) = \mathbf{P}(V \le t, \ V \le X), \quad P_{23}(t) = \mathbf{P}(X \le t, \ V > X),$$

$$P_{31}(t) = \mathbf{P}(V \le t), \quad P_{42}(t) = \mathbf{P}(U \le t).$$

It is clear that

$$P_{01}(t) = \mathbf{P}(X \le t, \ X < Y)$$

$$= \int_0^\infty \int_0^\infty \mathscr{I}_{\{u \le t, \ u < v\}} \lambda_A \lambda_B e^{-\lambda_A x} e^{-\lambda_B y} dx dy = \int_0^t \left(\int_u^\infty \lambda_A \lambda_B e^{-\lambda_A u} e^{-\lambda_B y} dv \right) du$$

$$= \int_0^t \lambda_A e^{-(\lambda_A + \lambda_B)u} du = 1 - \frac{\lambda_A}{\lambda_A + \lambda_B} e^{-(\lambda_A + \lambda_B)u}$$

then

$$P'_{01}(t) = \lambda_A e^{-(\lambda_A + \lambda_B)u}.$$

With the same computations we have

$$P'_{02}(t) = \lambda_B e^{-(\lambda_A + \lambda_B)u}, \quad P'_{10}(t) = \mu_A e^{-(\mu_A + \lambda_B)u}, \quad P'_{14}(t) = \mu_A e^{-(\mu_A + \lambda_B)u},$$

$$P'_{20}(t) = \mu_B e^{-(\lambda_A + \mu_B)u}, \quad P'_{23}(t) = \lambda_A e^{-(\lambda_A + \mu_B)u},$$

$$P'_{31}(t) = \mu_B e^{-\mu_B t}, \quad P'_{42}(t) = \mu_A e^{-\mu_A t}$$

and in other cases $P'_{ij}(t) = 0$. The transition rate matrix of the system

$$\mathbf{Q} = (q_{ij}) = \begin{pmatrix} -(\lambda_A + \lambda_B) & \lambda_A & \lambda_B & 0 & 0 \\ \mu_A & -(\mu_A + \lambda_B) & 0 & 0 & \lambda_B \\ \mu_B & 0 & -(\lambda_A + \mu_B) & \lambda_A & 0 \\ 0 & \mu_B & 0 & -\mu_B & 0 \\ 0 & 0 & \mu_A & 0 & -\mu_A \end{pmatrix}.$$

By the Kolmogorov forward differential equation

$$\boldsymbol{\Pi}'(t) = \boldsymbol{\Pi}(t)\mathbf{Q}, \ t \ge 0.$$

This ordinary differential equation is linear and has constant coefficient matrix $\mathbf{Q}$ with special structure, therefore (see, for example, [2, Ch. 14, §13.]) it has a unique solution $\Pi(t), t \geq t$ for all initial values $\boldsymbol{\Pi}(0) = (p_0, \ldots, p_4), \ p_i \geq 0, \ p_0 + \ldots + p_4 = 1$ and $\boldsymbol{\Pi}(t)$ determines a distribution for any $t \geq 0$. The stationary distribution of the system can be computed from the linear algebraic equations $\pi \, \boldsymbol{Q} = 0$, where $\pi = (\pi_0, \ldots, \pi_4)$:

$$(\lambda_A + \lambda_B)\pi_0 = \mu_A\pi_1 + \mu_B\pi_2,$$

$$(\mu_A + \lambda_B)\pi_1 = \lambda_A\pi_0 + \pi_3,$$

$$(\mu_B + \lambda_A)\pi_2 = \lambda_B\pi_0 + \pi_4,$$

$$\pi_3 = \lambda_A\pi_2,$$

$$\pi_4 = \lambda_B\pi_1,$$

and the normalizing equation

$$\pi_0 + \pi_1 + \pi_2 + \pi_3 + \pi_4 = 1.$$

Solving this system of equations, the quantity $\pi_0 + \pi_1 + \pi_2$ will be the probability that at least one of the machines works.

Second Solution The transition probability matrix of the embedding Markov chain is

$$P = \begin{pmatrix} 0 & p_{01} & 1 - p_{01} & 0 & 0 \\ p_{10} & 0 & 0 & 0 & 1 - p_{10} \\ p_{20} & 0 & 0 & 0 - p_{20} & 0 \\ 0 & 1 & 0 & 0 & 0 \\ 0 & 0 & 1 & 0 & 0 \end{pmatrix},$$

where $p_{01} = \mathbf{P}(X < Y), p_{02} = \mathbf{P}(Y \leq X) = 1 - p_{01}, p_{10} = \mathbf{P}(U < Y), p_{14} = \mathbf{P}(Y \leq U) = 1 - p_{10}, p_{20} = \mathbf{P}(V < X), p_{23} = \mathbf{P}(X \leq V) = 1 - p_{20}.$

Firstly, we compute the probabilities p_{01}, p_{01}, and p_{20}. It is clear that

$$p_{01} = \mathbf{P}(X < Y) = \int_0^\infty \int_0^\infty \mathscr{I}_{\{x<y\}}\lambda_A\lambda_B e^{-\lambda_A x}e^{-\lambda_B y}\mathrm{d}x\mathrm{d}y$$

$$= \int_0^\infty \int_0^y \lambda_A\lambda_B e^{-\lambda_A x}e^{-\lambda_B y}\mathrm{d}x\mathrm{d}y = \int_0^\infty \lambda_B[1 - e^{-\lambda_A y}]e^{-\lambda_B y}\mathrm{d}y$$

$$= 1 - \frac{\lambda_B}{\lambda_A + \lambda_B}\int_0^\infty (\lambda_A + \lambda_B)e^{-(\lambda_A+\lambda_B)y}\mathrm{d}y = 1 - \frac{\lambda_B}{\lambda_A + \lambda_B} = \frac{\lambda_A}{\lambda_A + \lambda_B}.$$

Analogously, with the change of parameters, we have

$$p_{10} = \frac{\mu_A}{\mu_A + \lambda_B}, \quad p_{20} = \frac{\mu_B}{\mu_B + \lambda_A}.$$

The stationary distribution of the embedded Markov chain is the solution of the system of equations

$$rP = r, \quad r = (r_0, \ldots, r_4), \quad r_i \geq 0 \quad \text{and} \quad \sum_{i=0}^{4} r_i = 1.$$

Let us use the method of stationary analysis based on the embedded MC (see p. 206). For this we need to determine the stationary distribution $r = (r_1, \ldots, r_4)$ and the mean times $\hat{\tau}_j$, $j = 0, \ldots, 4$ that the system spends in a state j. Then, by the proposed method, the stationary distribution of the process Z is

$$\pi_j = \frac{r_j \hat{\tau}_j}{\sum\limits_{j=0}^{4} r_j \hat{\tau}_j}, \quad j = 0, \ldots, 4.$$

The holding time of state 1 has exponential distributions with parameter $(\lambda_A + \lambda_B)$, because

$$\mathbf{P}(\tau_1 \leq t) = \mathbf{P}(\min(X, Y) < t) = 1 - \mathbf{P}(\min(X, Y) \geq t)$$
$$= 1 - \mathbf{P}(X \geq t, Y \geq t) = 1 - \mathbf{P}(X \geq t)\mathbf{P}(Y \geq t) = 1 - e^{-(\lambda_A + \lambda_B)t},$$

therefore $\hat{\tau}_0 = \frac{1}{\lambda_A + \lambda_B}$. With the same way we get

$$\hat{\tau}_1 = \frac{1}{\mu_A + \lambda_B}, \quad \hat{\tau}_2 = \frac{1}{\lambda_A + \mu_B}, \quad \hat{\tau}_3 = \frac{1}{\mu_b}, \quad \hat{\tau}_4 = \frac{1}{\mu_A}.$$

Solution 3.10 For $n = 1$, it is true

$$\frac{1}{2 - a - b}\boldsymbol{\Pi} + \frac{a + b - 1}{2 - a - b}(\boldsymbol{I} - \boldsymbol{P}) = -(\boldsymbol{I} - \boldsymbol{P}) + \frac{1}{2 - a - b}[\boldsymbol{\Pi} + (\boldsymbol{I} - \boldsymbol{P})]$$

$$= -\boldsymbol{I} + \boldsymbol{P} + \frac{1}{2 - a - b}\begin{bmatrix} (1 - b) + 1 - a & (1 - a) + 0 - (1 - a) \\ (1 - b) + 0 - (1 - b) & (1 - a) + 1 - b \end{bmatrix}$$

$$= -\boldsymbol{I} + \boldsymbol{P} + \boldsymbol{I} = \boldsymbol{P}.$$

It is easy to check that $\boldsymbol{\Pi}(\boldsymbol{I} - \boldsymbol{P}) = (\boldsymbol{I} - \boldsymbol{P})\boldsymbol{\Pi} = 0$ and $\boldsymbol{P} - \boldsymbol{\Pi} = (a + b - 1)\boldsymbol{I}$. Apply the method of induction. Suppose that the equation

$$\boldsymbol{P}^n = \frac{1}{2 - a - b}\boldsymbol{\Pi} + \frac{(a + b - 1)^n}{2 - a - b}(\boldsymbol{I} - \boldsymbol{P})$$

is true for $n \geq 2$, then we prove that it is true for $(n + 1)$. Thus

$$P^{n+1} = P P^n = \frac{1}{2 - a - b} P \boldsymbol{\Pi} + \frac{(a + b - 1)^n}{2 - a - b} P (I - P)$$

$$= \frac{1}{2 - a - b} \boldsymbol{\Pi} + \frac{(a + b - 1)^n}{2 - a - b} [(a + b - 1) I + \boldsymbol{\Pi}](I - P)$$

$$= \frac{1}{2 - a - b} \boldsymbol{\Pi} + \frac{(a + b - 1)^{n+1}}{2 - a - b} (I - P).$$

From this it follows

$$\lim_{n \to \infty} P^n = \frac{1}{2 - a - b} \begin{bmatrix} 1 - b & 1 - a \\ 1 - b & 1 - a \end{bmatrix}.$$

Note that the convergence rate is exponential because of the inequality $|a + b - 1| < 1$.

It can also be seen that for both initial values $X_0 = 0$ and $X_0 = 1$ of the Markov chain there exists the limit matrix of the n-step transition matrix of the chain, which does not depend on the initial value. Thus the Markov chain has limit distribution, for which

$$\lim_{n \to \infty} \mathbf{P}(X_n = 1 \mid X_0 = 0) = \lim_{n \to \infty} \mathbf{P}(X_n = 1 \mid X_0 = 1) = \frac{1 - a}{2 - a - b},$$

$$\lim_{n \to \infty} \mathbf{P}(X_n = 0 \mid X_0 = 0) = \lim_{n \to \infty} \mathbf{P}(X_n = 0 \mid X_0 = 1) = \frac{1 - b}{2 - a - b}.$$

References

1. Apostol, T.: Calculus II. Wiley, London (1969)
2. Bellman, R.: Introduction to Matrix Analysis. SIAM, Philadelphia (1997)
3. Bernstein, S.N.: The Theory of Probabilities. Leningrad, Moscow (1946)
4. Erdős, P., Feller, W., Pollard, H.: A theorem on power series. Bull. Amer. Math. Soc. **55**, 201–203 (1949)
5. Feller, W.: An Introduction to Probability Theory and Its Applications, vol. I. Wiley, New York (1968)
6. Foster, F.G.: On the stochastic matrices associated with certain queuing processes. Ann. Math. Stat. **24**, 355–360 (1953)
7. Gikhman, I., Skorokhod, A.: The Theory of Stochastic Processes, vol. I. Springer, Berlin (1974)
8. Gikhman, I.I., Skorokhod, A.V.: The Theory of Stochastic Processes, vol. II. Springer, Berlin (1975)
9. Haccou, P., Jagers, P., Vatutin, V.A.: Branching Processes: Variation, Growth, and Extinction of Populations, vol. 5. Cambridge University Press, Cambridge (2005)
10. Harris, T.E.: The Theory of Branching Processes. Courier Corporation, Chelmsford (2002)
11. Izsák, J., Szeidl, L.: Population dynamic models leading to logarithmic and Yule distribution. Acta Inform. Hung. **15**(1), 149–162 (2018)

12. Johnson, N.L., Kemp, A.M., Kotz, S.: Univariate Discrete Distributions, 3rd edn. Wiley, London (2005)
13. Kamke, E.: Differencialgleichungen. I. Gewoehnliche Differencialgleichungen, 10th edn. Springer, Berlin (1977)
14. Kendall, D.G.: On the generalized birth-and-death process. Ann. Math. Stat. **19**(1), 1–15 (1948)
15. Kendall, D.G.: Stochastic processes and population growth. J. R. Stat. Soc. Ser. B **11**(2), 230–282 (1949)
16. Klimov, G.P.: Extremal Problems in Queueing Theory. Energia, Moscow (1964)
17. Lange, K.: Applied Probability, 2nd edn. Springer, Berlin (2010)
18. Markov, A.A.: Rasprostranenie zakona bol'shih chisel na velichiny, zavisyaschie drug ot druga. Izvestiya Fiziko-matematicheskogo Obschestva pri Kazanskom Universitete **15**, 135–156 (1906)
19. Matveev, V.F., Ushakov, V.G.: Queueing Systems. MGU, Moscow (1984)
20. Meyn, S., Tweedie, R.: Markov Chains and Stochastic Stability. Springer, Berlin (1993)
21. Olver, P.J.: Introduction to Partial Differential Equations. Springer, Berlin (2014)
22. Seneta, E.: Non-negative Matrices and Markov Chains. Springer, Berlin (2006)
23. Serfozo, R.: Basics of Applied Stochastic Processes. Springer, Berlin (2009)
24. Shiryaev, A.N.: Probability. Springer, Berlin (1994)

Chapter 4
Renewal and Regenerative Processes

4.1 Basic Theory of Renewal Processes

Let $\{N(t), \ t \geq 0\}$ be a nonnegative-integer-valued stochastic process that counts the occurrences of a given event. That is, $N(t)$ is the number of events in the time interval $[0, t]$. For example, $N(t)$ can be the number of bulb replacements in a lamp that is continuously on, and the dead bulbs are immediately replaced (Fig. 4.1).

Let $0 \leq \tau_1 \leq \tau_2 \leq$ be the times of the occurrences of consecutive events and $\tau_0 = 0$ and $T_i = \tau i - \tau_{i-1}, i = 1, 2, 3, \ldots$ be the time intervals between consecutive events.

Definition 4.1 $\tau_1 \leq \tau_2 \leq \ldots$ is a **renewal process** if the time intervals between consecutive events $T_i = \tau_i - \tau_{i-1}, \ i = 2, 3, \ldots$ are independent and identically distributed (i.i.d.) random variables with CDF

$$F(x) = \mathbf{P}(T_k \leq x), \quad k = 1, 2, \ldots$$

The nth event time, $\tau_n, \ n = 1, 2, \ldots$, is referred to as the n**th renewal point** or renewal time. According to the definition, the first time interval might have a different distribution.

We assume that $F(0-) = 0$ and $F(0) = \mathbf{P}(T_k = 0) < 1$. In this case

$$\tau_0 = 0, \quad \tau_n = T_1 + \ldots + T_n, \quad n = 1, 2, \ldots,$$

$$N(0) = 0, \quad N(t) = \sup\{n : \ \tau_n \leq t, \ n \geq 0\} = \sum_{i=1}^{\infty} \mathscr{I}_{\{\tau_i \leq t\}}, \quad t > 0.$$

Remark 4.1 $\{N(t), \ t \geq 0\}$ and $\{\tau_n, \ n \geq 1\}$ mutually and univocally determine each other because for arbitrary $t \geq 0$ and $k \geq 1$ we have

$$N(t) \geq k \quad \Leftrightarrow \quad \tau_k \leq t.$$

© Springer Nature Switzerland AG 2019
L. Lakatos et al., *Introduction to Queueing Systems with Telecommunication Applications*, https://doi.org/10.1007/978-3-030-15142-3_4

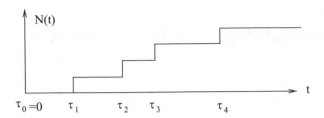

Fig. 4.1 Renewal process

Definition 4.2 When $\mathbf{P}(T_k \leq x) = F(x)$, $k = 2, 3, \ldots$, but $F_1(x) = \mathbf{P}(T_1 \leq x) \neq F(x)$, the process is referred to as a **delayed renewal process**.

Remark 4.2 $T_1, T_2, \ldots$ are i.i.d. random variables and from $\tau_n = T_1 + \ldots + T_n$, and we can compute the distribution of the time of the nth event $F^{(n)}(x) = \mathbf{P}(\tau_n \leq x)$ using the convolution formula

$$F^{(n)}(x) = \int_0^\infty F^{(n-1)}(x - y)\mathrm{d}F(y) = \int_0^x F^{(n-1)}(x - y)\mathrm{d}F(y), \quad n \geq 2, \quad x \geq 0,$$

$$F^{(n)}(x) \equiv 0, \quad \text{if } x \leq 0 \text{ and } n \geq 1.$$

Starting from $F^{(1)}(x) = F_1(x)$ the same formula applies in the delayed case.

Definition 4.3 The function $H(t) = \mathbf{E}(N(t))$, $t \geq 0$, is referred to as a **renewal function**.

One of the main goals of renewal theory is the analysis of the renewal function $H(t)$ and the description of its asymptotic behavior. Below we discuss the related results for regular renewal processes. The properties of delayed renewal processes are similar, and we do not provide details on them here. We will show that the law of large numbers and the central limit theorem hold for the renewal process (see also Ch. 5. in [4]).

Theorem 4.1 *If* $\{T_n, \ n = 1, 2, \ldots\}$ *is a series of nonnegative i.i.d. random variables and* $\mathbf{P}(T_1 = 0) < 1$, *then there exists* $\rho_0 > 0$ *such that for all* $\infty < \rho < \rho_0$ *and* $t \geq 0$

$$\mathbf{E}\left(e^{\rho N(t)}\right) < \infty$$

holds.

Proof (Proof 1 of Theorem 4.1) If we define $\rho_0 = (\log(\mathbf{E}(e^{-T_1})))^{-1}$, then obviously we can write $-\infty < \rho_0 < 0$. By the Markov inequality (Theorem 1.6) we have

$$\mathbf{E}\left(e^{\rho N(t)}\right) = \sum_{k=0}^{\infty} e^{\rho k} \mathbf{P}(N(t) = k) \le \sum_{k=0}^{\infty} e^{\rho k} \mathbf{P}(N(t) \ge k)$$

$$= \sum_{k=0}^{\infty} e^{\rho k} \mathbf{P}(\tau_k \le t) = \sum_{k=0}^{\infty} e^{\rho k} \mathbf{P}\left(1 \le e^{t - \tau_k}\right) \le \sum_{k=0}^{\infty} e^{\rho k} \mathbf{E}\left(e^{t - \tau_k}\right)$$

$$= \sum_{k=0}^{\infty} e^{\rho k} e^{t} \left(\mathbf{E}\left(e^{-T_1}\right)\right)^k = e^{t} \sum_{k=0}^{\infty} e^{-(\rho_0 - \rho)k} = e^{t}(1 - e^{-(\rho_0 - \rho)})^{-1},$$

which is finite if $-\infty < \rho < \rho_0$. □

Proof (Proof 2 of Theorem 4.1) According to the condition of the theorem, there exist ϵ and δ positive numbers such that $\mathbf{P}(T_k \ge \delta) > \epsilon$. We introduce $\{T'_k = \delta \mathscr{I}_{\{T_k \ge \delta\}}, \ k = 1, 2, \ldots\}$, which is a series of i.i.d. random variables. The related renewal process is $\{N'(t), \ t \ge 0\}$. According to the definition of $\{T'_k$ and $\{N'(t)$ we have that $\mathbf{P}(T'_k \le T_k) = 1, \ k \ge 1$, and consequently $\mathbf{P}(N'(t) \ge N(t)) = 1, \ t \ge 0$. The distribution of $N'(t)$ is negative binomial with the parameter $p = \mathbf{P}(T'_k \ge \delta)$ and order $r = \lfloor t/\delta \rfloor$,

$$\mathbf{P}(N'(t) = k + r) = \binom{k + r - 1}{r - 1} p^r (1 - p)^k, \quad k = 0, 1, 2, \ldots,$$

from which the statement of the theorem follows. □

Corollary 4.1 *All moments of $N(t)$ $(t \ge 0)$ are finite. Consequently, the renewal function $H(t)$ is also finite for all $t \ge 0$.*

Proof For any $n > 0$, according to Theorem 4.1 we have

$$\infty > \mathbf{E}\left(e^{\rho N(t)}\right) = \mathbf{E}\left(\sum_{i=0}^{\infty} \frac{\rho^i}{i!} N(t)^i\right) = \sum_{i=0}^{\infty} \frac{\rho^i}{i!} \mathbf{E}\left(N(t)^i\right) > \frac{\rho^n}{n!} \mathbf{E}\left(N(t)^i\right).$$

□

Before conducting an analysis of the renewal function we recall some properties of convolution.

Let $A(t)$ and $B(t)$ be monotonically nondecreasing right-continuous functions such that $A(0) = B(0) = 0$.

Definition 4.4 The **convolution** of $A(t)$ and $B(t)$ [denoted by $A * B(t)$] is

$$A * B(t) = \int_0^t B(t - y) dA(y), \quad t \ge 0.$$

Lemma 4.1 $A * B(t) = B * A(t)$.

Proof From $B(0) = 0$ we have $B(t - y) = \int\limits_0^{t-y} dB(s)$, and consequently

$$
A * B(t) = \int\limits_0^t \left\{ \int\limits_0^{t-y} dB(s) \right\} dA(y) = \int\limits_0^t \int\limits_0^t \mathscr{I}_{\{s<t-y\}} dA(y)dB(s)
$$

$$
= \int\limits_0^t \int\limits_0^t \mathscr{I}_{\{y<t-s\}} dA(y)dB(s) = \int\limits_0^t \left\{ \int\limits_0^{t-s} dA(y) \right\} dB(s)
$$

$$
= B * A(t).
$$

$\square$

Remark 4.3 The definition of the renewal function $H(t)$

$$
H(t) = \mathbf{E}(N(t)) = \mathbf{E}\left(\sum_{i=1}^\infty \mathscr{I}_{\{\tau_i \le t\}} \right) = \sum_{i=1}^\infty \mathbf{P}(T_1 + \ldots + T_i \le t)
$$

immediately determines the relation between the renewal function and the order k of the convolution of the event time distribution

$$
H(t) = \sum_{k=1}^\infty F^{(k)}(t).
$$

Theorem 4.2 *If* $\{T_n, \ n = 1, 2, \ldots\}$ *is a series of nonnegative i.i.d. random variables and* $\mathbf{P}(T_1 = 0) < 1$, *then* $H(t)$ *satisfies the* **renewal equation**

$$
H(t) = F(t) + \int\limits_0^t H(t - y)dF(y), \quad t \ge 0.
$$

Proof According to Remarks 4.2 and 4.3, the renewal function can be written as

$$
H(t) = F^{(1)}(t) + \sum_{k=1}^\infty \int\limits_0^t F^{(k)}(t - y)dF(y)
$$

$$
= F(t) + \int\limits_0^t \left(\sum_{k=1}^\infty F^{(k)}(t - y) \right) dF(y)
$$

$$
= F(t) + \int\limits_0^t H(t - y)dF(y),
$$

where the order of the summation and the integration are interchanged based on Corollary 4.1.   □

In the case of a delayed renewal process, the renewal function is denoted as $H_1(t)$, and the same composition holds as for the regular renewal process (Remark 4.3)

$$H_1(t) = \sum_{k=1}^{\infty} F^{(k)}(t), \ t \geq 0, \qquad (F^{(k)}(t) = \mathbf{P}(\tau_k \leq t)),$$

but in this case $F_1 \neq F$.

Theorem 4.3 *The renewal function can be written in the following forms:*

$$H_1(t) = F_1(t) + H_1 * F(t) = F_1(t) + F * H_1(t),$$
$$H_1(t) = F_1(t) + H * F_1(t) = F_1(t) + F_1 * H(t),$$
$$H(t) = F(t) + H * F(t) = F(t) + F * H(t).$$

Renewal Equations
Definition 4.5 An integral equation of type

$$A(t) = a(t) + \int_0^t A(t-x)dF(x), \ t \geq 0,$$

where $a(t)$ and $F(t)$ are known functions and $A(t)$ is unknown, is referred to as **renewal equation** (see also Theorem 4.1 of Ch. 5 in [4]).

Theorem 4.4 *If $a(t)$, $t \geq 0$, is a bounded real function that is Riemann–Stieltjes integrable according to $H(t)$ over any finite interval, then there uniquely exists the function $A(t)$, $t \geq 0$, which is finite over any finite interval and satisfies the renewal equation*

(i) $$A(t) = a(t) + \int_0^t A(t-x)dF(x), \quad t \geq 0,$$

and furthermore it satisfies

(ii) $$A(t) = a(t) + \int_0^t a(t-x)dH(x), \quad t \geq 0,$$

where $H(t) = \sum_{k=1}^{\infty} F^{(k)}(t)$, $t \geq 0$, is the renewal function.

Proof First we show that the function $A(t)$, $t \geq 0$, defined by equation (ii), is (a) bounded on the $[0, T]$ interval for all $T > 0$ and (b) satisfies (i). Next we prove that (c) all solutions of (i) that are bounded on $[0, T]$ can be given in form (ii), i.e., the solution is unique.

(a) Since $a(t)$ is bounded and $H(t)$ is monotonically nondecreasing, we have

$$\sup_{0 \leq t \leq T} |A(t)| \leq \sup_{0 \leq t \leq T} |a(t)| + \int_0^T [\sup_{0 \leq y \leq T} |a(y)| dH(x)$$

$$\leq \sup_{0 \leq t \leq T} |a(t)|(1 + H(T)) < \infty.$$

(b) Furthermore, we have

$$A(t) = a(t) + H * a(t) = a(t) + \left(\sum_{k=1}^{\infty} F^{(k)}\right) * a(t)$$

$$= a(t) + F * a(t) + \left(\sum_{k=2}^{\infty} F^{(k)}\right) * a(t)$$

$$= a(t) + F * [a(t) + \left(\sum_{k=1}^{\infty} F^{(k)}\right) * a(t)]$$

$$= a(t) + F * A(t).$$

(c) We prove this by successive approximation. According to equation (i), $A = a + F * A$. Substituting this into (i) we have

$$A = a(t) + F * (a + F * A) = a + F * a + F * (F * A)$$

$$= a + F * a + F^{(2)} * A.$$

Continuously substituting equation (i) we obtain for $n \geq 1$ that

$$A = a + F * a + F^{(2)} * (a + F * A) = \ldots = a + \sum_{k=1}^{n-1} (F^{(k)} * a) + F^{(n)} * A.$$

Since $A(t)$ is bounded on every finite interval according to (a), $F^{(n)}(0-) = 0$, $F^{(n)}(y)$ is monotonically nondecreasing, and $F^{(n)}(t) \to 0$, $n \to \infty$, for all fixed t, we have that for a fixed t

$$|F^{(n)} * A(t)| = |\int_0^t A(t-y) dF^{(n)}(y)| \leq \sup_{0 \leq y \leq t} |A(t-y)| F^{(n)}(t) \to 0, \quad n \to \infty.$$

From the fact that $a(t)$ is bounded it follows that

$$\lim_{n \to \infty} \left(\sum_{k=1}^{n-1} F^{(k)} \right) * a(t) = \left(\sum_{k=1}^{\infty} F^{(k)} \right) * a(t) = H * a(t),$$

and consequently

$$A(t) = a(t) + \lim_{n \to \infty} \left[\left(\sum_{k=1}^{n-1} F^{(k)} \right) * a(t) + F^{(n)} A(t) \right] = a(t) + H * a(t).$$

This means that if A is a bounded solution of (i), then it is identical with (ii).

$\square$

Analysis of the Renewal Function

One of the main goals of renewal theorem is the analysis of the renewal function. According to Theorem 4.3, in the case of delayed renewal processes the renewal function $H_1(t)$ can be obtained from $F_1(t)$ and $H(t)$. In the rest of this section we focus on the analysis of the renewal function of an ordinary renewal process, $H(t)$, that is, $F_k = F$, $k \geq 1$. During the subsequent analysis we assume that $F(t)$ is such that $F(0-) = 0$ and $F(0+) < 1$.

Theorem 4.5 (Elementary Renewal Theorem) *There exists the limit*

$$\lim_{t \to \infty} \frac{H(t)}{t} = \frac{1}{\mathbf{E}(T_1)}$$

and it is 0 if $\mathbf{E}(T_1) = \infty$.

Definition 4.6 The random variable X has a lattice distribution if there exists $d > 0$ and $r \in R$ such that the random variable $\frac{1}{d}(X - r)$ is distributed on the integer numbers, that is, $\mathbf{P}\left(\frac{1}{d}(X - r) \in \mathbf{Z} \right) = 1$. The largest d with that property is referred to as the step size of the distribution.

Remark 4.4 If X has a lattice distribution with step size d, then

$$d = \min\{s : |\psi(2\pi/s)| = 1\},$$

where $\psi(u) = \mathbf{E}\left(e^{iuX} \right)$, $u \in R$, denotes the characteristic function of X. In this case, $|\psi(u)| < 1$ if $0 < |u| < 2\pi/d$. If the distribution of X is not lattice, then $|\psi(u)| < 1$, if $u \neq 0$.

Theorem 4.6 (Blackwell's Theorem) *If $F(t)$ is a lattice distribution with step size d, then*

$$\lim_{n \to \infty} H(nd) - H((n-1)d) = \frac{d}{\mathbf{E}(T_1)}.$$

If $F(t)$ is not a lattice distribution, then for all $h > 0$

$$\lim_{t\to\infty} (H(t+h) - H(t)) = \frac{h}{\mathbf{E}(T_1)}$$

holds.

The following theorems require the introduction of *directly Riemann integrability.* which is more strict than Riemann integrability.

Let g be a nonnegative function on the interval $[0, \infty)$ and

$$s(\delta) = \delta \sum_{n=1}^{\infty} \inf\{g(x) : (n-1)\delta \leq x \leq n\delta\},$$

$$S(\delta) = \delta \sum_{n=1}^{\infty} \sup\{g(x) : (n-1)\delta \leq x \leq n\delta\}.$$

Definition 4.7 The function g is *directly Riemann integrable* if $s(\delta)$ and $S(\delta)$ are finite for all $\delta > 0$ and

$$\lim_{\delta\to 0} [S(\delta) - s(\delta)] = 0.$$

Remark 4.5 If the function g is directly Riemann integrable, then g is bounded, and the limit of $s(\delta)$ and $S(\delta)$ at $\delta \to 0$ is equal to the infinite Riemann integral, that is

$$\lim_{\delta\to 0} s(\delta) = \lim_{\delta\to 0} S(\delta) = \int_0^{\infty} g(x)\mathrm{d}x = \lim_{y\to\infty} \int_0^{y} g(x)\mathrm{d}x.$$

Sufficient and Necessary Condition for Direct Riemann Integrability
(a) There exists $\delta > 0$ such that $S(\delta) < \infty$.
(b) g is almost everywhere continuous along the real axes according to the Lebesgue measure (that is, equivalent to Riemann integrability on every finite interval).

Sufficient Condition for Direct Riemann Integrability
g is bounded and has a countable number of discontinuities, and at least either condition (a) and (b) holds:

(a) g equals 0 apart from a finite interval.
(b) g is monotonically decreasing and $\int_0^{\infty} g(x)\mathrm{d}x < \infty$.

Theorem 4.7 (Smith's Renewal Theorem) *If $g(x) \geq 0$, $x \geq 0$, is a nonincreasing directly Riemann integrable function on the interval $[0, \infty)$, then for $t \to \infty$ one of the following identities holds:*

(a) *If F is a nonlattice distribution, then*

$$\lim_{t \to \infty} H * g(t) = \lim_{t \to \infty} \int_0^t g(t - u) dH(u) = \frac{1}{E(T_1)} \int_0^\infty g(u) du.$$

(b) *If F is a lattice distribution with step size d, then*

$$\lim_{n \to \infty} H * g(x + nd) = \lim_{n \to \infty} \int_0^{x+nd} g(x + nd - u) dH(u) = \frac{d}{E(T_1)} \sum_{k=0}^\infty g(x + kd).$$

Remark 4.6 Blackwell's theorem (Theorem 4.6) follows from Smith's renewal theorem (Theorem 4.7) assuming that $g(u) = \mathscr{I}_{\{0 < u \leq h\}}$. The reverse direction is an implicit consequence of the proof of the Blackwell's theorem provided by Feller in [3].

Before proving Theorem 4.7 we collect some simple properties of the renewal function $H(t)$.

Lemma 4.2 *H is monotonically nondecreasing and continuous from the right.*

Proof $F^{(k)}(t)$ is monotonically nondecreasing and continuous from the right for all $k \geq 1$, and the series $\sum_{k=1}^\infty F^{(k)}(t)$ is uniformly convergent on every finite interval, from which the lemma follows. □

Lemma 4.3 *The function H is subadditive, that is,*

$$H(t + h) \leq H(t) + H(h) \tag{4.1}$$

for $t, h \geq 0$.

Proof Since $H(0) = 0$, it is enough to consider the case where $t, h > 0$. Let $n(t) = \inf\{n : \tau_n \geq t, n \geq 0\}$. If $\tau_n \leq t$ for all $n \geq 0$, then let $n(t) = \infty$. This case can occur only on a set with measure 0.

Due to the fact that $\mathbf{P}(T_1 = 0)$ might be positive, the relation of $n(t)$ and $N(t)$ is not deterministic. It holds that $n(t) \geq N(t) + 1$ and the right continuity of $N(t)$ implies $N(\tau_{n(t)}) = N(t)$, $t \geq 0$. Using that we have

$$N(t + h) - N(t) = N(t + h) - N(\tau_{n(t)}) \leq N(\tau_{n(t)} + h) - N(\tau_{n(t)}),$$

and using the total probability theorem, we obtain

$$\mathbf{E}(N(t+h) - N(t)) \le \mathbf{E}\big(N(\tau_{n(t)} + h) - N(\tau_{n(t)})\big)$$

$$= \sum_{k=1}^{\infty} \mathbf{E}\big(N(\tau_{n(t)} + h) - N(\tau_{n(t)}) | n(t) = k\big)\mathbf{P}(n(t) = k)$$

$$= \sum_{k=1}^{\infty} \mathbf{E}(N(\tau_k + h) - N(\tau_k) | n(t) = k)\mathbf{P}(n(t) = k).$$

Since τ_k is a renewal point, the conditional expected value in the last summation does not depend on the condition

$$\mathbf{E}(N(\tau_k + h) - N(\tau_k) | n(t) = k) = \mathbf{E}(N(h) - N(0)) = \mathbf{E}(N(h)),$$

and in this way we have

$$\mathbf{E}(N(t+h) - N(t)) \le \sum_{k=1}^{\infty} \mathbf{E}(N(h))\mathbf{P}(n(t) = k)$$

$$= \mathbf{E}(N(h)) \sum_{k=1}^{\infty} \mathbf{P}(n(t) = k) = \mathbf{E}(N(h)),$$

from which the lemma follows.   □

Lemma 4.4   *For the renewal function H the following inequality holds:*

$$H(t) \le H(1)(1 + t), \quad t \ge 0. \tag{4.2}$$

Proof From the previous statement and the monotonicity of H

$$H(t) \le H(\lfloor t \rfloor + 1) \le H(1) + H(\lfloor t \rfloor) \le H(1) + (H(1) + H(\lfloor t \rfloor - 1))$$

$$\le \dots \le H(1) + \lfloor t \rfloor H(1) \le H(1) + t H(1) = H(1)(1 + t).$$

□

Remark 4.7 The nonnegative subadditive functions can be estimated from the preceding expression by a linear function.

Lemma 4.5   *For arbitrary $\lambda > 0$ the Laplace–Stieltjes transform $H^{\sim}(\lambda) = \int_0^{\infty} e^{-\lambda t} dH(t), \lambda \ge 0$, of the function H can be represented in the Laplace–Stieltjes transform as*

$$H^{\sim}(\lambda) = (1 - \varphi^{\sim}(\lambda))^{-1},$$

where $\varphi^\sim(\lambda) = \mathbf{E}(e^{-\lambda T_1})$ is the Laplace–Stieltjes transform of the distribution function F.

Proof For $\lambda > 0$ there obviously exists $H^\sim(\lambda)$ since, according to Eqs. (1.3) and (4.2),

$$H^\sim(\lambda) = \lambda \int_0^\infty e^{-\lambda t} H(t) dt \le \lambda H(1) \int_0^\infty e^{-\lambda t}(1 + t) dt < \infty.$$

It is clear that

$$\int_0^\infty e^{-\lambda t} dN(t) = \sum_{k=0}^\infty e^{-\lambda \tau_k} = 1 + \sum_{k=1}^\infty \prod_{i=1}^k e^{-\lambda T_i}.$$

Using this equality we obtain

$$\mathbf{E}\left(\int_0^\infty e^{-\lambda t} dN(t)\right) = \mathbf{E}\left(\lambda \int_0^\infty N(t) e^{-\lambda t} dt\right)$$

$$= \lambda \int_0^\infty H(t) e^{-\lambda t} dt = \int_0^\infty e^{-\lambda t} dH(t) = H^\sim(\lambda)$$

$$= \mathbf{E}\left(1 + \sum_{k=1}^\infty \prod_{i=1}^k e^{-\lambda T_i}\right) = 1 + \sum_{k=1}^\infty (\varphi(\lambda))^k = \frac{1}{1 - \varphi(\lambda)},$$

where $0 < \varphi(\lambda) < 1$ if $\lambda > 0$. $\qquad\square$

Proof of Elementary Renewal Theorem First we prove the limit exists. If $t \ge 1$, then we have that $0 \le \frac{H(t)}{t} \le \frac{1+t}{t} H(1) \le 2H(1)$ is bounded. Let $c = \inf_{t \ge 1} \frac{\overline{H}(t)}{t}$. Then for arbitrary $\epsilon > 0$ there exists a number $t_0 > 0$ such that

$$\frac{H(\tau_0)}{t_0} < c + \epsilon.$$

Moreover, for all integers $k \ge 1$ and $\tau \ge 0$

$$\frac{H(kt_0 + \tau)}{kt_0 + \tau} \le \frac{kH(t_0) + H(\tau)}{kt_0} \le c + \epsilon + \frac{H(\tau)}{kt_0},$$

and consequently

$$\limsup_{t \to \infty} \frac{H(t)}{t} \leq c + \epsilon,$$

and

$$c \leq \liminf_{t \to \infty} \frac{H(t)}{t} \leq \limsup_{t \to \infty} \frac{H(t)}{t} \leq c$$

follows. We have proved the existence of the limit.

Using the preceding expression for the Laplace–Stieltjes transform $h(\lambda)$

$$\int_0^\infty e^{-\lambda t} H(t) \mathrm{d}t = \frac{1}{\lambda} \int_0^\infty e^{-\lambda t} \mathrm{d}H(t) = \frac{1}{\lambda} h(\lambda) = \frac{1}{\lambda} \frac{1}{1 - \varphi(\lambda)},$$

and we obtain

$$\frac{\lambda}{1 - \varphi(\lambda)} = \lambda^2 \int_0^\infty e^{-\lambda t} H(t) \mathrm{d}t = \int_0^\infty e^{-t} \lambda H\left(\frac{t}{\lambda}\right) \mathrm{d}t. \tag{4.3}$$

By means of the relation for the derivative of Laplace–Stieltjes transform

$$\lim_{\lambda \to 0+} \frac{\lambda}{1 - \varphi(\lambda)} = \lim_{\lambda \to 0+} \left(\mathbf{E}\left(\frac{1 - e^{-\lambda T_1}}{\lambda} \right) \right)^{-1} = \begin{cases} 0, & if \ \mathbf{E}(T_1) = \infty, \\ \frac{1}{\mathbf{E}(T_1)}, & if \ \mathbf{E}(T_1) < \infty. \end{cases}$$

On the other hand, in the case $0 < \lambda \leq 1$, we can give a uniform upper estimation for the integrand in Eq. (4.3):

$$e^{-t} \lambda H\left(\frac{t}{\lambda}\right) \leq e^{-t} \lambda \left(1 + \frac{t}{\lambda}\right) H(1) \leq e^{-t}(1 + t)H(1);$$

furthermore,

$$\lim_{\lambda \to 0+} \lambda H\left(\frac{t}{\lambda}\right) = t \lim_{\lambda \to 0+} \frac{H\left(\frac{t}{\lambda}\right)}{\frac{t}{\lambda}} = tc,$$

so from the Lebesgue majorated convergence theorem

$$\lim_{\lambda \to 0+} \int_0^\infty e^{-t} \lambda H\left(\frac{t}{\lambda}\right) \mathrm{d}t = \int_0^\infty e^{-t} ct \mathrm{d}t = c.$$

Summing up the previous results we obtain

$$c = \lim_{\lambda \to 0+} \frac{\lambda}{1 - \varphi(\lambda)} = \begin{cases} 0, & if \ \mathbf{E}(T_1) = \infty, \\ \frac{1}{\mathbf{E}(T_1)}, & if \ \mathbf{E}(T_1) < \infty. \end{cases}$$

4.1.1 Limit Theorems for Renewal Processes

Theorem 4.8 Let $0 < \mathbf{E}(T_1) = \mu < \infty$; then the following stochastic convergence holds:

$$\frac{N(t)}{t} \overset{P}{\to} \frac{1}{\mu}, \qquad t \to \infty.$$

Proof The proof of Theorem 4.8 is based on the relation

$$\{N(t) > k\} = \{\tau_k \leq t\}$$

from Comment 4.1. Let us estimate the probability $\mathbf{P}(|N(t)/t - 1/\mu| > \epsilon)$ for arbitrary $\epsilon > 0$. Let $n = n(t) = \lfloor t/\mu + \epsilon t \rfloor$; then

$$\mathbf{P}\left(\frac{N(t)}{t} - \frac{1}{\mu} > \epsilon\right) = \mathbf{P}\left(N(t) > \frac{t}{\mu} + \epsilon t\right) \leq \mathbf{P}(N(t) > n)$$

$$= \mathbf{P}(\tau_n \leq t) = \mathbf{P}\left(\frac{\tau_n}{n} \leq \frac{t}{\lfloor t/\mu + \epsilon t \rfloor}\right)$$

$$\leq \mathbf{P}\left(\frac{\tau_n}{n} \leq \frac{t}{t/\mu + \epsilon t - 1}\right)$$

$$= \mathbf{P}\left(\frac{\tau_n}{n} \leq \frac{1}{1/\mu + \epsilon - 1/t}\right)$$

$$\leq \mathbf{P}\left(\frac{\tau_n}{n} \leq \frac{\mu}{1 + \mu\epsilon/2}\right), \qquad if \ t \geq 2/\epsilon,$$

which by Bernoulli's law of large numbers tends to 0 for the sequence τ_n, $n = 1, 2, \ldots$, as $t \to \infty$. The probability $\mathbf{P}(N(t)/t - 1/\mu < -\epsilon)$ is estimated in a similar way. $\qquad\square$

Remark 4.8 By the strong law of large numbers $\frac{\tau_k}{k} \to \mu$, $k \to \infty$, with probability 1. Using this fact one can prove that with probability 1

$$\frac{N(t)}{t} \to \frac{1}{\mu}, \qquad t \to \infty.$$

The convergence with probability 1 remains valid for delayed renewal processes if the first time interval is finite with probability 1.

Theorem 4.9 *If* $E(T_1) = \mu > 0$, $D^2(T_1) = \sigma^2 < \infty$, *then as* $t \to \infty$

$$\lim_{t \to \infty} P\left(\frac{N(t) - t/\mu}{\sqrt{t\sigma^2/\mu^3}} \leq x\right) = \Phi(x) = \frac{1}{\sqrt{2\pi}} \int_{-\infty}^{x} e^{-u^2/2} du.$$

Proof Let x be a real number and denote

$$r(t) = \lfloor t/\mu + x\sqrt{t\sigma^2/\mu^3} \rfloor.$$

Note that $r(t) \geq 1$ if $\sqrt{t} + x\sigma/\sqrt{\mu} - \mu/\sqrt{t} \geq 0$. Since $r(t) \to \infty$ as $t \to \infty$, then from the central limit theorem it follows that for all $x \in \mathbb{R}$

$$P\left(\frac{\tau_{r(t)} - \mu r(t)}{\sigma\sqrt{r(t)}} \leq x\right) \to \Phi(x) = \frac{1}{\sqrt{2\pi}} \int_{-\infty}^{x} e^{-u^2/2} du, \quad t \to \infty. \tag{4.4}$$

Using the relation $\{N(t) \leq r(t)\} = \{\tau_{r(t)} > t\}$ we have

$$P\left(\frac{N(t) - t/\mu}{\sqrt{t\sigma^2/\mu^3}} \leq x\right)$$

$$= P\left(N(t) \leq t/\mu + x\sqrt{t\sigma^2/\mu^3}\right) = P(N(t) \leq r(t)) = P(\tau_{r(t)} > t)$$

$$= P\left(\frac{\tau_{r(t)} - \mu r(t)}{\sigma\sqrt{r(t)}} > \frac{t - \mu r(t)}{\sigma\sqrt{r(t)}}\right) = 1 - P\left(\frac{\tau_{r(t)} - \mu r(t)}{\sigma\sqrt{r(t)}} \leq \frac{t - \mu r(t)}{\sigma\sqrt{r(t)}}\right).$$

It can be easily checked that

$$\frac{t - \mu r(t)}{\sigma\sqrt{r(t)}} \to -x, \quad t \to \infty,$$

and the continuity of the standard normal distribution function implies the convergence

$$P\left(\frac{N(t) - t/\mu}{\sqrt{t\sigma^2/\mu^3}} \leq x\right) \to 1 - \Phi(-x) = \Phi(x), \quad t \to \infty.$$

The equation $1 - \Phi(-x) = \Phi(x)$ follows from the symmetry of the standard normal distribution. $\qquad\square$

The following results (without proof) concerning the mean value and variance of the renewal process $N(t)$ are a generalization of previous results and are valid for the renewal processes with delay, too.

Theorem 4.10 *If $\mu_2 = \mathbf{E}(T_1^2) < \infty$ and T_1 has a nonlattice distribution, then as $t \to \infty$ [[3], XIII-12§]*

$$\mathbf{E}(N(t)) - \frac{t}{\mu} = H(t) - \frac{t}{\mu} \to \frac{\mu_2}{2\mu^2} - 1,$$

$$\mathbf{D}^2(N(t)) = \frac{\mu_2 - \mu^2}{\mu^3} t + o(t).$$

If additionally, $\mu_3 = \mathbf{E}(T_1^3) < \infty$, then [3]

$$\mathbf{D}^2(N(t)) = \frac{\mu_2 - \mu^2}{\mu^3} t + \left(\frac{5\mu_2^2}{4\mu^4} - \frac{2\mu_3}{3\mu^3} - \frac{\mu_2}{2\mu^2} \right) + o(1).$$

4.2 Regenerative Processes

Many queueing systems can be described by means of regenerative processes. This property makes it possible to prove the limit and stability theorems in order to use the method of simulation.

Definition 4.8 Let T be a nonnegative random variable and $Z(t)$, $t \in [0, T)$ be a stochastic process. The pair $(T, Z(t))$, taking on values in the measurable space $(\mathcal{Z}, \mathcal{B})$, is called a cycle of length T.

Definition 4.9 The stochastic process $Z(t)$, $t \geq 0$, taking on values in the measurable space $(\mathcal{Z}, \mathcal{B})$, is called a **regenerative process** with moments of regeneration $\tau_0 = 0 < \tau_1 < \tau_2 < \dots$ if there exists a sequence of independent cycles $(T_k, Z_k(t))$, $k \geq 1$, such that

(1) $T_k = \tau_k - \tau_{k-1}$, $k \geq 1$;
(2) $\mathbf{P}(T_k > 0) = 1$, $\mathbf{P}(T_k < \infty) = 1$;
(3) All cycles are stochastically equivalent.
(4) $Z(t) = Z_k(t - \tau_{k-1})$, if $t \in [\tau_{k-1}, \tau_k)$, $k \geq 1$.

Definition 4.10 If the property (3) is fulfilled only starting with the second cycle (analogously to the renewal processes), then we have a **delayed regenerative process**.

Remark 4.9 τ_k, $k \geq 1$ is renewal process.

In the case of regenerative processes, an important task is to find conditions assuring the existence and possibility of determining the limit

$$\lim_{t \to \infty} \mathbf{P}(Z(t) \in B), \quad B \in \mathscr{B}.$$

It is also important to estimate the rate of convergence (especially upon examination of the stability problems of queueing systems and simulation procedures).

Let $\{Z(t), \ t \geq 0\}$ be a regenerative process taking on values in the measurable space $(\mathscr{Z}, \mathscr{B})$ with regeneration points $\tau_0 = 0 < \tau_1 < \tau_2 < \ldots, T_n = \tau_n - \tau_{n-1}, \quad n = 1, 2, \ldots$. Assume that $Z(t)$ is right continuous and there exists a limit from left. Then the cycles $\{T_n, \ \{Z(\tau_{n-1} + u) : \ 0 \leq u < T_n\}\}, \quad n = 1, 2, \ldots,$ are independent and stochastically equivalent; $\{\tau_n, \ n \geq 1\}$; and the corresponding counting process $\{N(t), \ t \geq 0\}$ is a renewal process. Let F denote the common distribution of random variables $\{T_n, \ n \geq 1\}$.

The most important application of Smith's theorem is the determination of limit values $\lim_{t \to \infty} \mathbf{E}(W(t))$ for the renewal and regenerative processes, where $W(t) = \Psi(t, N, Z)$ is the function of t, the renewal process N, and the regenerative process Z. The determination of the limit value is based on a more general theorem.

Theorem 4.11 *Let $\{V(t), \ t \geq 0\}$ be a real-valued stochastic process on the same probability space as the process $\{N(t), \ t \geq 0\}$, and for which the mean value $f(t) = \mathbf{E}(V(t))$ is bounded on each finite interval. Let*

$$g(t) = \mathbf{E}\big(V(t)\mathscr{I}_{\{T_1 > t\}}\big) + \int_0^t [\mathbf{E}(V(t)|T_1 = s) - \mathbf{E}(V(t - s))]\, \mathrm{d}F(s), \quad t \geq 0.$$

Assume that the positive and negative parts of g are directly Riemann integrable. If F is a nonlattice distribution, then

$$\lim_{t \to \infty} f(t) = \lim_{t \to \infty} \mathbf{E}(V(t)) = \frac{1}{\mu} \int_0^\infty g(x)\mathrm{d}x.$$

A similar result is valid if F is a lattice distribution.

Remark 4.10 In the theorem, the property of directly Riemann integrability was required separately for the positive and negative parts of function g. The reason is the property is defined only for nonnegative functions.

Proof It is clear that

$$f(t) = \mathbf{E}\big(V(t)\mathscr{I}_{\{T_1 > t\}}\big) + \mathbf{E}\big(V(t)\mathscr{I}_{\{T_1 \leq t\}}\big)$$

$$= \mathbf{E}\big(V(t)\mathscr{I}_{\{T_1 > t\}}\big) + \int_0^t \mathbf{E}(V(t)|T_1 = s)\mathrm{d}F(s).$$

Let us add and subtract $F * f(t)$; then we get the renewal equation

$$f = g + F * f.$$

The solution of the equation is $f(t) = g + H * g(t)$, which is because of the convergence $g(t) \to 0$, $t \to \infty$, and the elementary renewal theorem as a simple consequence of direct Riemann integrability tends to $\frac{1}{\mu} \int_0^\infty g(x)dx$ as $t \to \infty$.   □

Remark 4.11 From the proof it is clear that under the condition of Theorem 4.11 for an arbitrary process $V(t)$ there exists the representation $\mathbf{E}(V(t)) = H * g(t)$ and for the existence of the limit the direct Riemann integrability is required. This representation is interesting if $V(t)$ depends on $Z(t)$.

Special Case Let $h : \mathscr{L} \to R$ be a measurable function for which, for all t, $\mathbf{E}(|h(Z(t))|) < \infty$. $Z(t)$ is a regenerative process, and the part starting with the second cycle is independent of the first cycle of length T_1, so for arbitrary $0 < s < t$

$$\mathbf{E}((h(Z(t))|T_1 = s)) = \mathbf{E}(h(Z(t - s))).$$

Using the previous notation

$$g(t) = \mathbf{E}\big(h(Z(t))\mathscr{I}_{\{T_1 > t\}}\big).$$

Theorem 4.12 If g_+ and g_- are directly Riemann integrable, then

$$\lim_{t \to \infty} \mathbf{E}(h(Z(t))) = \mu^{-1} \int_0^\infty g(s)\, ds = \mu^{-1} \int_0^\infty \mathbf{E}\big(h(Z(s)\mathscr{I}_{\{T_1 > s\}}\big)\, ds$$

$$= \mu^{-1}\mathbf{E}\left(\int_0^{T_1} h(Z(s))\right)\, ds.$$

For arbitrary $A \in \mathscr{B}$ the following equality holds:

$$\lim_{t \to \infty} \mathbf{P}(Z(t) \in A) = \mu^{-1} \int_0^\infty \mathbf{P}(Z(s) \in A,\ T_1 > s)\, ds = \mu^{-1}\mathbf{E}\left(\int_0^{T_1} \mathscr{I}_{\{Z(s) \in A\}}\, ds\right).$$

Proof The first relation follows from the previous theorem, and for the second one it is necessary to mention that, since the trajectories of Z are right continuous and have left limits, the (integrable, bounded) function $\mathbf{P}(Z(s) \in A,\ T_1 > s)$ has a countable number of discontinuities and, consequently, it is directly Riemann integrable.   □

We give one more limit theorem (without proof) that is often useful in practice.

Theorem 4.13 Let F be a nonlattice distribution, and let at least one of the following conditions be fulfilled:

(a) $\mathbf{P}(Z(t) \in A)$ *is Riemann integrable on arbitrary finite interval, and* $\mu =$ $\int_0^{\infty} x \, dF(x) < \infty$ *holds.*

(b) *Starting with a certain integer* $n \geq 1$ *the distribution functions defined by* $F^{(1)} = F, \quad F^{(n+1)} = F^{(n)} * F$, *are absolute continuous and* $\mu = \int_0^{\infty} x \, dF(x) <$ ∞.

Then the following relation holds:

$$\lim_{t \to \infty} \mathbf{P}(Z(t) \in A) = \mu^{-1} \int_0^{\infty} \mathbf{P}(Z(s) \in A, T_1 > s) \, ds = \mu^{-1} \mathbf{E}\left(\int_0^{T_1} \mathscr{I}_{\{Z(s) \in A\}} \, ds \right).$$

Example 4.1 Let us consider the renewal process $\{N(t), \ t \geq 0\}$; the renewal moments are

$$\tau_0 = 0, \quad \tau_n = T_1 + T_2 + \ldots + T_n, \quad n \geq 1,$$

and, furthermore, $\mathbf{P}(T_k \leq x) = F(x), \ k \geq 1$, $\mu = \int_0^{\infty} x \, dF(x)$. For arbitrary $t > 0$ we define

$$
\begin{array}{lll}
\delta(t) = t - \tau_{N(t)}, & & \text{the age,} \\
\gamma(t) = \tau_{N(t)+1} - t, & & \text{the residual lifetime,} \\
\beta(t) = \gamma(t) - \delta(t) = \tau_{N(t)+1} - \tau_{N(t)}, & & \text{the total lifetime.}
\end{array}
$$

(For example, at instant t, $\delta(t)$ indicates how much time passed without a car arriving at the station, and $\gamma(t)$ indicates how long it was necessary to wait till the arrival of the next car, on the condition that the interarrival times are i.i.d. random variables with common distribution function F.)

Theorem 4.14 $\{\delta(t), \ t \geq 0\}$ *and* $\{\gamma(t), \ t \geq 0\}$ *are regenerative processes, and in the case of nonlattice distribution F,*

$$\lim_{t \to \infty} \mathbf{P}(\delta(t) \leq x) = \lim_{t \to \infty} \mathbf{P}(\gamma(t) \leq x) = \frac{1}{\mu} \int_0^x (1 - F(u)) \, du,$$

$$\lim_{t \to \infty} \mathbf{P}(\beta(t) \leq x) = \frac{1}{\mu} \int_0^x s \, dF(s).$$

Proof Both processes are obviously regenerative with common regeneration points $\tau_n, \ n \geq 1$. By our previous theorem,

$$\lim_{t\to\infty} \mathbf{P}(\delta(t) \le x) = \frac{1}{\mu} \int_0^\infty \mathbf{P}(\delta(s) \le x, T_1 > s)\,ds;$$

furthermore,

$$\mathbf{P}(\delta(s) \le x, T_1 > s) = \mathbf{P}(s \le x, T_1 > s) = \begin{cases} 1 - F(s), & \text{if } s < x, \\ 0, & \text{if } s \ge x, \end{cases}$$

so

$$\lim_{t\to\infty} \mathbf{P}(\delta(t) \le x) = \frac{1}{\mu} \int_0^x (1 - F(s))\,ds = \frac{1}{\mu} \int_0^x (1 - F(s))\,ds$$

using the identity $\mu = \int_0^\infty (1 - F(s))\,ds$ (Exercise 1.5). Similarly, for the process $\{\gamma(t),\ t \ge 0\}$ we obtain

$$\lim_{t\to\infty} \mathbf{P}(\gamma(t) \le x)$$

$$= \frac{1}{\mu} \int_0^\infty \mathbf{P}(\gamma(s) \le x, T_1 > s)\,ds = \frac{1}{\mu} \int_0^\infty \mathbf{P}(T_1 - s \le x, T_1 > s)\,ds$$

$$= \frac{1}{\mu} \int_0^\infty \mathbf{P}(s \le T_1 < s + x)\,ds = \frac{1}{\mu} \int_0^\infty (F(s+x) - F(s))\,ds$$

$$= -\frac{1}{\mu} \left(\int_x^\infty (1 - F(s))\,ds - \int_0^\infty (1 - F(s))\,ds \right) = \frac{1}{\mu} \int_0^x (1 - F(s))\,ds.$$

The statement for $\{\gamma(t),\ t \ge 0\}$ can be obtained analogously.   □

Similar to the renewal processes, the law of large numbers and the central limit theorem can be proved for the regenerative processes, too. Here we will not deal with these questions.

4.2.1 Convergence of Long-Run Averages of Regenerative Processes

Here we will give some statements which describe the limiting behavior of long-run averages $\frac{1}{t}\int_0^t h(Z(s))\,ds$ of a measurable function of a regenerative process

$Z(t)$, $t \geq 0$ as $t \to \infty$. We will apply these results later in the discussion of Little's Law.

Keeping the earlier notation, let $t_0 = 0 < t_1 < t_2 < \ldots$ be the renewal time points of the processes $Z(t)$. Denote by $N(t) = \max\{n : t_n \leq t\}$ the numbers of renewal points in time interval $[0, t]$ and define the integrals $H_0 = \int_{t_0}^{t_1} h(Z(s))ds$, $H_k = \int_{t_{k-1}}^{t_k} h(Z(s))ds$, $1 \leq k \leq N(t)$ and $H_{N(t)+1} = \int_{t_{N(t)}}^{t} h(Z(s))ds$. By the definition of regenerative processes it follows that the partial processes of $Z(t)$ given on the regenerative cycles $[t_{k-1}, t_k)$, $k \geq 1$ are independent and stochastically equivalent, but it may be possible, with the exception of the first part on $[0, t_1)$ if the process is delayed. It is clear that the process $h(Z(t))$, $t \geq 0$ also is a regenerative process with the same renewal points t_i, $i = 0, 1, 2, \ldots$. From this we can conclude that random variables H_k, $1 \leq k \leq N(t)$ are not only independent but are identically distributed. Moreover, the sequence $\{t_i, i \geq 0\}$ constitutes a renewal process, then by Remark 4.8 the convergence $\frac{N(t)}{t} \to \frac{1}{\mu}$ follows with probability 1 if $0 < \mu = \mathbf{E}(t_2 - t_1) < \infty$. This fact allows us to prove a strong law of large numbers for the long-run averages

$$\frac{1}{t} \int_0^t h(Z(s))ds = \frac{N(t)}{t} \frac{1}{N(t)} \left(H_0 + \sum_{k=1}^{N(t)} H_k + H_{N(t)+1} \right).$$

The following theorem holds (e.g., [8]).

Theorem 4.15 *Suppose that* $0 < \mathbf{E}(t_2 - t_1) < \infty$, $\mathbf{E}\left(\int_{t_1}^{t_2} |h(Z(s))| \, ds \right) < \infty$ *and in addition* $\mathbf{E}\left(\int_{t_0}^{t_1} |h(Z(s))| \, ds \right) < \infty$ *if the regenerative process* $Z(t)$ *is delayed. Then with probability* 1 *it is true that*

$$\lim_{t \to \infty} \frac{1}{t} \int_0^t h(Z(s))ds = \frac{\mathbf{E}\left(\int_{t_1}^{t_2} h(Z(s))ds \right)}{\mathbf{E}(t_2 - t_1)}$$

and

$$\lim_{t \to \infty} \frac{1}{t} \int_0^t \mathbf{E}(h(Z(s)))ds = \frac{\mathbf{E}\left(\int_{t_1}^{t_2} h(Z(s))ds \right)}{\mathbf{E}(t_2 - t_1)}.$$

Remark 4.12 Let us choose h to be indicator function as $h(Z(s)) = \mathscr{I}_{\{h(Z(s)) \in A\}}$, where A is a Borel set, then $\frac{1}{t} \int_0^t \mathscr{I}_{\{h(Z(s)) \in A\}}ds$ means the fraction of time when $h(Z(s))$ is in the set A. In this case

$$\lim_{t \to \infty} \frac{1}{t} \int_0^t \mathscr{I}_{\{h(Z(s)) \in A\}}ds = \lim_{t \to \infty} \frac{1}{t} \int_0^t \mathbf{P}\left(h(Z(s)) \in A \right)ds.$$

holds with probability 1.

4.2.2 Estimation of Convergence Rate for Regenerative Processes

For a wide class of regenerative processes (e.g., stochastic processes describing queueing systems) one can estimate the rate of convergence of distributions of certain parameters to a stationary distribution by means of the so-called coupling method (see [7]).

Lemma 4.6 (Coupling Lemma) *For the arbitrary random variables X and Y and arbitrary Borel set A of the real line the following statements hold:*

(i) $|\mathbf{P}(X \in A) - \mathbf{P}(Y \in A)| \leq \mathbf{P}(X \neq Y)$.
(ii) If $X = X_1 + \ldots + X_n$ and $Y = Y_1 + \ldots + Y_n$, then $|\mathbf{P}(X \in A) - \mathbf{P}(Y \in A)| \leq$
$$\sum_{k=1}^{n} \mathbf{P}(X_k \neq Y_k).$$

Proof If $\mathbf{P}(X \in A) = \mathbf{P}(Y \in A)$, then (i) is obviously true.

Suppose that $\mathbf{P}(X \in A) > \mathbf{P}(Y \in A)$ (if one changes notation, then this can always be done if the two probabilities differ). Then

$$|\mathbf{P}(X \in A) - \mathbf{P}(Y \in A)| = \mathbf{P}(X \in A) - \mathbf{P}(Y \in A)$$
$$\leq \mathbf{P}(X \in A) - \mathbf{P}(Y \in A, X \in A) = \mathbf{P}(X \in A, Y \in A^c) \leq \mathbf{P}(X \neq Y).$$

Proof of relation (ii). Since $\{X \neq Y\} \subset \bigcup_{k=1}^{n} \{X_k \neq Y_k\}$, we have

$$\mathbf{P}(X \neq Y) \leq \mathbf{P}\left(\bigcup_{k=1}^{n} \{X_k \neq Y_k\}\right) \leq \sum_{k=1}^{n} \mathbf{P}(X_k \neq Y_k).$$

□

Application of Coupling Lemma Let $Z = \{Z(j), \ j \geq 1\}$ be the discrete-time, real-valued regenerative process under consideration. Assume that there exists the weak stationary limit of the process $\tilde{Z} = \{Z(j + n), \ j \geq 1\}$ as $n \to \infty$ (its finite-dimensional distributions weakly converge to the finite-dimensional distributions of a stationary process), which is also regenerative, and let $Y = \{Y(j), \ j \geq 1\}$ be its realization, not necessarily different from Z on the same probability space. Let τ denote the first instant when the processes Z and Y are regenerated at the same time (in many concrete cases the distribution of τ can be easily estimated). Then the convergence rate of the distribution of $Z(j)$ can be estimated by means of the distribution of τ as follows: if after the regeneration point τ the process Z is replaced by the next part of process Y following the common regeneration point τ, then the finite-dimensional distributions of process Z do not change. It is clear that $\{\tau < j\} \subseteq \{Z(j) = Y(j)\}$, i.e., $\{Z(j) \neq Y(j)\} \subseteq \{\tau \geq j\}$, from which, using the coupling lemma for the arbitrary Borel set A of the real line, the estimation

$$|\mathbf{P}(Z(j) \in A) - \mathbf{P}(Y(j) \in A)| \leq \mathbf{P}(Z(j) \neq Y(j)) \leq \mathbf{P}(\tau \geq j)$$

holds.

4.3   Analysis Methods Based on Markov Property

Definition 4.11 A discrete-state, continuous-time stochastic process, $X(t)$, possesses the **Markov property** at time t_n if for all $n, m \geq 1, 0 \leq t_0 < t_1 < \ldots < t_n < t_{n+1} < \ldots < t_{n+m}$ and $x_0, x_1, \ldots, x_n, x_{n+1}, \ldots, x_{n+m} \in S$ we have

$$\begin{aligned}
\mathbf{P}(X(t_{n+m}) &= x_{n+m}, \ldots, X(t_{n+1}) = x_{n+1} | X(t_n) = x_n, \ldots, X(t_0) = x_0) \\
&= \mathbf{P}(X(t_{n+m}) = x_{n+m}, \ldots, X(t_{n+1}) = x_{n+1} | X(t_n) = x_n).
\end{aligned} \tag{4.5}$$

in this case t_n is referred to as regenerative point.

A commonly applied interpretation of the Markov property is as follows. Assuming that current time is t_n (present) which is a regenerative point, and we know the current state of the process $X(t_n)$, then the future of the stochastic process $X(t)$ for $t_n \leq t$ is independent of the past history of the process $X(t)$ for $0 \leq t < t_n$, and it only depends on the current state of the process $X(t_n)$. That is, if one knows the present state, the future is independent of the past.

In the case of discrete-time processes, it is enough to check if the one-step state transitions are independent of the past, i.e., it is enough to check the condition for $m = 1$.

Usually, we restrict our attention to stochastic processes with nonnegative parameters (positive half of the time axes), and in these cases we assume that $t = 0$ is a regenerative point.

4.3.1   Time-Homogeneous Behavior

Definition 4.12 The stochastic process $X(t)$ is *time-homogeneous* if the stochastic behavior of $X(t)$ is invariant for time shifting, that is, the stochastic behavior of $X(t)$ and $X'(t) = X(t + s)$ are identical in distribution $X(t) \overset{d}{=} X'(t)$.

Corollary 4.2 *If the time-homogeneous stochastic process* $X(t)$ *possesses the Markov property at time T and $X(T) = i$, then $X(t) \overset{d}{=} X(t - T)$ if $X(0) = i$.*

The corollary states that starting from two different Markov points with the same state results in stochastically identical processes.

4.4   Analysis of Continuous-Time Markov Chains

Definition 4.13 The discrete-state continuous-time stochastic process $X(t)$ is a *continuous-time Markov chain* (CTMC) if it possesses the Markov property for all $t \geq 0$.

Based on this definition and assuming time-homogeneous behavior we obtain the following properties.

Corollary 4.3 *An arbitrary finite-dimensional joint distribution of a CTMC is composed as the product of transition probabilities multiplied by an initial probability.*

Corollary 4.4 *For the time points $t < u < v$ the following Chapman–Kolmogorov equation holds:*

$$\hat{p}_{ij}(t, v) = \sum_{l \in S} \hat{p}_{il}(t, u)\hat{p}_{lj}(u, v); \quad \hat{\boldsymbol{\Pi}}(t, v) = \hat{\boldsymbol{\Pi}}(t, u)\hat{\boldsymbol{\Pi}}(u, v) \tag{4.6}$$

where $\hat{p}_{ij}(t, u) = \mathbf{P}(X(u) = j \mid X(t) = i)$, for all $i,\ j \in S, 0 \leq t \leq u, \hat{\boldsymbol{\Pi}}(t, u) = \left[\hat{p}_{ij}(t, u)\right]$. In the case of time-homogeneous processes the time shifts $u - t = \tau_1$ and $v - u = \tau_2$ play a role

$$p_{ij}(\tau_1 + \tau_2) = \sum_{l \in S} p_{il}(\tau_1)p_{lj}(\tau_2); \quad \boldsymbol{\Pi}(\tau_1 + \tau_2) = \boldsymbol{\Pi}(\tau_1)\boldsymbol{\Pi}(\tau_2), \tag{4.7}$$

where $\boldsymbol{\Pi}(\tau) = \left[p_{ij}(\tau)\right],\ p_{ij}(\tau) = \mathbf{P}(X(\tau) = j \mid X(0) = i)$, for all $i,\ j \in S$, $0 \leq \tau$.

Definition 4.14 The stochastic evolution of a CTMC is commonly characterized by an *infinitesimal generator* matrix (commonly denoted by $\boldsymbol{Q}$) which can be obtained from the derivative of the state-transition probabilities as follows:

$$\frac{d}{dt}\boldsymbol{\Pi}(t) = \lim_{\delta \to 0} \frac{\boldsymbol{\Pi}(t + \delta) - \boldsymbol{\Pi}(t)}{\delta} = \boldsymbol{\Pi}(t) \underbrace{\lim_{\delta \to 0} \frac{\boldsymbol{\Pi}(\delta) - \boldsymbol{I}}{\delta}}_{\boldsymbol{Q}} = \boldsymbol{\Pi}(t)\boldsymbol{Q}. \tag{4.8}$$

Corollary 4.5 *The sojourn time of a CTMC in a given state i is exponentially distributed with the parameter $q_i = -q_{ii}$. The probability that after state i the next visited state will be state j is q_{ij}/q_i, and it is independent of the sojourn time in state i.*

Remark 4.13 Based on Corollary 4.5 and the properties of the exponential distribution, the state transitions of a CTMC can also be interpreted in the following way. When the CTMC moves to state i, several exponentially distributed activities start, exactly one for each nonzero transition rate. The time of the activity associated with the state transition from state i to state j is exponentially distributed with the parameter q_{ij}. The CTMC leaves state i and moves to the next state when the first one of these activities completes. The next visited state is the state whose associated activity finishes first.

Corollary 4.6 (Short-Term Behavior of CTMCs)   *During a short time period Δ, a behavior of a CTMC is characterized by the following transition probabilities.*

(a) $\mathbf{P}(X(t + \Delta) = i | X(t) = i) = 1 - q_i \Delta + o(\Delta)$
(b) $\mathbf{P}(X(t + \Delta) = j | X(t) = i) = q_{ij} \Delta + o(\Delta)$ *for $i \neq j$*
(c) $\mathbf{P}(X(t + \Delta) = j, X(u) = k | X(t) = i) = o(\Delta)$ *for $i \neq k$, $j \neq k$ and $t < u < t + \Delta$.*

where $o(x)$ denotes the set of functions with the property $\lim_{x \to 0} o(x)/x = 0$.

According to the corollary, the two main things can happen with significant probability during a short time period:

- The CTMC stays in the initial state during the whole period [(a)].
- It moves from state i to j [(b)].

The event that more than one state transitions happen during a short time period [(c)] has a negligible probability as $\Delta \to 0$.

Corollaries 4.5 and 4.6 allow different analytical approaches for the description of the transient behavior of CTMCs.

4.4.1   Analysis Based on Short-Term Behavior

Let $X(t)$ be a CTMC with state space S, and let us consider the change in state probability $P_i(t + \Delta) = \mathbf{P}(X(t + \Delta) = i)$ ($i \in S$) considering the possible events during the interval $(t, t + \Delta)$. The following cases must be considered:

- There is no state transition during the interval $(t, t + \Delta)$. In this case $P_i(t) \to P_i(t + \Delta)$ and the probability of this event is $1 - q_i \Delta + o(\Delta)$.
- There is one state transition during the $(t, t + \Delta)$ interval from state k to state i. In this case $P_k(t) \to P_i(t + \Delta)$, and the probability of this event is $q_{ki} \Delta + o(\Delta)$.
- The process stays in state i at time $t + \Delta$ such that there is more than one state transition during the interval $(t, t + \Delta)$. The probability of this event is $o(\Delta)$.

Considering these cases we can compute $P_i(t + \Delta)$ from $P_k(t)$, $k \in S$, as follows:

$$P_i(t + \Delta) = (1 - q_i \Delta + o(\Delta)) P_i(t) + \sum_{k \in S, k \neq i} (q_{ki} \Delta + o(\Delta)) P_k(t) + o(\Delta)$$

$$= (1 - q_i \Delta) P_i(t) + \sum_{k \in S, k \neq i} (q_{ki} \Delta) P_k(t) + o(\Delta),$$

from which

$$\frac{P_i(t+\Delta)-P_i(t)}{\Delta} = -q_i P_i(t) + \sum_{k\in S, k\neq i} q_{ki} P_k(t) + \frac{o(\Delta)}{\Delta} = \sum_{k\in S} q_{ki} P_k(t) + \frac{o(\Delta)}{\Delta}.$$

Finally, setting the limit $\Delta \to 0$ we obtain that

$$\frac{dP_i(t)}{dt} = \sum_{k\in S} q_{ki} P_k(t).$$

Introducing the row vector of state probabilities $P(t) = \{P_i(t)\}, i \in S$, we obtain the vector-matrix form of the previous equation:

$$\frac{d}{dt} P(t) = P(t) Q. \tag{4.9}$$

A differential equation describes the evolution of a transient state probability vector. To define the state probabilities, we additionally need to have an initial condition. In practical applications, the initial condition is most often the state probability distribution at time 0, i.e., $P(0)$. The solution of (4.9) with initial condition $P(0)$ is [5]:

$$P(t) = P(0)e^{Qt} = P(0) \sum_{n=0}^{\infty} \frac{Q^n t^n}{n!}$$

Transform Domain Description The Laplace transform of the two sides of Eq. (4.9) gives

$$s P^*(s) - P(0) = P^*(s) Q,$$

from which we can express $P^*(s)$ in the following form:

$$P^*(s) = P(0)[sI - Q]^{-1}.$$

Comparing the time and transform domain expressions we have that e^{Qt} and $[sI - Q]^{-1}$ are Laplace transform pairs of each other.

Stationary Behavior If $\lim_{t\to\infty} P_i(t)$ exists, then we say that $\lim_{t\to\infty} P_i(t) = P_i$ is the stationary probability of state i. In this case, $\lim_{t\to\infty} dP_i(t)/dt = 0$, and the stationary probability satisfies the system of linear equations $\sum_{k\in S} q_{ki} P_k(t) = 0$ for all $k \in S$.

4.4.2   Analysis Based on First State Transition

Let $X(t)$ be a CTMC with state space S, and let $T_1, T_2, T_3, \ldots$ denote the time of the first, second, etc. state transitions of the CTMC. We assume that $T_0 = 0$, and $\tau_1, \tau_2, \tau_3, \ldots$ are the sojourn times spent in the consecutively visited states ($\tau_i = T_i - T_{i-1}$). We compute the state-transition probability $\pi_{ij}(t) = \mathbf{P}(X(t) = j \mid X(0) = i)$ assuming that $T_1 = h$, i.e., we are interested in

$$\pi_{ij}(t \mid T_1 = h) = \mathbf{P}(X(t) = j \mid X(0) = i, T_1 = h).$$

We have

$$\pi_{ij}(t \mid T_1 = h) = \begin{cases} \delta_{ij} & h \geq t \\ \displaystyle\sum_{k \in S, k \neq i} \frac{q_{ik}}{-q_{ii}} \pi_{kj}(t - h) & h < t, \end{cases} \tag{4.10}$$

where δ_{ij} is the Kronecker delta ($\delta_{ij} = 1$ if $i = j$ and $\delta_{ij} = 0$ if $i \neq j$), and $\dfrac{q_{ik}}{-q_{ii}}$ is the probability that after visiting state i the Markov chain moves to state k. In the case of general stochastic processes this probability might depend on the sojourn time in state i, but in case of CTMCs, it is independent.

Equation (4.10) contains two cases:

- If the time point of interest, t, is before the first state transition of the CTMC, $h \geq t$, then the conditional state-transition probability is either 1 (if the initial state and the final state are identical $i = j$) or 0 (if $i \neq j$).
- If the time point of interest, t, is after the first state transition of the CTMC, $T_1 < t$, then we can analyze the evolution of the process from T_1 to t using the fact that the process possesses the Markov property at time T_1. In this case we need to consider all possible states which might be visited at time $T_1, k \in S, k \neq i$, with the associated probability $\dfrac{q_{ik}}{-q_{ii}}$. The state-transition probabilities from T_1 to t are identical with the state-transition probabilities of the original process from 0 to $T_1 - t$ assuming that the original process starts from state k.

The distribution of T_1 is known. It is exponentially distributed with parameter $-q_{ii}$. Its cumulated and probability density functions are $F_{T_1}(x) = 1 - e^{q_{ii}x}$ and $f_{T_1}(x) = -q_{ii}e^{q_{ii}x}$, respectively. With that we can apply the total probability theorem to compute the (unconditional) state-transition probability $\pi_{ij}(t)$.

$$\pi_{ij}(t) = \int_{h=0}^{\infty} \pi_{ij}(t|T_1 = h) \, f_{T_1}(h) \, dh$$

$$= \int_{h=t}^{\infty} \delta_{ij} \, f_{T_1}(h) \, dh + \int_{h=0}^{t} \sum_{k \in S, k \neq i} \frac{q_{ik}}{-q_{ii}} \, \pi_{kj}(t - h) \, f_{T_1}(h) \, dh$$

$$= \delta_{ij} \, (1 - F_{T_1}(t)) + \int_{h=0}^{t} \sum_{k \in S, k \neq i} \frac{q_{ik}}{-q_{ii}} \, \pi_{kj}(t - h) \, f_{T_1}(h) \, dh \qquad (4.11)$$

$$= \delta_{ij} \, e^{q_{ii}t} + \sum_{k \in S, k \neq i} q_{ik} \int_{h=0}^{t} \pi_{kj}(t - h) \, e^{q_{ii}h} \, dh.$$

The obtained integral equation is commonly referred to as a Volterra integral equation. Its only unknown is the state-transition probability function $\pi_{ij}(t)$. The numerical methods developed for the numerical analysis of Volterra integral equations can be used to compute the state-transition probabilities of the CTMC.

Relation of Analysis Methods We can rewrite (4.11) in the following form

$$\pi_{ij}(t) = \delta_{ij} \, e^{q_{ii}t} + \sum_{k \in S, k \neq i} q_{ik} \int_{h=0}^{t} \pi_{kj}(t - h) \, e^{q_{ii}h} \, dh$$

$$= \delta_{ij} \, e^{q_{ii}t} + \sum_{k \in S, k \neq i} q_{ik} \int_{h=0}^{t} \pi_{kj}(h) \, e^{q_{ii}(t-h)} \, dh \qquad (4.12)$$

$$= \delta_{ij} \, e^{q_{ii}t} + \sum_{k \in S, k \neq i} q_{ik} \, e^{q_{ii}t} \int_{h=0}^{t} \pi_{kj}(h) \, e^{-q_{ii}h} \, dh \ .$$

The derivation of the two sides of (4.12) according to t is as follows

$$\pi'_{ij}(t) = \delta_{ij} q_{ii} e^{q_{ii}t} + \sum_{k \in S, k \neq i} q_{ik} \left(q_{ii} e^{q_{ii}t} \int_{h=0}^{t} \pi_{kj}(h) e^{-q_{ii}h} dh + e^{q_{ii}t} \pi_{kj}(t) e^{-q_{ii}t} \right)$$

$$= \sum_{k \in S, k \neq i} q_{ik} \pi_{kj}(t) + q_{ii} \underbrace{\left(\delta_{ij} e^{q_{ii}t} + \sum_{k \in S, k \neq i} q_{ik} e^{q_{ii}t} \int_{h=0}^{t} \pi_{kj}(h) e^{-q_{ii}h} dh \right)}_{\pi_{ij}(t)}$$

$$= \sum_{k \in S} q_{ik} \pi_{kj}(t),$$

where we used (4.11) for the substitution of the integral expression. The obtained differential equation is similar to the one provided by the analysis of the short-term behavior.

Transform Domain Description To relate the two transient descriptions of the CTMC, one with differential equation and the one with integral equation, we transform these descriptions into a Laplace transform domain. It is easy to take the Laplace transform from the last line of Eq. (4.11) because the second term of the right-hand side is a convolution integral. That is

$$\pi_{ij}^*(s) = \delta_{ij} \frac{1}{s - q_{ii}} + \sum_{k \in S, k \neq i} q_{ik} \, \pi_{kj}^*(s) \, \frac{1}{s - q_{ii}}.$$

Multiplying by the denominator and using that $-q_{ii} = \sum_{k \in S, k \neq i} q_{ik}$ we obtain

$$s \, \pi_{ij}^*(s) = \delta_{ij} + \sum_{k \in S} q_{ik} \, \pi_{kj}^*(s),$$

which can be written in the matrix form

$$s \, \boldsymbol{\Pi}^*(s) = \boldsymbol{I} + \boldsymbol{Q}\boldsymbol{\Pi}^*(s).$$

Finally, we have

$$\boldsymbol{\Pi}^*(s) = [s\boldsymbol{I} - \boldsymbol{Q}]^{-1},$$

which is identical with the Laplace transform expression obtained from the differential equation.

Embedded Markov Chain at State Transitions Let $X_i \in S, i = 0, 1, \ldots$ denote the ith visited state of the Markov chain $X(t)$, which is the state of the Markov chain in the interval (T_i, T_{i+1}) (see Fig. 4.2). The $X_0, X_1, \ldots$ series of random variables is a discrete-time Markov chain (DTMC) due to the Markov property of $X(t)$. This DTMC is commonly referred to as a Markov chain embedded at the state transitions or simply an *embedded Markov chain* (EMC). The state transition probability matrix of the EMC is

$$\Pi_{ij} = \begin{cases} \dfrac{q_{ij}}{-q_{ii}} & i \neq j \\ 0 & i = j. \end{cases}$$

Stationary Analysis Based on the EMC The stationary distribution of the EMC $\hat{P}$ (which is the solution of $\hat{P} = \hat{P}\Pi, \sum_i \hat{P}_i = 1$) defines the relative frequency of the visits to the state of the Markov chain. The higher the stationary probability is, the more frequently the state is visited. The stationary behavior of the CTMC $X(t)$ is characterized by two main factors: how often the state is visited (represented by $\hat{P}_i$) and how long a visit lasts. If state i is visited twice as frequently as state j but the mean time of a visit to state i is half the mean time of a visit to j, then the stationary probabilities of state i and j are identical. This intuitive behavior is summarized in the following general rule of renewal theory [6]:

$$P_i = \frac{\hat{P}_i \hat{\tau}_i}{\sum_j \hat{P}_j \hat{\tau}_j},$$

where $\hat{\tau}_j$ is the mean time spent in state j, which is known from the diagonal element of the infinitesimal generator, $\hat{\tau}_j = -1/q_{jj}$.

Discrete Event Simulation of CTMCs There are at least two possible approaches.

- When the CTMC is in state i, first draw an exponentially distributed random sample with parameter $-q_{ii}$ for the sojourn time in state i, then draw a discrete random sample for deciding the next visited state with distribution Π_{ij}, $j \in S$.
- When the CTMC is in state i, draw an exponentially distributed random sample with parameter q_{ij}, say τ_{ij}, for all positive transition rates of row i of the infinitesimal generator matrix. Find the minimum of these samples, $\min_j \tau_{ij}$. The sojourn time in state i is this minimum, and the next state is the one whose associated random sample is minimal.

4.5 Semi-Markov Process

Definition 4.15 The discrete state, continuous-time random process $X(t)$ is a semi-Markov process if it is time homogeneous and it possesses the Markov property at the state transition instants (Fig. 4.2).

The name semi-Markov process comes from the fact that such processes do not always possess the Markov property (during its sojourn in a state), but there are particular instances (state-transition instants) when they do. During a sojourn in state i, both the remaining time in that state and the next visited state depend on the elapsed time since the process is in state i.

Corollary 4.7 *The sojourn time in state i can be any general real-valued positive random variable. During a sojourn in state i, both the remaining time in both that*

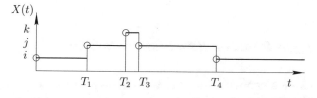

Fig. 4.2 Semi-Markov process, which possesses the Markov property at the indicated time points

*state and the next visited state depend on the elapsed time since the process entered
state* i.

Example 4.2 A two state (up/down) system fails at a rate λ (the up time of the
system is exponentially distributed with parameter λ) and gets repaired at a rate
μ. To avoid long down periods, the repair process is stopped and a replacement
process is initialized after a deterministic time limit d. The time of the replacement
is a random variable with a distribution $G(t)$. Define a system model and check if it
is a semi-Markov process.

Because a CTMC always possesses the Markov property, it follows that the
sojourn time in a state is exponentially distributed and that the distribution of the
next state is independent of the sojourn time. For example, considering the first
state transition and the sojourn time in the first state we have

$$\mathbf{P}(X_1 = j, T_1 = c | X_0 = i) = \mathbf{P}(X_1 = j | X_0 = i)\mathbf{P}(T_1 = c | X_0 = i).$$

This property does not hold for semi-Markov processes in general. The most
important consequences of the definition of semi-Markov processes are the following ones. The sojourn time in a state can have any positive distribution, and the
distribution of the next state and the time spent in a state are not independent in
general. Consequently, to define a semi-Markov process, this joint distribution must
be given. This is usually done by defining the kernel matrix of a process, whose i, j
element is

$$Q_{ij}(t) = \mathbf{P}(X(T_{i+1}) = j, \tau_{i+1} \leq t \mid X(T_i) = i).$$

Utilizing the time homogeneity of the process we further have for T_i that

$$Q_{ij}(t) = \mathbf{P}(X(T_{i+1}) = j, \tau_{i+1} \leq t \mid X(T_i) = i) = \mathbf{P}(X(T_1) = j, T_1 \leq t \mid X(0) = i).$$

The analysis of semi-Markov processes is based on the results of renewal theory
and the analysis of an EMC (of state-transition instances). The definition of a semi-
Markov process requires knowledge of the kernel matrix $Q(t) = \{Q_{ij}(t)\}$ (for
$t \geq 0$) and an initial distribution. It is commonly assumed that $X(t)$ possesses the
Markov property at time $t = 0$.

4.5.1   Analysis Based on First State Transitions

Let $X(t) \in S$ be a continuous-time semi-Markov process, $T_1, T_2, T_3, \ldots$ be the
state-transition instances, and $\tau_1, \tau_2, \tau_3, \ldots$ be the consecutive sojourn times ($\tau_i =
T_i - T_{i-1}$). We assume $T_0 = 0$. We intend to compute the state-transition probability
$\pi_{ij}(t) = \mathbf{P}(X(t) = j \mid X(0) = i)$ assuming that the sojourn in the first state finishes
at time h ($T_1 = h$), that is

$$\pi_{ij}(t|T_1 = h) = \mathbf{P}(X(t) = j \mid X(0) = i, T_1 = h).$$

In this case

$$\pi_{ij}(t|T_1 = h) = \begin{cases} \delta_{ij} & h \geq t \\ \displaystyle\sum_{k \in S} \mathbf{P}(X(T_1) = k \mid X(0) = i, T_1 = h) \, V_{kj}(t - h) & h < t, \end{cases}$$

(4.13)

where $\mathbf{P}(X(T_1) = j \mid X(0) = i, T_1 = h)$ is the probability that the process will start from state i at time 0 and it is in state j right after the state transition at time T_1 assuming $T_1 = h$. In contrast with CTMCs, this probability depends on the sojourn time in state i:

$$\mathbf{P}(X(T_1) = j \mid X(0) = i, T_1 = h)$$

$$= \lim_{\Delta \to 0} \frac{\mathbf{P}(X(T_1) = j, h < T_1 \leq h + \Delta \mid X(0) = i)}{\mathbf{P}(h < T_1 \leq h + \Delta \mid X(0) = i)}$$

(4.14)

$$= \lim_{\Delta \to 0} \frac{Q_{ij}(h + \Delta) - Q_{ij}(h)}{Q_i(h + \Delta) - Q_i(h)} = \frac{dQ_{ij}(h)}{dQ_i(h)},$$

where $Q_i(h)$ denotes the distribution of time spent in state i,

$$Q_i(t) = \mathbf{P}(T_1 \leq t \mid Z(0) = i) = \sum_j \mathbf{P}(Z(T_1) = j, T_1 \leq t \mid Z(0) = i) = \sum_j Q_{ij}(t).$$

It is commonly assumed that state transitions are real, which means that after staying in state i a state transition moves the process to a different state. It means that $Q_{ii}(t) = 0$, $\forall i \in S$. It is also possible to consider virtual state transitions from state i to state i, but this does not expand the set of semi-Markov processes and we do not consider it here. Note that the meaning of a diagonal element of a semi-Markov kernel matrix is completely different from that of a diagonal element of an infinitesimal generator of a CTMC. One of the technical consequences of this difference is the fact that we do not need to exclude the diagonal element from the summations over the set of states.

Two cases considered in Eq. (4.13):

- If the time point of interest, t, is before the first state transition of the process ($h \geq t$), then the conditional state-transition probability is either 0 or 1 depending on the initial and the final states. If the initial state i is identical with the final state j, then the transition probability is 1 because there is no state transition up to time t, otherwise it is 0.
- If the time point of interest, t, is after the first state transition of the process ($h < t$), then we need to evaluate the distribution of the next state k, assuming that the state transition occurs at time h, and after that the state transition probability

from the new state k to the final state j during time $t - h$, using the Markov property of the process at time h. The probability that the process moves to state k assuming it occurs at time h is $\dfrac{\mathrm{d}Q_{ij}(h)}{\mathrm{d}Q_i(h)}$, and the probability of its moving from state k to state j during an interval of length $t - h$ is $\pi_{ij}(t - h)$.

The distribution of the condition of Eq. (4.13) is known. The distribution of the sojourn time in state i is $Q_i(h)$. Using the law of total probability we obtain

$$
\begin{aligned}
\pi_{ij}(t) &= \int_{h=0}^{\infty} \pi_{ij}(t | T_1 = h) \, \mathrm{d}F_{T_1}(h) \\
&= \int_{h=t}^{\infty} \delta_{ij} \, \mathrm{d}Q_i(t) + \int_{h=0}^{t} \sum_{k \in S} \frac{\mathrm{d}Q_{ik}(h)}{\mathrm{d}Q_i(h)} \pi_{kj}(t - h) \, \mathrm{d}Q_i(h) \qquad (4.15) \\
&= \delta_{ij} (1 - Q_i(t)) + \int_{h=0}^{t} \sum_{k \in S} \pi_{kj}(t - h) \, \mathrm{d}Q_{ik}(h).
\end{aligned}
$$

Similar to the case of CTMCs, the analysis based on the first state transition resulted in a Volterra integral equation also in case of semi-Markov processes. The transient behavior of semi-Markov processes can be computed using the same numerical procedures.

Transform Domain Description We take the Laplace transform of both sides of the Volterra integral equation (4.15). The only nontrivial term is a convolution integral on the right-hand side:

$$
\pi_{ij}^*(s) = \delta_{ij} (1 - Q_i^*(s)) + \sum_{k \in S} q_{ik}^*(s) \, \pi_{kj}^*(s),
$$

where $q_{ik}(t) = \mathrm{d}Q_{ik}(t)/\mathrm{d}t$ and the transform domain functions are defined as $f^*(s) = \int_0^{\infty} f(t)e^{-st}\mathrm{d}t$.

Introducing the diagonal matrix $\boldsymbol{D}^*(s)$ composed of the elements $1 - Q_i^*(s)$, that is, $\boldsymbol{D}^*(s) = \mathrm{diag}\langle 1 - Q_i^*(s) \rangle$, the Laplace transform of the state transition probabilities are obtained in matrix form:

$$
\boldsymbol{\Pi}^*(s) = \boldsymbol{D}^*(s) + \boldsymbol{q}^*(s)\boldsymbol{\Pi}^*(s),
$$

from which

$$
\boldsymbol{\Pi}^*(s) = [\boldsymbol{I} - \boldsymbol{q}^*(s)]^{-1}\boldsymbol{D}^*(s).
$$

Stationary Behavior The stationary analysis of a semi-Markov process is very similar to the stationary analysis of a CTMC based on an EMC. Let the transition probability matrix of the EMC be $\boldsymbol{\Pi}$. It is obtained from the kernel matrix through the following relation:

$$\Pi_{ij} = \mathbf{P}(Z(T_1) = j \mid Z(0) = i)$$
$$= \lim_{t \to \infty} \mathbf{P}(Z(T_1) = j, T_1 \leq t \mid Z(0) = i) = \lim_{t \to \infty} Q_{ij}(t) = \lim_{s \to 0} q_{ij}^*(s).$$

The stationary distribution of the EMC $\hat{P}$ is the solution of the linear system $\hat{P} = \hat{P}\boldsymbol{\Pi}$, $\sum_i \hat{P}_i = 1$. The stationary distribution of the semi-Markov process is

$$P_i = \frac{\hat{P}_i \hat{\tau}_i}{\sum_j \hat{P}_j \hat{\tau}_j}, \qquad (4.16)$$

where $\hat{\tau}_i$ is the mean time spent in state i. It can be computed from the kernel matrix using $\hat{\tau}_i = \int_0^\infty (1 - Q_i(t)) \mathrm{d}t$.

Discrete Event Simulation of Semi-Markov Processes The initial distribution and the $Q_{ij}(t)$ kernel completely define the stochastic behavior of a semi-Markov process. As a consequence, it is possible to simulate the process behavior based on them.

The key step of the simulation is to draw dependent samples for the sojourn time and the next visited state. It can be done based on the marginal distribution of one of the two random variables and a conditional distribution of the other one. Depending on which random variable is sampled first, there are two ways to simulate a semi-Markov process:

- When the process is in state i, first draw a $Q_i(t)$ distributed sample for the sojourn time, denoted by τ, then draw a sample for the next state assuming that the sojourn is τ based on the discrete probability distribution $\mathbf{P}(X(T_1) = j \mid X(0) = i, T_1 = \tau)$ ($\forall j \in S$) given in Eq. (4.14).
- When the process is in state i, first draw a sample for the next visited state based on the discrete probability distribution $\Pi_{ij} = \mathbf{P}(X(T_1) = j \mid X(0) = i)$ ($\forall j \in S$), then draw a sample for the sojourn time given the next state with distribution

$$\mathbf{P}(T_1 \leq t \mid Z(0) = i, Z(T_1) = j) = \frac{Q_{ij}(t)}{\Pi_{ij}}. \qquad (4.17)$$

4.5.2 Transient Analysis Using the Method of Supplementary Variable

A semi-Markov process does not possess the Markov property during the sojourn in a state. For example the distribution of the time till the next state transition may depend on the amount of time that has passed since the last state transition. It is possible to extend the analysis of semi-Markov processes so that all information that makes the future evolution of the process conditionally independent of its past history is involved in the process description for $\forall t \geq 0$. It is indeed the Markov property for $\forall t \geq 0$. In the case of semi-Markov processes, it means that the discrete state of the process $X(t)$ and the time passed since the last state transition $Y(t) = t - \max(T_i \leq t)$ need to be considered together because the vector-valued stochastic process $\{X(t), Y(t)\}$ is already such that the future behavior of this vector process is conditionally independent of its past given the current value of the vector. That is, the $\{X(t), Y(t)\}$ process possesses the Markov property for $\forall t \geq 0$. The behavior of the $\{X(t), Y(t)\}$ process is depicted in Fig. 4.3.

This extension of a random process with an additional variable such that the obtained vector-valued process possesses the Markov property is referred to as the method of supplementary variables [2].

With $X(t)$, $Y(t)$, and the kernel matrix of the process we can compute the distribution of time till the next state transition at any time instant; this is commonly referred to as the remaining sojourn time in the given state. If at time t the process stays in state i for a period of τ $(X(t) = i, Y(t) = \tau)$ and the distribution of the total sojourn time in state i is $Q_i(t)$, then the distribution of the remaining sojourn time in state i, denoted as γ, is

$$\mathbf{P}(\gamma \leq t) = \mathbf{P}(\gamma_t \leq t + \tau \mid \gamma_t > \tau) = \frac{Q_i(t + \tau) - Q_i(\tau)}{1 - Q_i(\tau)},$$

where γ_t denotes the total time spent in state i during this visit in state i.

To analyze the $\{X(t), Y(t)\}$ process, we need to characterize the joint distribution of the following two quantities:

$$h_i(t, x) = \frac{\mathbf{P}(X(t) = i, x \leq Y(t) < x + \Delta)}{\Delta}.$$

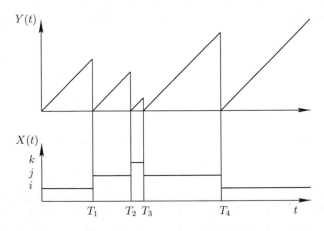

Fig. 4.3 Analysis of semi-Markov process with supplementary variable

It is possible to obtain $h_i(t, x)$ based on the analysis of the short-term behavior of CTMCs:

$$h_i(t + \Delta, x) = \mathbf{P}(\text{there is no state transition in the interval } (t, t + \Delta))$$
$$h_i(t + \Delta, x \mid \text{there is no state transition})$$
$$+ \mathbf{P}(\text{there is one state transition in the interval } (t, t + \Delta))$$
$$h_i(t + \Delta, x \mid \text{there is one state transition}) + o(\Delta),$$

where $h_i(t+\Delta, x \mid \text{condition})$ denotes $\dfrac{\mathbf{P}(X(t) = i, x \le Y(t) < x + \Delta \mid \text{condition})}{\Delta}$.

The probability of the state transition can be computed based on the distribution of the remaining sojourn time:

$$\mathbf{P}(\text{there is one state transition in the interval } (t, t + \Delta))$$
$$= \mathbf{P}(\text{remaining sojourn time} \le \Delta) = \frac{Q_i(x + \Delta) - Q_i(x)}{1 - Q_i(x)},$$

from which

$$\mathbf{P}(\text{there is no state transition in the interval } (t, t + \Delta)) = \frac{1 - Q_i(x + \Delta)}{1 - Q_i(x)}.$$

Immediately following a state transition $Y(t)$ is reset to zero. Consequently, the probability that $Y(t + \Delta) = x$ for a fixed $x > 0$ is zero when Δ is sufficiently small. That is,

$$h_i(t + \Delta, x \mid \text{there is one state transition in the interval } (t, t + \Delta)) = 0 \quad \text{if } x > 0.$$

It follows that

$$h_i(t + \Delta, x) = \mathbf{P}(\text{there is no state transition in the interval } (t, t + \Delta))$$
$$h_i(t + \Delta, x \mid \text{there is no state transition in the interval } (t, t + \Delta))$$
$$= \frac{1 - Q_i(x + \Delta)}{1 - Q_i(x)} \cdot h_i(t, x - \Delta).$$

Analysis of the process $\{X(t), Y(t)\}$ is made much simpler by the use of the transition rate of α_i instead of its distribution $Q_i(t)$. The transition rate is defined by

$$\lambda_i(t) = \lim_{\Delta \to 0} \frac{\mathbf{P}(\alpha_i \le t + \Delta \mid \alpha_i > t)}{\Delta} = \lim_{\Delta \to 0} \frac{Q_i(t + \Delta) - Q_i(t)}{\Delta(1 - Q_i(t))} = \frac{Q_i'(t)}{1 - Q_i(t)}.$$

It is also referred to as the hazard rate in probability theory. The probability of a state transition can be written in the following form:

$$\mathbf{P}(\text{there is one state transition in the interval } (t, t + \Delta)) = \frac{Q_i(x + \Delta) - Q_i(x)}{1 - Q_i(x)}$$

$$= \lambda_i(x)\Delta + o(\Delta),$$

from which

$$\mathbf{P}(\text{there is no state transition in the interval } (t, t + \Delta)) = 1 - \lambda_i(x)\Delta + o(\Delta).$$

Based on all of these expressions, $h_i(t, x)$ satisfies

$$h_i(t + \Delta, x) = \left(1 - \lambda_i(x)\Delta + o(\Delta)\right) h_i(t, x - \Delta).$$

From this difference equation we can go through the usual steps to obtain the partial differential equation for $h_i(t, x)$. First we move $h_i(t, x - \Delta)$ to the other side

$$h_i(t + \Delta, x) - h_i(t, x - \Delta) = \left(- \lambda_i(x)\Delta + o(\Delta)\right) h_i(t, x - \Delta),$$

then we add and subtract $h_i(t, x)$

$$h_i(t + \Delta, x) - h_i(t, x) + h_i(t, x) - h_i(t, x - \Delta) = \left(- \lambda_i(x)\Delta + o(\Delta)\right) h_i(t, x - \Delta)$$

and reorder the terms,

$$\frac{h_i(t + \Delta, x) - h_i(t, x)}{\Delta} + \frac{h_i(t, x) - h_i(t, x - \Delta)}{\Delta} = \left(- \lambda_i(x) + \frac{o(\Delta)}{\Delta}\right) h_i(t, x - \Delta).$$

Finally, the $\Delta \to 0$ transition results in

$$\frac{\partial h_i(t, x)}{\partial t} + \frac{\partial h_i(t, x)}{\partial x} = -\lambda_i(x) h_i(t, x). \tag{4.18}$$

This partial differential equation describes $h_i(t, x)$ for $x > 0$. The case of $x = 0$ requires a different treatment:

$$\mathbf{P}(X(t + \Delta) = i, Y(t) \le \Delta)$$

$$= \sum_{k \in S, k \neq i} \int_{x=0}^{\infty} \mathbf{P}(X(t) = k, Y(t) = x, \text{one transition to state } i \text{ in } (t, t + \Delta)) \, \mathrm{d}x.$$

The probability that in the interval $(t, t + \Delta)$ the process moves from state k to state i is

$\mathbf{P}($ there is one state transition in the interval $(t, t + \Delta)$ from k to $i)$
$= \mathbf{P}($one state transition in the interval $(t, t + \Delta))$
$\quad \mathbf{P}($state transition from k to $i \mid$ one state transition in the interval $(t, t + \Delta))$
$= \dfrac{Q_k(x + \Delta) - Q_k(x)}{1 - Q_k(x)} \cdot \dfrac{Q_{ki}(x + \Delta) - Q_{ki}(x)}{Q_k(x + \Delta) - Q_k(x)}$,

where the second term is already known from Eq. (4.14). We can also introduce the intensity of transition from k to i:

$$\lambda_{ki}(x) = \lim_{\Delta \to 0} \frac{\mathbf{P}(\text{there is a transition in the interval } (t, t + \Delta) \text{ from } k \text{ to } i)}{\Delta}$$

$$= \lim_{\Delta \to 0} \frac{Q_{ki}(x + \Delta) - Q_{ki}(x)}{\Delta(1 - Q_k(x))} = \frac{Q'_{ki}(x)}{1 - Q_k(x)} .$$

The transition probability can be written in the following form

$\mathbf{P}(\text{there is a transition in the interval } (t, t + \Delta) \text{ from } k \text{ to } i) = \lambda_{ki}(x)\Delta + o(\Delta) .$

Using this we can write

$$\mathbf{P}(X(t + \Delta) = i, Y(t) \le \Delta) = h_i(t + \Delta, 0)\Delta$$

$$= \sum_{k \in S, k \neq i} \int_{x=0}^{\infty} (\lambda_{ki}(x)\Delta + o(\Delta)) \, h_k(t, x) \, dx ,$$

from which a multiplication with Δ and the $\Delta \to 0$ transition result in

$$h_i(t, 0) = \sum_{k \in S, k \neq i} \int_{x=0}^{\infty} \lambda_{ki}(x) \, h_k(t, x) \, dx . \tag{4.19}$$

In summary, the method of supplementary variable allows the analysis of the process $\{X(t), Y(t)\}$ through the function $h_i(t, x)$, which is given by a partial differential equation (4.18) for $x > 0$ and a boundary equation (4.19) for $x = 0$. Based on these equations and the initial distributions of $h_i(0, x)$ for $\forall i \in S$ numerical partial differential solutions methods can be applied to compute the transient behavior of a semi-Markov process.

Stationary Behavior If the limit $\lim_{t \to \infty} h_i(t, x) = h_i(x)$ exists for all states $i \in S$, then we can evaluate the limit $t \to \infty$ of Eqs. (4.18) and (4.19)

$$\frac{dh_i(x)}{dx} = -\lambda_i(x) \, h_i(x) \tag{4.20}$$

$$h_i(0) = \sum_{k \in S, k \neq i} \int_{x=0}^{\infty} \lambda_{ki}(x)\, h_k(x)\, dx \ . \tag{4.21}$$

The solution of ordinary differential equation (4.20) is

$$h_i(x) = h_i(0) e^{\int_{u=0}^{x} -\lambda_i(u)\, du}$$

where the unknown quantity is $h_i(0)$. It can be obtained from Eq. (4.21) as follows:

$$h_i(0) = \sum_{k \in S, k \neq i} \int_{x=0}^{\infty} \lambda_{ki}(x)\, h_k(0) e^{\int_{u=0}^{x} -\lambda_k(u)\, du}\, dx$$

$$= \sum_{k \in S, k \neq i} h_k(0) \int_{x=0}^{\infty} \lambda_{ki}(x) e^{\int_{u=0}^{x} -\lambda_k(u)\, du}\, dx \ ,$$

where

$$\int_{x=0}^{\infty} \lambda_{ki}(x) e^{\int_{u=0}^{x} -\lambda_k(u)\, du}\, dx = \mathbf{P}(\text{after state } k \text{ the process moves to state } i) = \Pi_{ki} \ .$$

That is, we are looking for the solution of the linear system

$$h_i(0) = \sum_{k \in S, k \neq i} h_k(0)\, \Pi_{ki} \quad \forall i \in S$$

with the normalizing condition

$$\sum_{i \in S} \int_{x=0}^{\infty} h_i(x)\, dx = 1,$$

where the normalizing condition is the sum of the stationary-state probabilities. From

$$\sum_{i \in S} \int_{x=0}^{\infty} h_i(x)\, dx = \sum_{i \in S} h_i(0) \int_{x=0}^{\infty} e^{\int_{u=0}^{x} -\lambda_i(u)\, du}\, dx = \sum_{i \in S} h_i(0)\, \hat{\tau}_i = 1 \ ,$$

and Eq. (4.16) we have that the required solution is

$$h_i(0) = \frac{\hat{P}_i}{\sum_j \hat{P}_j \hat{\tau}_j} \ .$$

4.6   Markov Regenerative Process

Definition 4.16 The $X(t)$ discrete-state, continuous-time, time-homogeneous stochastic process is a **Markov regenerative process** if there exists a random time series $T_0, T_1, T_2, \ldots$ ($T_0 = 0$) such that the $X(t)$ process possesses the Markov property at time $T_0, T_1, T_2, \ldots$ [1, 6] (see Fig. 4.4).

Compared to the properties of semi-Markov processes, where the process possesses the Markov property at all state transition points, the definition of Markov regenerative processes is less restrictive. It allows that at some state-transition point the process does not possess the Markov property, but the analysis of Markov regenerative processes is still based on the occurrence of time points where the process possesses the Markov property.

Since Definition 4.16 does not address the behavior of the process between the consecutive time points $T_0, T_1, T_2, \ldots$, Markov regenerative processes can be fairly general stochastic processes. In practice, the use of renewal theorem for the analysis of these processes is meaningful only when the stochastic behavior between the consecutive time points $T_0, T_1, T_2, \ldots$ is easy to analyze.

A common method for analyzing Markov regenerative processes is based on the next time point with Markov property (T_1).

Definition 4.17 The series of random variables $\{Y_n, T_n; n \geq 0\}$ is a time-homogeneous **Markov renewal series** if

$$\mathbf{P}(Y_{n+1} = y, T_{n+1} - T_n \leq t \mid Y_0, \ldots, Y_n, T_0, \ldots, T_n)$$

$$= \mathbf{P}(Y_{n+1} = y, T_{n+1} - T_n \leq t \mid Y_n)$$

$$= \mathbf{P}(Y_1 = y, T_1 - T_0 \leq t \mid y_0)$$

for all $n \geq 0$, $y \in S$, and $t \geq 0$.

It can be seen from the definition of Markov renewal series that the series $Y_0, Y_1, \ldots$ is a DTMC. According to Definition 4.16, the sequence of states $X(T_i)$ of a Markov regenerative process at the time sequence T_i instants with the Markov

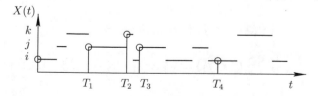

Fig. 4.4 Markov regenerative process—circles denote the points with Markov property

property and the time sequence T_i instants with Markov property form a Markov renewal sequence $\{X(T_i), T_i\}$ $(i = 0, 1, \ldots)$.

Analysis of Markov regenerative processes is based on this *embedded* Markov renewal series. To this end the joint distribution of the next time point and the state in that time point has to be known. In contrast with the similar kernel of semi-Markov processes, in the case of Markov regenerative processes, the kernel is denoted by

$$K_{ij}(t) = \mathbf{P}(X_1 = j, T_1 - T_0 \leq t \mid X_0 = i), \quad i, j \in S,$$

and the atrix $K(t) = \{K_{ij}(t)\}$ is referred to as the global kernel of the Markov regenerative process. The global kernel of the Markov regenerative process completely defines the stochastic properties of the Markov regenerative process at time points with the Markov property. The description of the process between those time points is complex, but for the transient analysis of the process (more precisely for computing transient-state probabilities) it is enough to know the transient-state probabilities between consecutive time points with the Markov property. This is given by the local kernel matrix of the Markov regenerative process $E(t) = \{E_{ij}(t)\}$ whose elements are

$$E_{ij}(t) = \mathbf{P}(X(t) = j, T_1 > t, \mid Z(0) = i),$$

where $E_{ij}(t)$ is the probability that the process will start in state i, the first point with Markov property will be later than t, and the process will stay in state j at time t.

4.6.1 Transient Analysis Based on Embedded Markov Renewal Series

Let the transient-state transition probability matrix be $\boldsymbol{\Pi}(t)$, whose elements are

$$\Pi_{ij}(t) = \mathbf{P}(X(t) = j \mid X(0) = i).$$

Assuming that $T_1 = h$, we can compute the conditional state-transition probability as follows:

$$\Pi_{ij}(t \mid T_1 = h) = \begin{cases} \mathbf{P}(X(t) = j \mid T_1 = h, X(0) = i), & h > t, \\ \displaystyle\sum_{k \in S} \mathbf{P}(X(T_1) = k \mid X(0) = i, T_1 = h) \cdot \Pi_{kj}(t - h), & h \leq t. \end{cases}$$

$$(4.22)$$

Similar to the transient analysis of semi-Markov processes, Eq. (4.22) describes two exclusive cases: $h \leq t$ and $h > t$. In case of semi-Markov processes, the $h > t$

case results in 0 or 1; in the case of a Markov regenerative process, the conditional probability for $h > t$ can be different from 0 or 1 because the process can have state transitions also before T_1.

Using the distribution of T_1 and the formula of total probability we obtain

$$
\begin{aligned}
\Pi_{ij}(t) = & \int_{h=t}^{\infty} \mathbf{P}(X(t) = j \mid T_1 = h, \, X(0) = i) \, \mathrm{d}K_i(h) \\
& + \int_{h=0}^{t} \sum_{k \in S} \frac{\mathrm{d}K_{ik}(t)}{\mathrm{d}K_i(t)} \, \Pi_{kj}(t - h) \, \mathrm{d}K_i(h) .
\end{aligned}
\tag{4.23}
$$

Let us consider the first term of the right-hand side:

$$
\begin{aligned}
& \int_{h=t}^{\infty} \mathbf{P}(X(t) = j \mid T_1 = h, \, X(0) = i) \, \mathrm{d}K_i(h) \\
& = \int_{h=t}^{\infty} \lim_{\Delta \to 0} \mathbf{P}(X(t) = j \mid h \leq T_1 < h + \Delta, \, X(0) = i) \, \mathrm{d}K_i(h) \\
& = \int_{h=t}^{\infty} \lim_{\Delta \to 0} \frac{\mathbf{P}(X(t) = j, \, h \leq T_1 < h + \Delta \mid X(0) = i)}{\mathbf{P}(h \leq T_1 < h + \Delta, \mid X(0) = i)} \, \mathrm{d}K_i(h) \\
& = \int_{h=t}^{\infty} \frac{\mathrm{d}_h \, \mathbf{P}(X(t) = j, \, T_1 < h \mid X(0) = i)}{\mathrm{d}K_i(h)} \, \mathrm{d}K_i(h) \\
& = \mathbf{P}(X(t) = j, \, t < T_1 \mid X(0) = i) = E_{ij}(t),
\end{aligned}
$$

from which

$$
\Pi_{ij}(t) = E_{ij}(t) + \sum_{k \in S} \int_{h=0}^{t} \Pi_{kj}(t - h) \, \mathrm{d}K_{ik}(h) .
\tag{4.24}
$$

Assuming that $\mathbf{K}(t)$ is continuous and $\mathrm{d}\mathbf{K}(t)/\mathrm{d}t = \mathbf{k}(t)$ we have

$$
\Pi_{ij}(t) = E_{ij}(t) + \sum_{k \in S} \int_{h=0}^{t} \Pi_{kj}(t - h) \, k_{ik}(h) \, \mathrm{d}h .
\tag{4.25}
$$

Similar to the transient analysis of CTMCs and semi-Markov processes we obtained a Volterra equation for the transient analysis of Markov regenerative processes.

Transform Domain Description The Laplace transform of Eq. (4.25) is

$$
\Pi_{ij}^*(s) = E_{ij}^*(s) + \sum_{k \in \Omega} k_{ik}^*(s) \, \Pi_{kj}^*(s),
\tag{4.26}
$$

which can be written in matrix form:

$$\boldsymbol{\Pi}^*(s) = \boldsymbol{E}^*(s) + \boldsymbol{k}^*(s)\, \boldsymbol{\Pi}^*(s). \tag{4.27}$$

The solution of Eq. (4.27) is

$$\boldsymbol{\Pi}^*(s) = [\boldsymbol{I} - \boldsymbol{k}^*(s)]^{-1}\, \boldsymbol{E}^*(s). \tag{4.28}$$

Based on Eq. (4.28), numerical inverse Laplace methods can also be used for the transient analysis of Markov regenerative processes.

Stationary Behavior Despite the differences between semi-Markov and Markov regenerative processes, their stationary analysis follows the same steps. The state-transition probability of the DTMC embedded in time points with Markov property is

$$\Pi_{ij} = \mathbf{P}(Z(T_1) = j \mid Z(0) = i)$$
$$= \lim_{t \to \infty} \mathbf{P}(Z(T_1) = j, T_1 \le t \mid Z(0) = i) = \lim_{t \to \infty} K_{ij}(t) = \lim_{s \to 0} k_{ij}^*(s).$$

The stationary distribution of the EMC is the solution of $\hat{P} = \hat{P}\boldsymbol{\Pi}, \sum_i \hat{P}_i = 1$. Now we need to compute the mean time spent in the different states during the interval (T_0, T_1). Fortunately, the local kernel carries the necessary information. Let τ_{ij} be the mean time the process spends in state j during the interval (T_0, T_1) assuming that it starts from state i $(X(T_0) = i)$. Then

$$\tau_{ij} = \mathbf{E}\left(\int_{t=0}^{\infty} \mathscr{I}_{\{X(t)=j, T_1 > t \mid X(0)=i\}} dt \right)$$
$$= \int_{t=0}^{\infty} \mathbf{P}(X(t) = j, T_1 > t \mid X(0) = i) dt$$
$$= \int_{t=0}^{\infty} E_{ij}(t) dt = \lim_{s \to 0} E_{ij}^*(s),$$

where $\mathscr{I}_{\{\bullet\}}$ is the indicator of event $\bullet$. The mean length of the interval (T_0, T_1) is

$$\tau_i = \sum_{j \in S} \tau_{ij}.$$

Finally, the stationary distribution of the process can be computed as

$$P_i = \frac{\displaystyle\sum_{j \in S} \hat{P}_j\, \tau_{ji}}{\displaystyle\sum_{j \in S} \hat{P}_j\, \tau_j}. \tag{4.29}$$

4.7 Exercises

Exercise 4.1 Applying Theorem 4.14 (on p. 196), find the limit (stationary) distributions of age, residual lifetime, and total lifetime ($\delta(t) = t - t_{N(t)}$, $\gamma(t) = t_{N(t)+1} - t$, $\beta(t) = t_{N(t)+1} - t_{N(t)}$), if the interarrival times are independent random variables having joint exponential distribution with parameter λ. Show the expected values for the limit distributions.

Exercise 4.2 (Ergodic Property of Semi-Markov Processes) Consider a system with finite state space $\mathscr{X} = \{1, \ldots, N\}$. The system begins to work at the moment $T_0 = 0$ in a state $X_0 \in \mathscr{X}$ and it changes the states at the random moments $0 < T_1 < T_2 < \ldots$ Denote by $X_1, X_2, \ldots$ the sequence of consecutive states of the system and suppose that it constitutes a homogeneous, irreducible and aperiodic Markov chain with initial distribution ($p_i = \mathbf{P}(X_0 = i)$, $1 \le i \le N$) and probability transition matrix $\Pi = (p_{ij})_{i,j=1}^{n}$. Define the process $X(t) = X_{n-1}$, $T_{n-1} \le t < T_n$, $n = 1, 2, \ldots$ and assume that the sequence of holding times $Y_k = T_k - T_{k-1}$, $k = 1, 2, \ldots$ depends only conditionally on the states $X_{k-1} = i$ and $X_k = j$ and denote $F_{ij}(x) = \mathbf{P}(Y_k \le x \mid X_{k-1} = i, X_k = j)$ if $p_{ij} > 0$, where
$$v_{ij} = \int_0^\infty x \, dF_{ij}(x) < \infty.$$
Find the limits for

(a) The average number of transitions/time;
(b) The relative frequencies of the states i in the sequence $X_0, X_1, \ldots$;
(c) The limit distribution $\mathbf{P}(X_t = i)$, $i \in X$;
(d) The average time which is spent in a state $i \in X$.

4.8 Solutions

Solution 4.1 By the use of Theorem 4.14 (on p. 196) we get the limit distributions in the following forms

$$G(x) = \lim_{t \to \infty} \mathbf{P}(\delta(t) \le x) = \lim_{t \to \infty} \mathbf{P}(\gamma(t) \le x) = \frac{1}{(1/\lambda)} \int_0^x [1 - F(s)] ds = \lambda \int_0^x e^{-\lambda s} ds$$

$$= 1 - e^{-\lambda x},$$

$$H(x) = \lim_{t \to \infty} \mathbf{P}(\beta(t) \le x) = \frac{1}{(1/\lambda)} \int_0^x s \, dF(s) = \frac{1}{(1/\lambda)} \int_0^x \lambda e^{-\lambda s} ds = \int_0^x \lambda^2 e^{-\lambda s} ds.$$

From these we can see that the limit distributions of $\delta(t)$ and $\gamma(t)$ as $t \to \infty$ coincide with the exponential distribution of parameter λ. The limit distribution of

$\beta(t)$ as $t \to \infty$ is gamma distribution with parameter $(2, \lambda)$, which coincides with the distribution of the sum of two independent exponentially distributed r.v.s with parameter λ. The expected values for the limit distributions are $1/\lambda$, $1/\lambda$, $2/\lambda$.

Solution 4.2 Since the Markov chain $(X_k, \ k = 0, 1, \ldots)$ with finite state space is homogeneous, irreducible, and aperiodic, then it is ergodic, consequently, the expected values of return times $\mu_i = \sum_{k=1}^{\infty} k f_{ii}(k) < \infty$, $1 \le i \le N$ are finite and its stationary distribution $\pi = (\pi_1, \ldots, \pi_N)$ can be given in the form $\pi_i = 1/\mu_i$, $1 \le i \le N$. From the ergodic property of the Markov chain it also follows that

$$\frac{1}{n} \sum_{k=1}^{n} \mathscr{I}_{\{X_k=i\}} \to \pi_i \ \text{ as } \ n \to \infty \ \text{ with probability 1.}$$

Let us introduce the notations

$$K_i(t) = \sum_{k=0}^{K(t)} \mathscr{I}_{\{X_k=i\}}, \quad K(t) = \max\{k : T_k \le t\}, \quad S_i(t) = \int_0^t \mathscr{I}_{\{X(s)=i\}} ds,$$

$$n_1^{(i)} = \min\{k : X_k = i, \ k \ge 1\}, \quad n_m^{(i)} = \min\{k : X_k = i, \ k > n_{m-1}^{(i)}\}, \ m = 2, 3, \ldots,$$

$$T_m^{(i)} = T_{n_m^{(i)}} = \sum_{k=1}^{n_m^{(i)}} Y_k, \ m = 1, 2, \ldots, \quad \tau_1^{(i)} = T_1^{(i)}, \quad \tau_m^{(i)} = T_m^{(i)} - T_{m-1}^{(i)}, \ m = 2, 3, \ldots$$

Note that $S_i(t)$ denotes the amount of time is spent by the process $X(t)$ on the interval $(0, t)$ in a state i.

Let us consider the process $(X(t), \ t \ge 0)$ for a fixed i. Since $(X_0, X_1, \ldots)$ is a Markov chain and the sequence $(Y_1, Y_2, \ldots)$ only conditionally depends on $(X_0, X_1, \ldots)$, then the cycles $(X(t), \ t \in [T_{m-1}^{(i)}, T_m^{(i)}))$, $m = 2, 3, \ldots$ are independent and stochastically equivalent. Consequently, the process $(X(t), \ t \ge 0)$ is regenerative under the condition $X_0 = i$, otherwise it is delayed regenerative. It is also clear that the r.v.s $\tau_m^{(i)}$, $m = 1, 2, \ldots$ are independent and moreover, $\tau_m^{(i)}$, $m = 2, 3, \ldots$ are identically distributed, which means that $(\tau_m^{(i)}, \ m = 1, 2, \ldots)$ forms a renewal or delayed renewal (in the case $X_0 \ne i$) process.

First we prove that $\mathbf{E}\left(\tau_1^{(i)} \mid X_0 = j\right)$, $1 \le j \le N$ are finite. Note that $\mathbf{E}\left(\tau_2^{(i)}\right) = \mathbf{E}\left(\tau_1^{(i)} \mid X_0 = i\right)$ and $\mathbf{E}\left(\tau_1^{(i)}\right) = \sum_{j=1}^{N} \mathbf{E}\left(\tau_1^{(i)} \mid X_0 = j\right) \mathbf{P}(X_0 = j)$. Denote $\nu = \max\{v_{ij} : 1 \le i, j \le N, \ p_{ij} > 0\}$ and $A_1^{(i)} = \{X_0 = i, X_1 = i\}$, $A_k^{(i)} = \{X_0 = i, X_1 \ne i, \ldots, X_{k-1} \ne i, X_k = i\}$, $k \ge 2$. Since for all $1 \le i, i_1, \ldots, i_{k-1} \le N$, $k = 1, 2, \ldots$ we have

$$\mathbf{E}\left(Y_1 + \ldots + Y_k \mid A_k^{(i)}\right) = v_{i,i_1} + v_{i_1,i_2} + \ldots + v_{i_{k-1},i} \le k\nu,$$

then

$$a_i = \mathbf{E}\left(\tau_2^{(i)}\right) = \mathbf{E}\left(\tau_1^{(i)} \mid X_0 = i\right) = \sum_{k=1}^{\infty} \mathbf{E}\left((Y_1 + \ldots + Y_k)\mathscr{I}_{\{A_k^{(i)}\}} \mid X_0 = i\right)$$

$$= \sum_{k=1}^{\infty} \mathbf{E}\left((Y_1 + \ldots + Y_k) \mid A_k^{(i)}\right)\mathbf{P}\left(A_k^{(i)}\right) \le \sum_{k=1}^{\infty} kv\mathbf{P}\left(A_k^{(i)}\right) = \sum_{k=1}^{\infty} kvf_{ii}(k) = v\mu_i < \infty.$$

Denote $m_{ij} = \min\{k : p_{ij}(k) > 0\}$, where $p_{ij}(k)$ is the k-step probability transition function. The definition of m_{ij} is correct because the Markov chain $(X_n, \ n \ge 0)$ is irreducible. From this it follows that there exist $i_1, \ldots, i_{m-1} \in \mathscr{X}$, $i_s \ne i$, $1 \le s \le m_{ij}$ such that $p = \mathbf{P}\left(B_{m_{ij}}^{(i,j)}\right) > 0$, where $B_{m_{ij}}^{(i,j)} = \{X_0 = i, X_1 = i_1, \ldots, X_{m_{ij}-1} = i_{m_{ij}-1}, X_{m_{ij}} = j\}$. Then

$$a_i \ge \mathbf{E}\left(\tau_1^{(i)}\mathscr{I}_{\left\{B_{m_{ij}}^{(i,j)}\right\}} \mid X_0 = i\right) = \mathbf{E}\left(\tau_1^{(i)} \mid B_{m_{ij}}^{(i,j)}\right)\mathbf{P}\left(B_{m_{ij}}^{(i,j)}\right)$$

$$\ge \mathbf{E}\left(Y_{m_{ij}} + \ldots + Y_{n_1^{(i)}} \mid B_{m_{ij}}^{(i,j)}\right)p = \mathbf{E}\left(Y_{m_{ij}} + \ldots + Y_{n_1^{(i)}} \mid X_{m_{ij}} = j\right)p$$

$$= \mathbf{E}\left(\tau_1^{(i)} \mid X_0 = j\right)p,$$

consequently

$$\mathbf{E}\left(\tau_1^{(i)} \mid X_0 = j\right) \le \frac{1}{p}a_i < \infty.$$

(a) By the Remark 4.8 (on p. 191) the strong law of the large number is also valid for the delayed renewal process $(\tau_m^{(i)}, \ m = 1, 2, \ldots)$, therefore with probability 1

$$\frac{K_i(t)}{t} \to \frac{1}{a_i}$$

and consequently with probability 1

$$\frac{K(t)}{t} = \sum_{i=1}^{N} \frac{K_i(t)}{t} \to a = \sum_{i=1}^{N} \frac{1}{a_i}.$$

(b) First, we prove that the convergence

$$\frac{K_i(t)}{K(t)} = \frac{1}{K(t)} \sum_{i=1}^{K(t)} \mathscr{I}_{\{X_{k-1}=i\}} \to \pi_i, \quad \text{as } t \to \infty$$

is true with probability 1. Note that from the convergence of $K(t)/t \overset{a.s.}{\to} a$ it follows that $K(t) \overset{a.s.}{\to} \infty$, as $t \to \infty$. Since the Markov chain $(X_k, \ k = 0, 1, \ldots)$ is ergodic with stationary distribution $(\pi_1, \ldots, \pi_N)$, therefore

$$\frac{1}{m} \sum_{i=1}^{m} \mathscr{I}_{\{X_{k-1}=i\}} \overset{a.s.}{\to} \pi_i, \ i = 1, \ldots, N \ \text{ as } m \to \infty.$$

On one hand

$$\frac{K_i(t)}{K(t)} = \frac{[at]}{K(t)} \frac{1}{[at]} \sum_{i=1}^{[at]} \mathscr{I}_{\{X_{k-1}=i\}} + \frac{[at]}{K(t)} \frac{1}{[at]} \left(\sum_{i=1}^{K(t)} \mathscr{I}_{\{X_{k-1}=i\}} - \sum_{i=1}^{[at]} \mathscr{I}_{\{X_{k-1}=i\}} \right)$$

and on the other hand

$$\frac{1}{[at]} \left| \sum_{i=1}^{K(t)} \mathscr{I}_{\{X_{k-1}=i\}} - \sum_{i=1}^{[at]} \mathscr{I}_{\{X_{k-1}=i\}} \right| \le \frac{1}{[at]} |K(t) - [at]| = \left| \frac{K(t)}{[at]} - 1 \right| \overset{a.s.}{\to} 0$$

because $\frac{[at]}{[t]} \frac{t}{K(t)} \overset{a.s.}{\to} 1$, $t \to \infty$, therefore $\frac{K_i(t)}{K(t)} \overset{a.s.}{\to} \pi_i$ as $t \to \infty$.

(c) The process $(X(t), \ t \ge 0)$ is a (delayed) regenerative one with regenerative cycles $\tau_k^{(i)}$ (the distribution of the cycle $\tau_1^{(i)}$ can differ in distribution from the distribution of other cycles), therefore the convergence with probability 1

$$\frac{S_i(t)}{t} = \frac{1}{t} \int_0^t \mathscr{I}_{\{X(s)=i\}} ds \overset{a.s.}{\to} \lim_{t \to \infty} \mathbf{P}(X(t) = i \mid X_0 = i)$$

$$= \frac{1}{a_i} \mathbf{E} \left(\int_0^{\tau_1^{(i)}} \mathscr{I}_{\{X(s)=i\}} ds \mid X_0 = i \right) = \frac{1}{a_i} \mathbf{E}(Y_1 \mid X_0 = i)$$

$$= \frac{1}{a_i} \sum_{i=1}^{N} \mathbf{E}(Y_1 \mid X_0 = i, X_1 = j) \mathbf{P}(X_1 = j \mid X_0 = i) = \frac{1}{a_i} \sum_{i=1}^{N} p_{ij} v_{ij} = \frac{v_i}{a_i}$$

is true, if $\tau_2^{(i)}$ has nonlattice distribution. (This condition is satisfied, for example, if $F_{ij}(x)$ are nonlattice distribution functions.) Note that the convergence $\frac{S_i(t)}{t} \overset{a.s.}{\to} \frac{v_i}{a_i}$ can be proved directly with the help of strong law of large numbers, because

$$\frac{K_i(t) - 1}{t} \frac{1}{K_i(t) - 1} \sum_{k=1}^{K_i(t)-1} Y_{n_k^{(i)}} \le \frac{S_i(t)}{t} \le \frac{K_i(t)}{t} \frac{1}{K_i(t)} \sum_{k=1}^{K_i(t)} Y_{n_k^{(i)}},$$

where $K_i(t) \xrightarrow{a.s.} \infty$, $t \to \infty$ and the r.v.s $Y_{n_k^{(i)}}$, $k = 1, 2, \ldots$ are independent and identically distributed with $E(Y_{n_k^{(i)}}) = \mathbf{E}(Y_1 \mid X_0 = i)$.

From the relations proved above it follows $\frac{K_i(t)}{K(t)} \frac{S_i(t)/t}{K_i(t)/t} \xrightarrow{a.s.} \pi_i v_i$ (if $\tau_1^{(i)} \le t$ then $K_i(t) \ge 1$) and the sum of average time which is spent in a state i equals to 1, thus

$$1 = \sum_{i=1}^N \frac{S_i(t)}{t} = \frac{K(t)}{t} \sum_{i=1}^N \frac{K_i(t)}{K(t)} \frac{S_i(t)/t}{K_i(t)/t},$$

where

$$\sum_{i=1}^N \frac{K_i(t)}{K(t)} \frac{S_i(t)/t}{K_i(t)/t} \xrightarrow{a.s.} \sum_{i=1}^N \pi_i v_i.$$

Then we get

$$\frac{K(t)}{t} \xrightarrow{a.s.} \left(\sum_{i=1}^N \pi_i v_i \right)^{-1}, \quad t \to \infty,$$

$$\frac{K_i(t)}{t} = \frac{K_i(t)}{K(t)} \frac{K(t)}{t} \xrightarrow{a.s.} \pi_i \left(\sum_{i=1}^N \pi_i v_i \right)^{-1} = \frac{1}{a_i}$$

and with probability 1

$$\lim_{t \to \infty} \frac{S_i(t)}{t} = \lim_{t \to \infty} \frac{S_i(t)}{K_i(t)} \frac{K_i(t)}{K(t)} \frac{K(t)}{t} = \frac{\pi_i v_i}{\sum_k \pi_k v_k}.$$

As a consequence, can be obtained the expected values of regenerative cycles

$$a_i = \frac{1}{\pi_i} \sum_{j=1}^N \pi_j v_j, \quad a = \sum_{i=1}^N \frac{1}{a_i} = \sum_{i=1}^N \frac{1}{\pi_i} \sum_{j=1}^N \pi_j v_j.$$

References

1. Cinlar, E.: Introduction to Stochastic Processes. Prentice-Hall, Englewood Cliffs (1975)
2. Cox, D.R.: The analysis of non-Markovian stochastic processes by the inclusion of supplementary variables. Proc. Cambridge Philo. Soc. **51**, 433–440 (1955)

3. Feller, W.: An Introduction to Probability Theory and Its Applications, vol. I. Wiley, New York (1968)
4. Karlin, S., Taylor, H.M.: A First Course in Stochastic Processes. Academic Press, New York (1975)
5. Kleinrock, L.: Queuing Systems, Volume 1: Theory. Wiley, New York (1975)
6. Kulkarni, V.G.: Modeling and Analysis of Stochastic Systems. Chapman & Hall, London (1995)
7. Lindwall, T.: Lectures on the Coupling Method. Wiley, New York (1992)
8. Sigman, K., Wolff, R.: A review of regenerative processes. SIAM Rev. **35**(2), 269–288 (1993)

Chapter 5
Markov Chains with Special Structures

The previous chapter presented analysis methods for stochastic models where some of the distributions were different from exponential. In these cases the analysis of the models is more complex than the analysis of Markov models. In this chapter we introduce a methodology to extend the set of models which can be analyzed by Markov models while the distributions can be different from exponential.

5.1 Phase Type Distributions

Combination of exponential distributions, such as convolution and probabilistic mixtures, was used for a long time to approximate nonexponential distributions such that the composed model remained to be a Markov model. The most general class of distributions fulfilling these requirement is the set of Phase type distributions (commonly abbreviated as PH distributions) [9, 10].

Definition 5.1 Time to absorption in a Markov chain with N transient and 1 absorbing state is *phase type* distributed (cf. Fig. 5.1).

5.1.1 Continuous-Time PH Distributions

Definition 5.1 is valid for both continuous and discrete-time Markov chains. In this subsection we focus on the case of continuous-time Markov chains.

It is possible to define a PH distribution by defining the initial probability vector p and the generator matrix Q of the Markov chain with $N + 1$ states. Let states of the Markov chain be numbered so that the first N states are transient and the $N + 1$th is absorbing and let $X(t)$ be the state of the Markov chain at time t. The distributions of the time to absorption, T, are related to the transient probabilities of

© Springer Nature Switzerland AG 2019
L. Lakatos et al., *Introduction to Queueing Systems with Telecommunication Applications*, https://doi.org/10.1007/978-3-030-15142-3_5

Fig. 5.1 Markov chain with
5 transient and an absorbing
states defines a PH
distribution

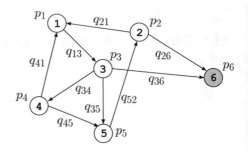

the Markov chain which can be computed from the initial probability vector and the
generator matrix as follows

$$\mathbf{P}(T < t) = \mathbf{P}(X(t) = N + 1) = pe^{Qt}e^T_{N+1},$$

where e_{N+1} is the row vector whose only nonzero element the $N + 1$th is one. A
multiplication of a row vector with e^T_{N+1} results in the $N + 1$th element of the row
vector.

Analysis of PH distributions based on this expression results in technical
difficulties in more complex cases. A more convenient expression can be derived
from the partitioned generator matrix, where the set of states is divided into transient
states and an absorbing one

$$Q = \begin{bmatrix} A & a \\ 0 & 0 \end{bmatrix},$$

where $a = -A\mathbb{1}$ and $\mathbb{1}$ is the column vector whose elements equal to one. The size
of $\mathbb{1}$ is always assumed to be such that the multiplication is valid. A multiplication
of a row vector with $\mathbb{1}$ results in the sum of the elements of the row vector. The
column vector a, which contains the transition rates to the absorbing state (Fig. 5.1)
can be computed from A due to the fact that the row sum of Q is zero. The last row
of Q is zero, because state $N + 1$ is absorbing.

Matrix A is called transient generator (or PH generator). It inherits its main
properties from matrix Q. The diagonal elements of A are negative, the nondiagonal
elements are nonnegative, and the row sums of A are nonpositive. Due to the fact
that the first N states are transient, matrix A is nonsingular, in contrast with matrix
Q, which is singular because $Q\mathbb{1} = 0$.

In this book we restrict our attention to the case when the Markov chain starts
from one of the transient states with probability one. In this case the partitioned
form of vector p is $p = [\alpha \mid 0]$. Based on the partitioned form of p and Q the
CDF of the PH distribution is

$$F_T(t) = \mathbf{P}(T \le t) = \mathbf{P}(T < t) = \mathbf{P}(X(t) = N + 1) = 1 - \mathbf{P}(X(t) < N + 1)$$

$$= 1 - [\alpha \mid 0] e^{Qt} \begin{bmatrix} \mathbb{1} \\ 0 \end{bmatrix} = 1 - [\alpha \mid 0] \sum_{i=0}^{\infty} \frac{t^i}{i!} \begin{bmatrix} A & a \\ 0 & 0 \end{bmatrix}^i \begin{bmatrix} \mathbb{1} \\ 0 \end{bmatrix}$$

$$= 1 - [\alpha \mid 0] \sum_{i=0}^{\infty} \frac{t^i}{i!} \begin{bmatrix} A^i & \mathscr{I}_{\{i>0\}} A^{i-1} a \\ 0 & 0 \end{bmatrix} \begin{bmatrix} 1 \\ 0 \end{bmatrix}$$

$$= 1 - [\alpha \mid 0] \begin{bmatrix} e^{At} & \bullet \\ 0 & 0 \end{bmatrix} \begin{bmatrix} 1 \\ 0 \end{bmatrix} = 1 - \alpha \, e^{At} 1,$$

where $\bullet$ denotes an irrelevant matrix block. Furthermore the PDF, the Laplace transform, and the moments of PH distribution can be computed as

$$f_T(t) = \frac{d}{dt} F_T(t) = -\frac{d}{dt} e^{At} 1 = -\alpha \sum_{i=0}^{\infty} \frac{d}{dt} \frac{t^i}{i!} A^i 1$$

$$= -\alpha \sum_{i=1}^{\infty} \frac{t^{i-1}}{(i-1)!} A^{i-1} A 1 = -\alpha \, e^{At} A 1 = \alpha \, e^{At} a,$$

$$f_T^*(s) = \int_{t=0}^{\infty} e^{-st} f_T(t) dt = \alpha \int_{t=0}^{\infty} e^{-st} e^{At} dt \, a$$

$$= \alpha \int_{t=0}^{\infty} e^{(-sI+A)t} dt \, a = \alpha \, (sI - A)^{-1} a,$$

$$\mathbf{E}(T^n) = \int_{t=0}^{\infty} t^n f_T(t) dt = \alpha \int_{t=0}^{\infty} t^n e^{At} dt \, a = \alpha \, n!(-A)^{-n-1} a$$

$$= \alpha \, n!(-A)^{-n-1} (-A)1 = \alpha \, n!(-A)^{-n} 1.$$

The infinite integrals of the above derivations are computed as follows

$$\int_{t=0}^{\infty} e^{(-sI+A)t} dt = \lim_{\tau \to \infty} \int_{t=0}^{\tau} e^{(-sI+A)t} dt = \lim_{\tau \to \infty} \int_{t=0}^{\tau} \sum_{i=0}^{\infty} \frac{t^i}{i!} (-sI + A)^i \, dt$$

$$= \lim_{\tau \to \infty} \sum_{i=0}^{\infty} \frac{\tau^{i+1}}{(i+1)!} (-sI + A)^{(i+1)} (-sI + A)^{-1}$$

$$= \lim_{\tau \to \infty} \left(e^{(-sI+A)\tau} - I \right) (-sI + A)^{-1} = (sI - A)^{-1},$$

$$(5.1)$$

where $e^{(-sI+A)\tau}$ vanishes in the convergence region of $f_T^*(s)$. The moments can also be computed from the Laplace transform

$$\mathbf{E}(T^n) = (-1)^n \frac{d^n}{ds^n} f_T^*(s) \Big|_{s=0} = (-1)^n \alpha \, \frac{d^n}{ds^n} (sI - A)^{-1} \Big|_{s=0} a$$

$$= (-1)^n \alpha \, (-1)^n n!(sI - A)^{-n-1} \Big|_{s=0} a = \alpha \, n!(-A)^{-n-1} a$$

$$= \alpha \, n!(-A)^{-n} 1.$$

The elements of $(-A)^{-1}$ have an important stochastic interpretation. Let T_{ij} be the time spent in state j before moving to the absorbing state when the process starts in state i.

$$
\mathbf{E}(T_{ij}) = \int_{t=0}^{\infty} \mathbf{E}\left(\mathscr{I}_{\{X(t)=j|X(0)=i\}}\right) dt = \int_{t=0}^{\infty} \mathbf{P}(X(t) = j | X(0) = i) dt
$$

$$
= \int_{t=0}^{\infty} \left(e^{At}\right)_{ij} dt = \left(\int_{t=0}^{\infty} e^{At} dt\right)_{ij} = \left((-A)^{-1}\right)_{ij}. \tag{5.2}
$$

Consequently, $(-A)^{-1}$ is nonnegative. Some characteristics properties of PH distributions can be seen from these expressions. From

$$
f^*(s) = \boldsymbol{\alpha}(sI - A)^{-1}\boldsymbol{a} = \boldsymbol{\alpha}\left[\frac{det(sI - A)_{ji}}{det(sI - A)}\right]\boldsymbol{a}
$$

we have that the Laplace transform is a rational function of s where the degree of the polynomial in the numerator is at most $N - 1$ and in the denominator is at most N, where N is the number of transient states and $det(sI - A)_{ji}$ denotes the sub-determinant associated with element i, j. The related properties of PH distributions in time domain can be obtained from the spectral decomposition of A. Let η be the number of eigenvalues of A, λ_i be the ith eigenvalue whose multiplicity is η_i. In this case

$$
f_T(t) = \boldsymbol{\alpha} \, e^{At} \, \boldsymbol{a} = \sum_{i=1}^{\eta} \sum_{j=1}^{\eta_i} a_{ij} t^{j-1} e^{\lambda_i t}.
$$

It means that in case of distinct eigenvalues ($\eta = N$, $\eta_i = 1$) $f_T(t)$ is a combination of exponential functions with possibly negative coefficients and in case of multiple eigenvalues $f_T(t)$ is a combination of exponential polynomial functions. As a consequence, as t goes to infinity, the exponential function associated with eigenvalue with maximal real part dominates the density. It means that PH distributions have asymptotically exponentially decaying tail behavior.

A wide range of positive distributions can be approximated with PH distributions of size N. A set of PH distributions approximating different positive distribution are depicted in Fig. 5.2. The exponentially decaying tail behavior is not visible in the figure, but there is another significant limitation of PH distributions of size N.

Theorem 5.1 ([1]) *The squared coefficient of variation of T ($cv^2(\tau) = E(T^2)/E(T)^2$) satisfies*

$$
cv^2(\tau) \geq \frac{1}{N}
$$

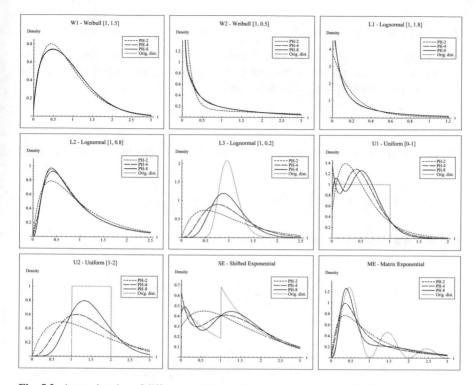

Fig. 5.2 Approximation of different positive distributions with $N = 2, 4, 8$ (figure is copied from [4])

and the only CPH distribution, which satisfies the equality is the Erlang(N) distribution:

Figure 5.2 shows several distributions with low coefficient of variation, whose approximation is poor due to this bound of the coefficient of variation. It is visible that PH distributions with larger N approximates significantly better these distributions. Theoretical results prove that as N tends to infinity any positive distribution can be approximated arbitrarily closely.

5.1.2   Discrete-Time PH Distributions

The majority of the analysis steps and the properties of discrete-time PH distributions are similar to the ones of the continuous-time PH distributions. Using a similar approach as for the continuous-time PH distributions, the state transition probability matrix can be partitioned as $P = \begin{bmatrix} B & b \\ 0 & 1 \end{bmatrix}$, where $b = \mathbb{1} - B\mathbb{1}$ and the initial probability vector p as $p = [\alpha \mid 0]$. B is a sub-stochastic matrix, whose elements are nonnegative and row sums are not greater than one. The probability that the chain moves to the absorbing state in the kth step is

$$r_k = \mathbf{P}(T = k) = \alpha B^{k-1} b,$$

which defines the probability mass function (PMF) of T. The CDF can be obtained as

$$F(k) = \mathbf{P}(T \le k) = \mathbf{P}(X_k = N + 1) = 1 - \mathbf{P}(X_k < N + 1) = 1 - \alpha B^k \mathbb{1},$$

and the z-transform or generator function of T is

$$\mathscr{F}(z) = \mathbf{E}\left(z^T\right) = \sum_{k=0}^{\infty} z^k r_k = z\, \alpha (I - zB)^{-1} b.$$

The factorial moments are

$$\gamma_n = \mathbf{E}(T(T - 1) \ldots (T - n + 1)) = \frac{\mathrm{d}^n}{\mathrm{d}z^n} \mathscr{F}(z)|_{z=1} = n!\, \alpha (I - B)^{-n} B^{n-1} \mathbb{1}$$

Similar to the continuous-time case the z-transform is a rational function of z

$$\mathscr{F}(z) = \mathbf{E}\left(z^T\right) = z\, \alpha (I - zB)^{-1} b = z\, \alpha \left[\frac{det(I - zB)_{ji}}{det(I - zB)} \right] b$$

and based on the spectral decomposition of B the PMF is a combination of geometric series. The coefficient of variation of discrete PH distributions is also bounded from below, but one of the most significant differences between the continuous and the discrete PH distributions is that the bound in this case also depends on the mean of the distribution, $\mu = \mathbf{E}(T)$.

Theorem 5.2 ([12]) *The squared coefficient of variation of T satisfies the inequality:*

$$cv^2(\tau) \ge \begin{cases} \dfrac{\langle \mu \rangle (1 - \langle \mu \rangle)}{\mu^2} & \text{if } \mu < N, \\[2ex] \dfrac{1}{N} - \dfrac{1}{\mu} & \text{if } \mu \ge N. \end{cases} \tag{5.3}$$

where $\langle x \rangle$ denotes the fraction part of x ($x = \lfloor x \rfloor + \langle x \rangle$). For $\mu \leq N$ CV_{min} is provided by the mixture of two deterministic distributions. Its DPH representation is

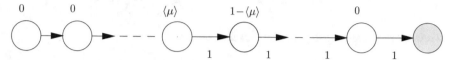

For $\mu > N$ CV_{min} is provided by the discrete Erlang distribution, whose DPH representation is

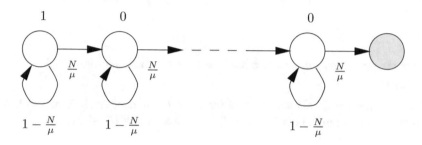

5.1.3 Special PH Classes

The set of PH distributions with N transient states is often too complex for particular practical applications (e.g., derivations by hand). There are special subclasses with restricted flexibility whose application is often more convenient. The most often used subclasses are

- Acyclic PH distributions,
- Hyper-Erlang distributions,
- Hyper-exponential distributions ("parallel," "$cv > 1$").

5.1.3.1 Acyclic PH Distributions

Definition 5.2 *Acyclic PH distributions are PH distributions whose generator is an upper triangular matrix.*

A direct consequence of the structural property of acyclic PH distributions is that the eigenvalues are explicitly given in the diagonal of the generator.

The practical applicability of acyclic PH distributions is due to the following result.

Theorem 5.3 ([5]) *Any acyclic PH distribution can be transformed into the following canonical form. In case of continuous-time acyclic PH distribution:*

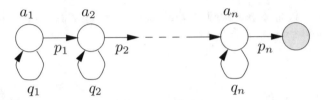

in case of discrete-time acyclic PH distribution:

where the transition rates and probabilities are ordered such that $\lambda_i \leq \lambda_{i+1}$ and $p_i \leq p_{i+1}$.

This essential result allows to consider only these canonical forms with $2N$ parameters to represent the whole acyclic PH class with N transient states.

5.1.3.2   Hyper-Erlang Distributions

Definition 5.3 A *hyper-Erlang distribution* is a probabilistic mixture of Erlang distributions.

Hyper-Erlang distributions are special acyclic PH distributions and even less than $2N$ parameters can define them. Let ϑ be the number of Erlang branches, p_i be the probability of taking branch i, and λ_i, n_i be the parameters of the ith Erlang branch. These 3ϑ parameters completely define the hyper-Erlang distribution

$$f(t) = \sum_{i=1}^{\vartheta} p_i \frac{\lambda_i^{n_i} t^{n_i-1} e^{-\lambda_i t}}{(n_i - 1)!} .$$

5.1.3.3   Hyperexponential Distributions

Definition 5.4 A *hyperexponential distribution* is a probabilistic mixture of Exponential distributions.

Hyperexponential distributions are special hyper-Erlang distributions where the order parameter of the Erlang distribution is one ($n_i = 1$). The PDF of hyperexponential distributions

$$f(t) = \sum_{i=1}^{\vartheta} p_i \lambda_i e^{-\lambda_i t}$$

is monotone decreasing due to the fact that it is the mixture of monotone decreasing exponential density functions.

5.1.4   Fitting with PH Distributions

As it is mentioned in the introduction of this chapter, PH distributions are often used to approximate experimental or exactly given but nonexponential positive distributions in order to analyze the obtained system behavior with discrete state Markov chains. The engineering description of the fitting procedure is rather straightforward: given a nonnegative distribution or a set of experimental data find a "similar" PH distribution, but for the practical implementation of this approach we need to answer several underlying questions. First we formalize the problem as an optimization problem:

$$\min_{PH\,parameters} \left\{ \text{Distance}(F_{PH}(t), \hat{F}_{Original}(t)) \right\},$$

that is, we optimize the parameters of the PH distribution such that the distance between the original distribution and the PH distribution is minimal. The two main technical problems are to find a proper distance measure and to solve the optimization problem. There are several proposals in the literature for the solutions of these problems, but there is still room for further improvement. Some of the typical distance measures are

- squared CDF difference: $\int_0^\infty (F(t) - \hat{F}(t))^2 dt,$

- density difference: $\int_0^\infty |f(t) - \hat{f}(t)| dt,$

- relative entropy: $\int_0^\infty f(t) \log\left(\dfrac{f(t)}{\hat{f}(t)}\right) dt.$

The optimization problems according to these distance measures are typically non-linear and numerically difficult. The close relation of the relative entropy measure with commonly applied statistical parameters (likelihood) makes this measure the most popular one in practice. It is worth mentioning that the complexity of the optimization procedures largely depends on the number of parameters of the PH distributions. That is why we discussed the number of parameters of the mentioned special PH subclasses. There are a few implemented fitting procedures available on the internet. An available fitting procedure using acyclic PH distributions is PhFit [7] and one using hyper-Erlang distributions is G-fit [13]. The literature of PH fitting is rather extended. Several other heuristic fitting approaches exist, e.g., combined with moment matching, that are left for the ambitions of interested readers.

5.2   Markov Arrival Process

Continuous-time Markov arrival process (MAP) is a generalization of Poisson process such that the interarrival times are PH distributed and can be dependent. One of the simplest interpretations of MAPs considers a continuous-time Markov

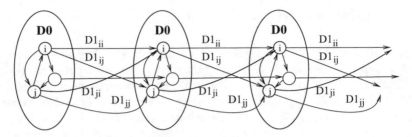

Fig. 5.3 Structure of the Markov chain describing the arrivals of a MAP

chain, $J(t)$, with N states and with generator D, which determines the arrivals in the following way. While the Markov chain stays in state i, it generates arrivals according to a Poisson process with rate λ_i. When the Markov chain experiences a state transition from state i to j, then an arrival occurs with probability p_{ij} and does not occur with probability $1 - p_{ij}$. Based on generator D, rates λ_i $(i = 1, \ldots, N)$ and probabilities p_{ij} $(i, j = 1, \ldots, N, i \neq j)$, one can easily simulate the behavior of the MAP. Due to technical convenience MAPs are most commonly defined by a pair of matrices D_0, D_1, which are obtained from the above introduced parameters in the following way:

$$D_{0ij} = \begin{cases} D_{ij}(1 - p_{ij}) & \text{if } i \neq j, \\ D_{ii} - \lambda_i & \text{if } i = j, \end{cases} \qquad D_{1ij} = \begin{cases} D_{ij} p_{ij} & \text{if } i \neq j, \\ \lambda_i & \text{if } i = j. \end{cases}$$

In this description matrix D_0 is associated with the events which do not result in arrival and matrix D_1 is associated with the events which result in arrivals. By these definitions we have $D_0 + D_1 = D$.

Based on these two matrices we can investigate the counting process of arrivals. Let $N(t)$ be the number of arrivals of a MAP and $J(t)$ be the state of the background Markov chain at time t. The $(N(t), J(t))$ $(N(t) \in \mathbb{N}, J(t) \in \{1, \ldots, N\})$ process is a CTMC. The transition structure of this Markov chain is depicted in Fig. 5.3. The set of states where the number of arrivals is n is commonly referred to as level n, and the state of the background Markov chain ($J(t)$) is commonly referred to as phase.

If the states are numbered in lexicographical order $((0, 1), \ldots, (0, N), (1, 1), \ldots, (1, N), \ldots)$, then the generator matrix has the form

$$Q = \begin{bmatrix} D_0 & D_1 & & & \\ & D_0 & D_1 & & \\ & & D_0 & D_1 & \\ & & & D_0 & \ddots \\ & & & & \ddots \end{bmatrix},$$

where the matrix blocks are of size N. Compared with the CTMC describing the number of arrivals of the Poisson process in (3.21) we have conspicuous similarities: only the diagonal elements/blocks and the first sub-diagonal elements/blocks are nonzero and the transition structure of the arrival process is independent of the number of arrivals.

It is commonly assumed that $N(0) = 0$, and this way the initial probability is 0 for all states (n, j) where $n > 0$. Let vector π_0 be the initial probability for the states with $n = 0$. The arrival instants are determined by $N(t)$ as follows $\Theta_n = \min(t : N(t) = n)$ and the nth interarrival time is $T_n = \Theta_n - \Theta_{n-1}$. Based on the simple block structure of the CTMC, we can analyze the properties of $N(t)$ and T_n. For example, the distribution of T_1 is

$$\mathbf{P}(T_1 \le t) = 1 - \mathbf{P}(T_1 > t) = 1 - \mathbf{P}(N(t) = 0) = 1 - \sum_{i=1}^{N} \mathbf{P}(N(t) = 0, J(t) = i)$$

$$= 1 - \pi_0 e^{D_0 t} \mathbb{1},$$

that is T_1 is PH distributed with initial vector π and generator D_0. For the analysis of the nth interarrival time we introduce the phase distributions vector after the $n-1$th arrivals, π_{n-1}. The ith element of this vector is the probability that after the $n-1$th arrivals the background Markov chain is in state i, that is $(\pi_{n_1})_i = \mathbf{P}(J(\Theta_{n-1}) = i)$. Based on π_{n-1} the distribution of T_n is

$$\mathbf{P}(T_n \le t) = 1 - \mathbf{P}(T_n > t) = 1 - \mathbf{P}(N(t + \Theta_{n-1}) = n - 1)$$

$$= 1 - \sum_{i=1}^{N} \sum_{j=1}^{N} \mathbf{P}(J(\Theta_{n-1}) = i)$$

$$\cdot \mathbf{P}(N(t + \Theta_{n-1}) = n - 1, J(t + \Theta_{n-1}) = j | J(\Theta_{n-1}) = i)$$

$$= 1 - \pi_{n-1} e^{D_0 t} \mathbb{1},$$

that is T_n is PH distributed with initial vector π_{n-1} and generator D_0. The π_n vectors can be computed recursively. The ith element of π_1 has the following stochastic interpretation:

$$(\pi_1)_i = \lim_{\Delta \to 0} \sum_{n=0}^{\infty} \sum_{j=1}^{N} \mathbf{P}(J(n\Delta) = j, T_1 > n\Delta)$$

$$\mathbf{P}(J((n+1)\Delta) = i, n\Delta < T_1 \le (n+1)\Delta)$$

$$= \lim_{\Delta \to 0} \sum_{n=0}^{\infty} \sum_{j=1}^{N} \left(\pi e^{D_0 n \Delta} \right)_j \left(D_{1 j,i} \Delta + o(\Delta) \right)$$

$$= \int_{t=0}^{\infty} \sum_{j=1}^{N} \left(\pi e^{D_0 t} \right)_j D_{1 j,i} \, dt \, ,$$

where the first term of the right-hand side of the first row is the probability that there is no arrival up to time $n\Delta$ and the background Markov chain is in state j at time $n\Delta$, and the second term of the right-hand side of the first row is the probability that there is an arrival between $n\Delta$ and $(n+1)\Delta$ such that the background Markov chain is in state i at time $(n+1)\Delta$. Using (5.1), we further have

$$\pi_1 = \pi \int_{t=0}^{\infty} e^{D_0 t}\, dt\, D_1 = \pi(-D_0)^{-1} D_1. \tag{5.4}$$

According to (5.4) we can compute the phase distribution after the first arrival from the initial distribution and the phase transition probability matrix $P = (-D_0)^{-1} D_1$. P is a stochastic matrix because from $(D_0 + D_1)\mathbb{1} = 0$ we have $-D_0 \mathbb{1} = D_1 \mathbb{1}$, from which $P\mathbb{1} = (-D_0)^{-1} D_1 \mathbb{1} = (-D_0)^{-1}(-D_0)\mathbb{1} = \mathbb{1}$, and $(-D_0)^{-1}$ is nonnegative according to Eq. (5.2). Applying the same analysis for the nth interval starting with initial phase distribution π_{n-1} we have $\pi_n = \pi_{n-1} P$.

5.2.1   Properties of Markov Arrival Processes

The basic properties of MAPs or the $(N(t), J(t))$ CTMC (with level process $N(t) \in \mathbb{N}$ and phase process $J(t) \in \{1, \ldots, N\}$) are as follows.

- The phase distribution at arrival instants form a DTMC with transition probability matrix $P = (-D_0)^{-1} D_1$. As a consequence, the phase distributions might be correlated at consecutive arrivals.
- The interarrival times are PH distributed with representation (π_0, D_0), (π_1, D_0), (π_2, D_0), ... The interarrival times can be correlated due to the correlation of the initial phases.
- The phase process $(J(t))$ is a CTMC with generator $D = D_0 + D_1$, from which $D\mathbb{1} = (D_0 + D_1)\mathbb{1} = 0$. Based on D, some properties of the phase process can be analyzed independent of the level process.
- The (time) stationary phase distribution α is the solution of $\alpha D = 0$, $\alpha \mathbb{1} = 1$.
- The (embedded) stationary phase distribution right after an arrival π is the solution of $\pi P = \pi$, $\pi \mathbb{1} = 1$.
- These stationary distributions are closely related. On one hand, the row vector of the mean time spent in the different phases during the stationary interarrival interval is $\pi(-D_0)^{-1}$ (cf. (5.2)), from which the portion of time spent in the phases is

$$\alpha = \frac{\pi(-D_0)^{-1}}{\pi(-D_0)^{-1}\mathbb{1}}.$$

On the other hand, when the phase process is (time) stationary, the arrival intensities resulting in different initial phases for the next interarrival period are given by αD_1 and after normalizing the result we have

$$\pi = \frac{\alpha D_1}{\alpha D_1 \mathbb{1}}.$$

- The stationary interarrival time (T) is PH distributed with representation (π, D_0), and its nth moment is $\mathbf{E}(T^n) = n!\pi(-D_0)^{-n}\mathbb{1}$.
- The stationary arrival intensity can be computed both from α and from π as follows

$$\lambda = \alpha D_1 \mathbb{1} = \frac{1}{\mathbf{E}(X)} = \frac{1}{\pi(-D_0)^{-1}\mathbb{1}}. \tag{5.5}$$

The first equality is based on the arrival intensities in the (time) stationary phase process. The second equality is based on the mean stationary interarrival time.

Further properties of stationary MAPs can be computed from the joint density functions of consecutive interarrivals:

$$f_{T_0,T_1,\dots,T_k}(x_0,\dots,x_k) = \pi e^{D_0 x_0} D_1 e^{D_0 x_1} D_1 \dots e^{D_0 x_k} D_1 \mathbb{1}.$$

This joint density function describes the probability density that the process starts in phase i with probability π_i at time 0, it does not generate arrival until time x_0 and an arrival occurs at x_0 according to the arrival intensities in D_1. This arrival results in that the second interarrival period starts in phase j, and so on. If the MAP starts from a different initial phase distribution, e.g., γ, then the stationary embedded phase distribution vector π needs to be replaced by γ and the same joint density function applies. For example, we can compute the joint PDF of T_0 and T_k as

$$f_{T_0,T_k}(x_0,x_k) = \int_{x_1} \dots \int_{x_{k-1}} f_{T_0,T_1,\dots,T_k}(x_0,\dots,x_k)\, dx_{k-1}\dots dx_1$$

$$= \pi e^{D_0 x_0} D_1 P^{k-1} e^{D_0 x_k} D_1 \mathbb{1},$$

where we used that $\int_x e^{D_0 x} dx = (-D_0)^{-1}$ according to (5.1). This expression indicates that T_0 and T_k are dependent due to their dependent initial phases. It is also visible that as k tends to infinity, this dependency vanishes according to the speed at which the Markov chain of the initial vectors with transition probability matrix P converges to its stationary distribution π.

The lag-k correlation of a MAP can be computed based on $f_{T_0,T_k}(x_0,x_k)$ as follows:

$$\mathbf{E}(T_0 T_k) = \int_{t=0}^{\infty} \int_{\tau=0}^{\infty} t\,\tau\, \pi e^{D_0 t} D_1 P^{k-1} e^{D_0 \tau} D_1 \mathbb{1}\, d\tau\, dt$$

$$= \pi(-D_0)^{-2} D_1 P^{k-1}(-D_0)^{-2} \underbrace{D_1 \mathbb{1}}_{-D_0 \mathbb{1}}$$

$$= \pi(-D_0)^{-1} P^k (-D_0)^{-1}\mathbb{1} = \frac{1}{\lambda}\alpha P^k (-D_0)^{-1}\mathbb{1}.$$

Since

$$\int_{t=0}^{\infty} t \, e^{\boldsymbol{D_0}t} \, \mathrm{d}t = \underbrace{\left[t \, (\boldsymbol{D_0})^{-1} e^{\boldsymbol{D_0}t} \right]_0^{\infty}}_{0} - \int_{t=0}^{\infty} (\boldsymbol{D_0})^{-1} \, e^{\boldsymbol{D_0}t} \, \mathrm{d}t$$

and

$$\int_{t=0}^{\infty} e^{\boldsymbol{D_0}t} \, \mathrm{d}t = \lim_{T\to\infty} \sum_{i=0}^{\infty} \frac{\boldsymbol{D_0}^i}{i!} \int_0^T t^i \, \mathrm{d}t = \lim_{T\to\infty} \sum_{i=0}^{\infty} \frac{\boldsymbol{D_0}^i}{i!} \frac{T^{i+1}}{i+1}$$

$$= \lim_{T\to\infty} (\boldsymbol{D_0})^{-1} \Big(\underbrace{e^{\boldsymbol{D_0}T}}_{\to 0} - I \Big) = (-\boldsymbol{D_0})^{-1} \, .$$

Based on $\mathbf{E}(T_0 T_k)$ the covariance is

$$\mathbf{Cov}(T_0, T_k) = \mathbf{E}(T_0 T_k) - \mathbf{E}(T)^2 = \frac{1}{\lambda} \, \alpha \, \boldsymbol{P}^k (-\boldsymbol{D_0})^{-1} \mathbb{1} - \frac{1}{\lambda^2},$$

and the coefficient of correlation is

$$\mathbf{Corr}(T_0, T_k) = \frac{\mathbf{Cov}(T_0, T_k)}{\mathbf{E}(T^2) - \mathbf{E}(T)^2} = \frac{\dfrac{\mathbf{E}(T_0 T_k)}{\mathbf{E}(T)^2} - 1}{\dfrac{\mathbf{E}(T^2)}{\mathbf{E}(T)^2} - 1} = \frac{\lambda \, \alpha \, \boldsymbol{P}^k (-\boldsymbol{D_0})^{-1} \mathbb{1} - 1}{2\lambda \, \alpha (-\boldsymbol{D_0})^{-1} \mathbb{1} - 1} \, .$$

Starting from the joint density function of consecutive interarrivals we compute any joint moment for arbitrary series of interarrivals in a similar way as the lag-k correlation. For the interarrival series $a_0 = 0 < a_1 < a_2 < \ldots < a_k$ we have

$$f_{T_{a_0}, T_{a_1}, \ldots, T_{a_k}}(x_0, x_1, \ldots, x_k)$$

$$= \pi e^{\boldsymbol{D_0}x_0} \boldsymbol{D_1} \boldsymbol{P}^{a_1 - a_0 - 1} e^{\boldsymbol{D_0}x_1} \boldsymbol{D_1} \boldsymbol{P}^{a_2 - a_1 - 1} \ldots e^{\boldsymbol{D_0}x_k} \boldsymbol{D_1} \mathbb{1} \, ,$$

and from that the joint moment $\mathbf{E}\Big(T_{a_0}^{i_0}, T_{a_1}^{i_0}, \ldots, T_{a_k}^{i_0} \Big)$ is

$$\mathbf{E}\Big(T_{a_0}^{i_0}, T_{a_1}^{i_0}, \ldots, T_{a_k}^{i_0} \Big) = \pi \, i_0! (-\boldsymbol{D_0})^{-i_0} \boldsymbol{P}^{a_1 - a_0} i_1! (-\boldsymbol{D_0})^{-i_1} \boldsymbol{P}^{a_2 - a_1} \ldots i_k! (-\boldsymbol{D_0})^{-i_k} \mathbb{1} \, .$$

One of the main performance measures of MAPs is the distribution of the number of arrivals in $(0, t)$. This measure can be obtained in several different ways. Intuitively, the probability that n arrivals happen in $(0, t)$ is the probability that the Markov chain depicted in Fig. 5.3 is at level n at time t such that it started from level zero at time 0. Since the Markov chain in Fig. 5.3 is characterized by the level, $N(t)$, and the phase process, $J(t)$, the related state transition probabilities of the Markov chain depend also on the initial and the final phase of the process $P(N(t) = n, J(t) = j | N(0) = 0, J(0) = i)$.

Considering the overall transition probability matrix of the $(N(t), J(t))$ Markov chain, e^{Qt}, the state transitions associated with n arrivals in $(0, t)$ are available in the $0, n$ block of the infinite matrix e^{Qt}.

For a block level solution of the number of arrivals we introduce the $N \times N$ matrix $P(t, n)$, whose i, j element is

$$P(t, n)_{ij} = \mathbf{P}(N(t) = n, J(t) = j | N(0) = 0, J(0) = i). \tag{5.6}$$

$P(t, 0)$ is indeed the $0, 0$ block of the infinite matrix e^{Qt} and can be obtained from the ODE

$$\frac{d}{dt} P(t, 0) = P(t, 0) D_0, \tag{5.7}$$

with initial value $P(t, 0) = I$. For $n \geq 1$, $P(t, n)$ can be obtained from

$$\frac{d}{dt} P(t, n) = P(t, n-1) D_1 + P(t, n) D_0,$$

with initial value $P(t, 0) = 0$. The matrix generating function of the number of arrivals in $(0, t)$, $\hat{P}(t, z) = \sum_{n=0}^{\infty} P(t, n) z^n$, can also be expressed from the block level description. Multiplying the ODE equation for $P(t, n)$ by z^n for $n \geq 1$ and summing up gives

$$\frac{d}{dt} \sum_{n=1}^{\infty} z^n P(t, n) = \sum_{n=1}^{\infty} z^n P(t, n-1) D_1 + \sum_{n=1}^{\infty} z^n P(t, n) D_0.$$

Additionally adding (5.7) gives

$$\frac{d}{dt} \hat{P}(t, z) = z \hat{P}(t, z) D_1 + \hat{P}(t, z) D_0 = \hat{P}(t, z)(z D_1 + D_0),$$

with initial value $\hat{P}(0, z) = I$, whose solution is

$$\hat{P}(t, z) = \hat{P}(0, z) e^{(z D_1 + D_0)t} = e^{(z D_1 + D_0)t}.$$

Assuming that the initial phase distribution is the stationary distribution of the phase process, α, where $\alpha(D_1 + D_0) = 0$, the mean number of arrivals in $(0, t)$ can be obtained as

$$\alpha \frac{d}{dz} \hat{P}(t, z)|_{z=1} \mathbb{1} = \alpha \frac{d}{dz} e^{(z D_1 + D_0)t}|_{z=1} \mathbb{1} = \alpha \frac{d}{dz} \sum_{i=0}^{\infty} \frac{t^i}{i!} (z D_1 + D_0)^i|_{z=1} \mathbb{1} =$$

$$\alpha \sum_{i=1}^{\infty} \sum_{j=0}^{i-1} \frac{t^i}{i!} (zD_1 + D_0)^j \frac{\mathrm{d}}{\mathrm{d}z}(zD_1 + D_0)(zD_1 + D_0)^{i-j-1}|_{z=1} \mathbb{1} =$$

$$\alpha \sum_{i=1}^{\infty} \sum_{j=0}^{i-1} \frac{t^i}{i!} (D_1 + D_0)^j D_1 (D_1 + D_0)^{i-j-1} \mathbb{1} =$$

$$t\alpha D_1 \mathbb{1} + \sum_{i=2}^{\infty} \sum_{j=0}^{i-1} \frac{t^i}{i!} \underbrace{\alpha (D_1 + D_0)^j}_{\mathbf{0}} D_1 \underbrace{(D_1 + D_0)^{i-j-1} \mathbb{1}}_{\mathbf{0}} = t\alpha D_1 \mathbb{1}$$

which verifies (5.5).

5.2.2 Examples of Simple Markov Arrival Processes

In this section we describe some basic arrival processes with MAP notations.

- PH renewal process:
  Consider the arrival process whose interarrival times are independent PH distributed with representation (α, A). This is a special MAP characterized by $D_0 = A$, $D_1 = a\alpha$.
- Interrupted Poisson process (IPP):
  Consider the arrival process determined by a background CTMC with two states, ON and OFF. The transition rate from ON to OFF is α and from OFF to ON is β. There is no arrival in state OFF and customers arrive according to a Poisson process with rate λ in state ON. The MAP description of the process is

$$D_0 = \begin{array}{|c|c|} \hline -\alpha-\lambda & \alpha \\ \hline 0 & -\beta \\ \hline \end{array}, \quad D_1 = \begin{array}{|c|c|} \hline \lambda & 0 \\ \hline 0 & 0 \\ \hline \end{array}.$$

- Markov modulated Poisson process (MMPP):
  Consider the arrival process determined by a background CTMC with generator Q. While the CTMS is in state i arrivals occur according to a Poisson process with rate λ_i. Let λ be the vector of arrival rates. This is a special MAP with representation $D_0 = Q - \mathrm{diag}\langle \lambda \rangle$, $D_1 = \mathrm{diag}\langle \lambda \rangle$.
- Filtered MAP:
  Consider a MAP with representation $\hat{D}_0$, $\hat{D}_1$. The arrivals of this MAP are discarded with probability p. The obtained process is a MAP with representation $D_0 = \hat{D}_0 + p\hat{D}_1$, $D_1 = (1 - p)\hat{D}_1$.
- Cyclically filtered MAP:
  In the previous example, every MAP arrival is discarded with probability p. Now we consider the same MAP such that only every second arrival is discarded with probability p. It requires that we keep track of odd and even arrivals of

the original MAP. It can be done by duplicating the phases, such that the first half of them represents odd arrivals of the original MAP and the second half of them the even arrivals of the original MAP. The obtained process is a MAP with representation

$$D_0 = \begin{array}{|c|c|} \hat{D}_0 & 0 \\ \hline p\hat{D}_1 & \hat{D}_0 \end{array}, \quad D_1 = \begin{array}{|c|c|} 0 & \hat{D}_1 \\ \hline (1-p)\hat{D}_1 & 0 \end{array}.$$

- Superposition of MAPs:
  Consider two MAPs with representation $\hat{D}_0, \hat{D}_1$ and $\tilde{D}_0, \tilde{D}_1$. The superposition of their arrival processes is a MAP with

$$D_0 = \hat{D}_0 \oplus \tilde{D}_0, \text{ and } D_1 = \hat{D}_1 \oplus \tilde{D}_1,$$

where the Kronecker product is defined as $A \otimes B = \begin{array}{|ccc|} A_{11}B & \dots & A_{1n}B \\ \vdots & & \vdots \\ A_{n1}B & \dots & A_{nn}B \end{array}$ and the Kronecker sum as $A \oplus B = A \otimes I_B + I_A \otimes B$. This example indicates one advantage of the D_0, D_1 description of MAPs. Using these matrices the description of the superposed process inherits the related property of the Cartesian product of independent Markov chains.

- Consider an arrival process where the interarrival time is either exponentially distributed with parameter λ_1 or with parameter λ_2 ($\lambda_1 \neq \lambda_2$). The arrivals are correlated such that an interarrival period with parameter λ_1 is followed by one with parameter λ_1 with probability p or one with parameter λ_2 with probability $1 - p$. The interarrival periods with parameter λ_2 follow the same behavior. The obtained process is a MAP with

$$D_0 = \begin{array}{|c|c|} -\lambda_1 & 0 \\ \hline 0 & -\lambda_2 \end{array} \quad D_1 = \begin{array}{|c|c|} p\lambda_1 & (1-p)\lambda_1 \\ \hline (1-p)\lambda_2 & p\lambda_2 \end{array}$$

Probability p has a very intuitive meaning in this model. If $p \to 1$ the correlation of the consecutive interarrivals is increasing and vice versa.

5.2.3 Batch Markov Arrival Process

Batch Markov arrival process (BMAP) is an extension of MAP with batch arrivals. It has an interpretation similar to that of MAP.

A CTMC with generator D determines the arrivals in the following way. While the Markov chain stays in state i, arrivals of batch size k occur according to a Poisson process with rate $\lambda_i^{(k)}$. When the Markov chain experiences a state transition from state i to j, then arrivals of batch size k occur with probability $p_{ij}^{(k)}$ and no arrival occurs with probability $1 - \sum_k p_{ij}^{(k)}$. Generator $\boldsymbol{D}$, rates $\lambda_i^{(k)}$ $(i = 1, \ldots, N)$ and probabilities $p_{ij}^{(k)}$ $(i, j = 1, \ldots, N, i \neq j)$ determine the stationary behavior of BMAP. Additionally, the initial distribution of the CTMC is needed for the analysis of the transient behavior. A BMAP is commonly described by matrices $\boldsymbol{D}_k$, which are obtained from the above introduced parameters in the following way:

$$
\boldsymbol{D}_{0ij} = \begin{cases} D_{ij}(1 - \sum_k p_{ij}^{(k)}) & \text{if } i \neq j, \\ D_{ii} - \sum_k \lambda_i^{(k)} & \text{if } i = j, \end{cases} \qquad \boldsymbol{D}_{kij} = \begin{cases} D_{ij} p_{ij}^{(k)} & \text{if } i \neq j, \\ \lambda_i^{(k)} & \text{if } i = j. \end{cases}
$$

Based on this description the $(N(t), J(t))$ $(N(t) \in \mathbb{N}, J(t) \in \{1, \ldots, N\})$ process is a CTMC with transition structure depicted in Fig. 5.4. If the states are numbered in lexicographical order $((0, 1), \ldots, (0, N), (1, 1), \ldots, (1, N), \ldots)$, then the generator matrix has the form

$$
\boldsymbol{Q} = \begin{array}{|c|c|c|c|c|}
\hline
\boldsymbol{D}_0 & \boldsymbol{D}_1 & \boldsymbol{D}_2 & \boldsymbol{D}_3 & \cdots \\
\hline
 & \boldsymbol{D}_0 & \boldsymbol{D}_1 & \boldsymbol{D}_2 & \\
\hline
 & & \boldsymbol{D}_0 & \boldsymbol{D}_1 & \ddots \\
\hline
 & & & \boldsymbol{D}_0 & \ddots \\
\hline
 & & & & \ddots \\
\hline
\end{array} \; .
$$

To avoid complex cases it is commonly assumed that the considered BMAPs are regular:

- The phase process $(\boldsymbol{D} = \sum_{i=0}^{\infty} \boldsymbol{D}_i)$ is irreducible.
- The mean interarrival time is positive and finite (the eigenvalues of $\boldsymbol{D}_0$ have negative real part, consequently $\boldsymbol{D}_0$ is nonsingular).
- The mean arrival rate, $\sum_{k=0}^{\infty} k \boldsymbol{D}_k \mathbb{1}$, is finite in each phase. This condition is obviously satisfied when there is a boundary K for which $\boldsymbol{D}_k = \boldsymbol{0}$ if $k > K$.

An example for BMAPs is the generalization of Markov modulated Poisson process (MMPP) with batch arrivals. Assuming that a background Markov chain with generator $\boldsymbol{Q}$ works such that, during a stay in state i, it generates arrival instants according to a Poisson process with rate λ_i and at an arrival instant the number of arriving customers is k with probability p_k the obtained arrival process is a BMAP with representation $\boldsymbol{D}_0 = \boldsymbol{Q} - diag\langle \boldsymbol{\lambda} \rangle$, and $\boldsymbol{D}_k = p_k \, diag\langle \boldsymbol{\lambda} \rangle$ for $k \geq 1$, where $diag\langle \boldsymbol{\lambda} \rangle$ is the diagonal matrix composed by the arrival rates λ_i.

The properties of BMAPs can be obtained in a similar approach as the ones of MAPs. In the next section we collect some properties and we refer to [8] for further details.

5.2.4 Properties of Batch Markov Arrival Processes

While a MAP generates a correlated sequence of interarrival times (T_k), a BMAP generates a correlated sequence of a pair of random variables (T_k, M_k), where T_k represents the kth interarrival time and M_k is the number of customers which arrive at the kth arrival instant. Still, the basic properties of BMAPs come from properties of the ($N(t), J(t)$) Markov chain depicted in Fig. 5.4.

- Similar to MAPs the inter arrival times are PH distributed with generator D_0 and the initial phase distribution might depend on various conditions.
- Irrespective of the number of arrivals at arrival instants the occurrence of the arrival instants follows a MAP with generator ($D_0, \sum_{i=1}^{\infty} D_i$). Consequently, irrespective of the number of arrivals the phase distribution at arrival instants forms a DTMC with transition probability matrix $P = (-D_0)^{-1} \sum_{i=1}^{\infty} D_i$.
- The consecutive interarrival times and number of arrivals can be correlated due to the correlation of the initial phases.
- The phase process ($J(t)$) is a CTMC with generator $D = \sum_{i=0}^{\infty} D_i$, where $D\mathbb{1} = 0$. Based on D, some properties of the phase process can be analyzed independent of the level process.
- The (time) stationary distribution of the phase process, α, is the solution of $\alpha D = 0, \alpha\mathbb{1} = 1$.
- Irrespective of the number of arrivals at arrival instants the (embedded) stationary phase distribution right after an arrival, π, is the solution of $\pi P = \pi, \pi\mathbb{1} = 1$, and the stationary interarrival time is PH distributed with representation (π, D_0).
- These stationary distributions are related by

$$\alpha = \frac{\pi(-D_0)^{-1}}{\pi(-D_0)^{-1}\mathbb{1}} \quad \text{and} \quad \pi = \frac{\alpha \sum_{i=1}^{\infty} D_i}{\alpha \sum_{i=1}^{\infty} D_i \mathbb{1}}.$$

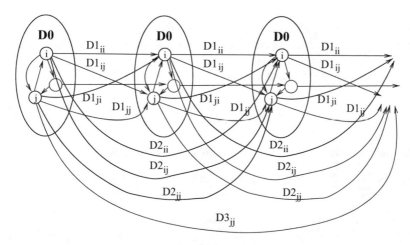

Fig. 5.4 Structure of the Markov chain describing the arrivals of a BMAP

- The stationary arrival intensity can be computed from

$$\lambda = \alpha \sum_{i=1}^{\infty} i \, D_i \mathbb{1}. \tag{5.8}$$

Further properties of BMAPs can be computed from the joint density function of consecutive interarrival times and number of arrivals $f_{T_0, M_0, \ldots, T_k, M_k}(x_0, m_0, \ldots, x_k, m_k)$, which is defined as

$$f_{T_0, M_0, \ldots, T_k, M_k}(x_0, m_0, \ldots, x_k, m_k)$$
$$= \frac{\partial}{\partial x_0} \cdots \frac{\partial}{\partial x_k} \mathbf{P}(T_0 < x_0, M_0 = m_0, \ldots, T_k < x_k, M_k = m_k).$$

The joint density function with initial phase distribution γ is

$$f_{T_0, M_0, T_1, M_1 \ldots, T_k, M_k}(x_0, m_0, \ldots, x_k, m_k) = \gamma e^{D_0 x_0} D_{m_0} e^{D_0 x_1} D_{m_1} \ldots e^{D_0 x_k} D_{m_k} \mathbb{1}. \tag{5.9}$$

For the stationary behavior of the BMAP γ needs to be replaced by π. Several measures can be derived from this joint density function in a similar manner as for MAPs. As a BMAP specific example, the probability that $M_0 = m_0$ and $M_k = m_k$ can be computed as

$$\mathbf{P}(M_0 = m_0, M_k = m_k) = \int_{x_0} \cdots \int_{x_k} \sum_{m_1} \cdots \sum_{m_{k-1}} f_{T_0, M_0, \ldots, T_k, M_k}$$

$$(x_0, m_0, \ldots, x_k, m_k) \, \mathrm{d}x_k \ldots \mathrm{d}x_0$$

$$= \gamma(-D_0)^{-1} D_{m_0} P^{k-1}(-D_0)^{-1} D_{m_k} \mathbb{1},$$

where we used that $\int_x e^{D_0 x} \mathrm{d}x = (-D_0)^{-1}$ and $P = (-D_0)^{-1} \sum_i D_i$.

For computing the distribution of the number of arrivals in $(0, t)$, $\mathbf{P}(t, n)$, defined in (5.6), we need to consider that level n can be reached from level $n - k$ by a batch arrival of k customers, from which

$$\frac{\mathrm{d}}{\mathrm{d}t} P(t, n) = \sum_{k=0}^{n} P(t, n - k) D_k, \tag{5.10}$$

with initial conditions $\mathbf{P}(0, 0) = I$, and $\mathbf{P}(0, n) = \mathbf{0}$ for $n \geq 1$. By defining $\hat{P}(t, z) = \sum_{n=0}^{\infty} z^n P(t, n)$ and $D(z) = \sum_{n=0}^{\infty} z^n D_n$, from (5.10) we have

$$\frac{\mathrm{d}}{\mathrm{d}t} \hat{P}(t, z) = \sum_{n=0}^{\infty} z^n \sum_{k=0}^{n} P(n-k, t) D_k = \sum_{k=0}^{\infty} z^k \sum_{n=k}^{\infty} z^{n-k} P(n-k, t) D_k = \hat{P}(t, z) D(z)$$

with initial condition $\hat{P}(0, z) = I$, whose solution is

$$\hat{P}(t, z) = \hat{P}(0, z)e^{D(z)t} = e^{D(z)t}. \tag{5.11}$$

With initial distribution α, where $\alpha D = \alpha D(1) = 0$, the mean number of arrivals in $(0, t)$ can be obtained as

$$\alpha \frac{d}{dz}\hat{P}(t, z)|_{z=1}\mathbb{1} = \alpha \frac{d}{dz}e^{D(z)t}|_{z=1}\mathbb{1} = \alpha \frac{d}{dz}\sum_{i=0}^{\infty}\frac{t^i}{i!}D(z)^i|_{z=1}\mathbb{1} =$$

$$\alpha \sum_{i=1}^{\infty}\sum_{j=0}^{i-1}\frac{t^i}{i!}D(z)^j\frac{d}{dz}D(z)D(z)^{i-j-1}|_{z=1}\mathbb{1} =$$

$$\alpha \sum_{i=1}^{\infty}\sum_{j=0}^{i-1}\frac{t^i}{i!}D(1)^j\left(\sum_{k=0}^{\infty}kD_k\right)D(1)^{i-j-1}\mathbb{1} =$$

$$t\alpha\sum_{k=0}^{\infty}kD_k\mathbb{1} + \sum_{i=2}^{\infty}\sum_{j=0}^{i-1}\frac{t^i}{i!}\underbrace{\alpha D(1)^j}_{\mathbf{0}}\left(\sum_{k=0}^{\infty}kD_k\right)\underbrace{D(1)^{i-j-1}\mathbb{1}}_{\mathbf{0}} = t\alpha\sum_{k=0}^{\infty}kD_k\mathbb{1},$$

which verifies (5.8).

5.2.5 Marked Markov Arrival Process

Marked Markov arrival processes (MMAPs) [6] are practically the same as BMAPs with a different interpretation. At an arrival instant of a BMAP the number of arrivals is the random variable denoted by N_i in (5.9). In a MMAP, N_i represents the *type* of the arrival, where the type can refer to any special modeling feature associated with the arrival, e.g., the type can refer to the priority class of the incoming customer. If type represents the number of customers arriving at an arrival instant, then MMAP is identical with BMAP.

For example, if a two-state background Markov chain with generator Q generates high priority (type 1) arrivals in state one with rate λ_1 and low priority (type 2) arrivals in state two with rate λ_2, then the generated arrival stream is an MMAP with representation

$$D_0 = \begin{array}{|c|c|} \hline q_{11} - \lambda_1 & q_{12} \\ \hline q_{21} & q_{22} - \lambda_2 \\ \hline \end{array}, \quad D_1 = \begin{array}{|c|c|} \hline \lambda_1 & 0 \\ \hline 0 & 0 \\ \hline \end{array}, \quad D_2 = \begin{array}{|c|c|} \hline 0 & 0 \\ \hline 0 & \lambda_2 \\ \hline \end{array}, \quad \text{if } Q = \begin{array}{|c|c|} \hline q_{11} & q_{12} \\ \hline q_{21} & q_{22} \\ \hline \end{array}.$$

5.3   Quasi-Birth–Death Process

There are very few Markov chain structures which ensure solutions with convenient analytical properties. One of these few Markov chain structures is the Quasi-Birth–Death (QBD) process.

Definition 5.5 A CTMC $\{N(t), J(t)\}$ with state space $\{n, j\}$ ($n \in \mathbb{N}, j \in \{1, \ldots, N\}$) is a *Quasi-Birth–Death process* if transitions are restricted to modify n by at most one and the transitions are homogeneous for different n values for $n \geq 1$. That is, the transition rate from $\{n, j\}$ to $\{n', j'\}$ is zero if $|n - n'| \geq 2$ and the transition rate from $\{n, j\}$ to $\{n', j'\}$ equals to the transition rate from $\{1, j\}$ to $\{n' - n + 1, j'\}$ (cf. Fig. 5.5).

These structural descriptions are relaxed below by considering various versions of this basic regular QBD model. Similar to the case of MAPs $N(t)$ is commonly referred to the "level" process (it represents, e.g., the number of customers in a queue), and $J(t)$ is commonly referred to the "phase" process (it represents, e.g., the state of a randomly changing environment). Hereafter we assume that the considered QBD processes are irreducible with irreducible phase process at the $n \geq 1$ levels (as detailed below).

Due to the structural properties of QBDs their state transitions can be classified as forward ($n \to n + 1$), local ($n \to n$), and backward ($n \to n - 1$). We apply the following notations:

- Matrix F of size $N \times N$ contains the rates of the forward transitions. The i, j element of F is the transition rate from $\{n, i\}$ to $\{n + 1, j\}$ ($n \geq 0$).
- Matrix L of size $N \times N$ contains the rates of the local transitions for $n \geq 1$.
- Matrix L' of size $N \times N$ contains the rates of the local transitions for $n = 0$. Level 0 is irregular, because there is no backward transition from level 0.
- Matrix B of size $N \times N$ contains the rates of the backward transitions. The i, j element of F is the transition rate from $\{n + 1, i\}$ to $\{n, j\}$ ($n \geq 0$).

Using these notations the structure of the generator matrix of a QBD process is

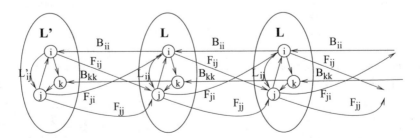

Fig. 5.5 The transition structure of Quasi-Birth–Death processes

$$Q = \begin{array}{|c|c|c|c|c|}
\hline
L' & F & & & \\
\hline
B & L & F & & \\
\hline
& B & L & F & \\
\hline
& & B & L & F \\
\hline
& & & \ddots & \ddots \\
\hline
\end{array} \, . \tag{5.12}$$

The name of QBD processes comes from the fact that on the matrix block level the generator matrix has a birth–death structure.

Condition of Stability

The phase process of a QBD in the regular part ($n > 1$) is a CTMC with generator matrix $A = F + L + B$. Let A be irreducible with stationary distribution α (that is, $\alpha A = 0, \alpha \mathbb{1} = 1$). The drift associated with the stationary distribution of the regular phase process is $d = \alpha F \mathbb{1} - \alpha B \mathbb{1}$. The sign of this drift indicates if the average tendency of the level process is increasing in the regular part. If $d < 0$, then the QBD process is positive recurrent [10]. That is, the condition of stability of QBD processes is $d = \alpha F \mathbb{1} - \alpha B \mathbb{1} < 0$ where α is the solution of $\alpha(F + L + B) = 0, \alpha \mathbb{1} = 1$.

5.3.1   Matrix Geometric Distribution

The stationary solution of a QBD with generator Q is the solution of the linear system of equations $\pi Q = 0, \pi \mathbb{1} = 1$, where π is the row vector of stationary probabilities. To utilize the regular structure of matrix Q we partition vector π according to the levels of the QBD process: $\pi = \{\pi_0, \pi_1, \pi_2, \ldots\}$. Using this partitioning the linear system of equations takes the following form

$$\pi_0 L' + \pi_1 B = 0, \tag{5.13}$$

$$\pi_{n-1} F + \pi_n L + \pi_{n+1} B = 0, \quad \forall n \geq 1, \tag{5.14}$$

$$\sum_{n=0}^{\infty} \pi_n \mathbb{1} = 1 . \tag{5.15}$$

Theorem 5.4 *The solution of* (5.13)–(5.15) *in case of a stable QBD is* $\pi_n = \pi_0 R^n$ *where matrix* R *is the only solution of the quadratic matrix equation*

$$F + RL + R^2 B = 0, \tag{5.16}$$

whose eigenvalues are inside the unit disk and vector π_0 *is the solution of linear system of size N*

$$\pi_0(L' + RB) = 0,$$

with normalizing condition

$$\pi_0(I - R)^{-1}\mathbb{1} = 1 \ .$$

Proof In case of stable irreducible CTMCs the solution of the linear system $\pi Q = 0$, $\pi \mathbb{1} = 1$ is unique and is identical with the stationary distribution of the CTMC. In this proof we only show that $\pi_n = \pi_0 R^n$ satisfies the linear system and do not discuss the properties of the solutions of the quadratic matrix equations. The details of the spectral properties of the solutions are discussed, e.g., in [8]. Substituting the $\pi_n = \pi_0 R^n$ solution into (5.14) gives:

$$\pi_0 R^{n-1} F + \pi_0 R^n L + \pi_0 R^{n+1} B = \pi_0 R^{n-1}(F + RL + R^2 B) = 0 \quad \forall n \geq 1,$$

which holds according to the definition of R. Due to the fact that the eigenvalues of R are inside the unit disk the infinite sum $\sum_{n=0}^{\infty} R^n$ is finite and we have $\sum_{n=0}^{\infty} R^n = (I - R)^{-1}$. Using this and substituting the $\pi_n = \pi_0 R^n$ solution into (5.13) and (5.15) gives:

$$\pi_0 L' + \pi_0 R B = 0$$

$$\sum_{n=0}^{\infty} \pi_0 R^n \mathbb{1} = \pi_0(I - R)^{-1}\mathbb{1} = 1,$$

which is the linear system defining π_0. $\square$

The stationary distribution, $\pi_n = \pi_0 R^n$, is commonly referred to as matrix geometric distributions. This terminology refers also to the relation of homogeneous birth and death processes and QBDs, since the stationary distribution of homogeneous birth and death processes is geometric. Similar to the relation of Poisson processes and MAPs, QBDs can be interpreted as an extension of birth and death processes such that their generator matrices have the same structure on the level of matrix blocks.

There is an extensive literature dealing with the properties of QBDs and the efficient computation of matrix R, therefore here we only present two computational methods for matrix R and we refer the interested readers to [2] for a recent survey and to [3] for a downloadable set of Matlab functions.

Linear Algorithm

$$R := 0;$$
$$\textbf{REPEAT}$$
$$R := F\,(-L - RB)^{-1}\,;$$
$$\textbf{UNTIL}||F + RL + R^2 B|| \leq \epsilon$$

Logarithmic Algorithm

$$H := F(-L)^{-1};$$
$$K := B(-L)^{-1};$$
$$R := H;$$
$$T := K;$$
REPEAT
$$U := HK + KH;$$
$$H := H^2(I-U)^{-1};$$
$$K := K^2(I-U)^{-1};$$
$$R := R + HT;$$
$$T := KT;$$
$$\textbf{UNTIL} \|F + RL + R^2B\| \le \epsilon$$

The input data of these algorithms are matrices F, L, B, and a predefined accuracy parameter ϵ. The main differences between the algorithms are that the linear algorithm has simpler iteration step and it is more sensitive to the drift d. When the drift is close to 0 the linear algorithm performs a huge number of iterations. The properties of the logarithmic algorithm are opposite of the respective ones. It has a more complex iteration step, but the number of iterations is tolerable also for drift values close to 0.

The next subsections present different QBD variants whose stationary distributions are different variants of the matrix geometric distribution.

5.3.2 Special Quasi-Birth–Death Processes

5.3.2.1 Quasi-Birth–Death Process with Irregular Level 0

There are a lot of practical examples where the system has a regular behavior when it is in normal operation mode in some sense, but it has a different behavior (e.g., different state transition structure or rates, or even different number of phases) when it is idle in some sense. Additionally, any CTMC, which exhibits a regular QBD structure from a given point on, can be considered to be a QBD with irregular level 0, where level 0 is defined such that it contains the whole irregular part of the state space.

In general, a QBD with irregular level 0 has the following block structure

$$Q = \begin{array}{|c|c|c|c|c|}
\hline
L' & F' & & & \\
\hline
B' & L & F & & \\
\hline
& B & L & F & \\
\hline
& & B & L & F \\
\hline
& & & \ddots & \ddots \\
\hline
\end{array}$$

where the sizes of the blocks are identical for levels $1, 2, \ldots$, but the sizes of the blocks at level 0 can be different from the regular block size. If N is the regular block size and N_0 is the block size at level 0, then matrices F, L, B are of size $N \times N$, matrix F' is of size $N_0 \times N$, matrix L' is of size $N_0 \times N_0$, and matrix B' is of size $N \times N_0$.

In this case the partitioned form of the linear system $\pi Q = 0$, $\pi \mathbb{1} = 1$ is

$$\pi_0 L' + \pi_1 B' = 0, \tag{5.17}$$

$$\pi_0 F' + \pi_1 L + \pi_2 B = 0, \tag{5.18}$$

$$\pi_{n-1} F + \pi_n L + \pi_{n+1} B = 0, \quad \forall n \geq 2, \tag{5.19}$$

$$\sum_{n=0}^{\infty} \pi_n \mathbb{1} = 1. \tag{5.20}$$

Theorem 5.5 *The solution of (5.17)–(5.20) in case of a stable QBD is π_0 and $\pi_n = \pi_1 R^{n-1}$ ($n \geq 1$) where matrix R is the only solution of the quadratic matrix equation*

$$F + RL + R^2 B = 0,$$

whose eigenvalues are inside the unit disk and vectors π_0, π_1 come from the solution of the linear system of size $N_0 + N$

$$\pi_0 L' + \pi_1 B' = 0,$$

$$\pi_0 F' + \pi_1 (L' + RB) = 0,$$

with normalizing condition

$$\pi_0 \mathbb{1} + \pi_1 (I - R)^{-1} \mathbb{1} = 1.$$

Proof The proof follows the same pattern as the one of Theorem 5.4. Substituting the matrix geometric solution into the partitioned form of the stationary equations indicates that the solution satisfies the stationary equations.   □

The linear system for π_0 and π_1 can be rewritten into the matrix from

$$[\pi_0 | \pi_1] \begin{array}{|c|c|} \hline L' & F' \\ \hline B' & L + RB \\ \hline \end{array} = [\, 0 \mid 0 \,]$$

5.3.2.2   Finite Quasi-Birth–Death Process

Another frequently applied variant of QBD processes is the case when the level process has an upper limit. When the upper limit is at level m the generator matrix takes the form:

$$Q = \begin{array}{|c|c|c|c|c|}
\hline
L' & F & & & \\
\hline
B & L & \ddots & & \\
\hline
& B & \ddots & F & \\
\hline
& & \ddots & L & F \\
\hline
& & & B & L'' \\
\hline
\end{array} \quad ,$$

and the partitioned form of the stationary equation is:

$$\pi_0 L' + \pi_1 B = 0, \tag{5.21}$$

$$\pi_{n-1} F + \pi_n L + \pi_{n+1} B = 0, \quad 1 \le n \le m - 1, \tag{5.22}$$

$$\pi_{m-1} F + \pi_m L'' = 0, \tag{5.23}$$

$$\sum_{n=0}^{m} \pi_n \mathbb{1} = 1 . \tag{5.24}$$

Theorem 5.6 *The solution of (5.21)–(5.24) in case of a finite QBD with $d < 0$ is $\pi_n = \alpha R^n + \beta S^{m-n}$ ($0 \le n \le m$) where matrix R is the only solution of the quadratic matrix equation*

$$F + RL + R^2 B = 0,$$

whose eigenvalues are inside the open unit disk, matrix S is the only solution of the quadratic matrix equation

$$B + SL + S^2 F = 0,$$

whose eigenvalues are on the closed unit disk and vectors α, β are the solution of linear system of size $2N$

$$\alpha \left(L' + RB \right) + \beta S^{m-1} \left(SL' + B \right) = 0,$$

$$\alpha R^{m-1} \left(F + RL'' \right) + \beta \left(SF + L'' \right) = 0,$$

with normalizing condition

$$\alpha \sum_{n=0}^{m} R^n \mathbb{1} + \beta \sum_{n=0}^{m} S^n \mathbb{1} = 1.$$

Proof The proof follows the same pattern as the one of Theorem 5.4. Substituting the solution into the partitioned form of the stationary equations indicates that the solution satisfies the stationary equations. □

The matrix form of the linear system for $\boldsymbol{\alpha}$ and $\boldsymbol{\beta}$ is

$$[\boldsymbol{\alpha}|\boldsymbol{\beta}] \begin{array}{|c|c|} \hline \boldsymbol{L'+RB} & \boldsymbol{R^{m-1}\left(F+RL''\right)} \\ \hline \boldsymbol{S^{m-1}\left(SL'+B\right)} & \boldsymbol{SF+L''} \\ \hline \end{array} = [\,\boldsymbol{0}\,|\,\boldsymbol{0}\,]\,.$$

Matrix $\boldsymbol{S}$ can be computed by the same linear or logarithmic procedures as matrix $\boldsymbol{R}$. In a finite QBD process the problem of stability does not exist and the drift could be positive, negative, or zero. If the drift is negative the eigenvalues of matrix $\boldsymbol{R}$ are inside the unit disk, and matrix $\boldsymbol{S}$ is such that it has an eigenvalue on the unit circle at one, while all other eigenvalues of matrix $\boldsymbol{S}$ are inside the unit disk. If the drift is positive, then the numbering of the levels needs to be inverted ($0 \rightarrow m, 1 \rightarrow m-1, \ldots, m \rightarrow 0$) and we obtain a new finite QBD process whose drift is negative and $\boldsymbol{R}$ and $\boldsymbol{S}$ are interchanged. If the drift is zero ($d=0$) then both, $\boldsymbol{R}$ and $\boldsymbol{S}$, are such that they have an eigenvalue on the unit circle at one, while all other eigenvalues are inside the unit disk [8].

We note that, if $\boldsymbol{R}$ or $\boldsymbol{S}$ has an eigenvalue on the unit circle, then $\sum_{n=0}^{\infty} \boldsymbol{R}^n$ or $\sum_{n=0}^{\infty} \boldsymbol{S}^n$ does not converge. Fortunately, it does not affect the applicability of Theorem 5.6 because we need to compute only the finite sum $\sum_{n=0}^{m} \boldsymbol{R}^n$ or $\sum_{n=0}^{m} \boldsymbol{S}^n$.

Computation of $\sum_{k=0}^{m} \boldsymbol{R}^k$ if the Drift Is Nonnegative

If the drift is nonnegative, then $\boldsymbol{R}$ has an eigenvalue on the unit circle at one, while all other eigenvalues of matrix $\boldsymbol{R}$ are inside the unit disk. Let the eigenvector associated with eigenvalue one be u and v, i.e., u and v are nonzero and satisfy $\boldsymbol{R}u = u$ and $v\boldsymbol{R} = v$, then

$$\sum_{k=0}^{m} \boldsymbol{R}^k = \left(\boldsymbol{I} - (\boldsymbol{R}-\boldsymbol{\Pi})^{m+1}\right)\left(\boldsymbol{I} - (\boldsymbol{R}-\boldsymbol{\Pi})\right)^{-1} + (m+1)\boldsymbol{\Pi} \qquad (5.25)$$

where $\boldsymbol{\Pi} = \dfrac{uv}{vu}$ and we used that $\boldsymbol{R}\boldsymbol{\Pi} = \boldsymbol{\Pi}\boldsymbol{R} = \boldsymbol{\Pi}$, $\boldsymbol{\Pi}^i = \boldsymbol{\Pi}$ for $i \geq 1$, and the eigenvalues of $\boldsymbol{R}-\boldsymbol{\Pi}$ are inside the unit disk. Based on these, for $k \geq 1$

$$\boldsymbol{R}^k = ((\boldsymbol{R}-\boldsymbol{\Pi})+\boldsymbol{\Pi})^k = (\boldsymbol{R}-\boldsymbol{\Pi})^k + \underbrace{(\boldsymbol{R}-\boldsymbol{\Pi})^{k-1}\boldsymbol{\Pi} + \ldots + (\boldsymbol{R}-\boldsymbol{\Pi})\boldsymbol{\Pi}^{k-1}}_{0} + \underbrace{\boldsymbol{\Pi}^k}_{\boldsymbol{\Pi}}$$

and $\sum_{k=0}^{\infty}(\boldsymbol{R}-\boldsymbol{\Pi})^k = \left(\boldsymbol{I} - (\boldsymbol{R}-\boldsymbol{\Pi})\right)^{-1}$.

5.3.2.3   Partially Homogeneous Quasi-Birth–Death Process

In this case, there is a homogeneous behavior on levels 0 to m and levels $m + 1$ to infinity. The generator matrix has the form

$$
Q =
\begin{array}{cccccccc}
 & 0 & 1 & \cdots & m{-}1 & m & m{+}1 & \cdots \\
\hline
L' & F & & & & & & \\
B & L & \ddots & & & & & \\
 & B & \ddots & F & & & & \\
 & & \ddots & L & F & & & \\
 & & & B & L'' & \hat{F} & & \\
 & & & & \hat{B} & \hat{L} & \hat{F} & \\
 & & & & & \hat{B} & \hat{L} & \ddots \\
 & & & & & & \ddots & \ddots \\
\end{array}
$$

and the associated stationary equations are

$$\pi_0 L' + \pi_1 B = 0,$$

$$\pi_{n-1} F + \pi_n L + \pi_{n+1} B = 0, \quad 1 \le n \le m - 1,$$

$$\pi_{m-1} F + \pi_m L'' + \pi_{m+1} \hat{B} = 0,$$

$$\pi_{n-1} \hat{F} + \pi_n \hat{L} + \pi_{n+1} \hat{B} = 0, \quad m + 1 \le n,$$

$$\sum_{n=0}^{\infty} \pi_n \mathbb{1} = 1.$$

Theorem 5.7 *The stationary solution is*

$$\pi_n = \alpha R^n + \beta S^{m-n}, \quad 0 \le n \le m - 1,$$

$$\pi_n = \pi_m \hat{R}^{n-m}, \quad m \le n,$$

where matrices R, S and $\hat{R}$ are the minimal solutions of the quadratic matrix equations

$$F + RL + R^2 B = 0$$

$$B + SL + S^2 F = 0$$

$$\hat{F} + \hat{R}\hat{L} + \hat{R}^2 \hat{B} = 0$$

and α, β and π_m, are the solution of the linear system

$$[\alpha \mid \beta \mid \pi_m] \cdot \begin{array}{|c|c|c|} \hline L' + RB & R^{m-2}(F + RL) & R^{m-1}F \\ \hline S^{m-2}(SL' + B) & SF + L & F \\ \hline 0 & B & L'' + \hat{R}\hat{B} \\ \hline \end{array} = [0 \mid 0 \mid 0],$$

$$\alpha \sum_{n=0}^{m-1} R^n \mathbb{1} + \beta \sum_{n=0}^{m-1} S^n \mathbb{1} + \pi_m (I - \hat{R})^{-1} \mathbb{1} = 1.$$

Proof The proof follows the same pattern as the one of Theorem 5.4. Substituting the solution into the stationary equations indicates that the solution satisfies the stationary equations. $\square$

Utilizing the regular (level homogeneous) structure of the transition matrix of the Markov chain the previous subsections provided examples where the stationary distribution is obtained from the solution of some quadratic matrix equations and a linear system.

5.3.2.4 Stationary Solution of Level Dependent QBD Process

In the previous subsections we considered Markov chains with homogeneous (level independent) QBD structure for different intervals of levels and obtained their stationary distribution utilizing the homogeneous QBD structure for different intervals. In this section we consider Markov chains with more general structure, where the QBD structure is present, but it is level dependent, and in general, it is different for all levels. The structure of the generator matrix is

$$Q = \begin{array}{|c|c|c|c|c|} \hline L_0 & F_0 & & & \\ \hline B_1 & L_1 & F_1 & & \\ \hline & B_2 & L_2 & F_2 & \\ \hline & & B_3 & L_3 & F_3 \\ \hline & & & \ddots & \ddots \\ \hline \end{array},$$

where the matrices of the transition rates to the neighboring and the same level are level dependent:

- F_n—(forward) transitions from level n to $n + 1$,
- L_n—(local) transitions in level n,
- B_n—(backward) transitions from level n to $n - 1$.

In this case the size of the consecutive levels can be different as well, in which case B_n and F_n are not square matrices.

The stationary equations are

$$0 = \pi_0 L_0 + \pi_1 B_1,$$

$$0 = \pi_{n-1} F_{n-1} + \pi_n L_n + \pi_{n+1} B_{n+1} \text{ for } n \geq 1. \tag{5.26}$$

Theorem 5.8 *The stationary solution has the form* $\pi_{i+1} = \pi_i R_i$, *from which* $\pi_i = \pi_0 \prod_{n=0}^{i-1} R_n$, *where* R_{n-1} *and* R_n *satisfy*

$$0 = F_{n-1} + R_{n-1} L_n + R_{n-1} R_n B_{n+1} \text{ for } n \geq 1. \tag{5.27}$$

Proof Substituting $\pi_{i+1} = \pi_i R_i$ into the stationary equation, (5.26), we obtain (5.27).

The numerical solution of (5.27) goes downwards, when it is possible to find a level boundary, N, such that $R_n \simeq R$ for $\forall n \geq N$. In that case R is obtained from the quadratic matrix equation

$$0 = F_N + R L_{N+1} + R^2 B_{N+2}$$

and for $n = N, N - 1, N - 2, \ldots, 1$, R_{n-1} is computed from

$$R_{n-1} = F_{n-1} (-L_n - R_n B_{n+1})^{-1},$$

where $R_N = R$ and for the computation of R_{n-1} for $n < N$, R_n is known. Finally, π_0 is computed from

$$0 = \pi_0 (L_0 + R_0 B_1),$$

with normalizing equation

$$1 = \pi_0 \sum_{i=0}^{N-1} \prod_{n=0}^{i-1} R_n \mathbb{1} + \pi_0 \prod_{n=0}^{N-1} R_n (I - R)^{-1} \mathbb{1}.$$

To apply the described numerical approach it is necessary that the sizes of the levels are identical for $\forall n \geq N$. If this necessary condition applies for $\forall n \geq N_S$ and the matrix blocks are such that $\lim_{n \to \infty} F_n = F$, $\lim_{n \to \infty} L_n = L$ and $\lim_{n \to \infty} B_n = B$, it is still a challenging problem to find $N \geq N_S$ above which $R_n \simeq R$ applies. A possible numerical approach for finding N is to start from $N = N_S$, complete the described numerical procedure and check the error of the obtained solution by

$$err = ||\pi_0 L_0 + \pi_1 B_1|| + \sum_{n=1}^{\infty} ||\pi_{n-1} F_{n-1} + \pi_n L_n + \pi_{n+1} B_{n+1}||.$$

If err is greater than the prescribed error limit increase N and repeat the procedure.

5.3.3   Analysis of the Level Process

For better understanding the stochastic behavior of regular QBD processes we evaluate some measures associated with some random intervals of the level process. Let γ_n be the time to the first visit to level n, i.e., $\gamma_n = \inf(t \,|\, t > 0, N(t) = n)$.

First Visit from Level n to $n - 1$

To characterize the process starting from level n until the first visit to level $n - 1$ we define $G_{ij}(t, n) = \mathbf{P}(J(\gamma_{n-1}) = j, \gamma_{n-1} < t \,|\, N(0) = n, J(0) = i)$, which is the joint distribution of the time to reach level $n - 1$ and the phase first visited in level $n - 1$ starting from phase i on level n. One of the most crucial property of QBD processes is the *level independence* of this measure, since the part of the Markov chain which plays role in the process from level n to $n - 1$ is identical with the one from level $n + 1$ to n. That is, $G_{ij}(t, n) = G_{ij}(t)$, for $\forall n \geq 1$. The Laplace Stieltjes transform of $G_{ij}(t)$ is $G_{ij}^{\sim}(s) = \int_{t=0}^{\infty} e^{-st} dG_{ij}(t)$.

Theorem 5.9 *Matrix $G^{\sim}(s)$, whose i, j element is $G_{ij}^{\sim}(s)$, satisfies*

$$sG^{\sim}(s) = B + LG^{\sim}(s) + FG^{\sim 2}(s). \tag{5.28}$$

Proof

$$G_{ij}^{\sim}(s) = \underbrace{\frac{-L_{ii}}{s - L_{ii}}}_{\text{time in }(n,i)} \left(\underbrace{\frac{B_{ij}}{-L_{ii}}}_{\text{prob. of moving to }(n-1,j)} + \sum_{k \in S, k \neq i} \underbrace{\frac{L_{ik}}{-L_{ii}}}_{\text{prob. of moving to }(n,k)} \underbrace{G_{kj}^{\sim}(s)}_{\text{time to }(n-1,j)} \right.$$

$$\left. + \sum_{k \in S} \underbrace{\frac{F_{ik}}{-L_{ii}}}_{\text{prob. of moving to }(n+1,k)} \sum_{\ell \in S} \underbrace{G_{k\ell}^{\sim}(s)}_{\text{time to }(n,\ell)} \underbrace{G_{\ell j}^{\sim}(s)}_{\text{time to }(n-1,j)} \right).$$

Multiplying both sides with $s - L_{ii}$ and adding $L_{ii} G_{ij}^{\sim}(s)$ result in the matrix form of the theorem.   □

Corollary 5.1 *Matrix $G = \lim_{t \to \infty} G(t) = \lim_{s \to 0} G^{\sim}(s)$, whose i, j element is $G_{ij} = \mathbf{P}(J(\gamma_{n-1}) = j \,|\, N(0) = n, J(0) = i)$ satisfies the quadratic matrix equation*

$$0 = B + LG + FG^2. \tag{5.29}$$

Sojourn in level n before moving to level $n - 1$

Let $V_{ij}(t) = \mathbf{P}(N(t) = n, J(t) = j, \gamma_{n-1} > t \,|\, N(0) = n, J(0) = i)$ and $V_{ij}^*(s) = \int_{t=0}^{\infty} e^{-st} V_{ij}(t) dt$ be the probability of staying in (n, j) before reaching level $n - 1$ starting from (n, i) and its Laplace transform.

Theorem 5.10 *Matrix* $V^*(s)$, *whose* i, j *element is* $V_{ij}^*(s)$, *satisfies*

$$V^*(s) = (sI - L - FG^{\sim}(s))^{-1}.$$

Proof Supposing that the sojourn time in state (n, i), denoted by H, is h we can compute the conditional probability of the sojourn in (n, j) as

$$V_{ij}(t|H = h)$$

$$= \begin{cases} \delta_{ij} & h > t \\ \sum_{k \neq i} \frac{L_{ik}}{-L_{ii}} V_{kj}(t - h) + \sum_k \sum_\ell \int_{\tau=0}^{t-h} \frac{F_{ik}}{-L_{ii}} V_{\ell j}(t - h - \tau) dG_{k\ell}(\tau) & h < t \end{cases}$$

If $h > t$, then the process stays in (n, j) at time t iff $i = j$. If $h < t$, then there is a state transition at t to either level n or to level $n + 1$. If the state transition was to level $n - 1$, then $\gamma_{n-1} > t$ is violated. If the state transition at h is inside level n, then $\frac{L_{ik}}{-L_{ii}}$ is the probability that the next state is (n, k), and $V_{kj}(t - h)$ gives the probability that the process is at (n, j) at time t such that $\gamma_{n-1} > t$. If the state transition at h is to level $n + 1$, then $\frac{F_{ik}}{-L_{ii}}$ is the probability that the next state is $(n + 1, k)$, $dG_{k\ell}(\tau)/d\tau$ is the probability density that the process returns back to level n at time $h + \tau$ in state (n, ℓ) and $V_{\ell j}(t - h - \tau)$ gives the probability that the process is at (n, j) at time t such that $\gamma_{n-1} > t$.

The sojourn time in state (n, i) is exponentially distributed with parameter $-L_{ii}$. Applying the law of total probability, $V_{ij}(t) = \int_{h=0}^{\infty} -L_{ii} e^{L_{ii} h} V_{ij}(t|H = h) dh$, for $V_{ij}(t)$ we have

$$V_{ij}(t) = e^{L_{ii} t} \delta_{ij} + \int_{h=0}^{t} e^{L_{ii} h} \left(\sum_{k \neq i} L_{ik} V_{kj}(t - h) \right.$$

$$\left. + \sum_k \sum_\ell \int_{\tau=0}^{t-h} F_{ik} V_{\ell j}(t - h - \tau) dG_{k\ell}(\tau) \right) dh,$$

whose Laplace transform, $V_{ij}^*(s) = \int_{t=0}^{\infty} e^{-st} V_{ij}(t) dt$, is

$$V_{ij}^*(s) = \frac{\delta_{ij}}{s - L_{ii}} + \frac{1}{s - L_{ii}} \left(\sum_{k \neq i} L_{ik} V_{kj}^*(s) + \sum_k \sum_\ell F_{ik} G_{k\ell}^{\sim}(s) V_{\ell j}^*(s) \right).$$

Multiplying both sides with $s - L_{ii}$ and adding $L_{ii} V_{ij}^*(s)$ result in

$$sV^*(s) = I + LV^*(s) + FG^{\sim}(s)V^*(s),$$

which is identical with the statement of the theorem. □

Corollary 5.2 *Starting from* (n, i), *the mean time spent in* (n, j) *before reaching level* $n - 1$ *is*

$$E\left(\int_{t=0}^{\infty} \mathscr{I}_{\{N(t)=n, J(t)=j, \gamma_{n-1}>t \mid N(0)=n, J(0)=i\}} dt\right)$$

$$= \int_{t=0}^{\infty} Pr(N(t) = n, J(t) = j, \gamma_{n-1} > t \mid N(0) = n, J(0) = i) dt$$

$$= \int_{t=0}^{\infty} V_{ij}(t) dt = \lim_{s \to 0} V_{ij}^{*}(s) = (-\boldsymbol{L} - \boldsymbol{F}\boldsymbol{G})_{ij}^{-1}.$$

That is, defining matrix $\boldsymbol{V}$ as $\boldsymbol{V} = \lim\limits_{s \to 0} \boldsymbol{V}^{*}(s)$ we have

$$\boldsymbol{V} = (-\boldsymbol{L} - \boldsymbol{F}\boldsymbol{G})^{-1}.$$

Theorem 5.11 $G_{ij}(t)$ *and* $V_{ij}(t)$ *are related by*

$$\frac{d}{dt} G_{ij}(t) = \sum_{k} V_{ik}(t) B_{kj}.$$

Proof By the definition of $G_{ij}(t)$ we have

$$\frac{d}{dt} G_{ij}(t) = \lim_{\Delta \to 0} \frac{1}{\Delta} P(J(\gamma_{n-1}) = j, t < \gamma_{n-1} < t + \Delta \mid N(0) = n, J(0) = i)$$

$$= \lim_{\Delta \to 0} \frac{1}{\Delta} \sum_{k} P(N(t) = n, J(t) = k, t < \gamma_{n-1} \mid N(0) = n, J(0) = i)$$

$$\cdot P(J(\gamma_{n-1}) = j, \gamma_{n-1} < t + \Delta \mid N(t) = n, J(t) = k)$$

$$= \lim_{\Delta \to 0} \frac{1}{\Delta} \sum_{k} V_{ik}(t)(\Delta B_{kj} + o(\Delta)),$$

from which the statement comes.   □

As a consequence of the theorem $G^{\sim}(s) = V^{*}(s)B$.

Sojourn Probability in Level $n + 1$ **Before Returning to Level** n
We define $R_{ij}(t)$ as follows

$$R_{ij}(t) = \tag{5.30}$$

$$\lim_{\Delta \to 0} \frac{1}{\Delta} P(N(t) = n + 1, J(t) = j, \gamma_n > t, N(0) = n + 1 \mid N(-\Delta) = n, J(-\Delta) = i)$$

That is $R_{ij}(t)$ is defined by the probability that starting from (n, i) at time $-\Delta$ a transition takes place to level $n+1$ in $(-\Delta, 0)$ and after that the process is at (n, j) at time t such that it never visits level n in $(0, t)$. In some sense, $R_{ij}(t)$ is a counterpart of $\frac{d}{dt}G_{ij}(t)$, because $R_{ij}(t)$ assumes a jump to level $n + 1$ at the beginning and a sojourn above level n after that, while $\frac{d}{dt}G_{ij}(t)$ assumes a sojourn above level $n - 1$ and a jump to level $n - 1$ at time t.

Theorem 5.12 $R_{ij}(t)$ and $V_{ij}(t)$ are related by

$$R_{ij}(t) = \sum_k F_{ik} V_{kj}(t).$$

Proof Based on the definition of $R_{ij}(t)$ in (5.30), we can write

$$R_{ij}(t) = \sum_k \lim_{\Delta \to 0} \frac{1}{\Delta} \mathbf{P}(N(0) = n + 1, J(0) = k \mid N(-\Delta) = n, J(-\Delta) = i)$$

$$\cdot \mathbf{P}(N(t) = n + 1, J(t) = j, \gamma_n > t \mid N(0) = n + 1, J(0) = k)$$

$$= \sum_k \lim_{\Delta \to 0} \frac{1}{\Delta} (F_{ik}\Delta + o(\Delta)) V_{kj}(t),$$

from which the statement comes. □

As a consequence of the theorem $\mathbf{R}^*(s) = \mathbf{F}\mathbf{V}^*(s)$, where $R_{ij}^*(s) = \int_{t=0}^{\infty} e^{-st} R_{ij}(t) dt$.

Summary of Transient Measures
Multiplying $\mathbf{V}^*(s)$ with $\mathbf{F}$ from the left and with $\mathbf{B}$ from the right and using the relation of the matrix transforms $\mathbf{V}^*(s)$, $\mathbf{F}^\star(s)$ and $\mathbf{G}^\sim(s)$ we have

$$\mathbf{F}\mathbf{V}^*(s)\mathbf{B} = \mathbf{R}^*(s)\mathbf{B} = \mathbf{F}\mathbf{G}^\sim(s).$$

Using this we obtain the following matrix equations for $\mathbf{V}^*(s)$ and $\mathbf{R}^*(s)$

$$\mathbf{V}^*(s) = \left(s\mathbf{I} - \mathbf{L} - \mathbf{F}\mathbf{V}^*(s)\mathbf{B}\right)^{-1},$$

$$\mathbf{R}^*(s) = \mathbf{F}\mathbf{V}^*(s) = \mathbf{F}\left(s\mathbf{I} - \mathbf{L} - \mathbf{F}\mathbf{V}^*(s)\mathbf{B}\right)^{-1}$$

that is

$$s\mathbf{R}^*(s) = \mathbf{F} + \mathbf{R}^*(s)\mathbf{L} + \mathbf{R}^{*2}(s)\mathbf{B},$$

and we have (5.28) for $\mathbf{G}^\sim(s)$. These three equations are such that the solution of one allows to compute the other two unknown matrices in a simple manner. $\mathbf{V}^*(s)$ can be computed from the other two as

$$V^*(s) = \left(sI - L - R^*(s)B\right)^{-1} = (sI - L - FG^\sim(s))^{-1},$$

$R^*(s)$ can be computed as

$$R^*(s) = FV^*(s) = F\,(sI - L - FG^\sim(s))^{-1},$$

and $G^\sim(s)$ can be computed as

$$G^\sim(s) = V^*(s)B = \left(sI - L - R^*(s)B\right)^{-1}B.$$

At $s \rightarrow 0$ the equations to compute the unknown matrices simplifies to (5.16), (5.29), and

$$V = (-L - FVB)^{-1},$$

and the relation of the unknown matrices to

$$V = (-L - RB)^{-1} = (-L - FG)^{-1},$$

$$R = FV = F\,(-L - FG)^{-1},$$

and

$$G = VB = (-L - RB)^{-1}B.$$

We remain to show the relation of the transient measures and the stationary state probabilities.

Theorem 5.13 *The stationary probabilities of the QBD satisfies*

$$\pi_{n+1} = \pi_n R,$$

where $R = \lim\limits_{s\to 0} R(s)$.

Proof To prove the theorem we interpret the $(N(t), J(t))$ Markov chain as a MRP whose embedded regenerative instants are the visits to level n. That is, let T_k be the kth time instant when the QBD starts a visit in a state of level n, as it is depicted in Fig. 5.6.

Since the states of the Markov chain are identified with 2 parameters, $N(t)$ and $J(t)$, the global and the local kernel of the MRP are

$$K_{n,i;m,j}(t) = \mathbf{P}(T_1 < t, N(T_1) = m, J(T_1) = j | N(0) = n, J(0) = i),$$

$$E_{n,i;m,j}(t) = \mathbf{P}(T_1 > t, N(t) = m, J(t) = j | N(0) = n, J(0) = i).$$

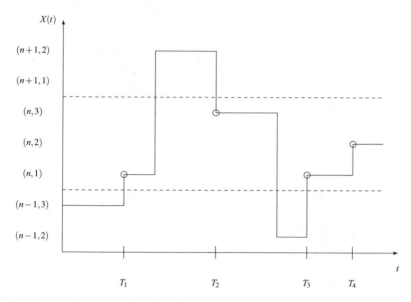

Fig. 5.6 Regenerative points of the MPR interpretation

An obvious consequence of the regeneration instant definition is that $N(T_k) = n$ for $\forall k > 0$, from which $K_{n,i;m,j}(t) = 0$ for $m \neq n$. Our main goal is to compute the stationary state probabilities, $\pi_{n,i}$ and $\pi_{n+1,j}$, based on the MRP interpretation using (4.29) and kernels $K_{n,i;m,j}(t)$ and $E_{n,i;m,j}(t)$. Let us suppose that the stationary distribution of the embedded Markov chains at regenerative instants is $\hat{\pi}$ that is

$$\hat{\pi}_j = \lim_{k \to \infty} \mathbf{P}(N(T_k) = n, J(T_k) = j) = \lim_{k \to \infty} \mathbf{P}(J(T_k) = j).$$

$\hat{\pi}$ can be computed from $K_{n,i;m,j}(t)$, but we do not need it. Instead, we need the behavior of $E_{n,i;m,j}(t)$ for $m = n$ and $m = n + 1$. For $m = n$, $E_{n,i;n,j}(t) = 0$ if $i \neq j$, because after visiting n, i the entrance to n, j results in a new regenerative instant. For $i = j$ we have the sojourn time in n, i which is exponentially distributed with parameter $-L_{ii}$, i.e.,

$$E_{n,i;n,i}(t) = e^{L_{ii}t} \quad \text{and} \quad E^*_{n,i;n,i}(s) = \frac{1}{s - L_{ii}}.$$

For $m = n + 1$, we need to consider the cases when after visiting n, i the process jumps to level $n + 1$. All other cases, a jump to level $n - 1$ and a jump inside level n, implies $E_{n,i;n+1,j}(t) = 0$ for $\forall j$. As a consequence,

$$E_{n,i;n+1,j}(t) = \int_{\tau=0}^{t} \underbrace{-L_{ii}e^{L_{ii}\tau}}_{\text{sojourn in } n,\, i} \sum_{k} \underbrace{\frac{F_{ik}}{-L_{ii}}}_{\text{jump to } n+1,\, k} \underbrace{V_{kj}(t-\tau)}_{\text{moving to } n+1,\, j \text{ in } t-\tau} \, d\tau,$$

from which

$$E^*_{n,i;n+1,j}(s) = \frac{1}{s - L_{ii}} \sum_k F_{ik} V^*_{kj}(s) = \frac{1}{s - L_{ii}} R^*_{ij}(s).$$

At the $s \to 0$ limit we have

$$\tau_{n,i;n,i} = \lim_{s \to 0} E^*_{n,i;n,i}(s) = \frac{1}{-L_{ii}},$$

$$\tau_{n,i;n+1,j} = \lim_{s \to 0} E^*_{n,i;n+1,j}(s) = \frac{1}{-L_{ii}} R_{ij}.$$

Now we can apply (4.29) and obtain

$$\pi_{n,i} = \frac{\hat{\pi}_i \frac{1}{-L_{ii}}}{c} \tag{5.31}$$

$$\pi_{n+1,j} = \frac{\sum_k \hat{\pi}_k \frac{1}{-L_{kk}} R_{kj}}{c}, \tag{5.32}$$

where $c = \sum_i \sum_m \sum_j \hat{\pi}_i \tau_{n,i:m,j}$ is a positive constant. Substituting (5.31) into (5.32) gives

$$\pi_{n+1,j} = \sum_k \pi_k R_{kj},$$

whose vector form is the statement of the theorem.   □

5.3.3.1   Characteristic Matrices of Inhomogeneous QBD Process

For the analysis of the level process of an inhomogeneous QBD we introduce the level dependent versions of the characteristic matrices:

- R_{nij}—"ratio of mean time spent in (n, i) and in $(n + 1, j)$ between two consecutive visits to level n, when the time spent in (n, i) is nonzero."
- G_{nij}—"probability that starting from (n, i) the process returns to level $n - 1$ by visiting $(n - 1, j)$"
- V_{nij}—"the mean time spent in (n, j) before leaving level n, starting from (n, i)."

These level dependent characteristic matrices satisfy

$$0 = B_n + L_n G_n + F_n G_{n+1} G_n$$

$$V_n = (-L_n - F_n V_{n+1} B_{n+1})^{-1}$$

and (5.27), and they mutually determine one another by the following relations

$$V_n = (-L_n - R_n B_{n+1})^{-1} = (-L_n - F_n G_{n+1})^{-1},$$

$$G_n = V_n B_n = (-L_n - R_n B_{n+1})^{-1} B_n,$$

$$R_{n-1} = F_{n-1} V_n = F_{n-1}(-L_n - F_{n+1} G_{n+1})^{-1}.$$

5.3.4   Discrete-Time Quasi-Birth–Death Process

When $\{N_n, J_n\}$ is a discrete-time QBD process (DTQBD), where

- N_n is the "level" process (e.g., number of customers in a queue after the nth step),
- J_n is the "phase" process (e.g., state of the environment after the nth step),

the structure of the transition probability matrix is

$$P = \begin{array}{|c|c|c|c|c|}
\hline
L' & F & & & \\
\hline
B & L & F & & \\
\hline
 & B & L & F & \\
\hline
 & & B & L & F \\
\hline
 & & & \ddots & \ddots \\
\hline
\end{array}, \tag{5.33}$$

where the element of matrix P are nonnegative state transition probabilities, and $P\mathbb{1} = \mathbb{1}$. Consequently, the block matrices $B + L + F$ and $L' + F$ are nonnegative and $(B + L + F)\mathbb{1} = \mathbb{1}$, $(L' + F)\mathbb{1} = \mathbb{1}$.

The transition probability matrix of the phase process is $P_J = F + L + B$, whose stationary solution, α, satisfies $\alpha P_J = \alpha$, $\alpha\mathbb{1} = 1$. The stationary drift of the level process is $d = \alpha F\mathbb{1} - \alpha B\mathbb{1}$, and the process is stable when the stationary drift is negative.

The stationary distribution of the DTQBD process with transition probability matrix P is the solution of the infinite linear system $\pi P = \pi$, with normalizing condition $\pi\mathbb{1} = 1$. Partitioning the stationary solution vector according to the levels $\pi = \{\pi_0, \pi_1, \pi_2, \ldots\}$ and using the special block structure of matrix P we obtain the following special, regular, and normalizing equations

$$\pi_0 L' + \pi_1 B = \pi_0, \tag{5.34}$$

$$\pi_{n-1} F + \pi_n L + \pi_{n+1} B = \pi_n, \quad \forall n \geq 1, \tag{5.35}$$

$$\sum_{n=0}^{\infty} \pi_n \mathbb{1} = 1. \tag{5.36}$$

Theorem 5.14 *In case of a stable DTQBD, the solution of (5.34)–(5.36) is* $\pi_n = \pi_0 R^n$ *where matrix R is the only solution of the quadratic matrix equation*

$$F + RL + R^2 B = R \tag{5.37}$$

whose eigenvalues are inside the unit disk and vector π_0 is the solution of linear system of size N

$$\pi_0(L' + RB) = \pi_0$$

with normalizing condition

$$\pi_0(I - R)^{-1}\mathbb{1} = 1 .$$

Proof Similar to the proof of Theorem 5.4 we only show that $\pi_n = \pi_0 R^n$ satisfies the linear system. Substituting $\pi_n = \pi_0 R^n$ into (5.35) gives

$$\pi_0 R^{n-1} F + \pi_0 R^n L + \pi_0 R^{n+1} B = \pi_0 R^n \quad \forall n \geq 1,$$

that is

$$\pi_0 R^{n-1}(F + RL + R^2 B) = \pi_0 R^{n-1} R \quad \forall n \geq 1,$$

which holds according to (5.37). Due to the fact that the eigenvalues of R are inside the unit disk we have $\sum_{n=0}^{\infty} R^n = (I - R)^{-1}$. Using this and substituting $\pi_n = \pi_0 R^n$ into (5.34) and (5.36) gives:

$$\pi_0 L' + \pi_0 RB = \pi_0$$

$$\sum_{n=0}^{\infty} \pi_0 R^n \mathbb{1} = \pi_0(I - R)^{-1}\mathbb{1} = 1,$$

which is the linear system defining π_0. $\square$

Analysis of the level process of a DTQBD is even more intuitive than the one of continuous-time QBD. Here we only discuss the analysis of matrix G and make a general comment on the relation of discrete and continuous-time QBDs.

Let γ_n be the time when the process visits level n for the first time. In case of a DTQBD, γ_n is a *positive* discrete r.v. defined as $\gamma_n = \min(u : N_u = n, u > 0)$. We define matrix G, whose i, j element is

$$G_{ij} = \mathbf{P}\big(J_{\gamma_{n-1}} = j | N_0 = n, J_0 = i\big).$$

We can compute G_{ij} based on the result of the first step of the discrete-time process as

$$G_{ij} = \sum_k B_{ik} \underbrace{\mathbf{P}\left(J_{\gamma_{n-1}} = j | N_1 = n-1, J_1 = k\right)}_{\delta_{kj}}$$

$$+ \sum_k L_{ik} \underbrace{\mathbf{P}\left(J_{\gamma_{n-1}} = j | N_1 = n, J_1 = k\right)}_{G_{kj}}$$

$$+ \sum_k F_{ik} \mathbf{P}\left(J_{\gamma_{n-1}} = j | N_1 = n+1, J_1 = k\right)$$

$$= B_{ij} + \sum_k L_{ik} G_{kj}$$

$$+ \sum_k F_{ik} \sum_\ell \mathbf{P}\left(J_{\gamma_n} = \ell | N_1 = n+1, J_1 = k\right) \mathbf{P}\left(J_{\gamma_{n-1}} = \ell | N_{\gamma_n} = n, J_{\gamma_n} = \ell\right)$$

$$= B_{ij} + \sum_k L_{ik} G_{kj} + \sum_k \sum_\ell F_{ik} G_{k\ell} G_{\ell j},$$

where δ_{kj} is the Kronecker delta. The matrix form of the scalar equation is

$$\mathbf{0} = \mathbf{B} + (\mathbf{L} - \mathbf{I})\mathbf{G} + \mathbf{F}\mathbf{G}^2. \tag{5.38}$$

There are several similarities between the measures of a DTMC with state transition probability matrix $\mathbf{P}$ and a CTMC with generator matrix $\mathbf{Q} = \mathbf{P} - \mathbf{I}$. Obviously the stationary distribution of the two processes is identical since $\pi \mathbf{Q} = \pi(\mathbf{P} - \mathbf{I}) = \mathbf{0}$ implies $\pi \mathbf{P} = \pi$, but there are several further measures of interest which are identical for these two processes. As an example, (5.29) and (5.38) indicate that matrix $\mathbf{G}$ of the DTQBD with state transition probability matrix (5.33) and the one of the continuous-time QBD with generator

$L' - I$	F			
B	$L - I$	F		
	B	$L - I$	F	
		B	$L - I$	F
			$\ddots$	$\ddots$

are identical. The same analogy applies for matrix $\mathbf{R}$ of the DTQBD process, whose i, j element is defined as

$$R_{ij} = \sum_{k=1}^\infty \mathbf{P}(N_k = n+1, J_k = j, \gamma_n > k | N_0 = n, J_0 = i).$$

The fact that γ_n is a positive r.v. has an important role in the definition of R_{ij}. R_{ij} accounts for the trajectories in which a level n to level $n + 1$ transition takes place in the first step. Matrix R is the solution of

$$F + R(L - I) + R^2 B = 0, \tag{5.39}$$

which can be obtained also from the analysis of the level process.

5.3.4.1   Level Process of Discrete-Time QBD

We define matrix $G(k)$, whose i, j element is

$$G_{ij}(k) = \mathbf{P}(J_k = j, \gamma_{n-1} = k | N_0 = n, J_0 = i).$$

$G_{ij}(k)$ defines the joint distribution of the time to reach level $n - 1$ and the state first visited on level $n - 1$. Similar to G_{ij}, $G_{ij}(k)$ can be analyzed based on the first state transition of the discrete-time process and the analysis contains a discrete convolution with respect to the time. For $k = 1$, $G_{ij}(1) = B_{ij}$, for $k > 1$

$$G_{ij}(k) = \sum_u L_{iu} G_{uj}(k-1) + \sum_u F_{iu} \sum_{\ell=1}^{k-2} \sum_v G_{uv}(\ell) G_{vj}(k - \ell - 1),$$

where for $k = 2$ the second term is zero. To avoid the discrete convolution we introduce

$$\hat{G}_{ij}(z) = \sum_{k=0}^{\infty} G_{ij}(k) z^k = \mathbf{E}\left(z^{\gamma_{n-1}} \mathscr{I}_{\{J_{\gamma_{n-1}} = j\}} \,\Big|\, N_0 = n, J_0 = i \right),$$

which satisfies

$$\hat{G}_{ij}(z) = z \left(B_{ij} + \sum_u L_{iu} \hat{G}_{uj}(z) + \sum_u F_{iu} \sum_v \hat{G}_{uv}(z) \hat{G}_{vj}(z) \right),$$

that is

$$\hat{G}(z) = z \left(B + L\hat{G}(z) + F\hat{G}^2(z) \right) \tag{5.40}$$

in matrix form.

   For $V(k)$, whose i, j element is

$$V_{ij}(k) = \mathbf{P}(J_k = j, N_k = n, \gamma_{n-1} > k | N_0 = n, J_0 = i),$$

we have a similar description with discrete convolution with respect to time. We avoid the time domain description and we introduce the z-transform $\hat{V}_{ij}(z) = \sum_{k=0}^{\infty} V_{ij}(k)z^k$, which satisfies

$$\hat{V}_{ij}(z) = \delta_{ij} + z\left(\sum_u L_{iu}\hat{V}_{uj}(z) + \sum_u F_{iu}\sum_v \hat{G}_{uv}(z)\hat{V}_{vj}(z)\right),$$

that is

$$\hat{V}(z) = I + z\left(L\hat{V}(z) + F\hat{G}(z)\hat{V}(z)\right) \tag{5.41}$$

in matrix form. The two measures are related as

$$G_{ij}(k) = \sum_u V_{iu}(k-1)B_{uj}, \quad \text{and} \quad G(k) = V(k-1)B,$$

which is identical with

$$\hat{G}_{ij}(z) = z\sum_u \hat{V}_{iu}(z)B_{uj}, \quad \text{and} \quad \hat{G}(z) = z\hat{V}(z)B. \tag{5.42}$$

Finally, for $R(k)$, whose i, j element is

$$R_{ij}(k) = \mathbf{P}(J_k = j, N_k = n+1, \gamma_n > k | N_0 = n, J_0 = i),$$

with z-transform $\hat{R}_{ij}(z) = \sum_{k=0}^{\infty} R_{ij}(k)z^k$, we have the following relation

$$\hat{R}_{ij}(z) = z\sum_u F_{iu}\hat{V}_{uj}(z), \quad \text{that is} \quad \hat{R}(z) = zF\hat{V}(z) \tag{5.43}$$

in matrix form.

Similar to the continuous-time case, these relations allow us to write matrix equations for $\hat{V}(z)$ and $\hat{R}(z)$. Substituting (5.42) into (5.41) we obtain

$$\hat{V}(z) = I + z\left(L\hat{V}(z) + zF\hat{V}(z)B\hat{V}(z)\right) = \left(I - zL - z^2F\hat{V}(z)B\right)^{-1}.$$

From (5.42) and (5.43) we have

$$\hat{R}(z)B = zF\hat{V}(z)B = F\hat{G}(z).$$

Using this and substituting $\hat{V}(z) = \left(I - zL - z\hat{R}(z)B\right)^{-1}$ into (5.43) we obtain

$$\hat{R}(z) = zF\left(I - zL - z\hat{R}(z)B\right)^{-1} = z\left(F + \hat{R}(z)L + \hat{R}^2(z)B\right). \tag{5.44}$$

From the definition of G_{ij} and $G_{ij}(k)$, we have $G = \lim_{z \to 1} \hat{G}(z)$ and similarly, $R = \lim_{z \to 1} \hat{R}(z)$. From (5.40) and (5.44), at $z \to 1$ we obtain (5.38) and (5.39).

5.4 G/M/1 Type Process

The G/M/1 type process [10, Chapter 3] is an extension of the level independent QBD process, where downward state transitions can occur to any lower levels. More precisely, an G/M/1 type process is a CTMC with two parameters, $\{N(t), J(t)\}$, where

- $N(t)$ is the "level" process (e.g., number of customers in a queue),
- $J(t)$ is the "phase" process (e.g., state of the environment),

and the upward transitions (which increases the level, $N(t)$) are restricted to one level up and there is no limit on downward transitions. The generator matrix of a G/M/1 type process has the following structure

$$
Q = \begin{array}{|c|c|c|c|c|c|}
\hline
L' & F & & & & \\
\hline
B'_1 & L & F & & & \\
\hline
B'_2 & B_1 & L & F & & \\
\hline
B'_3 & B_2 & B_1 & L & F & \\
\hline
B'_4 & B_3 & B_2 & B_1 & L & F \\
\hline
\vdots & & & \ddots & \ddots & \ddots \\
\hline
\end{array}
$$

where the matrix blocks contain

- F—transitions one level up (e.g., arrival),
- L—transitions in the same level,
- B_n—transitions n level down (e.g., n departures),
- L'—irregular block at level 0,
- $B'_n = \sum_{i=n}^{\infty} B_i$—irregular block down to level 0.

The related state transition graph is depicted in Fig. 5.7. The name of the process comes from the fact that on the block level the structure of the generator matrix is similar to the embedded Markov chain of an G/M/1 queue (c.f. Fig. 8.9 and (8.25)).

Condition of Stability

Asymptotically, as n tends to infinity, the phase process is a CTMC with generator

$$
A = F + L + \sum_{i=1}^{\infty} B_i.
$$

When A is irreducible and its stationary distribution is α,

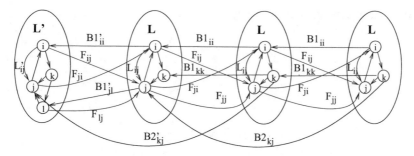

Fig. 5.7 State transitions in G/M/1 type process

where α is the unique solution of $\alpha A = 0, \alpha 1 = 1$, the asymptotic $(n \to \infty)$ stationary drift of the level process is $d = \alpha F 1 - \alpha \sum_{i=1}^{\infty} i \ B_i \ 1$.

Based on the drift, the condition of stability is the same as for QBD processes, i.e., $d < 0$, which means that the condition of stability is

$$d = \alpha F 1 - \alpha \sum_{i=1}^{\infty} i \ B_i \ 1 < 0. \tag{5.45}$$

Matrix Geometric Stationary Distribution
The block partitioned version of the stationary equation, $\pi Q = 0$, is

$$\pi_0 L' + \sum_{i=1}^{\infty} \pi_i B_i' = 0 \qquad \text{(special equation)}$$

$$\pi_{n-1} F + \pi_n L + \sum_{i=1}^{\infty} \pi_{n+i} B_i = 0 \quad \forall n \geq 1 \text{ (regular equation)}$$

where π_n is the stationary vector of level n and $\pi = \{\pi_0, \pi_1, \pi_2, \ldots\}$. The block partitioned version of the normalizing equation, $\pi 1 = 1$, is

$$\sum_{n=0}^{\infty} \pi_n 1 = 1.$$

Theorem 5.15 *The stationary distribution of the G/M/1 type process with generator Q is $\pi_n = \pi_0 R^n$ where matrix R is the minimal nonnegative solution of the matrix equation*

$$F + RL + \sum_{i=1}^{\infty} R^{i+1} B_i = 0 \tag{5.46}$$

and π_0 is the solution of the linear system

$$\pi_0 \left(L' + \sum_{i=1}^{\infty} R^i B_i \right) = 0$$

and

$$\pi_0 (I - R)^{-1} \mathbb{1} = 1.$$

Proof Substituting $\pi_n = \pi_0 R^n$ into the regular stationary equation indicates that R is defined such that the stationary solution satisfies all regular equations. The linear system for π_0 is defined such that the special equation and the normalizing conditions are satisfied as well.

Matrix R can be computed from similar iterative numerical procedures as in case of QBD processes. The following procedure linearly converges to R

$$R := 0;$$

REPEAT

$$R := F \left(-L - \sum_{i=1}^{\infty} R^i B_i \right)^{-1};$$

$$\textbf{UNTIL} \| F + RL + \sum_{i=1}^{\infty} R^{i+1} B_i \| \le \epsilon$$

Further computational methods and implemented Matlab functions can be obtained from [2, 3].

Properties of the Level Process
The matrix equation (5.46) has more than one solution. Similar to the QBD process when the G/M/1 type process is stable ($d < 0$), then the minimal nonnegative solution is such that the spectral radius of the minimal solution is less than one. The spectral radius of matrix R, ($SP(R)$), is the minimal radius of the circle around zero on the complex plain, which contains all eigenvalues of R. If the G/M/1-type process is not stable ($d \ge 0$), then the spectral radius of the minimal nonnegative solution of (5.46) is one and one is an eigenvalue of the solution with associated left eigenvector α.

The stochastic interpretation of matrix R is the same as for the QBD process. R_{ij} is the ratio of the mean time spent in (n, j) and the mean time spent in $(n-1, i)$ before the first return to level $n-1$ starting from $(n-1, i)$. This quantity is level dependent, because any jump below level $n-1$ results in a visit to level $n-1$, before visiting any level above $n-1$ again, since level process increases at most by one at a state transition.

Due to the same reasoning, matrix V is level independent and Theorem 5.12 remains valid, where similar to the QBD process, the i, j element of matrix V is the mean time spent in (n, j) before the first visit to level $n - 1$ starting from (n, i). Consequently, V satisfies

$$V = \left(-L - \sum_{i=1}^{\infty} (FV)^i B_i \right)^{-1} = \left(-L - \sum_{i=1}^{\infty} R^i B_i \right)^{-1}.$$

Unfortunately, due to the unbounded downward transitions in the level process, the level process from level n to level $n - 1$ may contain visits to levels below $n - 1$, which makes matrix G level dependent, where matrix G is the transition probability matrix from level n to level $n - 1$, as in the QBD case. Due to this level dependent property, matrix G, is not used in the analysis of G/M/1 type processes.

5.5 M/G/1 Type Process

The M/G/1 type process [11] is a counterpart of G/M/1 type process, where the downward jumps in the level process are restricted to one level, but the upward jumps are unbounded. The transition graph of an M/G/1 type process is depicted in Fig. 5.8 and the block structure of its generator matrix is

$$Q = \begin{array}{|c|c|c|c|c|}
\hline
L' & F'_1 & F'_2 & F'_3 & F'_4 \\
\hline
B & L & F_1 & F_2 & F_3 \\
\hline
 & B & L & F_1 & F_2 \\
\hline
 & & B & L & F_1 \\
\hline
 & & & \ddots & \ddots \\
\hline
\end{array},$$

where the matrix blocks contain

- L—transitions in the same level,
- B—transitions one level down (e.g., departure),
- F_n—transitions n level up (e.g., n arrivals),
- L'—irregular block at level 0,
- F'_n—irregular blocks starting from level 0.

The name of the process comes from the fact that on the block level the structure of the generator matrix is similar to the embedded Markov chain of an M/G/1 queue (c.f. Fig. 8.4 and (8.3)).

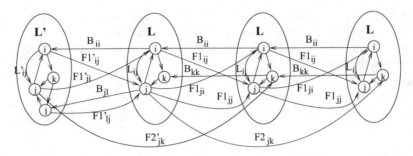

Fig. 5.8 State transitions in M/G/1 type process

Condition of Stability

For level $n \geq 2$, the phase process is a CTMC with generator matrix $A = B + L + \sum_{i=1}^{\infty} F_i$. When A is irreducible its stationary solution is α, which is the unique solution of the linear system $\alpha A = 0$, $\alpha \mathbb{1} = 1$. Based on this stationary distribution of the phase process, the condition of stability of the M/G/1 type process is that the stationary drift of the level process, d, is negative, that is

$$d = \alpha \sum_{i=1}^{\infty} i \, F_i \mathbb{1} - \alpha B \, \mathbb{1} < 0. \tag{5.47}$$

Stationary Solution of M/G/1 Type Process

The block partitioned version of the stationary and normalizing equations ($\pi Q = 0$, $\pi \mathbb{1} = 1$) are

$$\pi_0 L' + \pi_1 B = 0,$$

$$\pi_0 F'_n + \sum_{i=1}^{n-1} \pi_i F_{n-i} + \pi_n L + \pi_{n+1} B = 0 \quad \forall n \geq 1 \text{ (regular equation)},$$

$$\sum_{n=0}^{\infty} \pi_n \mathbb{1} = 1.$$

Due to the level dependent number of terms in the regular stationary equation the stationary solution is more complex than in the G/M/1 type case.

Theorem 5.16 *The stationary distribution of the M/G/1 type process with generator Q is*

$$\pi_i = \left(\pi_0 S_i' + \sum_{k=1}^{i-1} \pi_k S_{i-k}\right)(-S_0)^{-1}, \quad \forall i \geq 1, \tag{5.48}$$

where

$$S_i' = \sum_{k=i}^{\infty} F_k' G^{k-i}, \quad S_i = \sum_{k=i}^{\infty} F_k G^{k-i}, \text{ for } i \geq 1, \; S_0' = L' + S_1' G, \; S_0 = L + S_1 G,$$

matrix G is the minimal nonnegative solution of the matrix equation

$$B + LG + \sum_{i=1}^{\infty} F_i G^{i+1} = 0 \tag{5.49}$$

and the π_0 vector is the solution of the linear system:

$$\pi_0 S_0' = 0, \quad \pi_0 \left(S_0 + \sum_{i=1}^{\infty}(S_i - S_i')\right)\left(\sum_{j=0}^{\infty} S_j\right)^{-1} \mathbb{1} = 1. \tag{5.50}$$

Proof First, we define matrix G in the same way as for the QBD process. The i, j element of G is the probability that state $(n - 1, j)$ is the first state visited in level $n - 1$ starting from state (n, i). For the M/G/1 type process matrix G is level independent, since starting from level n, only levels greater than $n - 1$ can be visited before reaching level $n - 1$. Following the same reasoning as for the QBD case (Theorem 5.9 and Corollary 5.1), matrix G is the minimal nonnegative solution of (5.49).

We prove the rest of Theorem 5.16 using the concept of restricted process from Sect. 3.4.3. First we consider the restricted process on level 0. In the restricted process, a state transition from $(0, i)$ to $(0, j)$ (for $i \neq j$) can occur due to a direct state transition with rate L'_{ij}, or due to a state transition to (n, k) with rate F'_{nik}, which returns to level zero by first visiting state $(0, j)$, whose probability is G^n_{kj} (c.f., (3.55)). Altogether, in the restricted process the transition rate from $(0, i)$ to $(0, j)$ is

$$Q_{ij}^{(0)} = L'_{ij} + \sum_{n=1}^{\infty} \sum_k F'_{nik} G^n_{kj},$$

from which the generator of the restricted process is

$$Q^{(0)} = L' + \sum_{n=1}^{\infty} F'_n G^n = L' + \left(\sum_{n=1}^{\infty} F'_n G^{n-1} \right) G = L' + S'_1 G = S'_0,$$

which verifies the first equation of (5.50), since π_0 satisfies $\pi_0 Q^{(0)} = 0$.

Next, we consider the restricted process on level 0 and 1. Its generator is

$$Q^{(0,1)} = \begin{array}{|c|c|} \hline L' & \sum_{i=1}^{\infty} F'_i G^{i-1} \\ \hline B & L + \sum_{i=1}^{\infty} F_i G^i \\ \hline \end{array} = \begin{array}{|c|c|} \hline L' & S'_1 \\ \hline B & S_0 \\ \hline \end{array},$$

From $(\pi_0, \pi_1) Q^{(0,1)} = (0, 0)$ we have $\pi_0 L' + \pi_1 B = 0$ and $\pi_0 S'_1 + \pi_1 S_0 = 0$, and from this second equation we get

$$\pi_1 = \pi_0 S'_1 (-S_0)^{-1},$$

which is identical with (5.48) for $i = 1$.

Similarly, the generator of the restricted process on level 0, 1, and 2 is

$$Q^{(0,1,2)} = \begin{array}{|c|c|c|} \hline L' & F'_1 & \sum_{i=2}^{\infty} F'_i G^{i-2} \\ \hline B & L & \sum_{i=1}^{\infty} F_i G^{i-1} \\ \hline B & L + \sum_{i=1}^{\infty} F_i G^i \\ \hline \end{array} = \begin{array}{|c|c|c|} \hline L' & F'_1 & S'_2 \\ \hline B & L & S_1 \\ \hline B & S_0 \\ \hline \end{array},$$

and from the last block of $(\pi_0, \pi_1, \pi_2) Q^{(0,1,2)} = (0, 0, 0)$ we have

$$\pi_2 = \left(\pi_0 S'_1 + \pi_1 S_1 \right) (-S_0)^{-1},$$

which is identical with (5.48) for $i = 2$.

In general, for $n \geq 1$, the generator of the restricted process on levels $0, \ldots, n$ is

$$Q^{(0,\ldots,n)} = \begin{array}{|c|c|c|c|c|} \hline L' & F'_1 & \cdots & F'_{n-1} & \sum_{i=n}^{\infty} F'_i G^{i-n} \\ \hline B & L & \cdots & F_{n-2} & \sum_{i=n-1}^{\infty} F_i G^{i-n+1} \\ \hline \ddots & \ddots & & \vdots & \vdots \\ \hline & \ddots & & L & \sum_{i=1}^{\infty} F_i G^{i-1} \\ \hline & & & B & L + \sum_{i=1}^{\infty} F_i G^i \\ \hline \end{array} = \begin{array}{|c|c|c|c|c|} \hline L' & F'_1 & \cdots & F'_{n-1} & S'_n \\ \hline B & L & \cdots & F_{n-2} & S_{n-1} \\ \hline \ddots & \ddots & & \vdots & \vdots \\ \hline & \ddots & & L & S_1 \\ \hline & & & B & S_0 \\ \hline \end{array},$$

and from the last block of $(\pi_0, \ldots, \pi_n) Q^{(0,\ldots,n)} = (0, \ldots, 0)$ we have

$$\pi_n = \left(\pi_0 S'_1 + \sum_{i=1}^{n-1} \pi_i S_{n-i} \right) (-S_0)^{-1},$$

which is identical with (5.48).

To obtain the normalizing condition for $\boldsymbol{\pi}_0$, based on (5.48), we write

$$
\sum_{i=1}^{\infty} \boldsymbol{\pi}_i = \sum_{i=1}^{\infty} \boldsymbol{\pi}_0 S_i'(-S_0)^{-1} + \sum_{i=1}^{\infty} \sum_{k=1}^{i-1} \boldsymbol{\pi}_k S_{i-k}(-S_0)^{-1},
$$

$$
= \boldsymbol{\pi}_0 \sum_{i=1}^{\infty} S_i'(-S_0)^{-1} + \left(\sum_{k=1}^{\infty} \boldsymbol{\pi}_k \right) \left(\sum_{i=1}^{\infty} S_i \right) (-S_0)^{-1}.
$$

Multiplying both sides with $-S_0$ from the right it gives

$$
\sum_{i=1}^{\infty} \boldsymbol{\pi}_i \left(-S_0 - \sum_{i=1}^{\infty} S_i \right) = \boldsymbol{\pi}_0 \sum_{i=1}^{\infty} S_i'
$$

and

$$
\sum_{i=1}^{\infty} \boldsymbol{\pi}_i = \boldsymbol{\pi}_0 \sum_{i=1}^{\infty} S_i' \left(-S_0 - \sum_{i=1}^{\infty} S_i \right)^{-1} = -\boldsymbol{\pi}_0 \sum_{i=1}^{\infty} S_i' \left(\sum_{i=0}^{\infty} S_i \right)^{-1}
$$

from which we obtain the normalizing equation

$$
1 = \boldsymbol{\pi}_0 \mathbb{1} + \sum_{i=1}^{\infty} \boldsymbol{\pi}_i \mathbb{1} = \boldsymbol{\pi}_0 \mathbb{1} - \boldsymbol{\pi}_0 \left(\sum_{i=1}^{\infty} S_i' \right) \left(\sum_{j=0}^{\infty} S_j \right)^{-1} \mathbb{1}
$$

$$
= \boldsymbol{\pi}_0 \left[\left(\sum_{j=0}^{\infty} S_j \right) - \left(\sum_{i=1}^{\infty} S_i' \right) \right] \left(\sum_{j=0}^{\infty} S_j \right)^{-1} \mathbb{1},
$$

which is identical with the second equation of (5.50).

Remark 5.1 If $F_n' = F_n$ for $\forall n \geq 1$, then Theorem 5.16 simplifies to

$$
\boldsymbol{\pi}_i = \left(\sum_{k=0}^{i-1} \boldsymbol{\pi}_k S_{i-k} \right) (-S_0)^{-1}, \quad \forall i \geq 1, \tag{5.51}
$$

where $\boldsymbol{\pi}_0$ is the solution of the linear system:

$$
\boldsymbol{\pi}_0 S_0' = 0, \quad \boldsymbol{\pi}_0 S_0 \left(\sum_{j=0}^{\infty} S_j \right)^{-1} \mathbb{1} = 1.
$$

Level Process of the M/G/1 Type Process

Theorem 5.16 already indicated that matrix G is level dependent in case of an M/G/1 type process and plays fundamental role in the analysis of such processes. That is why it is often referred to as *fundamental matrix* of the M/G/1 type process.

If the drift of the process is negative, then G is a stochastic matrix, i.e., G is nonnegative and $G\mathbb{1} = \mathbb{1}$. Equivalently, it means that the spectral radius of G is 1, 1 is an eigenvalue of G and the associated right eigenvector is $\mathbb{1}$. If the drift is nonnegative, then G is sub-stochastic matrix, i.e., G is nonnegative and $G\mathbb{1} \leq \mathbb{1}$, such that the inequality is strict at least in one coordinate. Equivalently, it means that the spectral radius of G is less than 1. An iterative algorithm to calculate matrix G with linear convergence is

$$G := 0;$$
REPEAT
$$G := \left(-L - \sum_{i=1}^{\infty} F_i G^i\right)^{-1} B;$$
UNTIL $\left\lVert B + LG + \sum_{i=1}^{\infty} F_i G^{i+1}\right\rVert \leq \epsilon$

and we refer to [2, 3] for further computational methods and implemented Matlab functions.

Matrix V can also be defined similar to QBD case and due to the fact that Theorem 5.11 remains valid for M/G/1 type processes we have $G = VB$ and matrix V satisfies

$$V = \left(-L - \sum_{i=1}^{\infty} F_i G^i\right)^{-1} = (-S_0)^{-1} = \left(-L - \sum_{i=1}^{\infty} F_i (VB)^i\right)^{-1}.$$

The unbounded upward jumps in the level process inhibit the use of matrix R in the analysis of M/G/1 type processes.

5.6　Exercises

Exercise 5.1 X and Y are independent continuous PH distributed r.v. with representations (α, A) and (β, B), respectively. Define the distribution of the following r.v.

- $Z_1 = c_1 X$,
- Z_2 equals to X with probability p and to Y with probability $1 - p$,
- $Z_3 = c_1 X + c_2 Y$,
- $Z_4 = Min(X, Y)$,
- $Z_5 = Max(X, Y)$.

Exercise 5.2 X and Y are independent discrete-time PH distributed r.v. with representations (α, A) and (β, B), respectively. Define the distribution of the following r.v.

- $Z_1 = c_1 X$,
- Z_2 equals to X with probability p and to Y with probability $1 - p$,
- $Z_3 = c_1 X + c_2 Y$,
- $Z_4 = Min(X, Y)$,
- $Z_5 = Max(X, Y)$.

Exercise 5.3 There are two machines at a production site A and B. Their failure times are exponentially distributed with parameter λ_A and λ_B, respectively. Their repair times are also exponentially distributed with parameter μ_A and μ_B, respectively. There is a single repair man associated with the two machines, which can work on one machine at a time. At a given time both machines work. Compute the distribution and the moments of the time to the first complete breakdown, when both machines are failed.

5.7 Solutions

Solution 5.1 $Z_1, \ldots, Z_5$ are continuous PH distributed with the following representations. Denote the size of (α, A) by n and the size of (β, B) by m.

- PH representation of $Z_1 = c_1 X$ is of size n

$$(\gamma, G) = (\alpha, \frac{1}{c_1} A)$$

- PH representation of Z_2 is of size $n + m$

$$\gamma = (p\alpha, (1 - p)\beta), G = \begin{array}{|c|c|} \hline A & 0 \\ \hline 0 & B \\ \hline \end{array}.$$

- PH representation of $Z_3 = c_1 X + c_2 Y$ is of size $n + m$

$$\gamma = (\alpha, 0), G = \begin{array}{|c|c|} \hline \frac{1}{c_1} A & \frac{1}{c_1} a\beta \\ \hline 0 & \frac{1}{c_2} B \\ \hline \end{array}.$$

where $a = -A\mathbb{1}$ is the column vector of transition rates to the absorbing state.
- PH representation of $Z_4 = Min(X, Y)$ is of size nm

$$\gamma = \alpha \otimes \beta, G = A \oplus B,$$

- PH representation of $Z_5 = Max(X, Y)$ is of size $nm + n + m$

$$\gamma = (\alpha \otimes \beta, 0, 0), G = \begin{array}{|c|c|c|} \hline A \oplus B & I \otimes b & a \otimes I \\ \hline 0 & A & 0 \\ \hline 0 & 0 & B \\ \hline \end{array}.$$

where $a = -A\mathbb{1}$ and $b = -B\mathbb{1}$.

Solution 5.2 $Z_1, \ldots, Z_5$ are discrete PH distributed with the following representations. Denote the size of (α, A) by n and the size of (β, B) by m.

- When c_1 is a positive integer the PH representation of $Z_1 = c_1 X$ is of size $c_1 n$

$$\gamma = (\alpha, 0, \ldots, 0), G = \begin{array}{|c|c|c|c|} \hline 0 & I & \cdots & 0 \\ \hline \vdots & \ddots & \ddots & \vdots \\ \hline 0 & 0 & \cdots & I \\ \hline A & 0 & \cdots & 0 \\ \hline \end{array}.$$

- PH representation of Z_2 is of size $n + m$

$$\gamma = (p\alpha, (1 - p)\beta), G = \begin{array}{|c|c|} \hline A & 0 \\ \hline 0 & B \\ \hline \end{array}.$$

- When c_1 and c_2 are positive integers Z_3 is PH distributed with size $c_1 n + c_2 m$. In case of $c_1 = c_2 = 2$ the representation of Z_3 is

$$\gamma = (\alpha, 0, 0, 0), G = \begin{array}{|c|c|c|c|} \hline 0 & I & 0 & 0 \\ \hline A & 0 & a\beta & 0 \\ \hline 0 & 0 & 0 & I \\ \hline 0 & 0 & B & 0 \\ \hline \end{array}.$$

where $a = \mathbb{1} - A\mathbb{1}$ is the column vector of transition probabilities to the absorbing state. For $c_1, c_2 > 2$ the representation is obtained similarly.

- PH representation of $Z_4 = Min(X, Y)$ is of size nm

$$\gamma = \alpha \otimes \beta, G = A \otimes B,$$

- PH representation of $Z_5 = Max(X, Y)$ is of size $nm + n + m$

$$\gamma = (\alpha \otimes \beta, 0, 0), G = \begin{array}{|c|c|c|} \hline A \otimes B & A \otimes b & a \otimes B \\ \hline 0 & A & 0 \\ \hline 0 & 0 & B \\ \hline \end{array}.$$

where $a = \mathbb{1} - A\mathbb{1}$ and $b = \mathbb{1} - B\mathbb{1}$.

Solution 5.3 The time to the complete breakdown is continuous PH distributed with representation

$$\boldsymbol{\gamma} = (1, 0, 0), \boldsymbol{G} = \begin{array}{|c|c|c|} \hline -\lambda_A - \lambda_B & \lambda_A & \lambda_B \\ \hline \mu_A & -\lambda_B - \mu_A & 0 \\ \hline \mu_B & 0 & -\lambda_A - \mu_B \\ \hline \end{array}.$$

The distribution and the moments of the time to complete breakdown, denoted by T, can be obtained from this PH representation. For example, its cumulated density function is

$$F_T(t) = \mathbf{P}(T < t) = 1 - \boldsymbol{\gamma} \, e^{\boldsymbol{G}t} \mathbb{1},$$

and its moments are

$$\mathbf{E}(T^n) = n! \, \boldsymbol{\gamma}(-\boldsymbol{G})^{-n} \mathbb{1}.$$

References

1. Aldous, D., Shepp, L.: The least variable phase type distribution is Erlang. Stoch. Model. **3**, 467–473 (1987)
2. Bini, D., Latouche, G., Meini, B.: Numerical Methods for Structured Markov Chains. Oxford University Press, Oxford (2005)
3. Bini, D.A., Meini, B., Steffé, S., Van Houdt, B.: Structured Markov chains solver: software tools. In: Proceeding from the 2006 Workshop on Tools for Solving Structured Markov Chains, SMCtools '06. ACM, New York (2006). https://doi.org/10.1145/1190366.1190379
4. Bobbio, A., Telek, M.: A benchmark for PH estimation algorithms: results for Acyclic-PH. Stoch. Model. **10**, 661–677 (1994)
5. Cumani, A.: On the canonical representation of homogeneous Markov processes modelling failure-time distributions. Microelectron. Reliab. **22**, 583–602 (1982)
6. He, Q.M., Neuts, M.F.: Markov chains with marked transitions. Stoch. Process. Their Appl. **74**(1), 37–52 (1998)
7. Horváth, A., Telek, M.: PhFit: a general purpose phase type fitting tool. In: Tools 2002. Lecture Notes in Computer Science, vol. 2324, pp. 82–91. Springer, London (2002)
8. Latouche, G., Ramaswami, V.: Introduction to Matrix Analytic Methods in Stochastic Modeling. SIAM, Philadelphia (1999)
9. Neuts, M.F.: Probability distributions of phase type. In: Liber Amicorum Prof. Emeritus H. Florin, pp. 173–206. University of Louvain, Louvain-la-Neuve (1975)
10. Neuts, M.F.: Matrix Geometric Solutions in Stochastic Models. Johns Hopkins University Press, Baltimore (1981)
11. Neuts, M.F.: Structured Stochastic Matrices of M/G/1 Type and Their Applications. Marcel Dekker, New York (1989)
12. Telek, M.: Minimal coefficient of variation of discrete phase type distributions. In: Latouche, G., Taylor, P. (eds.) Advances in Algorithmic Methods for Stochastic Models, MAM3, pp. 391–400. Notable Publications Inc., Neshanic Station (2000)
13. Thümmler, A., Buchholz, P., Telek, M.: A novel approach for fitting probability distributions to trace data with the EM algorithm. IEEE Trans. Dependable Secure Comput. **3**(3), 245–258 (2006)

Part II
Queueing Systems

Chapter 6
Introduction to Queueing Systems

6.1 Queueing Systems

The theory of queueing systems dates back to the seminal work of A.K. Erlang (1878–1929), who worked for the telecom company in Copenhagen and studied the telephone traffic in the early twentieth century. To this today the terminology of queueing theory is closely related to telecommunications (e.g., channel, call, idle/busy, queue length, and utilization).

Due to the wide range of potential application fields (e.g., vehicular traffic, logistics, trade, banking, customer service, production lines, manufacturing systems, and stock-in-trade), queueing theory has attracted attention and developed quickly. This attention is also apparent in the number of queueing-related publications. The queueing theory book of Saaty [4], published in 1961, contained 896 references, and its Russian translation, published in 1965, contained 1115 references.

The early works of A.K. Erlang already contained the main elements of queueing theory: the (stochastic) arrival process of requests (calls), the (stochastic) service process of customers and, consequently, the departure process of customers, rejected/waiting customers, servers, etc. Later on real physical systems broke away from queueing and developed its own terminology, and the theory was applied in a wide range of application fields using the aforementioned basic terminology.

The mathematical description of queueing systems requires the description of the following elements:

- Arrival process: the stochastic description of customer arrivals, where customers might have any abstract or physical meaning depending on the considered system.

  Customer arrivals might depend on the current system's properties, e.g., the number of customers in the system. In the case of basic queueing models (where the interarrival times are independent), the arrival process is characterized by the interarrival time distribution.

© Springer Nature Switzerland AG 2019
L. Lakatos et al., *Introduction to Queueing Systems with Telecommunication Applications*, https://doi.org/10.1007/978-3-030-15142-3_6

- Service process: the stochastic description of customer service.

   Like customer arrival, customer service might also depend on the current system's properties, and in basic queueing models the service times are i.i.d. random variables.
- System structure: the resources of the queueing system, typically the number of servers and the size of the waiting room.
- Service discipline: a set of rules that determines the service order and service mode of customers. The most common orders are FCFS (first come, first served), FIFO (first in, first out), and LIFO (last in, first out). Service resources can also be used to serve all customers in parallel. This discipline is referred to as processor sharing (PS). Service order plays an important role when different types of customers arrive at a system. In this case, priority (with and without preemptions) can be used to provide faster service to one customer type.
- Performance parameters: to build an appropriately detailed model of a system, one should consider those performance parameters that must be computed. The most common performance parameters are system utilization, mean and distribution of waiting time, loss probability (the probability that a customer will be rejected by the system), etc. These measures are precisely defined later.

6.2   Classification of Basic Queueing Systems

The same queueing models might appear in completely different application fields. To avoid the parallel development of the same models in different fields, in 1953, Kendall proposed a classification and a standard notation of basic queueing systems. The current version of this set of notations is composed by six elements— A/B/c/d/e/x—where

A is the type of arrival process;
B is the type of service process;
c is the number of servers;
d is the system capacity, the maximum number of customers in the system;
e is the population of the set of customers (if it is finite, the arrival intensity decreases with an increasing number of customers in the system); and
x defines the service discipline (the most common service disciplines—e.g., FCFS, LCFS, PS—are defined above).

In basic queueing systems A and B take one of the following options:

M—memoryless, refers to exponentially distributed interarrival or service time;
E_r—order r Erlang distributed interarrival or service time;
H_r—order r hyperexponentially distributed interarrival or service time;
D—deterministic interarrival or service time,
G or GI—i.i.d. random interarrival or service time with any general distribution.

Fig. 6.1 Common
representation of queueing
systems

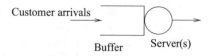

Customer arrivals

Buffer   Server(s)

The symbols d, e, and x are not indicated if they take their default values: $d = \infty$ infinite system capacity, $e = \infty$ infinite customer population, and $x = FCFS$ service in arrival order.

Additionally, most of the queueing systems have the following property. If there is an idle server when a waiting customer is present in the system or a new customer arrives at the system, then the service to the customer starts immediately. This property is referred to as *work-conserving* property, which we commonly assume in the sequel unless otherwise stated. Non-work-conserving queuing systems are resulted by, for instance, the presence of breakdowns or server vacations or sleeping modes.

It is assumed that $c \leq d$. If $c = d$ (the system capacity is identical to the number of servers), then there is no buffer position available for customers who arrive at the system when all servers are busy. In this case, the arriving customer leaves the system without service. These systems are also referred to as loss systems. If $c < d$, then the customers who arrive at the system when all servers are busy and there is still an available buffer position is not lost but waits until a server becomes available. The time period from the arrival of such customers to the beginning of the service is referred to as waiting time. The elements of queueing systems are commonly depicted as in Fig. 6.1.

6.3   Queueing System Performance Parameters

The optimal operation of queueing systems can be analyzed through several performance parameters, the most of which follow.

1. *Customer loss probability* (of queuing systems with finite capacity, $d < \infty$): Let $0 \leq t_1 \leq t_2 \leq \ldots$ be the arrival times of the first, second, $\ldots$ customers, and let m_n be the number of the first n customers who are lost. If $\lim_{n \to \infty} m_n/n$ converges to the deterministic value q almost surely, then q is referred to as the **loss probability**. In finite-capacity systems, it is also important to check if the $\lim_{n \to \infty} m_n/n$ limit exists at all.
2. *Waiting time distribution*: Let W_n, $n \geq 1$, be the waiting time of the nth customer; the number of the first n customers whose waiting time is less than x is

$$F_n(x) = \frac{1}{n} \sum_{i=1}^{n} \mathscr{I}_{\{W_i \leq x\}}, \quad x > 0.$$

$F_n(x)$ is the empirical distribution function of the waiting time based on the first n customers. If $\lim_{n \to \infty} F_n(x) = F(x)$ in a stochastic sense for $\forall x > 0$, then $F(x)$ is the CDF of the **waiting time distribution**.

3. *Mean waiting time*: If $\lim_{n \to \infty} \frac{1}{n}(W_1 + \ldots + W_n) = W^{(1)}$ in a stochastic sense, then $W^{(1)}$ is the mean waiting time. The higher moments of the waiting time are defined by stochastic convergence in a similar way:

$$W^{(k)} = \lim_{n \to \infty} \frac{1}{n}(W_1^k + \ldots + W_n^k), \quad k \geq 1.$$

In a wide range of practical cases,

$$W^{(k)} = \int_0^\infty x^k dF(x), \quad k \geq 1,$$

holds.

4. *Distribution of a server's busy period*: Consider one of the servers of a queueing system. Let $[a_n, b_n)$, $n \geq 1$, denote the consecutive intervals during which the server is busy (serving a customer). a_n and b_n are such that $a_n < b_n < a_{n+1}$, $n \geq 1$, and the server is idle (serving no customers) during the intervals $[b_n, a_{n+1})$, $n \geq 1$. In this case $[a_n, b_n)$ denotes the nth **busy period** of the given server. If

$$\lim_{n \to \infty} \frac{1}{n} \sum_{i=1}^{n} \mathscr{I}_{\{b_i - a_i \leq x\}} = G(x), \quad \forall x > 0,$$

in a stochastic sense, then $G(x)$ is the **distribution of the busy period** of the given server. The distribution of the idle period can be defined in an analogous way.

5. *Queue length distribution*: Let $L(t)$, $t \geq 0$, be the number of customers in the system (including those in the servers and those waiting in the buffer) at time t and $\bar{L}_k(t)$ be the portion of time in $(0, t)$ during which there were k customers in the system:

$$\bar{L}_k(t) = \frac{1}{t} \int_0^t \mathscr{I}_{\{L(s)=k\}} ds.$$

If

$$p_k = \lim_{t \to \infty} \bar{L}_k(t), \quad k \geq 0$$

exists in a stochastic sense, then (p_k, $k \geq 0$) defines the **queue length distribution**.

As was done previously, one can define the moments of the busy and idle periods and the queue length.

Remark 6.1 If the state (e.g., the number of customers in the system) of the queueing system can be described by the discrete-state $\mathcal{X} \subseteq \mathbb{N}^+$ homogeneous ergodic Markov chain $X(t)$, $t \geq 0$, and $f(i)$, $i \in \mathcal{X}$, is an arbitrary bounded function, then

$$\bar{f} = \lim_{t \to \infty} \frac{1}{t} \int_0^t f(v(s)) ds$$

and

$$\bar{f} = \sum_{i \in \mathcal{X}} f(i) \pi_i,$$

where $\{\pi_i, \; i \in \mathcal{X}\}$ denotes the stationary distribution of the Markov chain.

6.4   Little's Law

The famous relation $L = \lambda T$ was known as an experimental law for a long time. It was first proved by Little [2] in 1961 and was later commonly referred to as Little's law. In words, Little's law can be expressed as

mean number of customers in the *system*
= arrival intensity * mean time a customer spends in the *system*

and is independent of the definition of the system, the arrival and the service time distributions, the number of servers and buffer size, and the service discipline may also be arbitrarily chosen.

The importance of Little's law is that by knowing two of the three performance parameters of a queuing system the third parameter can be calculated. For example, depending on the definition of the *system*, one can obtain the following versions of Little's law:

(a) If the *system* is the buffer only, then

$$L_w = \lambda W,$$

where L_w is the mean number of waiting customers and W is the mean waiting time.

(b) If the *system* is the set of all servers, then

$$L_s = \lambda \bar{Y}$$

where L_s is mean number of busy servers and $\bar{Y}$ is the mean service time.

The relation described by Little's law is quite simple and general, and its extensions have stimulated many scientific works of theoretic and applied characteristics (see, e.g., [1, 3, 6, 8]).

To state Little's law formally, we will use the following notations. Let us consider a system, in which customers arrive at time moments $0 \leq t_1 \leq t_2 \leq \ldots$ and the nth customer spends T_n, $n \geq 1$ time in the system (i.e., the sum of its waiting time and of its service time) and after that leaves the system. Let us define

- $N(t)$, $t \geq 0$: **number of arrivals** (the number of customers arrive at the system in $[0, t]$);
- $M(t)$, $t \geq 0$: **number of departures** (the number of customers leaving system in $[0, t]$);
- $L(t) = N(t) - M(t)$, $t \geq 0$: **queue length** (the number of customers in the system at time t);
- $T(t) = \int_0^t [N(s) - M(s)]ds = \sum_{n=1}^{N(t)} T_n(t)$, $t > 0$: **aggregate time** the customers who arrived before time t spend in the system in $[0, t]$.

Based on the above definitions we can compute the following time average numbers:

- $\lambda(t) = \frac{N(t)}{t}$, $t > 0$: the **arrival intensity** at $t > 0$;
- $\overline{T}(t) = \frac{T(t)}{N(t)}$, $t > 0$: the **average time of a customer spends** in the system in $[0, t]$ considering the customers arrived before time t.
- $\overline{L}(t) = \frac{1}{t} \int_0^t L(s)ds = \frac{T(t)}{t}$, $t > 0$: the **mean number of customers** in the system in $[0, t]$.

Define the following quantities (when the limits exist)

$$\lambda = \lim_{t \to \infty} \lambda(t) = \lim_{t \to \infty} \frac{N(t)}{t}, \quad T = \lim_{n \to \infty} \frac{1}{n} \sum_{i=1}^n T_n, \quad L = \lim_{t \to \infty} \frac{1}{t} \int_0^t L(s)ds.$$
$$(6.1)$$

In the analysis of queueing systems we are usually interested in the long-term or stationary behavior. If we consider stochastic queueing systems, then the quantities t_n and T_n, $n \geq 1$ are random variables. In this case, the time average quantities $\lambda(t)$, $\overline{T}(t)$, and $\overline{L}(t)$ constitute stochastic processes which are determined by realizations of $t_n(\omega)$ and $T_n(\omega)$, $n \geq 1$ (from this point on, we omit writing ω in formulas).

There are two versions of Little's law. The sample-path version relates to the limiting time average numbers of $\lambda(t)$, $\overline{T}(t)$ and $\overline{L}(t)$ as $t \to \infty$ for particular realizations of t_n, T_n, $n \geq 1$. The stationary version relates to the system being in steady state. If supposing that the system is ergodic, then the limiting relations (6.1) hold for almost every trajectory of t_n, T_n, $n \geq 1$ (i.e., with probability 1) and Little's law says $\mathbf{E}(L_\infty) = \lambda \mathbf{E}(T_\infty)$, where L_∞ and T_∞ represent random variables with stationary distributions of queue length and the time that a customer spends in the system, respectively. Moreover, $\mathbf{E}(L_\infty) = L$, $\mathbf{E}(T_\infty) = T$.

The following theorem of [5] states the sample-path version of Little's law.

Theorem 6.1 (Little's Law) *Assume that $\lambda < \infty$ and $T < \infty$, then $L < \infty$ and*

$$L = \lambda T. \tag{6.2}$$

Proof (See, e.g., [8]) First we observe that the quantities $N(t)$, $M(t)$, $T(t)$, and $L(t)$ can be written more formally as

$$N(t) = \max\{n : t_n \leq t\} = \sum_{n=1}^{\infty} \mathscr{I}_{\{t_n \leq t\}}, \quad M(t) = \sum_{n=1}^{N(t)} \mathscr{I}_{\{t_n + T_n \leq t\}} \ t \geq 0,$$

$$T(t) = \sum_{n=1}^{N(t)} T_n(t) \quad \text{and} \quad L(t) = \sum_{n=1}^{\infty} \mathscr{I}_{\{t_n \leq t < t_n + T_n\}}.$$

Now note that the time the n-th customer spends in the system in $[0, t]$ is

$$T_n(t) = \int_0^t \mathscr{I}_{\{t_n \leq s \leq t_n + T_n\}} ds = \begin{cases} T_n, & \text{if } t_n + T_n \leq t, \\ t_n + T_n - t, & \text{if } t_n \leq t \leq t_n + T_n. \end{cases}$$

The following inequalities play a key role in establishing relations (6.2). Clearly,

$$\int_0^t L(s) ds = \int_0^t \sum_{n=1}^{\infty} \mathscr{I}_{\{t_n \leq s \leq t_n + T_n\}} ds = \sum_{n=1}^{\infty} \int_0^t \mathscr{I}_{\{t_n \leq s \leq t_n + T_n\}} ds$$

$$= \sum_{n: \ t_n + T_n < t} T_n + \sum_{n: \ t_n \leq t \leq t_n + T_n} (t_n + T_n - t) \leq \sum_{n: \ t_n < t} T_n.$$

From this, we get immediately a lower and upper bound for the mean number of customers in the system in $[0, t]$

$$\sum_{n: \ t_n + T_n < t} T_n \leq \int_0^t L(s) ds \leq \sum_{n: \ t_n < t} T_n.$$

Dividing by t throughout, we have

$$\frac{1}{t} \sum_{n: \ t_n + T_n < t} T_n \leq \overline{L}(t) \leq \frac{1}{t} \sum_{n: \ t_n < t} T_n. \tag{6.3}$$

We will first show that the last term in the preceding inequality converges to λT, as $t \to \infty$. Clearly, the fact that $N(t)/t$ has the finite limit λ as $t \to \infty$ implies that $N(t) \to \infty$, as $t \to \infty$. Moreover, if the arrival times t_n, $n \geq 1$ are distinct, then

$N(t_n) = n$, otherwise $N(t_n) \geq n$, consequently $t_n \to \infty$ and $N(t_n)/t_n \to \lambda$, as $n \to \infty$. Hence, T also exists and finite, which implies

$$\overline{\lim_{t \to \infty}} \frac{1}{t} \sum_{n:\, t_n < t} T_n = \lim_{t \to \infty} \frac{1}{t} \sum_{n:\, t_n < t} T_n = \lim_{t \to \infty} \frac{N(t_n)}{t} \frac{1}{N(t_n)} \sum_{n=1}^{N(t_n)} T_n = \lambda T. \qquad (6.4)$$

Now, we prove that $\lim_{t \to \infty} \frac{1}{t} \sum_{n:\, t_n + T_n < t} T_n = \lambda T$ with which (6.3) justifies (6.2). First, we verify that $T_n / t_n \to 0$, $n \to \infty$. It is obvious that

$$\frac{T_n}{n} = \frac{1}{n} \sum_{k=1}^{n} T_k - \frac{n-1}{n} \frac{1}{n-1} \sum_{k=1}^{n-1} T_k \to T - T = 0, \ n \to \infty.$$

Therefore,

$$\frac{T_n}{t_n} = \frac{T_n}{n} \frac{n}{t_n} \leq \frac{T_n}{n} \frac{N(t_n)}{t_n} \to 0 \cdot \lambda = 0.$$

By using this convergence we can state that for every positive ε there exists a number n_0 such that $\frac{T_n}{t_n} \leq \varepsilon$ if $n \geq n_0$. Then

$$\frac{1}{t} \sum_{n:\, t_n + T_n < t} T_n \geq \frac{1}{t} \sum_{n:\, n \geq n_0,\, t_n + T_n < t} T_n \geq \frac{1}{t} \sum_{n:\, n \geq n_0,\, t_n + \varepsilon t_n < t} T_n =$$

$$= \frac{1}{t} \sum_{n:\, 1 \leq n \leq n_0,\, t_n + \varepsilon t_n < t} T_n + \frac{1}{t} \sum_{n:\, (1+\varepsilon) t_n < t} T_n.$$

Clearly, for the finite sum $\sum_{n=1}^{n_0 - 1} T_n$

$$\frac{1}{t} \sum_{n:\, 1 \leq n \leq n_0,\, t_n + \varepsilon t_n < t} T_n \leq \frac{1}{t} \sum_{n=1}^{n_0 - 1} T_n \to 0, \ t \to \infty$$

is true. Moreover,

$$\frac{1}{t} \sum_{n:\, (1+\varepsilon) t_n < t} T_n = \frac{1}{t} \sum_{n:\, t_n < t/(1+\varepsilon)} T_n = \frac{1}{t} \sum_{n=1}^{N(t/(1+\varepsilon))} T_n$$

$$= \frac{N(t/(1+\varepsilon))}{t} \frac{1}{N(t/(1+\varepsilon))} \sum_{n=1}^{N(t/(1+\varepsilon))} T_n \to \frac{\lambda}{1+\varepsilon} T, \ n \to \infty,$$

thus

$$\lim_{t \to \infty} \frac{1}{t} \sum_{n:\, (1+\varepsilon) t_n < t} T_n \geq \frac{\lambda}{1+\varepsilon} T.$$

Since ε is an arbitrary positive number, consequently,

$$\lim_{t\to\infty} \frac{1}{t} \sum_{n:\,t_n<t} T_n = \overline{\lim_{t\to\infty}} \frac{1}{t} \sum_{n:\,t_n<t} T_n = \lambda T.$$

As a direct application of Theorem 4.15 we can conclude the following statement, which is valid in statistical equilibrium for a large class of queueing systems.

Theorem 6.2 *Suppose that* $Z(t) = (N(t), M(t))$, $t \geq 0$ *is a regenerative process with renewal time points* $0 \leq t_0 < t_1 < \dots$ *If* $\mathbf{E}(t_2 - t_1) < \infty$, $\mathbf{E}\left(\int_{t_1}^{t_2}(N(t) - M(t))dt\right) < \infty$ *and* $\mathbf{E}\left(\int_{t_0}^{t_1}(N(t) - M(t))dt\right) < \infty$, *then Little' law has the form*

$$\mathbf{E}(L_\infty) = \lambda \mathbf{E}(T_\infty)$$

6.5 Poisson Arrivals See Time Averages (PASTA)

In many models of queueing systems customers arrive according to a Poisson process and induce state transitions in the systems. The results are known as PASTA (see [7]) which states that in steady state the probability of a state indexed by k ($k \geq 0$) equals the probability that the system is in state k just before a customer arrives.

More formally, let $\{N(t),\ t \geq 0\}$ be a Poisson process at rate $\lambda > 0$ and suppose that the arrival times of customers $0 < t_1 < t_2 < \dots$ are the jumping points of the process $N(t)$. Denote by $L(t)$, $t \geq 0$ the number of customers in the system at t and assume that the following property (Lack of Anticipation Assumption, [7]) holds: for each $t \geq 0$ $\{N(t + s) - N(t),\ s \geq 0\}$ and $\{L(s),\ 0 \leq s \leq t\}$ are independent. Introduce the following notation: let $E_k(t) = \{$an arrival occurred just after time $t\}$, $E_k(t + \delta) = \{$an arrival occurred in $(t, t + \delta)\}$, $t > 0$, $\delta > 0$, $k \geq 0$ and $p_k(t) = \mathbf{P}(L(t) = k)$, $r_k(t) = \mathbf{P}(L(t) = k \mid E_k(t))$, $t > 0$.

Theorem 6.3 *The limiting values*

$$p_k = \lim_{t\to\infty} \mathbf{P}(L(t) = k) \quad and \quad r_k = \lim_{t\to\infty} \mathbf{P}(L(t) = k \mid E_k(t)),\ k \geq 0$$

are the same.

Proof Clearly, by the use of the theorem of multiplication and lack of anticipation assumption we obtain

$$r_k = \lim_{t\to\infty} \frac{\mathbf{P}(L(t) = k,\ E_k(t))}{\mathbf{P}(E_k(t))}$$

$$= \lim_{t\to\infty}\lim_{\delta\to 0} \frac{\mathbf{P}(L(t) = k,\ E_k(t, t + \delta))}{\mathbf{P}(E_k(t, t + \delta))}$$

$$= \lim_{t\to\infty} \lim_{\delta\to 0} \frac{\mathbf{P}(L(t) = k)\mathbf{P}(E_k(t, t+\delta) \mid L(t) = k)}{\mathbf{P}(E_k(t, t+\delta))}$$

$$= \lim_{t\to\infty} \lim_{\delta\to 0} \frac{\mathbf{P}(L(t) = k)\mathbf{P}(E_k(t, t+\delta))}{\mathbf{P}(E_k(t, t+\delta))}$$

$$= \lim_{t\to\infty} \lim_{\delta\to 0} \mathbf{P}(L(t) = k)$$

$$= p_k.$$

We mention here a consequence of the result given in [7] related to the path-wise convergence of long-run averages with probability 1, which is an important property for simulation of queueing systems. Note that the first relation is true if the process $L(t)$ is regenerative with finite mean of regenerative cycles (see Theorem 4.15).

Theorem 6.4 *Under the lack of anticipation assumption the following statement holds for all $k \geq 0$:*

$$\lim_{t\to\infty} \frac{1}{t} \int_0^t \mathscr{I}_{\{L(s)=k\}} \mathrm{d}s = p_k$$

with probability 1 *if and only if*

$$\lim_{t\to\infty} \frac{1}{N(t)} \sum_{n=1}^{N(t)} \mathscr{I}_{\{L(t_n)=k\}} = p_k$$

with probability 1.

6.6   Exercises

Exercise 6.1   Interpret the following Kendall's notations

- M/M/1/∞/∞-FIFO, M/M/1,
- M/M/2//4,
- M/M/1/m-PS,
- M/M/m-LIFO.

Exercise 6.2   In a single server infinite buffer queueing model the arrival rate is λ and the service time is exponentially distributed with parameter μ.

- Define the Little's law for the whole queueing system, for the buffer and for the server.
- Which one of these expressions define the server utilization?
- What is the utilization?

Exercise 6.3 Which ones of the following queueing systems are lossless?

- M/M/1,
- M/M/2/5/4,
- M/M/1/2-PS,
- M/M/m/m,
- M/M/m.

Exercise 6.4 Which ones of the following queueing systems provide immediate service for the customers?

- M/M/1,
- M/M/4/5/3,
- M/M/1/2-PS,
- M/M/m/m,
- M/M/m.

6.7 Solutions

Solution 6.1 The default values are usually eliminated from the Kendall's notations.

- M/M/1/∞/∞-FIFO and M/M/1 refer to the same queueing system with (time homogeneous) memoryless arrival and service processes, single server, infinite buffer, and infinite population. The first version of the notation is the extended version of the second one.
- M/M/2//4 refers to the queueing system with (time homogeneous) memoryless arrival and service processes, two server units, infinite buffer, and a finite population of 4 customers.
- M/M/1/m-PS refers to the queueing system with (time homogeneous) memoryless arrival and service processes, single server, finite buffer of $m - 1$ positions, and processor sharing service discipline. In case of processor sharing the server serves as many customers, as many are present in the system at the same time. The service capacity is uniformly distributed among the customers. With this service discipline the size of the buffer does not play role.
- M/M/m-LIFO refers to the queueing system with (time homogeneous) memoryless arrival and service processes, m servers, infinite buffer and infinite population, and last-in-first-out service discipline. It means that an arriving customer always enters a server independent of the number of customers in the system. If all servers are busy upon the arrival of the new customer, the new customer moves one of the customers under service to the waiting queue and its server starts the service of the newly arrived customer.

Solution 6.2 It is an M/M/1 queuing system.

- Let T, W, and $\bar{Y}$ be the mean system time, the mean waiting time, and the mean service time, and L, L_w, and L_s be the mean number of customers in the system, in the buffer, and in the server, respectively, and λ be the mean arrival rate. The application of Little's law for the whole queueing system, for the buffer, and for the server results in

$$L = \lambda T,$$

$$L_w = \lambda W,$$

$$L_s = \lambda \bar{Y}.$$

- The Little's law applied for the server is related to the utilization, because the mean number of customers in the server queue (L_s) define it.
- The mean number of customers in a single server queue (L_s) is the utilization of the queue, $\rho = L_s$. In case of m servers L_s/m defines the utilization, $\rho = L_s/m$.

Solution 6.3

- M/M/1—it is lossless, because there is an infinite buffer,
- M/M/2/5/4—it is lossless, because the population is 4 customers and there are 2 server and 3 buffer positions for customers.
- M/M/1/2-PS—it is lossless, because the server serves all customers at the same time and buffer is not used.
- M/M/m/m—it is a lossy queue, because the customers arrive when all servers are busy are lost.
- M/M/m—it is lossless, because there is an infinite buffer.

Solution 6.4

- M/M/1—it is a waiting system, because the customers arrive when the server is busy wait in the buffer.
- M/M/2/5/4—it is a waiting system, because the customers arrive when both servers are busy wait in the buffer.
- M/M/1/2-PS—it is an immediate service system, because the server serves all customers at the same time with a portion of the server capacity.
- M/M/m/m—it is an immediate service system, because the customers arriving when all servers are busy are lost and the ones which are not lost start the service immediately at arrival.
- M/M/m—it is a waiting system, because the customers arrive when all servers are busy wait in the buffer.

References

1. Harchol-Balter, M.: Performance Modeling and Design of Computer Systems: Queueing Theory in Action. Cambridge University Press, Cambridge (2013)
2. Little, J.D.C.: A proof of the queuing formula: L =AW. Oper. Res. **9**, 383–387 (1961)
3. Little, J.D.C.: Little's law as viewed on its 50th anniversary. Oper. Res. **59**(3), 536–549 (2011)
4. Saaty, T.: Elements of Queueing Theory. McGraw-Hill, New York (1961)
5. Stidham, S.: A last word on l= λ w. Oper. Res. **22**(2), 417–421 (1974)
6. Whitt, W.: A review of $L = \lambda W$ and extensions. Queueing Syst. **9**(3), 235–268 (1991)
7. Wolff, R.W.: Poisson arrivals see time averages. Oper. Res. **30**(2), 223–231 (1982)
8. Wolff, R.W.: Little's law and related results. In: Wiley Encyclopedia of Operations Research and Management Science, vol. 4, pp. 2828–2841. Wiley, New York (2011)

Chapter 7
Markovian Queueing Systems

Queueing systems whose underlying stochastic process is a continuous-time Markov chain (CTMCs) are the simplest and most often used class of queueing systems. The analysis of these systems is based on the essential results available for the analysis of CTMCs. As a consequence, several interesting properties of these queueing systems can be described by simple closed-form analytical expressions both in transient (as a function of time and initial state) and in steady state.

The most often studied property of basic queueing systems is the queue length process, $N(t)$ $(t \geq 0)$, which represents the number of customers in the system at time t. If customers belong to K different customer classes, then the vector-valued function $N(t) = (N_1(t), \ldots, N_K(t))$ $t \geq 0$, describes the queue length process. In this case the ith component of $N(t)$, $N_i(t)$, is the number of class i customers in the system.

The queue length process of basic queueing systems with single class of customers is the birth–death process, which is a special CTMC. We will utilize the previously introduced results of CTMCs and birth–death processes for the analysis of queueing systems.

7.1 M/M/1 Queue

The most basic queueing system is the M/M/1 queue, which is composed of a single server and an infinite buffer (see Fig. 7.1). Identical customers arrive according to a time-homogenous Poisson process at rate λ, and the service time of a customer is exponentially distributed at rate μ. The customers are served in the order of their arrivals (FCFS). The server is always busy when there is at least one customer in the system, i.e., the service process is *work-conserving*.

Let $N(t)$ be the number of customers in a system (either being served or waiting in the buffer) at time t. Due to the memoryless property of the arrival and the service

© Springer Nature Switzerland AG 2019
L. Lakatos et al., *Introduction to Queueing Systems with Telecommunication Applications*, https://doi.org/10.1007/978-3-030-15142-3_7

Customer arrivals: Service time distribution:
                          Exponetial with parameter μ

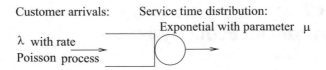

Fig. 7.1 M/M/1(/∞) queueing system

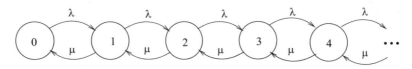

Fig. 7.2 Markov chain of number of customers in M/M/1 queue

process, $N(t)$ is a CTMC (Fig. 7.2). Its (nonvanishing) state-transition probabilities
are

$$p_{i,i+1}(\Delta) = \lambda\Delta + o(\Delta), \quad (i = 0, 1, \ldots),$$

$$p_{i,i-1}(\Delta) = \mu\Delta + o(\Delta), \quad (i = 1, 2, \ldots),$$

$$p_{ii}(\Delta) = 1 - (\lambda + \mu)\Delta + o(\Delta), \quad (i = 1, 2, \ldots),$$

$$p_{0,0}(\Delta) = 1 - \lambda\Delta + o(\Delta),$$

where $p_{i,j}(t) = \mathbf{P}(N(t) = j \mid N(0) = i)$. That is, $N(t)$ is an infinite state birth–
death process with $\lambda_i = \lambda$ $(i = 0, 1, \ldots)$ and $\mu_i = \mu$ $(i = 1, 2, \ldots)$. The Markov
chain is irreducible, and from its stationary equations we have

$$\pi_i = \left(\frac{\lambda}{\mu}\right)^i \quad (i = 0, 1, \ldots).$$

According to Eq. (3.27) this Markov chain is stable if

$$\frac{\lambda}{\mu} < 1.$$

The intuitive explanation of this relation is straightforward. It means that the queue
is stable if the mean service time $(1/\mu)$ is less than the mean interarrival time $(1/\lambda)$.
Introducing $\rho = \frac{\lambda}{\mu}$ from Eqs. (3.29) and (3.28) we have

$$p_i = \lim_{t\to\infty} \mathbf{P}(N(t) = i) = p_0\rho^i \quad (i = 0, 1, \ldots),$$

where

$$p_0 = \frac{1}{\displaystyle\sum_{j=0}^{\infty} \rho^j} = \frac{1}{\frac{1}{1-\rho}} = 1 - \rho$$

and

$$p_i = (1 - \rho)\rho^i \quad (i = 0, 1, \ldots).$$

Consequently, the stationary number of customers in the system is geometrically distributed on $\{0, 1, \ldots\}$ with parameter $(1 - \rho)$. The mean of this distribution is

$$\bar{N} = \sum_{i=0}^{\infty} i\, p_i = \frac{\rho}{1 - \rho} = \frac{\lambda}{\mu - \lambda}. \tag{7.1}$$

Now we compute the mean time a customer spends in the system in stationary state. Let us consider that the customer arrives at time t. The number of customers in the system at this time instant is $N(t)$. According to the FCFS service order, the new customer must wait while all of the customers present at the system at time t are served. Due to the memoryless property of the exponentially distributed service time, the remaining service for the customer being served at time t (if any) is also exponentially distributed with the same parameter. Furthermore, the service times of the $N(t) - 1$ customers waiting in the buffer at time t [if $N(t) \geq 1$] and the service time of the newly arrived customer are also exponentially distributed with parameter μ. Summing up all these, the total time the customer spends in the system is

$$D(t) = \sum_{i=1}^{N(t)} Y_i + Y, \tag{7.2}$$

where $Y_1, \ldots, Y_{N(t)}$ and Y are i.i.d. exponentially distributed random variables with parameter μ. In stationary state, neither the distribution of $N(t)$ nor the distribution of $D(t)$ depends on t, that is, $\mathbf{E}(D(t)) = \bar{D}$. $\bar{D}$ can be computed using the Wald's lemma (see Exercise 2.1), Eqs. (7.1) and (7.2) as follows

$$\bar{D} = \mathbf{E}(Y_1) \cdot \bar{N} + \mathbf{E}(Y) = \frac{1}{\mu} \cdot \frac{\lambda}{\mu - \lambda} + \frac{1}{\mu} = \frac{1}{\mu - \lambda}.$$

We can also evaluate the stationary system time (time spent in the system by a customer) distribution $D(t) \equiv D$ because

$$P(D \leq s) = \sum_{i=0}^{\infty} P(D \leq s|N(t) = i) \cdot P(N(t) = i),$$

where $P(D \leq s|N(t) = i)$ is the distribution of the sum of $(i+1)$ i.i.d. exponentially distributed random variables with parameter μ according to Eq. (7.2). Thus,

$$
\begin{aligned}
P(D < s) &= \sum_{i=0}^{\infty} \left(\int_0^s \mu \frac{(\mu u)^i}{i!} e^{-\mu u} du \right) (1 - \rho)\rho^i \\
&= \int_0^s \left(\sum_{i=0}^{\infty} \mu \frac{\mu^i u^i}{i!} e^{-\mu u} \left(1 - \frac{\lambda}{\mu}\right) \left(\frac{\lambda}{\mu}\right)^i \right) du \\
&= \mu \left(1 - \frac{\lambda}{\mu}\right) \int_0^s \left(e^{-\mu u} \cdot \sum_{i=0}^{\infty} \frac{(\lambda u)^i}{i!} \right) du \\
&= \mu \left(1 - \frac{\lambda}{\mu}\right) \int_0^s e^{-\mu u} \cdot e^{\lambda u} du \\
&= \mu \left(1 - \frac{\lambda}{\mu}\right) \int_0^s e^{-(\mu-\lambda)u} du \\
&= \mu \cdot \frac{\mu - \lambda}{\mu} \cdot \frac{1}{\mu - \lambda} \left(1 - e^{-(\mu-\lambda)s}\right) \\
&= 1 - e^{-(\mu-\lambda)s} \quad (s \geq 0),
\end{aligned}
$$

where we used that the sum of $(i + 1)$ i.i.d. exponentially distributed random variables with parameter μ is Gamma (or Erlang) distributed with parameters μ and $i + 1$. We obtained that the system time is exponentially distributed with parameter $(\mu - \lambda)$ and its mean is

$$\bar{D} = \frac{1}{\mu - \lambda},$$

as we saw previously.

The departure process of a stationary M/M/1 queue (the point process of the consecutive departure instants) is a Poisson process with parameter λ. This property is referred to as Burke's theorem and plays an important role in the analysis of queueing networks, as discussed in Sect. 10.1.

In single-server systems, the probability of finding the server busy is referred to as utilization.

$$P(\text{the server is busy}) = \sum_{k=1}^{\infty} p_k = 1 - p_0 = 1 - (1 - \rho) = \rho.$$

Let X_s be the number of customers being served in stationary state. In this case, $E(X_s) = 0 \cdot p_0 + 1 \cdot (1 - p_0)$, whence

$$E(X_s) = \rho. \tag{7.3}$$

According to Little's law, if $\rho < 1$ (i.e., the system is stable), then

$$E(X_s) = \bar{\lambda}\bar{x} = \frac{\lambda}{\mu} = \rho \tag{7.4}$$

because $\bar{\lambda} = \lambda$ and $\bar{x} = \frac{1}{\mu}$.

The mean number of customers in the system is

$$E(X) = \sum_{k=0}^{\infty} k \, p_k = \sum_{k=0}^{\infty} k \, (1 - \rho)\rho^k = \rho(1 - \rho) \sum_{k=0}^{\infty} k \, \rho^{k-1}$$

$$= \rho(1 - \rho) \sum_{k=0}^{\infty} \frac{d}{d\rho}\rho^k = \rho(1 - \rho) \frac{d}{d\rho} \sum_{k=0}^{\infty} \rho^k = \rho(1 - \rho) \frac{1}{(1 - \rho)^2},$$

whence

$$E(X) = \frac{\rho}{1 - \rho}. \tag{7.5}$$

The mean number of waiting customers in the system is

$$E(X_w) = \sum_{k=1}^{\infty} (k - 1) \, p_k = \sum_{k=1}^{\infty} k \, p_k - \sum_{k=1}^{\infty} p_k$$

$$= \sum_{k=1}^{\infty} k \, (1 - \rho)\rho^k - (1 - p_0) = \frac{\rho}{1 - \rho} - \rho,$$

whence

$$E(X_w) = \frac{\rho^2}{1 - \rho}, \tag{7.6}$$

which also verifies the relation $X = X_w + X_s$.

A customer's system time, waiting time, and service time (T, W, and x, respectively) fulfill

$$T = W + x, \tag{7.7}$$

and

$$\bar{T} = \bar{W} + \bar{x}. \tag{7.8}$$

According to Little's law,

$$\bar{T} = \frac{\mathbf{E}(X)}{\lambda} = \frac{1}{\mu(1 - \rho)}, \tag{7.9}$$

$$\bar{W} = \frac{\mathbf{E}(X_w)}{\lambda} = \frac{\rho}{\mu(1 - \rho)}, \tag{7.10}$$

$$\bar{x} = \frac{1}{\mu}, \tag{7.11}$$

which confirms Eq. (7.8).
    $\bar{T}$ can also be computed as

$$\bar{T} = \frac{1}{\mu(1 - \rho)} = \frac{1}{\mu} + \sum_{k=0}^{\infty} k \, \frac{1}{\mu} \, p_k,$$

which has an intuitive interpretation. Upon arrival, a customer finds k customers in the queue with probability p_k. Its mean system time is composed by the mean of the k service times of the customers in the system at its arrival, $k \frac{1}{\mu}$, and its own mean service time, $\frac{1}{\mu}$.
    The probability that there are at least k customers in the system is

$$\mathbf{P}(X \geq k) = \sum_{i=k}^{\infty}(1 - \rho)\rho^i = (1 - \rho)\rho^k \sum_{j=0}^{\infty} \rho^j = \rho^k. \tag{7.12}$$

7.2   Transient Behavior of M/M/1 Queueing System

Let $A(x) = 1 - e^{-\lambda x}$ and $B(x) = 1 - e^{-\mu x}$ be the CDF of the interarrival and the service time distribution, and $N(t)$, $t \geq 0$ be the number of customers in the system at time t. From the fact that $N(t)$ is a birth–death process (Sect. 3.4.1) we will derive the following characteristics (Fig. 7.3):

(a) The parameters of $\{N(t), \ t \geq 0\}$.
(b) The distribution of $N(t)$ at an arbitrary $t \geq 0$ instant (using point 2 of Theorem 3.28).
(c) The distribution of the length of the busy period of the server (based on Theorem 3.30).

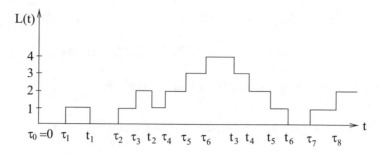

Fig. 7.3 Number of customers in the M/M/1(/∞) queueing system

(d) The distribution of the stationary virtual waiting time (the time required to serve the customers who are in the system at an arbitrary time instant).

(a) Parameters of birth–death process:

Applying the notations of Sect. 3.4.1 we have $N(t)$ is a birth–death process with parameters

$$a_k = \lambda, \quad k \geq 0, \quad b_k = \mu, \quad k \geq 1;$$

$$p_0 = 1, \quad p_k = \frac{\lambda}{\lambda+\mu}, \quad q_k = \frac{\mu}{\lambda+\mu}, \quad k \geq 1.$$

(b) Distribution of $N(t)$:

We assume that the system is idle at time 0 $[L(0) = 0]$ with probability 1 ($\varphi_0 = 1$, $\varphi_k = 0$, $k \geq 1$) and compute the distribution of $N(t)$, $t \geq 0$. More precisely, we evaluate

$$P_k(t) = \mathbf{P}(N(t) = k), \quad p_k^*(s) = \int_0^\infty e^{-st} P_k(t)\mathrm{d}t, \quad \mathrm{Re}\, s > 0.$$

The $p_k^*(s)$ functions are given by point 2 of Theorem 3.28:

$$sp_0^*(s) - 1 = -\lambda p_0^*(s) + \mu p_1^*(s), \tag{7.13}$$

$$sp_k^*(s) = \lambda p_{k-1}^*(s) - (\lambda + \mu)p_k^*(s) + \mu p_{k+1}^*(s), \quad k \geq 1. \tag{7.14}$$

Following the approach proposed in [2] we define the probability generating function

$$p^*(z, s) = \sum_{k=0}^\infty p_k^*(s)\, z^k.$$

Multiplying both sides of Eq. (7.14) by z^k ($0 < |z| \leq 1$, $k \geq 1$) and summing up the terms $k = 1, 2, \ldots$ we obtain

$$sp^*(z, s) - sp_0^*(s) = \lambda z p^*(z, s) - (\lambda + \mu)[p^*(z, s) - p_0^*(s)]$$

$$+ \frac{1}{z} \mu [p^*(z, s) - p_0^*(s) - z p_1^*(s)].$$

Further adding (7.13) and rearranging the terms we obtain

$$p^*(z, s) \left[s - \lambda z + (\lambda + \mu) - \frac{\mu}{z} \right] = 1 + \mu p_0^*(s) - \frac{\mu}{z} p_0^*(s),$$

whence

$$p^*(z, s) = \frac{z - \mu(1 - z) p_0^*(s)}{s z - (1 - z)(\mu - \lambda z)}. \tag{7.15}$$

The function $p^*(z, s)$ is the Laplace transform of a generator function. It is analytic and bounded for $|z| \leq 1$ and fixed Re $s > 0$. Consequently, on the right-hand side of Eq. (7.15) the numerator must have a root at the root of the denominator. The denominator has a root at

$$z = \gamma_1(s) = \frac{\lambda + \mu + s - \sqrt{(\lambda + \mu + s)^2 - 4\lambda\mu}}{2\lambda}.$$

Using that the numerator also has a root at $z = \gamma_1(s)$ we obtain that $|\gamma_1(s)| < 1$, if Re $s > 0$. It is easy to see for real $s > 0$ because

$$\lambda + \mu + s - \sqrt{(\lambda + \mu + s)^2 - 4\lambda\mu} < 2\lambda$$

holds for $\mu - \lambda + s < 0$ since in this case $\mu - \lambda + s < \sqrt{(\lambda + \mu + s)^2 - 4\lambda\mu}$. If $\mu - \lambda + s \geq 0$, then the squares of the two sides of the equation remain equal and we obtain a simple identity.

The numerator of $p^*(z, s)$ in Eq. (7.15) must be zero at $z = \gamma_1(s)$, that is,

$$\gamma_1(s) - \mu(1 - \gamma_1(s)) p_0^*(s) = 0, \quad \text{that is,} \quad p_0^*(s) = \frac{\gamma_1(s)}{\mu(1 - \gamma_1(s))}.$$

Thus

$$p^*(z, s) = \frac{z - (1 - z)\frac{\gamma_1(s)}{1 - \gamma_1(s)}}{s z - (1 - z)(\mu - \lambda z)} = \frac{z - \gamma_1(s)}{(1 - \gamma_1(s))(s z - (1 - z)(\mu - \lambda z))}.$$

Introducing

$$\gamma_2(s) = \frac{\lambda + \mu + s + \sqrt{(\lambda + \mu + s)^2 - 4\lambda\mu}}{2\lambda}$$

and using

$$s\,z - (1-z)(\mu - \lambda z) = -\lambda(z - \gamma_1(s))(z - \gamma_2(s))$$

we modify the expression in the following way:

$$p^*(z, s) = \frac{1}{\lambda(1 - \gamma_1(s))(\gamma_2(s) - z)}.$$

It can be seen that $|\gamma_2(s)| > 1$ $[\gamma_2(s) \neq z, \ \ |z| \leq 1$, since $p^*(z, s)$ is bounded there] and

$$\gamma_1(s)\gamma_2(s) = \frac{\mu}{\lambda}.$$

The series expansion of the fraction $\frac{1}{\gamma_2(s) - z}$, according to z, gives

$$p^*(z, s) = [\lambda(1 - \gamma_1(s))\gamma_2(s)]^{-1} \sum_{k=0}^{\infty} \left[\frac{z}{\gamma_2(s)}\right]^k.$$

Comparing the coefficients of the z^k terms and using the series expansion of the fraction $\frac{1}{1-\gamma_1(s)}$ we have

$$p_k^*(s) = [\lambda(1 - \gamma_1(s))(\gamma_2(s))^{k+1}]^{-1} = \frac{1}{\lambda(\gamma_2(s))^{k+1}} \sum_{j=0}^{\infty} (\gamma_1(s))^j$$

$$= \frac{1}{\lambda[\gamma_1(s)\gamma_2(s)]^{k+1}} \sum_{j=k+1}^{\infty} (\gamma_1(s))^j = \frac{1}{\lambda}\left(\frac{\lambda}{\mu}\right)^{k+1} \sum_{j=k+1}^{\infty} (\gamma_1(s))^j$$

$$= \frac{1}{\lambda}\left(\frac{\lambda}{\mu}\right)^{k+1} \sum_{j=k+1}^{\infty} \left(\frac{\mu}{\lambda}\right)^j (\gamma_2(s))^{-j}.$$

The last expression allows the explicit description of $P_k(t)$. Let $I_m(z)$ be the modified first-order Bessel function, i.e.,

$$I_m(z) = \sum_{k=0}^{\infty} \frac{1}{k!\Gamma(m + k + 1)} \left(\frac{z}{2}\right)^{m+2k}.$$

Since the Laplace transform of

$$\int_0^{\infty} e^{-sx} x^{-1} I_m(cx)\,dx$$

is (see [3])

$$\frac{c^m}{m}(s + \sqrt{s^2 - c^2})^{-m},$$

the inverse Laplace transform of

$$\left(\frac{s + \sqrt{s^2 - 4\lambda\mu}}{2\lambda}\right)^{-m}$$

is

$$m\left(\frac{\lambda}{\mu}\right)^{m/2} t^{-1} I_m(2\sqrt{\lambda\mu t}).$$

Additionally, using $e^{-st} e^{-(\lambda+\mu)t} = e^{-(s+\lambda+\mu)t}$ the inverse Laplace transform of

$$\gamma_2(s)^{-m} = \left(\frac{s + \lambda + \mu + \sqrt{(s + \lambda + \mu)^2 - 4\lambda\mu}}{2\lambda}\right)^{-m}$$

is

$$e^{-(\lambda+\mu)t} m\left(\frac{\lambda}{\mu}\right)^{m/2} t^{-1} I_m(2\sqrt{\lambda\mu t}),$$

and finally we obtain

$$P_k(t) = \frac{1}{\lambda}\left(\frac{\lambda}{\mu}\right)^{k+1} e^{-(\lambda+\mu)t} \sum_{j=k+1}^{\infty} j t^{-1} \left(\frac{\mu}{\lambda}\right)^{j/2} I_j(2\sqrt{\lambda\mu t}).$$

(c) Distribution of the busy intervals:

Let Π_k random variables be the length of the busy period starting from state k, and let

$$\Pi_k(t) = \mathbf{P}(\Pi_k \le t), \quad \text{and} \quad \pi_k^*(s) = \mathbf{E}(e^{-s\Pi_k})$$

be its CDF and Laplace transform, respectively. To compute $\Pi_k(t)$ and $\pi_k^*(s)$, we assume that state 0 is an absorbing state and put

$$\varphi_k = 1, \quad a_0 = 0, \quad p_0 = 0, \quad a_n = \lambda, \quad p_n = \frac{\lambda}{\lambda + \mu}, \quad n \ge 1.$$

We note (see Remark 3.8) that

$$P_0(t) = \mathbf{P}(N(t) = 0) = \mathbf{P}(\Pi_k \le t) = \Pi_k(t), \quad P_0'(t) = \Pi_k'(t).$$

According to Theorem 3.30 for $p_n^*(s)$, $n \ge 1$, we have

$$sp_0^*(s) = \mu p_1^*(s), \tag{7.16}$$

$$sp_1^*(s) - \delta_{1,k} = -(\lambda + \mu)p_1^*(s) + \mu p_2^*(s), \tag{7.17}$$

$$sp_n^*(s) - \delta_{n,k} = \lambda p_{n-1}^*(s) - (\lambda + \mu)p_n^*(s) + \mu p_{n+1}^*(s), \quad n \geq 2. \tag{7.18}$$

Furthermore, according to point 1 of Theorem 3.30 $\Pi_k'(t) = P_0'(t) = \mu P_1(t)$, and consequently

$$\pi_k^*(s) = \int_0^\infty e^{-sx} d\Pi_k(x) = \int_0^\infty e^{-sx} \Pi_k'(x) dx = \int_0^\infty e^{-sx} b P_1(x) dx = b p_1^*(s).$$

Multiplying Eq. (7.18) by z^n, summing it up for $n \geq 2$, and adding Eq. (7.17) z times, we obtain

$$s\, p^*(z, s) - z^k = \lambda z\, p^*(z, s) - (\lambda + \mu)\, p^*(z, s) + \frac{\mu}{z}\, p^*(z, s) - \mu\, p_1^*(s),$$

where $p^*(z, s) = \sum_{n=1}^\infty p_n^*(s) z^n$ and $z^n \delta_{n,k} = z^k I (n = k)$. Further rearranging the expression gives

$$[s\, z - (1 - z)(\mu - \lambda z)] \frac{p^*(z, s)}{z} = z^k - \mu p_1^*.(s). \tag{7.19}$$

As was shown previously, the roots of $s\, z - (1 - z)(\mu - \lambda z) = 0$ are $z = \gamma_1(s)$ and $z = \gamma_2(s)$. Since $|\gamma_1(s)| < 1$, if Re $s > 0$, then $p^*(z, s)/z$ is bounded for $|z| \leq 1$ and $\neq 0$; if $z \neq 0$, then from Eq. (7.19) we get for $\mu p_1^*(s)$ that [because in the case of $|z| \leq 1$ the only root of $z^k - \mu p_1^*(s) = 0$ is $z = \gamma_1(s)$]

$$\mu p_1^*(s) = \gamma_1(s)^k = \left(\frac{\mu}{\lambda}\right)^k \gamma_2(s)^{-k}.$$

Using this and $\int_0^\infty e^{-sx} \Pi_k'(x) dx = \mu p_1^*(s)$ we have that

$$\Pi_k'(t) = \left(\frac{\mu}{\lambda}\right)^k e^{-(\lambda+\mu)t} \left(\sqrt{\frac{\lambda}{\mu}}\right)^k k t^{-1} I_k(2\sqrt{\lambda\mu t})$$

$$= \left(\sqrt{\frac{\mu}{\lambda}}\right)^k \frac{k}{t} e^{-(\lambda+\mu)t} I_k(2\sqrt{\lambda\mu t}).$$

In the special case where $k = 1$, we have

$$\Pi_1'(t) = \sqrt{\frac{\mu}{\lambda}} \frac{1}{t} e^{-(\lambda+\mu)t} I_1(2\sqrt{\lambda\mu t}).$$

(d) Distribution of the virtual waiting time:

At time t the virtual waiting time, $W(t)$, satisfies

$$W(t) \overset{d}{=} \begin{cases} 0, & \text{if } N(t) = 0, \\ \displaystyle\sum_{i=1}^{k} \xi_i, & \text{if } N(t) = k, \end{cases}$$

where $\overset{d}{=}$ denotes the equality in distribution and $\xi_1, \ldots, \xi_k$ are the i.i.d. service times of the waiting customers (ξ_i, $i = 2, \ldots, k$) and the remaining service time of the customer being served (ξ_1). $W(t)$ is also referred to as *workload process* in the literature.

Due to the memoryless property of the exponential service time distribution, all of these random variables are exponentially distributed with parameter μ, and their CDF is $1 - e^{-\mu x}$. According to the law of total probability, this gives

$$W(x,t) = \mathbf{P}(W(t) \le x) = \sum_{k=0}^{\infty} \mathbf{P}(W(t) \le x | N(t) = k) \mathbf{P}(N(t) = k)$$

$$= P_0(t) + \sum_{k=1}^{\infty} P_k(t) \mathbf{P}\left(\sum_{i=1}^{k} \xi_i \le x\right), \tag{7.20}$$

where $P_k(t) = \mathbf{P}(N(t) = k)$.

Introducing the Laplace transforms $W^*(x,s) = \int_0^{\infty} e^{-st} W(x,t) dt$ and $p_k^*(s) = \int_0^{\infty} e^{-st} P_k(t) dt$ from Eq. (7.20) we obtain

$$W^*(x,s) = p_0^*(s) + \sum_{k=1}^{\infty} p_k^*(s) \mathbf{P}\left(\sum_{i=1}^{k} \xi_i < x\right).$$

Since ξ_i, $i = 1, 2, \ldots, k$ are independent exponentially distributed random variables with parameter μ we have

$$\mathbf{P}\left(\sum_{i=1}^{k} \xi_i \le x\right) = \mathbf{P}\left(\sum_{i=1}^{k} (\mu \xi_i) \le \mu x\right) = \int_0^{\mu x} \frac{u^{k-1}}{(k-1)!} e^{-u} du$$

and according to point (b) we also have

$$p_k^*(s) = [\lambda(1 - \gamma_1(s))(\gamma_2(s))^{k+1}]^{-1}, \quad k \ge 0.$$

Using all these we obtain the Laplace transform of $W(x,t)$:

$$W^*(x,s) = [\lambda(1 - \gamma_1(s))\gamma_2(s)]^{-1} \left(1 + \sum_{k=1}^{\infty} [\gamma_2(s)]^{-k} \int_0^{\mu x} \frac{u^{k-1}}{(k-1)!} e^{-u} du\right)$$

$$= [\lambda(1 - \gamma_1(s))\gamma_2(s)]^{-1} \left(1 + \sum_{k=1}^{\infty}(\gamma_2(s))^{-1} \int_0^{\mu x} \frac{[u / \gamma_2(s)]^{k-1}}{(k-1)!} e^{-u} du \right)$$

$$= [\lambda(1 - \gamma_1(s))\gamma_2(s)]^{-1} \left(1 + \frac{1}{\gamma_2(s) - 1}(1 - e^{-\mu(1-\gamma_2^{-1}(s))x}) \right).$$

According to Eq. (7.20), in case of stable system ($\lambda/\mu < 1$) there exists the limit

$$\bar{W}(x) = \lim_{t \to \infty} W(x, t) = \pi_0 + \sum_{k=1}^{\infty} \pi_k \mathbf{P}\left(\sum_{i=1}^{k} \xi_i < x \right),$$

where according to Theorem 3.28

$$\pi_k = \lim_{t \to \infty} P_k(t) = \left(1 - \frac{\lambda}{\mu} \right) \left(\frac{\lambda}{\mu} \right)^k, \quad k \geq 0.$$

Thus

$$\bar{W}(x) = \left(1 - \frac{\lambda}{\mu} \right) \left(1 + \sum_{k=1}^{\infty} \left(\frac{\lambda}{\mu} \right)^k \int_0^{\mu x} \frac{u^{k-1}}{(k-1)!} e^{-u} du \right)$$

$$= \left(1 - \frac{\lambda}{\mu} \right) \left(1 + \frac{\lambda}{\mu} \int_0^{\mu x} e^{-(1-\lambda/\mu)u} du \right) = 1 - \frac{\lambda}{\mu} e^{-(\mu-\lambda)x}, \quad x > 0,$$

and

$$\lim_{t \to \infty} \mathbf{P}(W(t) = 0) = \lim_{t \to \infty} \mathbf{P}(N(t) = 0) = \lim_{t \to \infty} P_0(t) = \pi_0 = 1 - \frac{\lambda}{\mu}.$$

7.3   M/M/m Queueing System

The arrival process (Poisson process at rate λ) and the service time distribution (exponential with parameter μ) are the same as before, but there are m servers in the service unit of this queueing system. While there is at least one idle server, an arriving customer is assigned to one of the idle servers upon arrival, and service of this customer starts immediately. If all the servers are busy at an arrival, then the arriving customer waits in the buffer. When i ($1 \leq i \leq m$) servers are busy, the i service processes go on in parallel. Due to the memoryless property of the service time distribution, the remaining service times are also independent exponentially distributed random variables. The minimum of i independent exponentially distributed random variables with parameter μ is exponentially distributed with parameter $i\mu$. Another intuitive interpretation of the service process is through the service rate. A single server serves a customer at rate μ, i.e., the probability that an ongoing service will be completed in the next interval of length Δ is $\mu\Delta + o(\Delta)$. If i servers are working in parallel, then they serve customers at a rate $i\mu$, i.e., the probability that one of the i ongoing services will be completed in the next Δ long interval is $i\mu\Delta + o(\Delta)$. The transitions of the birth–death process describing the number of customers in the system are as follows:

$$p_{i,i+1}(\Delta) = \lambda\Delta + o(\Delta), \quad (i = 0, 1, \ldots),$$

$$p_{i,i-1}(\Delta) = i\mu\Delta + o(\Delta), \quad (0 < i \leq m),$$

$$p_{ii}(\Delta) = 1 - (\lambda + i\mu)\Delta + o(\Delta), \quad (0 \leq i \leq m),$$

$$p_{i,i}(\Delta) = 1 - (\lambda + m\mu)\Delta + o(\Delta), \quad (i \geq m).$$

The Markov chain is stable if $0 < \lambda < m\mu < \infty$. In this case the stationary equations are

$$p_{k-1}\lambda + p_{k+1}(k+1)\mu = p_k(\lambda + k\mu), \quad 0 < k < m,$$

$$p_{k-1}\lambda + p_{k+1}m\mu = p_k(\lambda + m\mu), \quad k \geq m,$$

$$p_1\mu = p_0\lambda.$$

The solution of this set of equations is

$$p_k = \frac{\lambda}{k\mu}p_{k-1} = \left(\frac{\lambda}{\mu}\right)^k \frac{1}{k!}p_0 \quad \text{if } k = 1, 2, \cdots, m, \tag{7.21}$$

$$p_k = \frac{\lambda}{m\mu}p_{k-1} \quad \text{if } k = m+1, m+2, \cdots, \tag{7.22}$$

whence

$$p_{m+i} = \left(\frac{\lambda}{m\mu}\right)^i p_m \quad i \geq 1. \tag{7.23}$$

Combining the two cases we have

$$p_j = \begin{cases} \left(\dfrac{\lambda}{\mu}\right)^j \dfrac{1}{j!} p_0, & j = 1, 2, \cdots, m, \\[3mm] \left(\dfrac{\lambda}{\mu}\right)^j \dfrac{1}{m!}\dfrac{1}{m^{j-m}} p_0, & j > m, \end{cases} \tag{7.24}$$

from which the normalized solution of p_0 is

$$p_0 = \cfrac{1}{1 + \sum_{j=1}^{m} \left(\frac{\lambda}{\mu}\right)^{j} \frac{1}{j!} + \sum_{j=m+1}^{\infty} \left(\frac{\lambda}{\mu}\right)^{j} \frac{1}{m!} \frac{1}{m^{j-m}}}. \tag{7.25}$$

The second term of the denominator can be rewritten as

$$\frac{m^m}{m!} \sum_{j=m}^{\infty} \left(\frac{\lambda}{m\mu}\right)^{j} = \frac{\left(\dfrac{\lambda}{m\mu}\right)^{m}}{\left(1 - \dfrac{\lambda}{m\mu}\right)} \frac{m^m}{m!}. \tag{7.26}$$

The mean system time can be computed as

$$\bar{T} = \bar{x} + \bar{W} = \frac{1}{\mu} + \sum_{k=m}^{\infty} \mathbf{E}(W \mid k)\, p_k^{(a)}, \tag{7.27}$$

where $p_k^{(a)}$ denotes the queue length distribution at arrival instants. In the case of M/M/m queue, $p_k^{(a)} = p_k$. The mean system time can be expressed as

$$\mathbf{E}(T) = \frac{1}{\mu} + \sum_{k=m}^{\infty} \frac{k - m + 1}{m\mu}\, p_k, \tag{7.28}$$

whence

$$\bar{T} = \frac{1}{\mu} + \sum_{k=m}^{\infty} \frac{k - m + 1}{m\mu}\, p_m \left(\frac{\lambda}{m\mu}\right)^{k-m} = \frac{1}{\mu} + \frac{p_m}{m\mu} \sum_{k=m}^{\infty} (k - m + 1) \left(\frac{\lambda}{m\mu}\right)^{k-m}. \tag{7.29}$$

Using that the arrival process is a Poisson process we further have

$$\bar{T} = \frac{1}{\mu} + \frac{p_m}{m\mu} \sum_{i=1}^{\infty} i \left(\frac{\lambda}{m\mu}\right)^{i-1} = \frac{1}{\mu} + \frac{m\mu p_m}{(m\mu - \lambda)^2}. \tag{7.30}$$

With the help of Little's law we can also compute the mean number of customers in the system:

$$\mathbf{E}(X) = \lambda \bar{T} = \frac{\lambda}{\mu} + \frac{m\lambda\mu p_m}{(m\mu - \lambda)^2}. \tag{7.31}$$

The probability that all servers will be busy and an arriving customer will have to wait is

$$\mathbf{P}(\text{waiting}) = \sum_{k=m}^{\infty} p_k = \sum_{k=m}^{\infty} p_0 \left(\frac{\lambda}{\mu}\right)^k \frac{1}{m!} \frac{m^m}{m^k} = \frac{m^m}{m!} p_0 \sum_{k=m}^{\infty} \left(\frac{\lambda}{m\mu}\right)^k$$

$$= p_0 \frac{m^m}{m!} \frac{\left(\frac{\lambda}{m\mu}\right)^m}{1 - \frac{\lambda}{m\mu}} = p_0 \frac{1}{m!} \left(\frac{\lambda}{\mu}\right)^m \frac{1}{1 - \rho}$$

$$= \frac{\dfrac{1}{m!} \left(\dfrac{\lambda}{\mu}\right)^m \dfrac{1}{1-\rho}}{\displaystyle\sum_{k=0}^{m-1} \left(\dfrac{\lambda}{\mu}\right)^k \dfrac{1}{k!} + \dfrac{1}{m!} \left(\dfrac{\lambda}{\mu}\right)^m \dfrac{1}{1-\rho}},$$

where $\rho = \dfrac{\lambda}{m\mu}$.

This expression is known as the C (or waiting probability) formula of Erlang [1]. The parameters of this formula are m the number of servers and the λ/μ ratio, which is also referred to as *traffic*. The shorthand notation of the C formula is $C(m, \lambda/\mu)$.

7.4   M/M/∞ Queueing System

An M/M/∞ queueing system is obtained as the number of servers in the M/M/m queueing system tends to infinity. Obviously, no waiting is possible at the limiting case because there is always an idle server in the system. The M/M/∞ queue does not occur in practice, but this model can be used efficiently to approximate the behavior of high-capacity service units. The condition of stability is $0 < \lambda, \mu < \infty$.

The analysis of M/M/∞ queueing system can be carried out in an analogous way to the analysis of the M/M/1 system. In an M/M/∞ system, the number of customers, $N(t)$, is also a birth–death process with the following parameters:

$$p_0 = 1, \quad a_k = \lambda, \quad k \geq 0, \quad b_k = k\mu, \quad k \geq 1,$$

$$p_k = \frac{\lambda}{\lambda + k\mu}, \quad q_k = \frac{k\mu}{\lambda + k\mu}, \quad k \geq 1.$$

According to Theorem 3.28, the stationary and the transient solution, assuming $P_0(0) = 1$, $P_k(0) = 0$ for $k \geq 1$, are

$$\pi_k = \frac{1}{k!} \left(\frac{\lambda}{\mu}\right)^k e^{-\lambda/\mu}, \quad k \geq 0.$$

and

$$P_k(t) = \exp\left\{-\frac{\lambda}{\mu t}(1 - e^{-\mu t})\right\} \frac{1}{k!}\left(\frac{\lambda}{\mu t}(1 - e^{-\mu t})\right)^k, \quad k \geq 0,$$

It is interesting to notice that both of them follow a Poisson distribution. A stochastic interpretation of the transient behavior is based on property (f) and (g) of the Poisson process in Sect. 2.7.3. If a customer arrives to the system in $(0, t)$ its arrival time is uniformly distributed in $(0, t)$ (property (f)) and consequently is present in the system at time t with probability

$$\int_0^t \frac{1}{t}e^{-\mu y}dy = \frac{1}{\mu t}(1 - e^{-\mu t}).$$

Customers arrive to the system according to a Poisson process with rate λ and are filtered (are present at time t) with probability $\frac{1}{\mu t}(1 - e^{-\mu t})$. According to property (g), the number of customers present in the system at time t is Poisson distributed with parameter $\frac{\lambda}{\mu t}(1 - e^{-\mu t})$.

7.5   M/M/m/m Queueing System

An M/M/m/m queueing system contains m servers but it does not contain a buffer for waiting customers. Thus customers that arrive while the servers are busy are lost. It can be interpreted as a finite-state variant of the M/M/m queueing system because the number of customers in the system cannot exceed m. The infinitesimal generator of the Markov chain describing the number of customers in the system contains the following nonzero elements:

$$q_{ij} = \begin{cases} \lambda_i = \lambda & \text{if } 0 \leq i < m, j = i + 1, \\ \mu_i = i\mu & \text{if } 1 \leq i \leq m, j = i - 1, \\ -\lambda_i - \mu_i = -\lambda - i\mu & \text{if } 0 \leq i < m, j = i, \\ -\mu_m = -m\mu & \text{if } i = m, j = i. \end{cases}$$

The stationary distribution is

$$p_k = p_0 \left(\frac{\lambda}{\mu}\right)^k \frac{1}{k!} \quad k = 1, 2, \cdots, m, \qquad (7.32)$$

where

$$p_0 = \frac{1}{\sum_{k=0}^m \left(\frac{\lambda}{\mu}\right)^k \frac{1}{k!}}. \qquad (7.33)$$

An M/M/m/m system is stable ($p_k > 0$ $\forall k$) if $0 < \lambda, \mu < \infty$. The mean service time, the mean customer arrival intensity, and the mean number of customers are

$$\bar{x} = \frac{1}{\mu}, \quad \bar{\lambda} = \sum_{k=0}^{m-1} \lambda_k p_k = \lambda(1 - p_m), \quad \text{and } \mathbf{E}(X) = \frac{\lambda}{\mu}(1 - p_m), \quad (7.34)$$

which fulfills Little's law. M/M/m/m queueing systems are referred to as loss systems in telecommunications. The probability that an arriving customer will be lost is

$$\mathbf{P}(\text{loss}) = p_m^{(a)} = p_m = \frac{\left(\dfrac{\lambda}{\mu}\right)^m \dfrac{1}{m!}}{\displaystyle\sum_{k=0}^{m} \left(\dfrac{\lambda}{\mu}\right)^k \dfrac{1}{k!}} = B(m, \lambda/\mu), \quad (7.35)$$

which is known as the B (loss) formula of Erlang [1]. The dimensioning of switched telephone networks was based on this formula for several decades in the twentieth century.

7.6 M/M/1//N Queueing System

All previous queueing systems have infinite populations and a state-independent Poisson customer arrival process. In an M/M/1//N queueing system, the population is finite and the customer arrival intensity depends on the state of the system because the customers in the system do not contribute to the new arrivals. For example, if all customers of the population are in the system, then the new customer arrival intensity reduces to 0. The infinitesimal generator of a Markov chain describing the possible changes in the system contains:

$$q_{ij} = \begin{cases} \lambda_i = (N - i)\lambda & \text{if } 0 \le i < N, j = i + 1, \\ \mu_i = i\mu & \text{if } 1 \le i \le N, j = i - 1, \\ -\lambda_i - \mu_i = -(N - i)\lambda - \mu & \text{if } 0 \le i \le N, j = i. \end{cases}$$

The Markov chain is stable if $0 < \lambda, \mu < \infty$. In this case the stationary distribution is

$$p_k = p_0 \left(\frac{\lambda}{\mu}\right)^k [N(N - 1) \cdots (N - k + 1)] = p_0 \left(\frac{\lambda}{\mu}\right)^k \frac{N!}{(N - k)!}$$

$$k = 1, 2, \cdots, N, \quad (7.36)$$

where

$$p_0 = \frac{1}{1 + \sum\limits_{j=1}^{N} \left(\dfrac{\lambda}{\mu}\right)^j \dfrac{N!}{(N-j)!}} \qquad (7.37)$$

and the utilization is $\rho = 1 - p_0$. The mean arrival intensity of this system is

$$\bar{\lambda} = \sum_{i=0}^{N} \lambda_i p_i = \sum_{i=0}^{N-1} (N-i) \lambda\, p_i . \qquad (7.38)$$

According to Little's law,

$$\rho = \bar{\lambda} E(x) = \bar{\lambda}/\mu,$$

from which we obtain an expression for the mean arrival intensity:

$$\bar{\lambda} = \frac{\rho}{E(x)} = \mu\rho = \mu(1 - p_0). \qquad (7.39)$$

There is another way to express the mean arrival intensity. We can interpret the life cycle of a customer such that it stays outside the system for an exponentially distributed amount of time with parameter λ and after that it enters the system and spends a system time (waiting time + service time) there. Thus, the cycle time of a customer is $1/\lambda + E(T)$, and a customer generates a new arrival at the system once in every cycle. Consequently, a customer generates arrivals at an average rate $(1/\lambda + E(T))^{-1}$, and the N members of the population generate arrivals at a rate of

$$\bar{\lambda} = \frac{N}{1/\lambda + E(T)}.$$

From this expression we have

$$E(T) = \frac{N}{\bar{\lambda}} - \frac{1}{\lambda}, \qquad (7.40)$$

and using Little's law again we have

$$E(X) = \bar{\lambda} E(T) = N - \frac{\bar{\lambda}}{\lambda} \qquad (7.41)$$

and $\mathbf{E}(W) = \mathbf{E}(T) - \dfrac{1}{\mu}$. The probability that a member of the population is in the system is

$$\mathbf{P}(\text{in system}) = \frac{\mathbf{E}(T)}{1/\lambda + \mathbf{E}(T)}.$$

7.7   Exercises

Exercise 7.1 Compute the mean and the variance of the waiting time in an M/M/1 queue based on the Wald's identity.

Exercise 7.2 Consider a data packet transmission unit, which receives data packets from a set of terminals and transmits them to a destination unit through a transmission line. The packets arrive according to a Poisson process. In average one packet arrives in every 4 ms. The packet transmission time is exponentially distributed. The mean packet transmission time is 3 ms.

- What is the mean number of packets in the transmission unit if it has an infinite buffer?
- What is the mean system time of a customer?
- How much has to increase the arrival rate to double the mean system time?

Exercise 7.3 Customers arrive to an infinite buffer queueing system according to a Poisson process with rate $K\lambda$ and the service time is exponentially distributed. The mean service time of a high capacity server is $1/(K\mu)$ and the mean service time of a low capacity server is $1/\mu$. Compare the performance of the single queue using one high capacity server with the performance of K parallel queues having low capacity servers. In the latter case customers arrive to each queue according to a Poisson process with rate λ (c.f. decomposition of a Poisson process in Sect. 2.7.3).

Exercise 7.4 Two kinds of customers arrive to a queueing system with 3 servers. Type 1 customers arrive according to a Poisson process with rate λ_1. A type 1 customer occupies one server for an exponentially distributed amount of time with parameter μ_1. Type 2 customers arrive according to a Poisson process with rate λ_2. A type 2 customer occupies two servers for an exponentially distributed amount of time with parameter μ_2. Compute the loss probability of type 2 customers if there is no buffer in the system.

Exercise 7.5 There are N terminals in a computer system. Each terminal infinitely repeats the following steps:

- Generates a task in an exponentially distributed amount of time with parameter λ.
- Submits the task to the central processing unit.
- Waits for the answer.

The central processing unit processes a task in an exponentially distributed amount of time with parameter μ. Approximate the task completion rate and the system time of this system assuming the two extreme cases where the system is heavily loaded (N, λ/μ are small) and when the system load is light (N, λ/μ are large).

Exercise 7.6 One shop assistant serves the customers in a shop with exponentially distributed service time with parameter μ. The shop assistant wants to smoke after an exponentially distributed time with parameter α. If the shop is idle leaves for smoking immediately. If he is busy when he wants to smoke, then he serves the customers while shop is not idle and then he leaves for smoking. The length of the smoke break is exponentially distributed with parameter β. The customers arrive according to a Poisson process with rate λ. Compute the mean shopping time of customers if at most 3 customers can enter the shop. (Compute the same measure if infinitely many customers can enter the shop.)

Exercise 7.7 There is a queueing system with two servers and two types of customers. Type i customers arrive according to a Poisson process with rate λ_i and their service time is exponentially distributed with parameter μ_i, $i = 1, 2$. Server i is typically assigned with type i customers. If there is a type i customer in the system when server i is idle, then it serves a type i customer. If there is no type i customer in the system when server i is idle, then it can serve a customer of the other type. The arrival of a new customer does not interrupt the ongoing service. Compute the loss probability of type i customers if the buffer size is 3.

Exercise 7.8 Two kinds of customers arrive at a discrete-time queueing system. In every time slot a type i customer arrives with probability p_i, $i = 1, 2$, and no customer arrives with probability $1 - p_1 - p_2$. There is a single server. The service time of a type 1 customer is geometrically distributed with parameter q_1. The service time of a type 2 customer is k time slot and the buffer size is b. Compute the mean system time of type i customers for $i = 1, 2$, if $k = 1, 2$ and $b = 0, 3, \infty$.

Exercise 7.9 To improve the energy efficiency of a discrete-time queueing system the server is switched off (goes on vacation) for a geometrically distributed amount of time with parameter r if the system is idle at the end of a time slot. At the end of the vacation period the server starts serving the arrived customers (if any) or goes for an other vacation (if none). In every time slot one customer arrives with probability p and no customer arrives with probability $1 - p$. The service time is geometrically distributed with parameter q and the buffer size is b. Compute the mean system time, the mean vacation time and the mean idle time of the server for $b = 3, \infty$.

Exercise 7.10 Compute the stationary number of customers in an M/M/2/3/4 queue if $\lambda = 1$ and $\mu = 2$.

Exercise 7.11 There are 4 leased telephone lines between two sites of a company. Phone call requests arrive according to a Poisson process at rate $1/2$ [calls/min]. The lengths of the calls is exponentially distributed. The mean call holding time 4 [min]. If all lines are busy when a call arrives the call is waiting while a telephone line becomes available. What is the probability that a call has to wait?

Exercise 7.12 Compare the performance of the M/M/1 and the M/M/2 queueing systems if $\lambda = 1/2$ in both systems, the service rates of the M/M/1 and the M/M/2 systems are $\mu_1 = 1$ and $\mu_2 = 1/2$, respectively.

Exercise 7.13 Compute the loss probability of the M/M/m/m/K system for $K > m$.

Exercise 7.14 Consider the same system as in Exercise 7.11 and assume that the calls which arrive when all lines are busy are lost. Compute the parameters of this loss system and compare them with the one of the waiting system from Exercise 7.11.

Exercise 7.15 Compare the probability of waiting in an M/M/m queue with the loss probability in an M/M/m/m queue for $m = 1, 2, 3$, where the arrival and service intensities are identical. Interpret the relation of the results.

Exercise 7.16 A complex system is composed by two main units. The failure and the repair time of unit i, $i = 1, 2$, are exponentially distributed with parameter λ_i and μ_i, respectively. The units are maintained by a single repairman. Define the Markov chain of the system behavior if the service discipline of the repairman is FIFO, preemptive LIFO, processor sharing, if the repair of unit 1 has a preemptive priority over the one of unit 2, if the repair of unit 1 has a nonpreemptive priority over the one of unit 2.

Exercise 7.17 Customers of a discrete-time queueing system (under service and waiting) can be lost. Each customer is lost with probability r in each time slot. One customer arrives with probability p (and with $1-p$ no customer arrives) in each time slot and the service time is geometrically distributed with parameter q. Compute the probability of successful service completion if the buffer size is 3.

7.8 Solutions

Solution 7.1 The waiting of a customer in an M/M/1 queue, W, is the sum of the service times of the customers, X_i, which are in the system at its arrival. If the number of customers in the system is N at the arrival of the customer, then its waiting time is

$$W = \sum_{i=1}^{N} X_i,$$

from which

$$\mathbf{E}(W) = \mathbf{E}\left(\sum_{i=1}^{N} X_i\right) = \sum_{n=0}^{\infty} \mathbf{E}\left(\sum_{i=1}^{n} X_i\right) \mathbf{P}(N = n) = \sum_{n=0}^{\infty} n\mathbf{E}(X)\mathbf{P}(N = n) = \mathbf{E}(X)\mathbf{E}(N),$$

and

$$\sigma_W = \sum_{n=0}^{\infty} \sigma_{\sum_{i=1}^{n} X_i} P(N = n) = \sum_{n=0}^{\infty} n\sigma_X P(N = n) = \sigma_X E(N).$$

Solution 7.2

- The mean number of packets in the transmission unit:

$$\rho = \frac{1}{4} \cdot 3 = \frac{3}{4},$$

$$E(X) = \frac{\rho}{1 - \rho} = 3.$$

- Mean system time of a customer:

$$E(T) = \frac{E(X)}{\lambda} = \frac{3}{1/4 \, 1 \, \text{ms}} = 12 \, \text{ms}.$$

- To double the mean system time the arrival rate has to be:

$$E(T') = 24 \, \text{ms} = \frac{1/\mu}{1 - \rho'} = \frac{3 \, \text{ms}}{1 - \rho'},$$

$$\rho' = 1 - \frac{1}{8} = 7/8,$$

from which

$$\lambda' = \rho'\mu = \frac{7}{8 \cdot 3} = \frac{7}{24}.$$

It means that a small (17%) increase of the arrival rate doubles the mean system time.

Solution 7.3 Performance of the single high capacity queue:

$$\rho = \frac{K\lambda}{K\mu} = \frac{\lambda}{\mu},$$

$$E(T) = \frac{E(X)}{1 - \rho} = \frac{1}{K\mu(1 - \rho)},$$

Performance of the K low capacity queues:

$$\rho = \frac{\lambda}{\mu}$$

per servers, and

$$\mathbf{E}(T') = \frac{\mathbf{E}(X)}{1-\rho} = \frac{1}{\mu(1-\rho)} = K \cdot \mathbf{E}(T).$$

Consequently, the system time is K times longer in the latter case.

The result demonstrates that aggregating the resources and the demands in service systems increases the system performance.

Solution 7.4 The stochastic process describing the number of type 1 (first coordinate) and type 2 (second coordinate) customers in the system is a CTMC with the following transition graph.

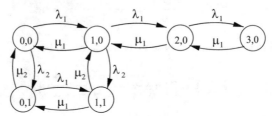

Compute the stationary probability distribution, $p_{i,j}$, of this CTMC and the loss probability is

$$p_{\text{loss}} = \frac{\lambda_2}{\lambda_1 + \lambda_2} (p_{0,2} + p_{1,0}) + p_{1,1} + p_{0,3}.$$

In state $p_{0,2}$ and $p_{1,0}$ only the type 2 customers are lost.

Solution 7.5

- In case of a light load:

$$\mathbf{E}(N) \approx \frac{1}{\mu},$$

$$\bar{\lambda} = \frac{N}{1/\lambda + \mathbf{E}(T)} \approx \frac{N}{1/\lambda + 1/\mu}.$$

- In case of a heavy load:

$$\bar{\lambda} \approx \mu,$$

$$\mathbf{E}(T) \approx \frac{N}{\mu} - \frac{1}{\lambda}.$$

Solution 7.6 The stochastic process describing the state of the shop assistant (first coordinate) and the number of customers in the shop (second coordinate) is a CTMC with the following transition graph. The state of the shop assistant is 0 if it works and does not miss a cigarette, is 1 if it works and misses a cigarette, is 2 if it smokes.

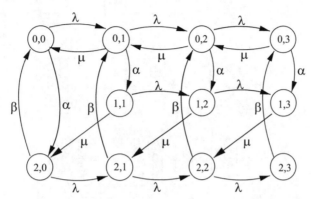

We compute the mean shopping time of customers with the help of Little's law. For that we need the mean number of customers in the shop, L, and the mean customer arrival rate, $\bar{\lambda}$. Based on the stationary probability distribution, $p_{i,j}$, of the CTMC the mean number of customers in the shop is

$$L = \sum_{j=1}^{3} j \sum_{i=0}^{2} p_{i,j}$$

and mean customer arrival rate is

$$\bar{\lambda} = \lambda p_{\text{loss}}$$

where

$$p_{\text{loss}} = p_{03} + p_{13} + p_{23}.$$

Finally, the mean shopping time of customers is

$$T = L/\bar{\lambda}.$$

If infinitely many customers can enter the shop then the Markov chain is infinite according to the second coordinate and we obtain a quasi-birth-death process of the form

$$
Q = \begin{array}{|c|c|c|c|}
\hline
L' & F' & \mathbf{0} & \cdots \\
\hline
B' & L & F & \ddots \\
\hline
\mathbf{0} & B & L & \ddots \\
\hline
\vdots & \ddots & \ddots & \ddots \\
\hline
\end{array}.
$$

where

$$
L = \begin{array}{|c|c|c|}
\hline
\bullet & \alpha & 0 \\
\hline
0 & \bullet & 0 \\
\hline
\beta & 0 & \bullet \\
\hline
\end{array}, \quad
B = \begin{array}{|c|c|c|}
\hline
\mu & 0 & 0 \\
\hline
0 & 0 & \mu \\
\hline
0 & 0 & 0 \\
\hline
\end{array}, \quad
F = \begin{array}{|c|c|c|}
\hline
\lambda & 0 & 0 \\
\hline
0 & \lambda & 0 \\
\hline
0 & 0 & \lambda \\
\hline
\end{array}.
$$

$$
L' = \begin{array}{|c|c|}
\hline
\bullet & \alpha \\
\hline
\beta & \bullet \\
\hline
\end{array}, \quad
B' = \begin{array}{|c|c|}
\hline
\mu & 0 \\
\hline
0 & \mu \\
\hline
0 & 0 \\
\hline
\end{array}, \quad
F' = \begin{array}{|c|c|c|}
\hline
\lambda & 0 & 0 \\
\hline
0 & 0 & \lambda \\
\hline
\end{array}.
$$

and $\bullet$ are the negative diagonal elements which are set such that all row sums of Q are zero. Assuming that process is positive recurrent and the stationary solution of this QBD process with irregular level zero is vector p_0, and vectors $p_i = p_1 R^{i-1}$ for $i \geq 1$, we can compute the mean number of customers in the shop as

$$
L = \sum_{i=1}^{\infty} i p_1 R^{i-1} \mathbb{1} = p_1 \left(\sum_{i=1}^{\infty} (i-1) R^{i-1} + \sum_{i=1}^{\infty} R^i \right) \mathbb{1}
$$
$$
= p_1 \left(R(I - R)^{-2} + R(I - R)^{-1} \right) \mathbb{1} = p_1 \left(R(I - R)^{-1} \left((I - R)^{-1} + I \right) \right) \mathbb{1}.
$$

Due to the fact that there is no loss in case of infinite shop capacity ($\bar{\lambda} = \lambda$) the mean shopping time of a customer is

$$
T = L / \bar{\lambda} = L / \lambda.
$$

Solution 7.7 A finite CTMC describes the behavior of the queueing system, where the states are identified by the triple (i, j, k). i indicates the state of server 1. $i = 0$ when server 1 is idle, $i = 1$ when server 1 serves a type 1 customer, and $i = 2$ when server 1 serves a type 2 customer. j indicates the state of server 2 in a similar manner, and k indicates the number of customers in the buffer.

As it is visible from the (i, j, k) state description the type of the customers in the buffer is not identified in the system state. It is a widely applicable trick to reduce the complexity of the Markov chains describing the behavior of queueing systems. When the servers are busy customers arrive to the buffer with rate $\lambda = \lambda_1 + \lambda_2$ and a given customer in the buffer is of type 1 with probability λ_1 / λ and of type 2 with probability λ_2 / λ.

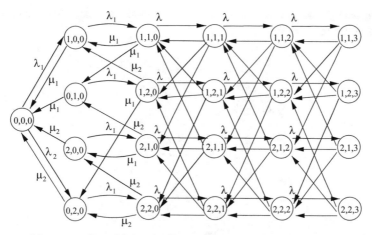

The transition rates from higher buffer occupancy to lower one are as follows

- from $(1, 1, i)$ to $(1, 1, i-1)$ at rate $2\mu_1\lambda_1/\lambda$ because either server 1 completes the service at rate μ_1 and the next customer is type 1 with probability λ_1/λ or server 2 completes the service at rate μ_1 and the next customer is type 1 with probability λ_1/λ,
- from $(1, 1, i)$ to $(1, 2, i-1)$ at rate $2\mu_1\lambda_2/\lambda$ because server 2 completes the service at rate μ_1 and the next customer is type 2 with probability λ_2/λ,
- ...

The loss probability of both customers type is identical and is associated with the states when the buffer is full.

$$P_{loss1} = P_{loss2} = p_{1,1,3} + p_{1,2,3} + p_{2,1,3} + p_{2,2,3}.$$

Further type related measures can be computed based on the stationary distribution of this Markov chain. E.g., the probability that a type 1 customer is served by server 2 is

$$P_{type1\text{-}server2} = \frac{\displaystyle\sum_{i=0}^{2}\sum_{k=0}^{3} p_{i,1,k}}{\displaystyle\sum_{i=0}^{2}\sum_{k=0}^{3} p_{i,1,k} + \sum_{j=0}^{2}\sum_{k=0}^{3} p_{1,j,k}}.$$

Solution 7.8 We present the solution for $k = 2$. The solution for $k = 1$ is straightforward based on that. A finite DTMC describes the behavior of the queueing system, where the states are identified by the couple (i, j). i indicates the state of the server. $i = 0$ when the server is idle, $i = 1$ when the server serves a type 1 customer, $i = 2$ when the server serves a type 2 in the first time slot, and $i = 3$ when the server serves a type 2 in the second time slot. j indicates the number of

customers in the buffer. The following Markov chain describes the behavior when
there is no buffer ($b = 0$).

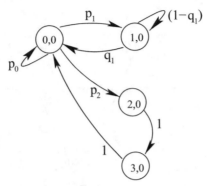

The Markov chain also indicates that the customers that receive service, because
the system is idle at their arrival, start service immediately after arrival. Consequently, the total system time is the service time for both types of customers and
their means are

$$T_1 = \frac{1}{q_1}, \quad T_2 = 2.$$

For the case when there is buffer in the system we follow the same approach as
in Exercise 7.7 and the state of the Markov chain does not identify the type of the
customers in the buffer only the type of the customer at the server. That is, when
the server is busy customers arrive to the buffer with probability $p = p_1 + p_2$ and
a given customer in the buffer is of type 1 with probability p_1/p and of type 2 with
probability p_2/p. In case of buffer capacity 3 ($b = 3$) the following Markov chain
describes the queueing system, where $p_1^* = p_0 p_1/p$ and $p_2^* = p_0 p_2/p$.

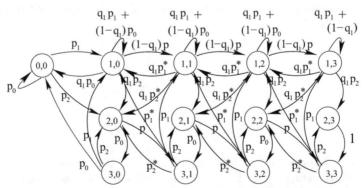

When $b = 3$ the computation of the system time requires the stationary solution
of the Markov chain ($p_{i,j}$ denote the stationary probabilities) and the application
of the Little's law in a similar way as in Exercise 7.6. The loss probability of both
customers type is identical and is associated with the states when the buffer is full

and the arriving customer is lost.

$$P_{loss} = P_{loss1} = P_{loss2} = \frac{p_{1,3}(1-q_1)p + p_{2,3}p}{p} = p_{1,3}(1-q_1) + p_{2,3},$$

and the mean arrival rate of type 1 (type 2) customers is $\bar{\lambda}_1 = p_1 P_{loss}$ ($\bar{\lambda}_2 = p_2 P_{loss}$). The mean number of type 1 and type 2 customers in the queue is

$$L_1 = \sum_{j=0}^{3} p_{1,j}\left(1+\frac{jp_1}{p}\right)+\sum_{i=2}^{3}\sum_{j=0}^{3} p_{i,j}\frac{jp_1}{p}, \quad L_2 = \sum_{j=0}^{3} p_{1,j}\frac{jp_2}{p}+\sum_{i=2}^{3}\sum_{j=0}^{3} p_{i,j}\left(1+\frac{jp_2}{p}\right).$$

Finally, from the Little's law we have

$$T_1 = L_1/\bar{\lambda}_1 . T_2 = L_2/\bar{\lambda}_2.$$

The case when the buffer is infinite results in a quasi-birth–death process with special level zero. Both the special structure of level zero and the regular structure of the higher levels are readable from the Markov chain of buffer capacity 3. For example, the matrices describing the regular part are

$$L = \begin{array}{|ccc|} q_1 p_1 + (1-q_2)p_0 & q_1 p_2 & 0 \\ 0 & 0 & p_0 \\ p_1 & p_2 & 0 \end{array}, \quad B = \begin{array}{|ccc|} q_1 p_1^* & q_1 p_2^* & 0 \\ 0 & 0 & 0 \\ p_1^* & p_2^* & 0 \end{array}, \quad F = \begin{array}{|ccc|} (1-q_1)p & 0 & 0 \\ 0 & 0 & p \\ 0 & 0 & 0 \end{array}.$$

Note that $(B+L+F)\mathbb{1} = \mathbb{1}$ holds, that is, the sum of the exit probabilities of each state is one. The stability condition of the QBD process can also be obtained by work load consideration. The service of a type 1 customer takes $1/q_1$ time slots in average and the service of a type 2 customer takes 2 time slots. This way $p_1/q_1 + p_2$ 2 workload arrive to the server in a time slot, which has to be less than 1 in a stable system.

When the buffer is infinite we can compute the system time based on the stationary solution of the QBD type Markov chain. In this case $\bar{\lambda}_1 = p_1$ and $\bar{\lambda}_2 = p_2$, because there is no loss due to the infinite buffer. The mean number of type 1 and type 2 customers in the queue is

$$L_1 = \sum_{j=0}^{\infty} p_{1,j}\left(1+\frac{jp_1}{p}\right)+\sum_{i=2}^{\infty}\sum_{j=0}^{3} p_{i,j}\frac{jp_1}{p}, \quad L_2 = \sum_{j=0}^{\infty} p_{1,j}\frac{jp_2}{p}+\sum_{i=2}^{3}\sum_{j=0}^{\infty} p_{i,j}\left(1+\frac{jp_2}{p}\right)$$

and similarly

$$T_1 = L_1/\bar{\lambda}_1 . T_2 = L_2/\bar{\lambda}_2.$$

Solution 7.9 When $b = 3$ the following finite DTMC describes the behavior of the queueing system. The states are identified by the couple (i, j). i indicates the state of the server. $i = 0$ when the server is on vacation, $i = 1$ when the server is active and serving a customer. j the number of customers in the buffer. In the figure $\bar{q} = 1 - q$, $\bar{p} = 1 - p$, and $\bar{r} = 1 - r$.

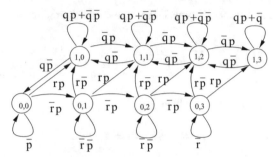

We compute the mean system time of customers, T, based on the stationary solution of the Markov chain (denoted as $p_{i,j}$) and the Little's law in a similar way as in Exercise 7.6. The loss probability is

$$p_{\text{loss}} = \frac{p_{1,3}(1-q)p + p_{0,3}(1-r)p}{p} = p_{1,3}(1-q) + p_{0,3}(1-r),$$

and the mean arrival rate of customers is $\bar{\lambda} = p p_{\text{loss}}$. The mean number of customers in the queue is

$$L = \sum_{i=0}^{1} \sum_{j=0}^{3} (i+j) p_{i,j} \text{ and } T = L/\bar{\lambda}.$$

The idle time and the vacation time of the server are identical because the server immediately starts the vacation when it becomes idle and it finishes the vacation only when there is customer to serve. The vacation time of the server is discrete PH distributed and its representation can be extracted from the Markov chain by interpreting the transition from the lower row of states (server in vacation) to the upper ones (server busy) as transitions to an absorbing state and recognizing that the vacation always starts in state $(0, 0)$.

$$\beta = \{1, 0, 0, 0\}, \quad \mathbf{B} = \begin{array}{|cccc|} \hline 1-p & (1-r)p & 0 & 0 \\ 0 & (1-p)(1-r) & (1-r)p & 0 \\ 0 & 0 & (1-p)(1-r) & (1-r)p \\ 0 & 0 & 0 & 1-r \\ \hline \end{array}.$$

The mean of the vacation time can be computed as $\beta \mathbf{B}^{-1} \mathbb{1}$.

In the case when the buffer is infinite the regular structure of the above Markov chain follows for all buffer levels. There are more than one way to define a QBD process based on this regular Markov chain. It is possible to define a QBD such that level j is composed by states $(0, j)$ and $(1, j)$, in this case the structure is regular for level zero; and it is also possible to define a QBD such that level j is composed by states $(0, j)$ and $(1, j - 1)$, in this case the structure is special for level zero. In the second case level zero has different dimension, it is composed by a single state, state $(0, 0)$.

The regular matrix blocks of the QBD whose level j is composed by states $(0, j)$ and $(1, j)$ are

$$
L = \begin{array}{|c|c|} \hline (1-r)(1-p) & rp \\ \hline 0 & pq + (1-p)(1-q) \\ \hline \end{array}, \quad
B = \begin{array}{|c|c|} \hline 0 & r(1-p) \\ \hline 0 & q(1-p) \\ \hline \end{array}, \quad
F = \begin{array}{|c|c|} \hline (1-r)p & 0 \\ \hline 0 & (1-q)p \\ \hline \end{array}.
$$

Note again that $(B + L + F)\mathbb{1} = \mathbb{1}$ holds.

When the buffer is infinite and the system is stable we can compute the mean system time of customers based on the stationary solution of the QBD type Markov chain. In this case $\bar{\lambda} = p$, because there is no loss due to the infinite buffer. The mean number of customers in the queue is

$$
L = \sum_{i=0}^{1} \sum_{j=0}^{\infty} (i + j) p_{i,j} \quad \text{and} \quad T = L/\bar{\lambda}.
$$

We compute the mean vacation time with infinite buffer based on the stochastic interpretation of the system behavior. The vacation starts in state $(0, 0)$ and there are two conditions to finish the vacation. There is an arrival and at the same time or after the arrival a transition with probability r occurs. This stochastic interpretation allows a simpler PH representation of the vacation time

$$
\gamma = \{1, 0\}, \quad G = \begin{array}{|c|c|} \hline 1 - p & (1 - r)p \\ \hline 0 & 1 - r \\ \hline \end{array}.
$$

Note that (β, B) and (γ, G) define the same distribution.

Solution 7.10 The Markov chain of M/M/2/3/4 queue is

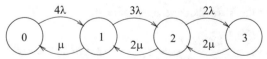

With $\lambda = 1$ and $\mu = 2$ the stationary probabilities satisfy the following local balance and normalizing equations.

$$p_1 = \frac{4}{2}p_0, \; p_2 = \frac{3}{4}p_1 = \frac{3}{2}p_0, \; p_3 = \frac{2}{4}p_2 = \frac{3}{4}p_0, \; \sum_{i=0}^{3} p_i = 1,$$

from which

$$p_0 = \frac{4}{21}, \; p_1 = \frac{8}{21}, \; p_2 = \frac{6}{21}, \; p_3 = \frac{3}{21},$$

and the mean number of customers is

$$L = \sum_{i=0}^{3} i p_i = \frac{29}{21}.$$

Solution 7.11 We have

$$\lambda = 1/2, \quad 1/\mu = 4, \quad a = \lambda/\mu = 2, \quad \rho = a/m = 2/4 = 0.5,$$

from which

$$p_0 = \frac{1}{1 + 2 + 2^2/2 + 2^3/6 + 16/24(1/(1 - 0.5))} = 3/23,$$

and

$$C(4, 2) = \frac{2^4/4!}{1 - 0.5} \frac{3}{23} = 4/23 = 0.17.$$

Solution 7.12 The parameters of the M/M/1 queueing system are

$$\rho = \frac{\lambda}{\mu_1} = \frac{1/2}{1} = 0.5,$$

$$E(W) = \frac{\rho/\mu}{1 - \rho} = 1 \, \text{s},$$

$$E(T) = \frac{1/\mu}{1 - \rho} = 2 \, \text{s}.$$

The parameters of the M/M/2 queueing system are

$$a = \frac{\lambda}{\mu_2} = \frac{1/2}{1/2} = 1, \quad \rho = a/m = 1/2 = 0.5,$$

$$p_0 = \frac{1}{1 + 2 + \frac{a^2/2}{1 - 0.5}} = 1/3,$$

from which

$$C(2, 1) = \frac{a^2/2}{1 - 0.5} p_0 = 1/3,$$

$$E(W') = \frac{1/\mu_2}{1 - \rho} C(2, 1) = 2/3,$$

$$E(T) = 2/3 + 1/\mu_2 = 8/3sec.$$

Consequently, the system time of the M/M/1 system is lower, in spite of the fact that its waiting time is higher.

Solution 7.13 The Markov chain of M/M/m/m/K queue is

The stationary probabilities satisfy the following local balance and normalizing equations.

$$p_i = \frac{(K - i + 1)\lambda}{i\mu} p_{i-1}, \quad i = 1, \ldots, m, \quad \sum_{i=0}^{m} p_i = 1,$$

from which

$$p_i = p_0 \prod_{j=1}^{i} \frac{(K - j + 1)\lambda}{j\mu} = p_0 \frac{K! \lambda^i}{K - i! \, i! \, \mu^i} = p_0 \binom{K}{i} \left(\frac{\lambda}{\mu}\right)^i, \quad i = 0, \ldots, m,$$

$$p_i = \frac{\binom{K}{i} \left(\frac{\lambda}{\mu}\right)^i}{\sum_{j=0}^{m} \binom{K}{j} \left(\frac{\lambda}{\mu}\right)^j}, \quad i = 0, \ldots, m.$$

The loss probability of the M/M/m/m/K queue is $p_{loss} = p_m$.

Solution 7.14

$$p_{loss} = B(4, 2) = \frac{16/24}{1 + 2 + 2^2/2 + 2^3/6 + 16/24} = \frac{2/3}{5 + 4/3 + 2/3} = 2/21 = 0.095$$

That is, $B(4, 2) = 0.095 < C(4, 2) = 0.17$. This relation can be explained by the load of the two systems. In case of waiting system all arriving customers have to be served, while in case of loss systems the load of the servers is reduced by the lost customers.

Solution 7.15 In an M/M/1 queue the probability of waiting is $p_{\text{wait}} = \frac{\lambda}{\mu}$ and in an M/M/1/1 queue (Markov chain with 2 states) the loss probability is $p_{\text{loss}} = \frac{\frac{\lambda}{\mu}}{1+\frac{\lambda}{\mu}}$.

Next we compute the probability of waiting and the loss probability for $m = 3$ only. The stationary state probabilities of the M/M/3 queue satisfy

$$p_i = \frac{\lambda}{i\mu}p_{i-1}, \ i = 1, 2, \quad p_i = \frac{\lambda}{3\mu}p_{i-1}, \ i = 3, 4, \dots, \quad \sum_{i=0}^{\infty} p_i = 1,$$

$$p_i = p_0 \frac{\lambda^i}{i!\mu^i}, \ i = 1, 2, \quad p_i = p_2 \left(\frac{\lambda}{3\mu}\right)^{i-2}, \ i = 2, \dots, m, \quad \text{and} \quad \sum_{i=2}^{\infty} p_i = p_2 \frac{3\mu}{3\mu - \lambda},$$

from which

$$p_{\text{wait}} = \frac{\sum_{i=3}^{\infty} p_i}{p_0 + p_1 + \sum_{i=2}^{\infty} p_i} = \frac{\frac{\lambda^3}{3!\mu^3}\frac{3\mu}{3\mu-\lambda}}{1 + \frac{\lambda}{\mu} + \frac{\lambda^2}{2\mu^2}\frac{3\mu}{3\mu-\lambda}}.$$

The stationary state probabilities of the M/M/3/3 queue (Markov chain with 4 states) satisfy

$$p_i = \frac{\lambda}{i\mu}p_{i-1}, \ i = 1, 2, 3, \quad \sum_{i=0}^{3} p_i = 1,$$

from which

$$p_{\text{loss}} = \frac{p_3}{p_0 + p_1 + p_2 + p_3} = \frac{\frac{\lambda^3}{3!\mu^3}}{1 + \frac{\lambda}{\mu} + \frac{\lambda^2}{2\mu^2} + \frac{\lambda^3}{3!\mu^3}}.$$

In both cases the waiting probability is larger than the loss probability because the Markov chains spend the same amount of time in the $0, \dots, m - 1$ part of the state space, but in case of M/M/m queue the Markov chain spends more time in the other part of the state space, which is composed by infinitely many states, than in the case of M/M/m/m queue, when the other part of the state space is composed by a single state.

Solution 7.16

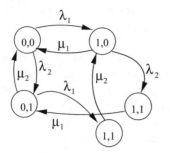

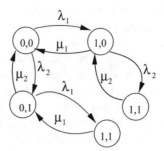

FIFO service   Preemptive LIFO

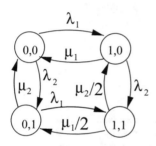

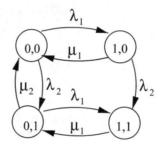

Processor sharing   Preemptive priority of unit1

The nonpreemptive priority of unit 1 is identical with the FIFO case, because the repairman does not interrupt the ongoing service process and at the time when the ongoing service process is completed there is only one failed unit to repair.

Solution 7.17 The following DTMC describes the system behavior,

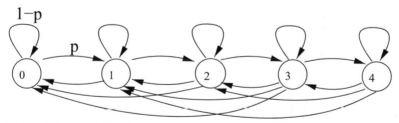

with transition probabilities

$$p_{i,i+1} = (1-q)pr(i,0),$$

$$p_{i,i} = qpr(i,0) + (1-q)(1-p)r(i,0) + (1-q)pr(i,1),$$

$$p_{i,i-1} = qpr(i,1) + q(1-p)r(i,0) + (1-q)(1-p)r(i,1) + \underbrace{(1-q)pr(i,2)}_{\text{if } i \geq 2},$$

$$\ldots,$$

where $r(i, j)$ denotes the probability that j customers are lost in a time slot, when there are i customers in the system at the beginning of the time slot. The number of lost customers is binomially distributed with parameters i, r, that is, $r(i, j) = \binom{i}{j} r^j (1 - r)^{i-j}$.

The transition probabilities are determined by the distribution of the number of served customers, Y, the number of arrived customers, V, and the number of lost customers, Z. Y is Bernoulli distributed with parameter q, V is Bernoulli distributed with parameter p and in case of i customers, and V is binomially distributed with parameters i, r. When there are i customers $2 \times 2 \times (i + 1)$ cases need to be evaluated to obtain the transition probabilities. The evolution equation describes the DTMC in a compact way. Let X_n be the number of customers in the system in time slot n and Y_n, Z_n, V_n the number of served, lost and arrived customers in time slot n, than

$$X_n = \min(\max(X_{n-1} - Y_n - Z_n, 0) + V_n, 3).$$

References

1. Kleinrock, L.: Queuing Systems, Volume 1: Theory. Wiley, New York (1975)
2. Matveev, V.F., Ushakov, V.G.: Queueing Systems (in Russian). MGU, Moscow (1984)
3. Prudnikov, A.P., Brychkov, Y.A., Marichev, O.I.: Integrals and Series, vol. 2. Gordon and Breach Science Publishers, New York (1986). Special functions

Chapter 8
Non-Markovian Queueing Systems

8.1 M/G/1 Queueing System

The M/G/1 queueing system (Fig. 8.1) is similar to the M/M/1 queueing system and the only difference is that the service time is not exponential. First we mention some ideas, most of which were described in the previous chapter in connection with an M/M/1 system.

8.1.1 Description of M/G/1 System

8.1.1.1 Conditions of Functioning

1. At the starting moment τ_0 the system is empty. For the sake of simplicity we generally assume $\tau_0 = 0$ (Fig. 8.2).
2. $\{N(t), \ t \geq \tau_0\}$ describes the number of entering customers; this is a Poisson process with intensity $\lambda > 0$.
3. There is one server functioning without breakdowns; after having served a customer it immediately starts serving of next one. If a customer enters the system and the server is busy, the customer joins the waiting queue. There is no limitation on the queue's size.
4. The service discipline is FCFS (FIFO).
5. The service times are independent identically distributed (i.i.d.) random variables with distribution function $\mathbf{P}(Y < x) = B(x)$ and mean $\mathbf{E}(Y) = \mu_B < \infty$, and they do not depend on the arrival process.

The main characteristic of the system is the queue length $\{L(t), \ t \geq 0\}$, i.e., how many customers are in the system at moment t. Let $L(t)$ be continuous from the right, i.e., $L(t) = L(t+), \ t \geq 0$.

© Springer Nature Switzerland AG 2019
L. Lakatos et al., *Introduction to Queueing Systems with Telecommunication Applications*, https://doi.org/10.1007/978-3-030-15142-3_8

Fig. 8.1 M/G/1 system

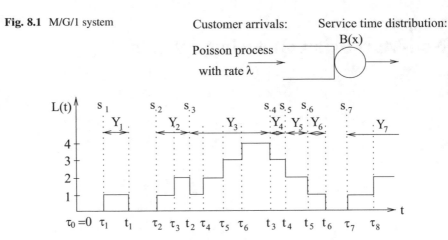

Fig. 8.2 Number of customers in M/G/1 system

In the case of queueing systems, the basic issues concern the distribution of queue length, whether there exists a limit distribution and how it can be found, the average number of customers in the system, etc. In this chapter we will deal with the asymptotic behavior of queue length $L(t)$ as $t \to \infty$.

We introduce the following notations:

- $\tau_0 + X_1$: moment of entry of first customer; X_n: interarrival time between $(n-1)$st and nth customers;
- $\tau_n = \tau_0 + X_1 + \ldots + X_n$ $(n \geq 1)$: moment of entry of nth customer;
- Y_n: service time of nth customer;
- s_n, $n \geq 1$: starting moment of service of nth customer;
- t_n, $n \geq 1$: moment when nth customer leaves system (service in system is completed at this moment).

According to these assumptions, $\{(X_n, Y_n), \ n \geq 1\}$ is a sequence of i.i.d. random variables, where the components of vectors are independent, too. Furthermore, the intervals between consecutive arrivals X_n, $n \geq 1$, have exponential distribution with parameter λ, the service times Y_n, $n \geq 1$, have distribution function $B(x)$. It is also clear that $\{\tau_n, \ n \geq 1\}$, are moments of jumps of the Poisson process $N(t)$ (Fig. 8.2).

8.1.2 Main Differences Between M/M/1 and M/G/1 Systems

For the M/M/1 system both the interarrival and service times are independent exponentially distributed random variables. These distributions have the memoryless property, so one can derive that $\{L(t),\ \ t \geq 0\}$ is a Markov (birth–death) process. This fact simplifies the investigation of system.

In the M/G/1 queueing system the examined processes (queue length, waiting time, etc.) are not necessarily of the Markov type since the service time distribution may not have the memoryless property, so their investigation requires different methods.

The foregoing conditions do not guarantee that $L(t)$ is a Markov process, but by means of an supplementary variable one can make it a Markov process with an extended state vector.

If $U(t)$ denotes the service time passed till t [$U(t)$ is right continuous], then the vector process $\{(L(t), U(t)),\ \ t \geq 0\}$ is already Markov and can be considered as the state of queueing system.

Generally, the vector $W(t) = \{L(t), U(t)\}$ ($t \geq \tau_0$) describing (from a certain viewpoint) the functioning of a system is called the *state vector* of the system if at arbitrary $t_1 > t$ one can determine the vector $W(t_1)$ in a stochastic sense based on the value of $W(t)$ and the arrivals for $(t, t_1]$.

Compared with the M/M/1 system the difference is not only that the system state is characterized by a vector process, but—and this is an important feature—the state space will not be discrete since the possible values of $U(t)$ are not discrete and take on values from the set $R_+ = [0, \infty)$ (or its subset).

8.1.3   Main Methods for Investigating M/G/1 System

1. **Method of embedded Markov chains,** also called Kendall's method because its wide use is connected with Kendall [7]. This method appeared in the 1950s, but the possibility of such approach was noted by Khinchin [8] (see also Palm [13]). We will consider this method in detail.
2. **Lindley's integral equation** [11]: can be derived for the more general G/G/1 systems and, hence, so is applicable in our case, too. This approach leads to a special Wiener–Hopf type integral equation for the limit distribution of customer waiting time.
3. **Method of supplementary variables** [4, 7]: based on the fact that the system may be investigated via the state vector $\{L(t), U(t),\ t \geq 0\}$ with the help of supplementary variable $U(t)$ (Fig. 8.3). Instead of the time interval from the beginning of service, one can use the time interval till the end of service (Henderson [6]).
4. **Method of random walk** and **combinatorial approach** [16].
5. **Method of recurrent processes** [1, 2].

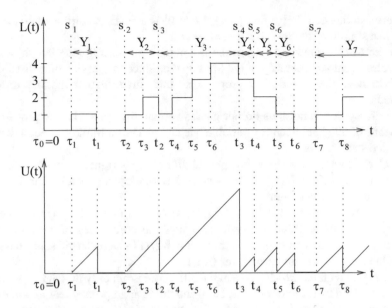

Fig. 8.3 $L(t)$ and $U(t)$ process of M/G/1 queue

In the following sections we investigate the M/G/1 queue using these approaches.

8.2 Embedded Markov Chains

The method includes the following steps:

(A) Choose random $(\tau_0 = 0 <) t_1 < t_2 < \ldots$ moments when the process describing the evolution of system is of a Markov type.

(B) Prove the ergodicity of the Markov chain $L_n = L(t_n)$, $n \geq 1$.

(C) Determine the ergodic distribution

$$\pi_k = \lim_{n \to \infty} \mathbf{P}(L_n = k), \quad k \geq 0,$$

of the Markov chain.

(D) Prove the coincidence of limiting values $\lim_{t \to \infty} \mathbf{P}(L(t) = k) = \pi_k, k \geq 0$.

8.2.1 Step (A): Determining Queue Length

Step (A) of the Embedded Markov chain method in Sect. 8.2: As earlier, t_n ($n = 1, 2, \ldots$) denotes the moment when the service of the nth customer is completed.

Let $L_n = L(t_n)$, $n \geq 1$ ($L_0 = 0$). Since $Y_{t_n} = 0$, $n \geq 0$, at moments t_n, the behavior of the state-vector process $\{L(t), Y_t), \ t \geq 0\}$ is described by the sequence $\{L_n, n \geq 1\}$. In our case the main idea of application of embedded Markov chains is to consider the process $\{(L(t), Y_t), \ t \geq 0\}$ at moments t_n, $n = 1, 2, \ldots$. Using this method we come to a Markov chain $\{L_n, n \geq 1\}$ with countable state space $\mathcal{X} = \{0, 1, 2, \ldots\}$, and so we obtain the final result [see (D) for the method]. In practice this means that the states of the system (the number of customers in the system) are considered at moments just after having served a customer. In this restricted view of the process, every state transition between consecutive service completion moments (e.g., customer arrival) is considered at the service completion moments.

The process $L(t)$ is Markov regenerative; this fact will be used at the proof of step (D).

We prove the following theorem, fulfilling the tasks formulated in steps (A) and (B).

Theorem 8.1 *The stochastic process $\{L_n, n \geq 1\}$ is homogeneous, irreducible, aperiodic Markov chain with state space $\mathcal{X} = \{0, 1, 2, \ldots\}$. If the condition $\rho = \lambda \mu_B < 1$ is fulfilled, then the Markov chain $\{L_n, n \geq 1\}$ is ergodic.*

Proof First we prove that the stochastic process $\{L_n, n \geq 1\}$ is a Markov chain. Let Δ_n, $n = 1, 2, \ldots$, denote the number of customers entering the system for the service time Y_n of the nth customer, i.c.,

$$\Delta_n = N(t_n) - N(t_n - Y_n) = N(s_n + Y_n) - N(s_n), \quad n \geq 1.$$

Service to the nth customer may start at $s_n = t_n - Y_n > t_{n-1}$ if the system is empty at t_{n-1}, i.e., $L_{n-1} = 0$. In this case $s_n = \tau_n$, and consequently $\Delta_n = N(t_n) - N(t_{n-1})$. Then

$$L_n = \begin{cases} L_{n-1} - 1 + \Delta_n, & \text{if } L_{n-1} > 0, \\ \Delta_n, & \text{if } L_{n-1} = 0, \end{cases}$$

or

$$L_n = \mathscr{I}_{\{L_{n-1}>0\}}(L_{n-1} - 1) + \Delta_n = (L_{n-1} - 1)^+ + \Delta_n. \tag{8.1}$$

Since the arrival process $\{N(t), t \geq 0\}$ is Poisson, independent of service times $\{Y_n, n \geq 1\}$, the number of customers Δ_n entering at service time Y_n is independent of $L_1, \ldots, L_{n-1}$, and consequently the sequence L_n constitutes a Markov chain.

Now we determine the distribution and mean value of random variable Δ_n ($n \geq 1$).

Using the fact that the behavior of $N(t)$ is independent of past events, the distribution of Δ_n can be written by the total mean value formula

$$a_k = \mathbf{P}(\Delta_n = k) = \int_0^\infty \mathbf{P}(\Delta_1 = k|Y_1 = x)\mathrm{d}B(x) = \int_0^\infty \frac{(\lambda x)^k}{k!} e^{-\lambda x}\mathrm{d}B(x), \quad k \geq 0.$$

Excluding the degenerate case $\mathbf{P}(Y = 0) = 1$, the inequality $a_k > 0$, $k \geq 0$, is always valid.

For the mean value of Δ_n we obtain

$$\mathbf{E}(\Delta_n) = \sum_{k=1}^{\infty} k a_k = \sum_{k=1}^{\infty} \int_0^{\infty} \frac{(\lambda x)^k}{(k-1)!} e^{-\lambda x} dB(x)$$

$$= \int_0^{\infty} \lambda x \sum_{k=0}^{\infty} \frac{(\lambda x)^k}{k!} e^{-\lambda x} dB(x) = \lambda \int_0^{\infty} x \, dB(x) = \lambda \mu_B = \rho. \quad (8.2)$$

Since $\lambda \mu_B < 1$, by Eq. (8.1), the Foster criterion is fulfilled (Theorem 3.15).   □

The possibility of changing the order of summation and integration in the previous formula follows from the Fubini theorem but can also be proved in an elementary way. Since for the function

$$Q(A, n) = \int_0^A \sum_{k=1}^n \frac{(\lambda x)^k}{(k-1)!} e^{-\lambda x} dB(x), \quad A \in R_+, \quad n \in N,$$

there exists a limit as $A \to \infty$ and $n \to \infty$, and moreover $\mu_B = \int_0^{\infty} x \, dB(x) < \infty$, then as $A \to \infty$ uniformly in n

$$|Q(\infty, n) - Q(A, n)| = \int_A^{\infty} \sum_{k=1}^n \frac{(\lambda x)^k}{(k-1)!} e^{-\lambda x} dB(x)$$

$$\leq \int_A^{\infty} \lambda x \sum_{k=0}^{\infty} \frac{(\lambda x)^k}{k!} e^{-\lambda x} dB(x) = \int_A^{\infty} \lambda x \, dB(x) \to 0,$$

from which the interchangeability follows.

Proof of Homogeneity Let

$$p_{ij}(n) = \mathbf{P}(L_{n+1} = j | L_n = i), \quad i, j \geq 0, \ n \geq 0,$$

be one-step transition probabilities. Then, using Eq. (8.1)

$$p_{ij} = \mathbf{P}\big(\mathscr{I}_{\{i>0\}}(i-1) + \Delta_{n+1} = j\big),$$

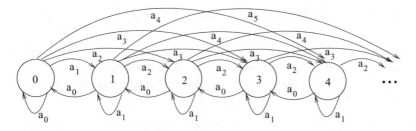

Fig. 8.4 Embedded Markov chain of M/G/1 queue

so

$$p_{ij}(n) = p_{ij} = \begin{cases} a_j & \text{if } i = 0, 1, \ j = 0, 1, 2, \ldots, \\ 0 & \text{if } i \geq 2, \ j \leq i - 2, \\ a_{j+1-i} & \text{if } i \geq 2, \ j \geq i - 1, \end{cases}$$

i.e., the sequence $\{L_n, n \geq 0\}$ is a homogeneous Markov chain. This behavior is depicted in Fig. 8.4, and the associated matrix of one-step transition probabilities may be written in the form

$$\boldsymbol{P} = (p_{ij})_{i,j=0}^{\infty} = \begin{pmatrix} a_0 & a_1 & a_2 & a_3 & \cdots \\ a_0 & a_1 & a_2 & a_3 & \cdots \\ 0 & a_0 & a_1 & a_2 & \cdots \\ 0 & 0 & a_0 & a_1 & \cdots \\ \vdots & \vdots & \vdots & \vdots & \ddots \end{pmatrix}. \tag{8.3}$$

In this matrix, a_i gives the probability that i customers arrive at the system while a customer is being served. Fixing the initial state, adding the arriving customers, we get the next state. We can descend one level if no new customers appear, remain at the same level if one new customer arrives, and go up if at least two new customers arrive. This explains the structure of the matrix. In this matrix the first two rows coincide. In the case of one already present customer, the foregoing reasoning is valid, but in the zero state there is a special situation. We arrive at the zero state when the last customer in a busy period is served. After a free period the first customer of the next busy period arrives, and we will consider the system state after this customer has been served. This new state will be determined by the number of customers arriving while this customer is being served. The coincidence of two rows is explained by the fact that in both cases we must consider the number of new customers arriving for the service of one customer. In the first case it is within a busy period, while in another case it is at the beginning of a busy period.

8.2.2   Proof of Irreducibility and Aperiodicity

Both properties may be derived from the matrix of one-step transition probabilities, but they may also be obtained from the following considerations.

Since the interarrival times have an exponential distribution with the parameter λ, it is clear that

- From arbitrary state $i \in \mathscr{X}$ for i services (steps) with positive probability we arrive at the state 0; this is enough so that no new customers enter.
- From state 0 with positive probability we can get to any state $j \in \mathscr{X}$ in one step.

The i and one-step transition probabilities (in the case of arbitrary $i, j \in \mathscr{X}$) are $p_{i0}^{(i)} > 0$, $p_{0j}^{(1)} > 0$, and consequently $p_{ij}^{(i+1)} > 0$, from which it follows that the Markov chain $\{L_n, n \geq 0\}$ is irreducible [for all $i, j \in \mathscr{X}$ exists such n that $p_{ij}^{(n)} > 0$].

Obviously, for arbitrary $i \in \mathscr{X}$ $p_{ii}^{(1)} > 0$ (since for all $i \geq 1$ for the service of a customer with positive probability a new customer enters and there is no entry at $i = 0$). So the Markov chain $\{L_n, n \geq 0\}$ is aperiodic [if $d(i)$ is the period of state i, i.e., $d(i) = \{$greatest common divisor (g.c.d.) of n for which $p_{ii}^{(n)} > 0\}$, in our case $d(i) = 1$].

8.2.3   Step (B): Proof of Ergodicity

Step (B) of the Embedded Markov chain method in Sect. 8.2: One way to prove the ergodicity is to show that all states of the Markov chain are recurrent nonzero ones (with probability 1 it returns to all states and the mean value of return time is finite), i.e.,

$$F_{ii} = \sum_{n=1}^{\infty} f_{ii}(n) = 1, \quad m_i = \sum_{n=0}^{\infty} n f_{ii}(n) < \infty,$$

where $f_{ij}(n)$ is the probability that the Markov chain which starts from state i goes to state j for the first time in the nth step. This approach requires a lot of computation, so we use the sufficient condition for the ergodicity of Markov chains obtained by Klimov (Theorem 3.14).

We check the conditions of Theorem 3.14 in the case $\rho < 1$. It is enough to find such function $g(i), i \in \mathscr{X}$, for which its conditions are fulfilled.

Let $\varepsilon = 1 - \rho$ (> 0) and $g(i) = i$, $i \geq 0$ (this case is known in the literature as Foster's criterion). From Eq. (8.1) it follows that

$$\mathbf{E}(g(L_{n+1})|L_n = i) = \mathbf{E}(i - 1 + \Delta_{n+1}) = i - 1 + \lambda \mu_B = i - \varepsilon, \quad i \geq 1$$

and

$$\mathbf{E}(g(L_{n+1})|L_n = 0) = \mathbf{E}(\Delta_{n+1}) = \lambda \mu_B = 1 - \varepsilon, \quad i = 0,$$

i.e., the conditions of Klimov's theorem are fulfilled, and we have proved the ergodicity of the Markov chain $\{L_n, n \geq 0\}$.

8.2.4   Pollaczek–Khinchin Mean Value Formula

Equation (8.1) makes it possible to find the moments of ergodic distribution. We present it for the case of mean value; the computations are similar for other moments. The derivation requires less computation than the Pollaczek–Khinchin transform equation, but in that case we automatically obtain the necessary conditions ($\rho < 1$ and the service time has finite second moment).

Assume that the following finite limits exist:

$$\lim_{n \to \infty} \mathbf{E}(L_n) = m_1 \quad \text{és} \quad \lim_{n \to \infty} \mathbf{E}\left(L_n^2\right) = m_2 \tag{8.4}$$

(we do not deal with the conditions of existence).

Equation (8.4) follows from (8.1) if $\mathbf{E}(\Delta_n^2) < \infty$, and the service time also has finite second moment. Taking on both sides of Eq. (8.1) limit as $n \to \infty$

$$m_1 = \lim_{n \to \infty} \mathbf{E}(L_n) = \lim_{n \to \infty} [\mathbf{E}\left(\mathscr{I}_{\{L_{n-1}>0\}}(L_{n-1} - 1)\right) + \mathbf{E}(\Delta_n)]$$

$$= \lim_{n \to \infty} [\mathbf{E}(L_{n-1}) - \mathbf{E}\left(\mathscr{I}_{\{L_{n-1}>0\}}\right) + \rho] = m_1 - \lim_{n \to \infty} \mathbf{E}\left(\mathscr{I}_{\{L_{n-1}>0\}}\right) + \rho,$$

whence

$$\lim_{n \to \infty} \mathbf{P}(L_n > 0) = \lim_{n \to \infty} \mathbf{E}\left(\mathscr{I}_{\{L_{n-1}>0\}}\right) = \rho$$

and

$$\pi_0 = \lim_{n \to \infty} \mathbf{P}(L_n = 0) = 1 - \lim_{n \to \infty} \mathbf{E}\left(\mathscr{I}_{\{L_{n-1}>0\}}\right) = 1 - \rho.$$

Though this procedure leads to important results, it does not produce the desired mean value. Repeating it for the second moments we meet our objective. Using the independence of L_{n-1} and Δ_n, we obtain

$$m_2 = \lim_{n \to \infty} \mathbf{E}\left(L_n^2\right)$$

$$= \lim_{n \to \infty} \mathbf{E}\left((L_{n-1}^2 - 2L_{n-1} + 1)\mathscr{I}_{\{L_{n-1}>0\}} + 2(L_{n-1} - 1)\mathscr{I}_{\{L_{n-1}>0\}} \Delta_n + \Delta_n^2\right)$$

$$= m_2 - 2m_1 + \rho + 2m_1\rho - 2\rho^2 + \mathbf{E}\left(\Delta_1^2\right),$$

whence

$$m_1 = \frac{\rho - 2\rho^2 + \mathbf{E}\left(\Delta_1^2\right)}{2(1-\rho)} = \rho + \frac{\mathbf{E}\left(\Delta_1^2\right) - \rho}{2(1-\rho)}.$$

Later, by means of the generating function, we obtain the equality $\mathbf{E}\left(\Delta_1^2\right) = \lambda^2 \mathbf{E}\left(Y_1^2\right) + \rho$; using it from the last equation we come to the Pollaczek–Khinchin mean value formula:

$$m_1 = \rho + \frac{\lambda^2 \mathbf{E}\left(Y_1^2\right)}{2(1-\rho)}. \tag{8.5}$$

8.2.5   *Proof of Equality* $\mathbf{E}\left(\Delta_1^2\right) = \lambda^2 \mathbf{E}\left(Y_1^2\right) + \rho$

Let $B^\sim(s) = \int_0^\infty e^{-sx} dB(x)$, $s \geq 0$, be the Laplace–Stieltjes transform of distribution function $B(x)$. The generating function of entering customers for one service will be

$$\mathbf{E}\left(z^{\Delta_1}\right) = A(z) = \sum_{i=0}^\infty a_i z^i = \sum_{i=0}^\infty \int_0^\infty \frac{(\lambda x z)^i}{i!} e^{-\lambda x} dB(x) = B^\sim(\lambda(1-z)). \tag{8.6}$$

Similar to the derivation of mean value $\mathbf{E}(\Delta_1)$ we get

$$A'(1) = \mathbf{E}(\Delta_1) = \sum_{k=1}^\infty k a_k = \sum_{k=1}^\infty k \int_0^\infty \frac{(\lambda x)^k}{k!} e^{-\lambda x} dB(x)$$

$$= \int_0^\infty \lambda x \sum_{k=0}^\infty \frac{(\lambda x)^k}{k!} e^{-\lambda x} dB(x) = \lambda \int_0^\infty x \, dB(x) = \rho. \tag{8.7}$$

There exists a second moment of service time, so the Laplace–Stieltjes transform is twice continuously differentiable from the right, and for the right derivatives

$$B^{\sim\prime}(0) = -\int_0^\infty x\,dB(x) = -\mathbf{E}(Y_1) = -\mu_B,$$

$$B^{\sim\prime\prime}(0) = \int_0^\infty x^2\,dB(x) = \mathbf{E}\left(Y_1^2\right).$$

From here (and taking the left-side derivatives at point 1)

$$\mathbf{E}(\Delta_1) = A'(1) = -\lambda B^{\sim\prime}(0) = \lambda\mu_B = \rho,$$

$$\mathbf{E}\left(\Delta_1^2\right) = (zA'(z))'_{z=1} = -\lambda B^{\sim\prime}(0) + \lambda^2 B^{\sim\prime\prime}(0) = \lambda\mu_B + \lambda^2\mathbf{E}\left(Y_1^2\right)$$

$$= \rho + \lambda^2\mathbf{E}\left(Y_1^2\right).$$

8.2.6 Step (C): Ergodic Distribution of Queue Length

Step (C) of the Embedded Markov chain method in Sect. 8.2: From the ergodicity of the Markov chain $\{L_n,\ n \ge 0\}$ follows the existence of the ergodic distribution

$$\pi_k = \lim_{n\to\infty} \mathbf{P}(L_n = k), \quad k = 0, 1, 2, \ldots,$$

which can be obtained as the solution of system of equations

$$\pi_k = \sum_{j=0}^\infty \pi_j p_{jk}, \quad k = 0, 1, 2, \ldots,$$

$$\sum_{k=0}^\infty \pi_k = 1.$$

The matrix P has a special structure [see Eq. (8.3)], and the stationary equations take the form

$$\pi_k = \pi_0 a_k + \pi_1 a_k + \pi_2 a_{k-1} + \ldots + \pi_{k+1} a_0$$

$$= \sum_{i=0}^k \pi_{k-i+1} a_i + \pi_0 a_k. \tag{8.8}$$

We solve this system of equations by the method of generating functions. Let us introduce the notation

$$\pi(z) = \sum_{k=0}^{\infty} \pi_k z^k, \quad A(z) = \sum_{k=0}^{\infty} a_k z^k, \quad |z| \le 1.$$

First, $\pi(1) = A(1) = 1$, and, according to our previous computations, $A'(1) = \lim_{z \to 1-0} A'(z) = \sum_{k=1}^{\infty} k a_k \ (= \mathbf{E}(\Delta_1)) = \rho$. Multiplying both sides of Eq. (8.8) by z^k and summing up by k for $k \ge 0$, we obtain

$$\pi(z) = \sum_{k=0}^{\infty} z^k \sum_{m=0}^{k} a_m \pi_{k-m+1} + \pi_0 A(z)$$

$$= \sum_{m=0}^{\infty} a_m z^m \sum_{k=m}^{\infty} z^{k-m} \pi_{k-m+1} + \pi_0 A(z)$$

$$= A(z) \frac{\pi(z) - \pi_0}{z} + \pi_0 A(z),$$

whence

$$\pi(z)[1 - A(z)/z] = -\pi_0 A(z)(1/z - 1),$$

and so

$$\pi(z) = \pi_0 \frac{(1-z)A(z)}{A(z) - z}, \quad |z| < 1. \tag{8.9}$$

This includes the unknown probability π_0, which will be found from the condition $\pi(1) = \sum_{k=0}^{\infty} \pi_k = 1$. In the derivation of the Pollaczek–Khinchin mean value formula under special conditions we already found the value of π_0, and here it will come from Eq. (8.9) when $\int_0^{\infty} x^2 dB(x) < \infty$.

$\pi(z)$ is continuous from left at point 1, so at $z = 1$ the numerator and denominator of Eq. (8.9) disappear. By L'Hospital's rule

$$\pi(1) = \lim_{z \to 1-0} \pi(z) = \lim_{z \to 1-0} \pi_0 \frac{-A(z) + A(z)(1-z)}{A(z) - 1} = \frac{-\pi_0}{A'(1) - 1} = \frac{\pi_0}{1 - \rho} = 1,$$

and we obtain $\pi_0 = 1 - \rho$.

Earlier we proved (8.6), i.e.,

$$A(z) = B^{\sim}(\lambda(1-z)), \quad |z| \le 1.$$

From it and Eq. (8.9) we get the Pollaczek–Khinchin transform equation (or, more precisely, one of its forms):

$$\pi(z) = \frac{(1 - \rho)(1 - z)B^{\sim}(\lambda(1 - z))}{B^{\sim}(\lambda(1 - z)) - z}. \tag{8.10}$$

Recall that this gives the generating function of ergodic distribution for the embedded Markov chain $\{L_n, \ n \geq 0\}$.

Corollary 8.1 *The inversion of the Pollaczek–Khinchin transform equation generally is not simple, but the moments may be obtained from it without inversion.*

Taking into account $B^{\sim\prime}(0) = -E(Y) = -\mu_B$, $B^{\sim\prime\prime}(0) = E(Y^2) = \int_0^\infty x^2 dB(x)$, and using the L'Hospital's rule twice we obtain the mean value of the number of customers in the system (Pollaczek–Khinchin mean value formula):

$$\sum_{k=0}^{\infty} k\pi_k = \pi'(1) = \lim_{z \to 1-} \pi(z)$$

$$= \lim_{z \to 1-} (1-\rho)\frac{\lambda z B^{\sim\prime}(\lambda(1-z)) - \lambda z^2 B^{\sim\prime}(\lambda(1-z)) + B^{\sim}(\lambda(1-z)) - B^{\sim 2}(\lambda(1-z))}{[B^{\sim}(\lambda(1-z)) - z]^2}$$

$$= \lim_{z \to 1-} (1 - \rho)\frac{\lambda^2 B^{\sim\prime\prime}(\lambda(1-z)) - 2\lambda B^{\sim\prime}(\lambda(1-z)) - 2\lambda^2[B^{\sim\prime}(\lambda(1-z))]^2}{2\lambda^2[B^{\sim\prime}(\lambda(1-z))]^2 + 4\lambda B^{\sim\prime}(\lambda(1-z)) + 2}$$

$$= (1 - \rho)\frac{\lambda^2 B^{\sim\prime\prime}(0) + 2\rho - 2\rho^2}{2(1 - \rho)^2}$$

$$= \rho + \frac{\lambda^2 E(Y^2)}{2(1 - \rho)}.$$

The variance of stationary queue length can be computed on a similar way:

$$\sigma^2 = \frac{\lambda^3 E(Y^3)}{3(1 - \rho)} + \frac{\lambda^4 E(Y^2)}{4(1 - \rho)^2} + \frac{\lambda^2 E(Y^2)(3 - 2\rho)}{2(1 - \rho)} + \rho(1 - \rho),$$

where $E(Y^i)$, $i = 2, 3$, denotes the ith moment of service time [3].

Example 8.1 (Inversion in the Case of M/M/1 System) In this case the intensity of arrivals is $\lambda > 0$, and the intensity of service $\mu > 0$ (the interarrival and service times are independent exponentially distributed random variables with parameters λ and μ, respectively). Then

$$B^{\sim}(s) = \frac{\mu}{s + \mu}, \quad \text{Re } s > -\mu,$$

$$\pi(z) = \frac{\mu}{\lambda - \lambda z + \mu} \cdot \frac{(1 - \rho)(1 - z)}{[\mu/(\lambda - \lambda z + \mu)] - z}.$$

Since $\rho = \lambda\tau = \lambda/\mu$,

$$\pi(z) = \frac{1-\rho}{1-\rho z},$$

and for the stationary distribution we obtain

$$\pi_k = (1-\rho)\rho^k, \quad k \ge 0.$$

8.2.7 Investigation of Busy/Free Intervals in M/G/1 Queueing System

Observing a queueing system we see that there are periods during which it is empty or occupied. The time interval when the server is occupied is called the busy period. It begins with the arrival of a customer at the empty system and is finished when the last customer leaves the system (Fig. 8.5).

If (ξ_i, η_i), $i = 1, 2, \ldots$, denote consecutive free and busy periods, then (ξ_i, η_i) is a sequence of i.i.d. random variables, where the components ξ_i and η_i are also independent of each other. The sequence $(\xi_i + \eta_i)$, $i = 1, 2, \ldots$ is a renewal process, ξ_i has an exponential distribution with the parameter λ. Finding the distribution of busy periods η_i is more complicated and will be considered later.

Let $\Psi(x) = \mathbf{P}(\eta_i \le x)$. Assume that at moment $t = 0$ a customer enters the system and a busy period begins. Its service time is $Y = y$. There are two cases:

1. During service no new customer enters the system and the busy period ends, i.e., its duration is $Y = y$.
2. For $y, n \ge 1$, customers enter the system and the busy period continues (Fig. 8.6).

In the last case n successive service times are denoted by $Y_1, Y_2, \ldots, Y_n$. Assume that the service is realized in inverse order, i.e., according to the LCFS discipline [14], then according to our assumptions the interarrival and service times are independent. Their distributions are exponential with parameter λ and $B(x)$, and

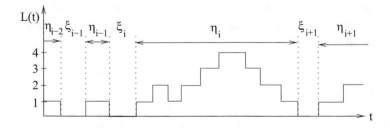

Fig. 8.5 Busy periods

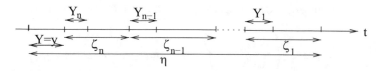

Fig. 8.6 Length of busy period

the distribution of busy periods remains the same (Ψ). The whole busy period η can be divided into intervals $Y, \zeta_n, \zeta_{n-1}, \ldots, \zeta_1$ (if $n = 0$, then $\eta = Y$), where $\zeta_n, \zeta_{n-1}, \ldots, \zeta_1$ mean the busy periods generated by the different customers, they are

1. Independent,
2. Identically distributed, and
3. Their distribution coincides with that of η.

By the formula of total probability ($\zeta_n + \cdots + \zeta_1 = 0$, if $n = 0$)

$$\Psi(x) = \mathbf{P}(\eta \leq x)$$

$$= \mathbf{P}(Y + \zeta_n + \cdots + \zeta_1 \leq x)$$

$$= \sum_{j=0}^{\infty} \mathbf{P}(Y + \zeta_n + \cdots + \zeta_1 \leq x, n = j)$$

$$= \int_0^{\infty} \sum_{j=0}^{\infty} \mathbf{P}\big(Y + \zeta_j + \cdots + \zeta_1 \leq x, n = j \mid Y = y\big)\, dB(y)$$

$$= \int_0^{\infty} \sum_{j=0}^{\infty} \mathbf{P}\big(y + \zeta_j + \cdots + \zeta_1 \leq x\big)\frac{(\lambda y)^j}{j!} e^{-\lambda y}\, dB(y)$$

$$= \int_0^{\infty} \sum_{j=0}^{\infty} \Psi_j(x - y)\frac{(\lambda y)^j}{j!} e^{-\lambda y}\, dB(y)$$

(the order of summation and integration may be changed), where

$$\Psi_j(x) = \mathbf{P}\big(\zeta_1 + \cdots + \zeta_j \leq x\big).$$

This functional equation will be simpler if we use the Laplace–Stieltjes transforms. Let

$$B^{\sim}(s) = \int_0^{\infty} e^{-sx}\, dB(x), \qquad \Psi^{\sim}(s) = \int_0^{\infty} e^{-sx}\, d\Psi(x).$$

The ζ_j are independent and have the same distribution Ψ, so

$$
\begin{aligned}
\Psi^\sim(s) &= \int_0^\infty \left\{ \sum_{j=0}^\infty \frac{(\lambda y)^j}{j!} \int_0^\infty e^{-sx}\, \mathrm{d}x \Psi_j(x-y) \right\} e^{-\lambda y}\, \mathrm{d}B(y) \\
&= \int_0^\infty \left\{ \sum_{j=0}^\infty \frac{(\lambda y)^j}{j!} \left[e^{-sy}(\Psi^\sim(s))^j \right] e^{-\lambda y} \right\} \mathrm{d}B(y) \\
&= \sum_{j=0}^\infty \frac{(\lambda \Psi^\sim(s))^j}{j!} \int_0^\infty y^j e^{-(\lambda+s)y}\, \mathrm{d}B(y) \\
&= \sum_{j=0}^\infty (-1)^j \frac{(\lambda \Psi^\sim(s))^j}{j!} \frac{\mathrm{d}^j}{\mathrm{d}s^j}(B^\sim(\lambda+s)),
\end{aligned}
$$

which corresponds to the Taylor expansion of the function $B^\sim(\lambda + s - \lambda\Psi^\sim(s))$ in the neighborhood of $\lambda\Psi^\sim(s)$; consequently,

$$\Psi^\sim(s) = B^\sim(\lambda + s - \lambda\Psi^\sim(s)). \tag{8.11}$$

The next theorem deals with the solution of the functional equation Eq. (8.11).

Theorem 8.2 *Equation* (8.11) *has a unique solution at Re $s > 0$, $|\Psi^\sim(s)| \le 1$ and $\Psi^\sim(s)$ is real for all $s > 0$. Let p^* ($0 \le p^* \le 1$) denote the least positive number for which $B^\sim(\lambda(1 - p^*)) = p^*$. Then*

$$\Psi(\infty) = p^*.$$

If $\rho = \lambda\tau \le 1$, then $p^* = 1$ and $\Psi(x)$ is a (nondegenerate) distribution function; if $\rho > 1$, then $p^* < 1$ and the busy period may be infinite with probability $1 - p^*$.

Remark 8.1 Since $B^\sim(\lambda(1 - p))$, $0 \le p \le 1$ is a continuous and strictly monotonically function of p, and if $p = 1$, then $B^\sim(\lambda(1 - p^*)) = p^*$, p^* is well defined.

Proof First we show that Eq. (8.11) has a unique solution $\Psi^\sim(s)$ for which, at arbitrary $s > 0$, $|\Psi^\sim(s)| \le 1$. The proof uses the Rouché's theorem. □

Theorem 8.3 (Rouché) *Let $G(z)$ and $g(z)$ regular functions on the domain D and continuous on its closure. If on the boundary of D $|g(z)| < |G(z)|$, then $G(z)+g(z)$ and $G(z)$ have the same number of roots in D (with multiplicities).*

Let s be an arbitrary complex number, $Re\ s > 0$, and consider the equation $z = B^\sim(s + \lambda - \lambda z)$. The right and left sides are analytical functions of z on a domain that contains the unit circle $|z| \le 1$. If $|z| = 1$, then, because of $Re\ s > 0$

$Re(s + \lambda - \lambda z) > 0,$

$$|B^{\sim}(s + \lambda - \lambda z)| \le \int_0^\infty |e^{-(s+\lambda-\lambda z)x}| \, dB(x)$$

$$= B^{\sim}(Re(s + \lambda - \lambda z))$$

$$< 1 = |z|.$$

By Rouché's theorem, z and $(z - B^{\sim}(s + \lambda - \lambda z))$ have the same number of roots on the domain $|z| < 1$, i.e., one.

Now let us examine Eq. (8.11) on the positive real half-line. Let $s + \lambda - \lambda \Psi^{\sim}(s) = x$ and consider the solution of the equation

$$(s + \lambda - x)/\lambda = B^{\sim}(x) \tag{8.12}$$

at $s > 0$. We will see that in this case there exists a unique solution x_0 for which $s < x_0 < s + \lambda$. Figure 8.7 helps to understand the problem.

$B^{\sim}(x)$ is convex from below and continuous; consequently the root of Eq. (8.12)—and so $\Psi^{\sim}(s)$ also—for all $s > 0$ is uniquely determined on the whole $(0, \infty)$ half-line.

We remark that $B^{\sim}$ is a regular function, as is $\Psi^{\sim}$ [for all points $(0, \infty)$ there exists a neighborhood with radius $r > 0$, where it can be expanded]; consequently, it can be analytically continued for the right half-plane. (This means that Eq. (8.12) has an analytical inverse for x.)

If $s \to 0$, then $B_s \to C$, while the tangent of $B_s A_s$ remains $-\frac{1}{\lambda}$. At the same time

$$B^{\sim\prime}(0) = -\mu_B = -\int_0^\infty x \, dB(x),$$

Fig. 8.7 Solution of
Eq. (8.12)

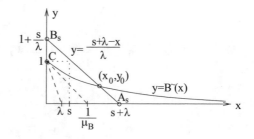

so, by using the fact $B^\sim(x)$ is convex from below:

1. If $\rho > 1$ $(1/\mu_B < \lambda)$, then $B_s A_s$ (in case $s \to 0+$) for a certain $x_* > 0$ intersects $B^\sim(x)$. Then $\lim_{s \to 0} x_0(s) = x_*$, $p^* = \Psi^\sim(0) = \frac{\lambda - x_*}{\lambda} < 1$ (in this case the busy period can be infinite with positive probability).
2. If $\rho \le 1$ $(1/\mu_B \ge \lambda)$, then the limit of $B_s A_s$ intersects $B^\sim(x)$ at the only point $x_0 = 0$, when $p^* = 1$. Consequently, $\Psi(\infty) = \Psi^\sim(0) = 1$.

Corollary 8.2 *Assume that $\rho < 1$. Differentiating Eq. (8.11) at $s = 0$ we obtain a linear equation for the mean value $\mathbf{E}(\eta) = -\Psi^\sim(0)$:*

$$\Psi^{\sim\prime}(0) = B^{\sim\prime}(0)(1 - \lambda\Psi^{\sim\prime}(0)).$$

From it the mean value of busy period

$$\mathbf{E}(\eta) = -\Psi^{\sim\prime}(0) = \frac{\mu_B}{1 - \rho}.$$

The other moments can be computed on a similar way, e.g.,

$$\mathbf{E}\left(\eta^2\right) = \frac{\mathbf{E}\left(Y^2\right)}{(1 - \rho)^3}.$$

Results concerning the distribution function of a busy period's length may be derived from other considerations. With one customer have been served, n ones remain in the system; let H_n denote the time period till the moment when there will be $n - 1$ customers. Furthermore, let Q_n denote the number of served customers for this period. The structure of this period (while we descend one level) coincides with the structure of the busy period and is independent of n.

Let the service time of a customer $Y = y$, then (since we have a Poisson process with parameter λ) the length of busy period is

$$\{\eta | Y = y\} = \begin{cases} y & \text{with probability } e^{-\lambda y}, \\ y + H_1 & \text{with probability } \lambda y e^{-\lambda y}, \\ y + H_1 + H_2 & \text{with probability } \frac{(\lambda y)^2}{2!}e^{-\lambda y}, \\ \cdots \end{cases}$$

We have $\Psi^\sim(s) = \mathbf{E}\left(e^{-s\eta}\right) = \Psi_1^\sim(s) = \Psi_2^\sim(s) = \ldots$, so

$$\mathbf{E}\left(e^{-s\eta} | Y = y\right) = \sum_{i=0}^{\infty} \frac{(\lambda y)^i}{i!} e^{-\lambda y} e^{-sy} \left(\Psi^\sim(s)\right)^i$$

$$= e^{-\lambda y} e^{-sy} e^{\lambda y \Psi^\sim(s)} = e^{-y(s + \lambda - \lambda\Psi^\sim(s))}$$

and

$$\Psi^{\sim}(s) = \int\limits_0^\infty \mathbf{E}\big(e^{-s\eta}|y\big)dB(y) = \int\limits_0^\infty e^{-y(s+\lambda-\lambda\Psi^{\sim}(s))}dB(y)$$

$$= B^{\sim}(s+\lambda-\lambda\Psi^{\sim}(s)).$$

The number of customers served in a busy period is

$$\{Q|Y=y\} = \begin{cases} 1 & \text{with probability } e^{-\lambda y}, \\ 1+Q_1 & \text{with probability } \lambda y e^{-\lambda y}, \\ 1+Q_1+Q_2 & \text{with probability } \frac{(\lambda y)^2}{2!}e^{-\lambda y}, \\ \dots \end{cases}$$

Let $Q(z) = \mathbf{E}\big(z^Q\big) = Q_1(z) = Q_2(z) = \dots$,

$$\mathbf{E}\big(z^Q|y\big) = z\sum_{i=0}^\infty \frac{(\lambda y)^i}{i!}e^{-\lambda y}Q^i(z) = ze^{-\lambda y}e^{\lambda yQ(z)} = ze^{-y(\lambda-\lambda Q(z))},$$

and using this result we obtain

$$Q(z) = \int\limits_0^\infty \mathbf{E}\big(z^Q|y\big)dB(y) = \int\limits_0^\infty ze^{-y(\lambda-\lambda Q(z))}dB(y)$$

$$= zB^{\sim}(\lambda(1-Q(z))).$$

We have already computed the moments for the length of the busy period; the mean value of customers served for the busy period is

$$\mathbf{E}(Q) = \frac{1}{1-\rho}.$$

8.2.8 Investigation on the Basis of the Regenerative Process

The functioning of an M/G/1 system may be considered a regenerative process. Our aim now is to derive the Pollaczek–Khinchin transform equation on its basis.

We introduce the following notations:

$\mathbf{E}(\eta) = \dfrac{\mu_B}{1-\rho}$: the mean value of busy period;

ω_i : mean value of time spent above ith level during a busy period;

η_i : mean value of time spent on the ith level for a busy period.

Theorem 8.4 *Let us consider an M/G/1 queueing system with arrival rate λ and service time distribution $B(x)$. If the service time of a customer has a finite mean μ_B, $\lambda\mu_B < 1$, then there exists an equilibrium distribution in the system. These probabilities are determined by the fractions $p_i = \eta_i/\mathbf{E}(\eta)$ $(i = 0, 1, \ldots)$, where $\mathbf{E}(\eta)$ is the mean value of the busy period and η_i is the mean value of time spent on the ith level during a busy period.*

Proof The proof of the theorem is a direct consequence of Theorem 4.13 (see also [17, Theorems 1.3.2 and 1.3.3]). The mean values appearing in the theorem are given by the following lemma.   □

Lemma 8.1 *In the M/G/1 system*

$$\eta_0 = \mu_B, \qquad \eta_1 = \frac{1 - a_0}{a_0}\eta_0, \qquad \eta_2 = \frac{1 - a_0 - a_1}{a_0}(\eta_0 + \eta_1),$$

and η_k $(k \geq 3)$ satisfy the recurrence relation

$$\eta_k = \sum_{i=1}^{k-2} \frac{1 - a_0 - a_1 - \ldots - a_i}{a_0}\eta_{k-i} + \frac{1 - a_0 - a_1 - \ldots - a_{k-1}}{a_0}(\eta_0 + \eta_1).$$

Proof Let j customers be present in the system, with one of them being served. An actual customer having been served, the number of present customers does not change with probability a_1. The number of present customers changes with probability $1 - a_1$, we come to another level, with probability $\frac{a_0}{1-a_1}$ to $j - 1$, and with probability $\frac{1-a_0-a_1}{1-a_1}$ to a level above j.

Let us consider a busy period and intervals in it where one or more customers stay in the system. When we used the embedded Markov chain technique the states of the system were identified by the number of customers remaining in the system after a customer had been served. Now it will be better to regard the number of customers at the beginning of service. The difference will be clear from the following reasoning. If one considers service periods of customers when at the starting moment there are no other customers, then each of them corresponds to state 1, excluding two cases. The first case is when we jump to a level above the first one, then the service of the last customer from the viewpoint of states corresponds to the new level (from the viewpoint of the number of present customers to the first level). But the whole duration does not change because coming from the second level to the first the inverse situation takes place. The situation will be similar for all levels above the first. The second case is the service of the last customer in the busy period: it corresponds to a zero state (after this customer is served there will be no customers in the system), so it must be excluded from the number of customers served on the first level.

We determine the mean value of a period during which there is only one customer in the system. For the service of a customer a new one enters with probability a_1, so this state is continued with probability a_1 and terminated with probability $1 - a_1$ (there is no entry or more than one customer appears). For such a period with probabilities $1 - a_1$ is served one, $a_1(1 - a_1)$ are served two,..., with probability $a_1^{k-1}(1 - a_1)$ are served k customers. The mean value of the number of customers served is

$$\sum_{k=1}^{\infty} k a_1^{k-1}(1 - a_1) = \frac{1}{1 - a_1}.$$

Now let us determine the mean value of a period above the first level (in this case we will have the aforementioned deviation concerning the states and number of customers, but finally we obtain the correct value). Assume that at the beginning of this period there are k customers in the system [while the last customer on the first level is being served, with probability $1 - a_0 - a_1$ at least two customers arrived, with probabilities $\frac{a_k}{1-a_0-a_1}$ $(k = 2, 3, \ldots)$ we will have k ones]. To return to the first level, we have to complete $k - 1$ present and all further customers entered for their services. (The structure of a period during which one customer is served with the generated ones coincides with that of the busy period.) The mean value of a busy period is $\frac{\mu_B}{1-\rho}$; consequently, the length of such an interval is

$$\sum_{k=2}^{\infty} \frac{a_k}{1 - a_0 - a_1}(k - 1)\frac{\mu_B}{1 - \rho} = \frac{\mu_B}{(1 - \rho)(1 - a_0 - a_1)}\left(\rho - a_1 - (1 - a_0 - a_1)\right)$$

$$= \frac{\rho - 1 + a_0}{(1 - \rho)(1 - a_0 - a_1)}\mu_B,$$

where we used the equalities

$$\rho = \sum_{k=1}^{\infty} k a_k \quad \text{and} \quad \sum_{k=0}^{\infty} a_k = 1.$$

For the busy period we have a certain number of intervals with one present customer; such an interval is finished either without entry (meaning the end of the busy period) or with the entry of more than one customer. With probabilities

$$\frac{a_0}{1 - a_1}, \quad \frac{1 - a_0 - a_1}{1 - a_1}\frac{a_0}{1 - a_1}, \ldots, \frac{(1 - a_0 - a_1)^k}{(1 - a_1)^k}\frac{a_0}{1 - a_1}, \ldots$$

we will have $0, 1, \ldots, k, \ldots$ intervals with the presence of more than one customers. Thus the mean values of intervals of two types are

$$\sum_{k=1}^{\infty} k \frac{(1 - a_0 - a_1)^{k-1}}{(1 - a_1)^{k-1}} \frac{a_0}{1 - a_1} \frac{\mu_B}{1 - a_1} = \frac{\mu_B}{a_0},$$

$$\sum_{k=1}^{\infty} k \frac{(1 - a_0 - a_1)^{k}}{(1 - a_1)^{k}} \frac{a_0}{1 - a_1} \frac{\rho - 1 + a_0}{(1 - \rho)(1 - a_0 - a_1)} \mu_B = \frac{\rho - 1 + a_0}{a_0(1 - \rho)} \mu_B.$$

The sum of these two values obviously gives the busy period's mean value:

$$\frac{\mu_B}{a_0} + \frac{\rho - 1 + a_0}{a_0(1 - \rho)} \mu_B = \frac{\mu_B}{1 - \rho}.$$

We derive the mean value of time spent above the kth level for a busy period. First let us consider the case of second level. We have two possibilities:

1. From the first level we arrive at the second one.
2. From the first level we arrive at least at the third one.

If the period under consideration begins at the second level, then we are in the same situation as in the case of the first level. We serve a certain number of customers on the second level, then we go either to the first level or above the second one. In the first case, intervals on and above the second level will change, and spending on average ω_1 time above it we come to the first one. In the second case the period above the second level begins with a jump from the first level immediately to a level above the second, and the mean value of time to return to the second one is equal to

$$\sum_{k=3}^{\infty} \frac{a_k}{1 - a_0 - a_1 - a_2}(k - 2)\frac{\mu_B}{1 - \rho} = \frac{\rho - 2 + 2a_0 + a_1}{(1 - \rho)(1 - a_0 - a_1 - a_2)} \mu_B = \varepsilon_2.$$

Now we are in the same situation as in the previous case, i.e., we spend above the second level ω_1 time. The probabilities of the two cases are

$$\frac{a_2}{1 - a_0 - a_1} \quad \text{and} \quad \frac{1 - a_0 - a_1 - a_2}{1 - a_0 - a_1},$$

so for a period beginning and ending on the first level we spend above the second level on average

$$\frac{a_2}{1 - a_0 - a_1} \omega_1 + \frac{1 - a_0 - a_1 - a_2}{1 - a_0 - a_1}(\omega_1 + \varepsilon_2) = \omega_1 + \varepsilon'_2,$$

where

$$\varepsilon'_2 = \frac{\rho - 2 + 2a_0 + a_1}{(1 - \rho)(1 - a_0 - a_1)} \mu_B.$$

For a busy period we have i such intervals with probability $\frac{(1-a_0-a_1)^i}{(1-a_1)^i}\frac{a_0}{1-a_1}$; consequently,

$$\omega_2 = \sum_{i=1}^{\infty} i\, \frac{(1-a_0-a_1)^i}{(1-a_1)^i}\, \frac{a_0}{1-a_1}(\omega_1+\varepsilon_2') = \frac{1-a_0-a_1}{a_0}\omega_1 + \frac{1-a_0-a_1-a_2}{a_0}\varepsilon_2.$$

Let us assume that our formula is valid for the $k-1$st level and compute ω_k. We consider again an interval starting and ending on the first level. ω_k may be written in the form

$$\omega_k : \quad \omega_{k-1}$$

$$\omega_{k-2} + \omega_{k-1}$$

$$\dots\dots\dots\dots\dots\dots$$

$$\omega_{k-i} + \omega_{k-i+1} + \dots + \omega_{k-2} + \omega_{k-1}$$

$$\dots\dots\dots\dots\dots\dots\dots\dots\dots$$

$$\omega_1 + \omega_2 + \dots + \omega_{k-2} + \omega_{k-1} + \varepsilon_k$$

From the first level we can come to the second, third,...,$k-1$st, kth, or one above the kth level. The first possibility is the second level. We are in the same situation as in the case with the time spent above the $k-1$st level from the viewpoint of the first one; the mean value is ω_{k-1}. In the case of the third level, first we have an interval starting with three and ending with two customers. This corresponds to the situation where one considers the time above the $k-2$nd level from the viewpoint of the first one; the mean value is ω_{k-2}. Now we are in the previous situation (two customers), and the mean value of the remaining part is ω_{k-1}. So under condition that from the first level we come at once to the third level, the desired mean value is equal to $\omega_{k-2} + \omega_{k-1}$.

Let us consider the last case, which takes place when from the first level we jump to a level above k. The mean value of time to reach the kth level is

$$\sum_{i=k+1}^{\infty} \frac{a_i}{1-a_0-a_1-\dots-a_k}(i-k)\frac{\mu_B}{1-\rho}$$

$$= \frac{\rho - k + ka_0 + (k-1)a_1 + \dots + 2a_{k-2} + a_{k-1}}{(1-\rho)(1-a_0-a_1-\dots-a_k)}\mu_B = \varepsilon_k.$$

After this period we will be at the kth level, and according to our previous reasoning, spending on average ω_1 time above the kth level we come to the $k-1$st level, spending ω_2 above the kth level we come to the $k-2$nd,..., and finally starting from the second level and spending ω_{k-1} above the kth one we reach the first level. So, in the last case, the desired mean value is $\omega_1 + \omega_2 + \dots + \omega_{k-1} + \varepsilon_k$. The probability of the first case is $\frac{a_2}{1-a_0-a_1}$, the probability of the second one is $\frac{a_3}{1-a_0-a_1}$,..., the probability of the last case is $\frac{1-a_0-a_1-...-a_k}{1-a_0-a_1}$. Multiplying the conditional mean values by the corresponding probabilities we obtain

$$\omega_{k-1} + \frac{1-a_0-a_1-a_2}{1-a_0-a_1}\omega_{k-2} + \ldots$$
$$+ \frac{1-a_0-a_1-\ldots-a_{k-1}}{1-a_0-a_1}\omega_1 + \frac{1-a_0-a_1-\ldots-a_k}{1-a_0-a_1}\varepsilon_k.$$

For the busy period we will stay above the first level i times with probability $\frac{(1-a_0-a_1)^i}{(1-a_1)^i}\frac{a_0}{1-a_1}$, so the mean value of time spent above the kth level for a busy period equals

$$\omega_k = \sum_{i=1}^{\infty} i \frac{(1-a_0-a_1)^i}{(1-a_1)^i} \frac{a_0}{1-a_1} \left(\omega_{k-1} + \frac{1-a_0-a_1-a_2}{1-a_0-a_1}\omega_{k-2} \right.$$
$$+\ldots+ \frac{1-a_0-a_1-\ldots-a_{k-1}}{1-a_0-a_1}\omega_1 + \left. \frac{1-a_0-a_1-\ldots-a_k}{1-a_0-a_1}\varepsilon_k \right)$$
$$= \sum_{i=1}^{k-1} \frac{1-a_0-a_1-\ldots-a_i}{a_0}\omega_{k-i} + \frac{1-a_0-a_1-\ldots-a_k}{a_0}\varepsilon_k.$$

In a similar way

$$\omega_{k-1} = \sum_{i=1}^{k-2} \frac{1-a_0-a_1-\ldots-a_i}{a_0}\omega_{k-i-1} + \frac{1-a_0-a_1-\ldots-a_{k-1}}{a_0}\varepsilon_{k-1}.$$

The mean value of time spent on the kth level for the busy period is

$$\eta_k = \omega_{k-1} - \omega_k$$
$$= \frac{1-a_0-a_1}{a_0}(\omega_{k-2} - \omega_{k-1}) + \frac{1-a_0-a_1-a_2}{a_0}(\omega_{k-3} - \omega_{k-2}) + \ldots$$
$$+ \frac{1-a_0-\ldots-a_{k-2}}{a_0}(\omega_1 - \omega_2) - \frac{1-a_0-\ldots-a_{k-1}}{a_0}\omega_1$$
$$+ \frac{\rho - (k-1) + (k-1)a_0 + (k-2)a_1 + \ldots + 2a_{k-3} + a_{k-2}}{a_0(1-\rho)}\mu_B$$
$$- \frac{\rho - k + ka_0 + (k-1)a_1 + (k-2)a_2 + \ldots + 2a_{k-2} + a_{k-1}}{a_0(1-\rho)}\mu_B$$
$$= \frac{1-a_0-a_1}{a_0}\eta_{k-1} + \frac{1-a_0-a_1-a_2}{a_0}\eta_{k-2} + \ldots + \frac{1-a_0-\ldots-a_{k-2}}{a_0}\eta_2.$$

$$-\frac{1 - a_0 - a_1 - \ldots - a_{k-1}}{a_0}\omega_1 + \frac{1 - a_0 - a_1 - \ldots - a_{k-1}}{a_0}\frac{\mu_B}{1 - \rho}$$

$$= \sum_{i=1}^{k-2}\frac{1 - a_0 - \ldots - a_i}{a_0}\eta_{k-i} + \frac{1 - a_0 - \ldots - a_{k-1}}{a_0}(\eta_0 + \eta_1).$$

The lemma is proved.   □

We show that from these mean values one can derive the Pollaczek–Khinchin transform equation. Let us multiply the expression for η_i in the lemma by z^i and sum up them from the row η_2, excluding the last term (containing η_0). Then

$$\frac{1 - a_0 - a_1}{a_0}z(\eta_1 z + \eta_2 z^2 + \ldots) + \frac{1 - a_0 - a_1 - a_2}{a_0}z^2(\eta_1 z + \eta_2 z^2 + \ldots)$$

$$+\frac{1 - a_0 - a_1 - a_2 - a_3}{a_0}z^3(\eta_1 z + \eta_2 z^2 + \ldots) + \ldots$$

$$= \left(\sum_{i=1}^{\infty}\eta_i z^i\right)\left(\frac{1 - a_0 - a_1}{a_0}z + \frac{1 - a_0 - a_1 - a_2}{a_0}z^2\right.$$

$$\left.+\frac{1 - a_0 - a_1 - a_2 - a_3}{a_0}z^3 + \ldots\right) \qquad (8.13)$$

$$= \left(\sum_{i=1}^{\infty}\eta_i z^i\right)\frac{1}{a_0}\left(\frac{z}{1 - z} - \frac{a_0 z}{1 - z} - \frac{a_1 z}{1 - z} - \frac{a_2 z^2}{1 - z} - \frac{a_3 z^3}{1 - z} - \ldots\right)$$

$$= \left(\sum_{i=1}^{\infty}\eta_i z^i\right)\frac{1}{a_0(1 - z)}\left(z(1 - a_0) - (A(z) - a_0)\right)$$

$$= (\overline{P}(z) - \eta_0)\frac{1}{a_0(1 - z)}\left(z(1 - a_0) - (A(z) - a_0)\right),$$

where $\overline{P}(z) = \sum_{i=0}^{\infty}\eta_i z^i$. For the terms containing η_0

$$\eta_0 z\sum_{i=1}^{\infty}\frac{1 - a_0 - \ldots - a_i}{a_0}z^i = \eta_0 z\frac{1}{a_0(1 - z)}\left(z(1 - a_0) - (A(z) - a_0)\right). \qquad (8.14)$$

Summing up Eqs. (8.13) and (8.14), the formula for η_0 and η_1 multiplied by z, we obtain

$$\overline{P}(z) = (\overline{P}(z) - \eta_0)\frac{1}{a_0(1 - z)}\left(z(1 - a_0) - (A(z) - a_0)\right)$$

$$+\eta_0 z \frac{1}{a_0(1-z)}\left(z(1-a_0)-(A(z)-a_0)\right)+\eta_0+\frac{1-a_0}{a_0}\eta_0 z,$$

whence

$$\overline{P}(z)=\frac{(1-z)A(z)}{A(z)-z}\eta_0.$$

Dividing this by the mean value of busy period $\frac{\mu_B}{1-\rho}$ and taking into account $\eta_0 = \mu_B$, we finally get the well-known formula

$$P(z)=\frac{(1-\rho)(1-z)A(z)}{A(z)-z}.$$

For details see in [10].

8.2.9 Proof of Relation (D) (Khinchin [8])

Step (D) of the Embedded Markov chain method in Sect. 8.2: Using the embedded Markov chain technique we found the ergodic distribution of the number of customers at moments just after having served the individual customers. Actually, our objective is to show that this stationary distribution holds not only for the service completion moments but also for the continuous-time $L(t)$ process. We prove the equality

$$\lim_{t\to\infty}\mathbf{P}(L(t)=k)=\pi_k, \quad k \geq 0,$$

i.e., the same formula (8.10) is valid for the generating function of the limiting distribution of $L(t)$.

The proof consists of three parts. First we show $p_0 = \pi_0$, then compute the distributions of backward and forward stationary distributions of service time; using it we show the coincidence of p_k and π_k in the general case ($k \geq 1$).

1. Let ξ_i, η_i, $i = 1, 2, \ldots$, denote the successive empty/busy periods, which are independent and separately identically distributed. The empty periods have an exponential distribution with parameter λ, and the corresponding mean value is $\mathbf{E}(\xi_i) = 1/\lambda$.

   The sequence $(\xi_i + \eta_i)$, $i = 1, 2, \ldots$, is a renewal process; at the same time $(\xi_i + \eta_i)$ are the regenerative cycles of process $L(t)$.

   Earlier we derived a functional equation for the Laplace–Stieltjes transform of a busy period's distribution function; from this for the mean value we obtained $\mathbf{E}(\eta_i) = \frac{\mu_B}{1-\rho}$, so the mean value of a regenerative cycle is

$$\kappa = \mathbf{E}(\xi_1+\eta_1)=\frac{1}{\lambda}+\frac{\mu_B}{\lambda(1-\rho)}=\frac{1}{\lambda(1-\rho)}.$$

$L(t)$ is a regenerative process, and from the limit theorem for the regenerative processes

$$\lim_{t\to\infty} \mathbf{P}(L(t) > 0) = \kappa^{-1}\mathbf{E}\left(\int_0^{T_1} \mathscr{I}_{\{L(t)>0\}}\,dt\right) = \kappa^{-1}\mathbf{E}(\eta_1)$$

$$= \frac{\mu_B}{1-\rho}\lambda(1-\rho) = \rho,$$

so, using the earlier proved relation $\pi_0 = 1 - \rho$, we get

$$p_0 = \lim_{t\to\infty} \mathbf{P}(L(t) = 0) = 1 - \rho = \pi_0.$$

By the repeated use of theorem for the regenerative processes one can show the existence of the limits

$$p_n = \lim_{t\to\infty} \mathbf{P}(L(t) = n), \quad n \geq 1,$$

but finding them in explicit form appears to be a difficult problem.

2. We find the limit distributions of backward and forward service times, δ_t and γ_t, of a customer being served at moment t as $t \to \infty$

$$F(y) = \lim_{t\to\infty} \mathbf{P}(\delta_t < y), \quad \text{and} \quad G(y) = \lim_{t\to\infty} \mathbf{P}(\gamma_t < y).$$

Let $y > 0$. Using the aforementioned theorem for regenerative processes

$$F(y) = \lambda(1-\rho)\mathbf{E}\left(\int_0^{T_1} \mathscr{I}_{\{0\leq\delta_s<y\}}\,ds\right)$$

$$= \lambda(1-\rho)\mathbf{E}\left(\xi_1 + \int_{\xi_1}^{\xi_1+\eta_1} \mathscr{I}_{\{0<\delta_s<y\}}\,ds\right)$$

$$= 1 - \rho + \lambda(1-\rho)\mathbf{E}\left(\sum_{j=1}^{K} \int_{\xi_1+Y_1+\ldots+Y_{j-1}}^{\xi_1+Y_1+\ldots+Y_j} \mathscr{I}_{\{0<\delta_s<y\}}\,ds\right)$$

$$= 1 - \rho + \lambda(1-\rho)\mathbf{E}\left(\sum_{j=1}^{K} \min(y, Y_j)\right),$$

where K is a random variable, the number of customers served in the first regenerative cycle T_1, and it coincides with the number of customers served in the first busy period of the system. Integrating by parts, we get

$$E\big(\min(y, Y_j)\big) = \int_0^\infty \min(y, x) dB(x)$$

$$= \int_0^y x dB(x) + \int_y^\infty y dB(x)$$

$$= -\int_0^y x d(1 - B(x)) + y(1 - B(y))$$

$$= -y(1 - B(y)) + \int_0^y (1 - B(x)) dx + y(1 - B(y))$$

$$= \int_0^y (1 - B(x)) dx.$$

On the other hand, since K is a regenerative point for the sequence Y_j, $j = 1, 2, \ldots$, using the Wald identity

$$E\left(\sum_{j=1}^K \min(y, Y_j)\right) = E(K) \cdot E\big(\min(y, Y_j)\big) = E(K) \int_0^y (1 - B(x)) dx.$$

Similarly,

$$E(\eta_1) = E\left(\sum_{j=1}^K Y_j\right) = E(K) \cdot E\big(Y_j\big) = E(K) \cdot \mu_B = \frac{\mu_B}{1 - \rho},$$

whence $E(K) = \frac{1}{1-\rho}$, and on the basis of these expressions we get the limiting distribution of δ_t:

$$F(y) = 1 - \rho + \lambda \int_0^y (1 - B(x)) dx.$$

We mention that $F(0+) = 1 - \rho$, $F(+\infty) = 1 - \rho + \lambda \mu_B = 1$.

The limiting distribution of γ_t may be obtained in a similar way.

$$1 - G(y) = \mu^{-1} \mathbf{E} \left(\int_0^{T_1} \mathscr{I}_{\{y < \gamma_s\}} ds \right)$$

$$= \lambda (1 - \rho) \mathbf{E} \left(\int_{\xi_1}^{\xi_1 + \eta_1} \mathscr{I}_{\{y < \gamma_s\}} ds \right)$$

$$= \lambda (1 - \rho) \mathbf{E} \left(\sum_{j=1}^{K} \int_{\xi_1 + Y_1 + \ldots + Y_{j-1}}^{\xi_1 + Y_1 + \ldots + Y_j} \mathscr{I}_{\{y < \gamma_s\}} ds \right)$$

$$= \lambda (1 - \rho) \mathbf{E} \left(\sum_{j=1}^{K} (Y_j - y)^+ \right)$$

$$= \lambda (1 - \rho) \mathbf{E}(K) \cdot \mathbf{E} \big((Y_j - y)^+ \big)$$

$$= \lambda \int_y^{\infty} (x - y) dB(x)$$

$$= \lambda \int_y^{\infty} (1 - B(x)) dx,$$

and so

$$G(y) = 1 - \lambda \int_y^{\infty} (1 - B(x)) dx.$$

From it follows that $G(0+) = 1 - \lambda \mu_B = 1 - \rho = \pi_0$ and $G(+\infty) = 1$.

3. Now, let us prove that for the stationary distribution $p_n = \lim_{t \to \infty} \mathbf{P}(L(t) = n)$ holds $p_n = \pi_n$, $n \geq 1$ (for the case $n = 0$ we have proved the equality). We will follow the reasoning by Khinchin [8].

In the stationary case the event that at the completion of a service n customers remain in the system has probability π_n, at an arbitrary moment p_n, and the remaining part of the service has distribution $G(x)$. For a small service time δ_t j customers enter the system with probability

$$a_j = \lambda \int_0^{\infty} \frac{(\lambda x)^j}{j!} e^{-\lambda x} (1 - B(x)) dx, \quad j \geq 0.$$

If the number of customers in the system at moment t is $L(t) = n > 0$, then $L(t - \delta_t)$ gives the possible number of customers there at the previous departure moment ($k = 1, \ldots, n$), or a new customer entered the empty system at $t - \delta_t$. Using the formula of total probability in the case $n > 0$ we obtain

$$
p_n = \mathbf{P}(L(t) = n) = \sum_{k=0}^{n} \mathbf{P}(L(t) = n \mid L(t - \delta_t) = k)\mathbf{P}(L(t - \delta_t) = k)
$$

$$
= \pi_n a_0 + \pi_{n-1} a_1 + \ldots + \pi_1 a_{n-1} + \pi_0 a_{n-1}, \quad n > 0.
$$

Let

$$
h_k(x) = \frac{(\lambda x)^k}{k!} e^{-\lambda x}, \quad k \geq 0,
$$

and in the case $k \geq 1$,

$$
g_{k-1}(x) = \pi_k h_0(x) + \pi_{k-1} h_1(x) + \ldots + \pi_1 h_{k-1}(x) + \pi_0 h_{k-1}(x). \quad (8.15)
$$

Then

$$
p_n = \lambda \int_0^\infty (1 - B(x)) g_{n-1}(x) \mathrm{d}x. \tag{8.16}
$$

One can directly check that the functions $h_k(x)$ satisfy the difference-differential equation

$$
h_k'(x) = \lambda[h_{k-1}(x) - h_k(x)],
$$

so for the functions $g(x)$ we have

$$
g_k'(x) = \lambda[g_{k-1}(x) - g_k(x)], \quad k \geq 1.
$$

Since $h_0(0) = 1$ and $h_k(0) = 0$, if $k \geq 1$, then $g_n(0) = \pi_{n+1}$. On the other hand, the recurrence relation (8.8) is valid for π_k, and taking into account Eqs. (8.15) and (8.16) for $k \geq 0$ we have

$$
\pi_k = \int_0^\infty h_k(x) \mathrm{d}B(x) = h_k(0) + \int_0^\infty h_k'(x)(1 - B(x)) \mathrm{d}x
$$

$$
= \pi_{k+1} + \lambda \int_0^\infty [h_{k-1}(x) - h_k(x)](1 - B(x)) \mathrm{d}x = \pi_{k+1} + p_k - p_{k+1}.
$$

Using this result and the equality proved earlier, $p_0 = \pi_0$, we obtain

$$\pi_k - p_k = \pi_{k+1} - p_{k+1} = \text{const},$$

i.e.,

$$\pi_k = p_k, \quad k \geq 0.$$

8.3 Limit Distribution of Virtual Waiting Time

Let $\gamma(t)$, $t \geq 0$, be the *virtual waiting time* at moment t (customers that entered up to moment t leave the system up to $t + \gamma(t)$). If the system is empty at moment t, then $\gamma(t) = 0$. The notion of virtual waiting time was introduced and investigated by Takács [15].

Assume that $\gamma(0) = x_0$. Our aim is to determine the distribution function of $\gamma(t)$. Let $t_1, t_2, \ldots$ ($t_0 = 0$) denote the arrival process. According to our previous assumptions $t_j - t_{j-1}$, $j \geq 1$, are independent exponentially distributed random variables with parameter λ. In this case if $t_n < t < t_{n+1}$, then

$$\gamma(t) = \begin{cases} 0, & \text{if } \gamma(t_n) < t - t_n, \\ \gamma(t_n) - (t - t_n), & \text{if } \gamma(t_n) \geq t - t_n. \end{cases}$$

If $t = t_n$, then $\gamma(t_n+) = \gamma(t_n-) + Y_n$, where Y_n is the service time of a customer entering at moment t_n (Fig. 8.8).

Theorem 8.5 $\gamma(t)$, $t \geq 0$ *is a Markov process.*

Proof The arrival process $N(t)$ has independent increments (it is a Poisson process), so the number of customers and the associated service times of customers appearing by $[t, t + s)$ are independent of the number and service times of those appearing before t. $\gamma(t+s)$ is determined by the value of $\gamma(t)$ and the customers entering after

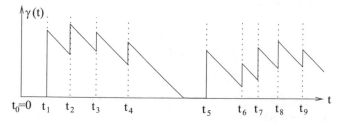

Fig. 8.8 Evolution of remaining service time

t; they do not depend on those entering before t (the service times are independent of one another and the arrival process), so our statement is valid.   □

Example 8.2 At moments $t_1, t_2, \ldots$ random amounts of water $Y_1, Y_2, \ldots$ flow to a reservoir. The outflow is uniform. In this case $\gamma(t)$ gives the actual amount of water in the reservoir.

8.3.1   Takács' Integrodifferential Equation

In previous sections we considered the number of customers in an M/G/1 system at special points. Here we intend to give a full description of its behavior. For the sake of simplicity let us denote the distribution function $F(t, x; x_0) = \mathbf{P}(\gamma(t) \leq x \mid \gamma(0) = x_0)$ by $F(t, x)$; we assume that the continuous partial derivatives of $F(t, x)$ by t and x on the set $t > 0, \ x \geq 0$ exist, and

$$\lim_{t \to 0+} F(t, x) = F(0, x) = \mathscr{I}_{\{x \geq x_0\}}.$$

Theorem 8.6 (Takács [16]) *Under these conditions the distribution function* $F(t, x)$ *satisfies the integrodifferential equation*

$$\frac{\partial F(t, x)}{\partial t} = \frac{\partial F(t, x)}{\partial x} - \lambda F(t, x) + \lambda \int_0^x B(x - y) \, d_y F(t, y). \qquad (8.17)$$

Proof $\{\gamma(t + \Delta) < x\}$ is the union of three disjoint events (we take into account that the arrival process is Poisson and independent of service times):

- $\gamma(t) < x + \Delta$ and for $(t, t + \Delta)$ no customer enters the system. The probability of this event is $(1 - \lambda\Delta)F(t, x + \Delta) + o(\Delta)$.
- At moment t, $\quad 0 \leq \gamma(t) < x$, and for $(t, t + \Delta)$ one new customer enters [the corresponding probability is $\lambda\Delta$ independently of $\gamma(t)$ and the service time], and the customer's service time is $Y \ < \ x - \gamma(t)$. The probability of this event [Y and $\gamma(t)$ are independent, and the distribution of $Y + \gamma(t)$ is $B * F$: $\lambda\Delta\mathbf{P}(Y + \gamma(t) < x) + o(\Delta)$] is

$$\lambda\Delta \int_0^x B(x - y) \, d_y F(t, y) + o(\Delta).$$

- $0 \leq \gamma(t) < x$, and for $(t, t + \Delta)$ more than one customer enters the system, and its probability is $o(\Delta)$.

Then

$$F(t + \Delta, x) = (1 - \lambda\Delta)F(t, x + \Delta) + \lambda\Delta \int_0^x B(x - y)\, d_y F(t, y) + o(\Delta),$$

which can be rewritten as

$$\frac{1}{\Delta}\left(F(t + \Delta, x) - F(t, x)\right)$$

$$= \frac{1}{\Delta}\left(F(t, x + \Delta) - F(t, x)\right) - \lambda F(t, x + \Delta) + \lambda \int_0^x B(x - y)\, d_y F(t, y) + o(1).$$

If $\Delta \to 0$, then we obtain Eq. (8.17).   □

Takács derived this theorem in the case of an inhomogeneous Poisson arrival process with intensity $\lambda(t)$. He proved that this integrodifferential equation holds for all $t, x \geq 0$ for which $\frac{\partial}{\partial x} F(t, x)$ exists.

With the help of the previous theorem we prove the following one giving an integrodifferential equation for the stationary distribution.

Theorem 8.7 *If $\mu_B = \int_0^\infty x\, dB(x) < \infty$, $\rho = \lambda\tau < 1$, then there exists*

$$\lim_{t\to\infty} F(t, x) = F(x),$$

and it is independent of the initial distribution $F(0, x)$. It satisfies the equation

$$F'(x) = \lambda F(x) - \lambda \int_0^x B(x - y)\, dF(y), \quad x > 0, \tag{8.18}$$

and $F(0+) = 1 - \rho$.

Proof The proof is based on results for regenerative processes. We can use the fact that the distribution of cycles for a given index is absolute continuous, or the process $\gamma(t)$, $t \geq 0$, is right continuous and has a limit from left. From both conditions it follows that the process has a limit distribution and can be written in the form given previously in Theorem 8.6.

Let $0 < \tau_1 < \tau_2 < \cdots$ be successive moments when free periods begin. Then $\{\gamma(t), t \geq 0\}$ is a regenerative process with regeneration points τ_k, $k = 1, 2, \ldots$; the intervals $Z_k = \tau_k - \tau_{k-1}$, $k = 1, 2, \ldots$ ($\tau_0 = 0$), whose lengths are the sums of free and busy periods (perhaps excluding Z_1), are a (delayed) renewal process.

Let

$$G_1(x) = \mathbf{P}(Z_1 \leq x), \quad G(x) = \mathbf{P}(Z_k \leq x), \quad k \geq 2,$$

and

$$G^{(n+1)}(x) = \mathbf{P}(\tau_k \leq x) = \int_0^x G^{(n)}(x - y) \, dG(x), \ n \geq 1.$$

Since the free and busy periods are i.i.d. random variables (the free periods have an exponential distribution with parameter λ), the distribution function G, and thus $G^{(n)}$, $n \geq 2$, is absolutely continuous (this is sufficient condition for the existence of a limit distribution).

The mean value of a regenerative cycle is

$$\int_0^\infty x \, dG(x) = \frac{1}{\lambda} + \frac{\mu_B}{1 - \rho} < \infty,$$

and for arbitrary x there exists the limit distribution

$$F(x) = \lim_{t \to \infty} F(t, x) = \lim_{t \to \infty} \mathbf{E}\left(\mathscr{I}_{\{\gamma(t) \leq x\}}\right) = \lim_{T \to \infty} \frac{1}{T} \mathbf{E}\left(\int_0^T \mathscr{I}_{\{\gamma(s) \leq x\}} \, ds\right).$$

If in (8.17) $t \to \infty$, then we obtain Eq. (8.18). $\square$

If we take the initial distribution $F(0, x) = F(x)$, then the distribution function $F(t, x) = F(x)$ satisfies Eq. (8.17). It is clear that

$$F(0+) = \lim_{t \to \infty} F(t, 0+) = \lim_{t \to \infty} \mathbf{P}(\gamma(t) = 0) = \lim_{t \to \infty} \mathbf{P}(L(t) = 0) = 1 - \rho.$$

One can see that [5]), if $\rho \geq 1$, then

$$\lim_{t \to \infty} F(t, x) = 0, \quad x \in \mathbb{R}.$$

Equation (8.18) may be solved by means of the Laplace–Stieltjes transforms. Let

$$F^\sim(s) = \int_0^\infty e^{-sx} \, dF(x) = 1 - \rho + \int_0^\infty e^{-sx} F'(x) \, dx,$$

where substituting F' from Eq. (8.18) yields

$$F^\sim(s) = 1 - \rho + \frac{\lambda}{s} F^\sim(s) - \frac{\lambda}{s} F^\sim(s) B^\sim(s)$$

$$= 1 - \rho + \frac{\lambda}{s} F^\sim(s)(1 - B^\sim(s)),$$

whence

$$F^\sim(s) = \frac{1 - \rho}{1 - \frac{\lambda}{s}(1 - B^\sim(s))}. \tag{8.19}$$

This expression is called the Pollaczek–Khinchin formula for the waiting time. The inversion of the Laplace–Stieltjes transform gives the probability of an event in a stationary regime; the waiting time is less than x (see, e.g., [12]).

Example 8.3 Let us consider the case where the distribution function $B(x)$ is exponential with parameter μ, i.e.,

$$B(x) = 1 - e^{-\mu x}, \quad x \geq 0.$$

Then $B^\sim(s) = \frac{\mu}{s+\mu}$, according to the Pollaczek–Khinchin formula (8.19), for the Laplace–Stieltjes transform $F^\sim$ of distribution function F we obtain

$$F^\sim(s) = 1 - \rho + \lambda \frac{\mu - \lambda}{s + \mu - \lambda}.$$

The inversion of the Laplace–Stieltjes transform gives

$$F(x) = 1 - \rho + \rho(1 - e^{-(\mu - \lambda)x}),$$

whence

$$F(0+) = \lim_{x \to 0+} F(x) = 1 - \rho.$$

8.4   Queue Length in Discrete-Time System

Previously, we characterized the M/G/1 system with continuously distributed interarrival and service times. Sometimes, it may be useful to apply the discrete analogue of such system.

Let us consider a discrete-time queueing system where for a time unit (slot) a new customer arrives with probability r and does not appear with probability $1 - r$, i.e., the interarrival time has geometrical distribution with probabilities

$$\mathbf{P}(X = i) = (1 - r)^{i-1} r \qquad i = 1, 2, \ldots;$$

the service time has general distribution

$$\mathbf{P}(Y = i) = q_i \qquad i = 1, 2, \ldots.$$

We are interested in the number of present customers at moments t_n-, where t_n denotes the moment of beginning to serve the nth customer. In this case for the queue length L_n at this moment we have the recurrence relation

$$L_n = \begin{cases} L_{n-1} - 1 + \Delta_n, & \text{if } L_{n-1} > 0, \\ \Delta_n - 1, & \text{if } L_{n-1} = 0, \end{cases}$$

Here Δ_n is the number of customers arriving in the interval $[t_{n-1}, t_n)$. The generating function of service time is

$$Q(z) = \sum_{i=1}^{\infty} q_i z^i,$$

the generating function of the arrival of a customer for a slot is $1 - r + rz$; consequently, the generating function of arriving customers Δ_n for a service is

$$A(z) = \sum_{i=1}^{\infty} q_i (1 - r + rz)^i = Q(1 - r + rz).$$

The matrix of transition probabilities will be

$$P = \begin{pmatrix} a_0 & a_1 & a_2 & a_3 & \cdots \\ a_0 & a_1 & a_2 & a_3 & \cdots \\ 0 & a_0 & a_1 & a_2 & \cdots \\ 0 & 0 & a_0 & a_1 & \cdots \\ \vdots & \vdots & \vdots & \vdots & \ddots \end{pmatrix}.$$

The ergodic probabilities may be obtained as the solution of the system of equations

$$\pi_0 = \pi_0 a_0 + \pi_1 a_0,$$

$$\pi_k = \pi_0 a_k + \sum_{i=1}^{k+1} \pi_i a_{k-i+1} \qquad k = 1, 2, \ldots.$$

Introducing the notation

$$\pi(z) = \sum_{k=0}^{\infty} \pi_k z^k,$$

one obtains

$$\pi(z) = \frac{(z-1)A(z)}{z - A(z)} \pi_0 = \frac{(z-1)Q(1-r+rz)}{z - Q(1-r+rz)} \pi_0.$$

From the condition $\pi(1) = 1$, we have

$$\pi_0 = \lim_{z \to 1-0} \frac{z - Q(1-r+rz)}{(z-1)Q(1-r+rz)}$$

$$= \left. \frac{1 - Q'(1-r+rz)}{Q(1-r+rz) + zQ'(1-r+rz)r - rQ'(1-r+rz)} \right|_{z=1}$$

$$= 1 - rQ'(1) = 1 - r \sum_{i=1}^{\infty} iq_i = 1 - r\tau = 1 - \rho,$$

where τ is the mean value of the service time of a customer. Finally,

$$\pi(z) = \frac{(1-\rho)(1-z)Q(1-r+rz)}{Q(1-r+rz) - z}.$$

We find the mean value of the queue length, and it is equal to $\pi'(1)$. Introducing the notations

$$S = Q(1-r+rz) - zQ(1-r+rz),$$

$$N = Q(1-r+rz) - z,$$

it equals

$$(1-\rho)\frac{S''(1)N'(1) - S'(1)N''(1)}{2N'^2(1)}.$$

Substituting the values

$$S'(1) = -1, \qquad\qquad S''(1) = -2rQ'(1),$$

$$N'(1) = rQ'(1) - 1, \qquad\qquad N''(1) = r^2 Q''(1),$$

one gets

$$\pi'(1) = rQ'(1) + \frac{r^2 Q''(1)}{2[1 - rQ'(1)]} = \rho + \frac{r^2 Q''(1)}{2(1-\rho)}.$$

This expression remembers the Pollaczek–Khinchin mean value formula, but instead of the second moment of the service time here appears the second derivative of the service time generating function whose value is less. This is explained by the fact that in this case the number of arriving customers is limited by the length of service time.

8.5 G/M/1 Queue

In the case of a G/M/1 queue, customers arrive according to a renewal process. The service times are independent exponentially distributed with parameter μ. There is one server, and the waiting room is infinite. The analysis methods available for a G/M/1 queue are very similar to those available for the M/G/1 queue. In this section we analyze the G/M/1 queue with the method of embedded Markov chain. In our analysis we will follow the classical way usual in the theory of Markov chains. One finds the generating function of n step transition probabilities and using the Tauber theorem the limiting distribution and the stability condition will be obtained. Another approach based on the existence of stability condition and the solution of an equation is presented in Kleinrock's book (see Kleinrock [9]).

Let T_n be the time between the $(n-1)$st and nth arrivals (in the case of the M/G/1 queue it had another meaning). The arrivals constitute a renewal process, so $\{T_n\}$ is a sequence of i.i.d. random variables, and let T have the same distribution. For the sake of simplicity we assume that T is continuous with density function $a(x)$ and has a finite mean. Let $\lambda = 1/T$, i.e.,

$$\mathbf{P}(T \leq x) = \int_0^x a(u)\, du \quad (x \geq 0)$$

and

$$\frac{1}{\lambda} = \int_0^\infty x\, a(x)\, dx.$$

$L(t)$ denotes the number of customers in the system at moment t. Similar to the M/G/1 queue, the process $\{L(t) : t \geq 0\}$ generally is not a Markov chain, and the future behavior at t depends not only on $L(t)$ but also on time elapsed from the moment of the last arrival. We will use the embedded Markov chain technique, and the embedded points will be the moments just before the arrivals.

8.5.1 Embedded Markov Chain

Let X_n be the number of customers in a G/M/1 system at the moment just before the entry of nth one, formally

$$X_n = \lim_{\Delta \to 0+} N(\textstyle\sum_{i=1}^n T_i - \Delta).$$

We show that $\{X_n\}$ is a homogeneous Markov chain. Let V_n' denote the number of customers served between the arrivals of $(n-1)$st and nth customers (i.e., for T_n). $\{X_n\}$ satisfies the equation

$$X_{n+1} = X_n + 1 - V_{n+1}'. \tag{8.20}$$

It is not simple to work with the recursion Eq. (8.20) since V_n' depends on X_{n-1}, so $\{V_n'\}$ is not an identically distributed sequence (e.g., $V_1' \equiv 0$). Let V_n be the number of customers that the system would have served had it not become empty. $\{V_n\}$ are i.i.d., which is a consequence of the fact that $\{T_n\}$ and the service times are independent. Equation (8.20) can be written in the form

$$X_{n+1} = (X_n + 1 - V_{n+1})^+, \tag{8.21}$$

from which it follows that $\{X_n\}$ is a homogeneous Markov chain.

We compute the transition probabilities of this chain:

$$\begin{aligned} p_{ij} &= \mathbf{P}(X_{n+1} = j | X_n = i) \\ &= \mathbf{P}\big((X_n + 1 - V_{n+1})^+ = j | X_n = i\big) \\ &= \mathbf{P}\big((i + 1 - V_{n+1})^+ = j | X_n = i\big) \\ &= \mathbf{P}\big((i + 1 - V_{n+1})^+ = j\big) \end{aligned} \tag{8.22}$$

because of the independence of X_n and V_{n+1}. Obviously, if $j > i+1$, then $p_{ij} = 0$. Let $0 < j \le i+1$, then in Eq. (8.22) we can cancel the sign of the positive part and

$$\begin{aligned} p_{ij} &= \mathbf{P}(V_{n+1} = i - j + 1) \\ &= \int_0^\infty \mathbf{P}(V_{n+1} = i - j + 1 | T_{n+1} = x)\, a(x)\, dx \tag{8.23} \\ &= \int_0^\infty \frac{(\mu x)^{i-j+1}}{(i-j+1)!} e^{-\mu x} a(x)\, dx \quad (0 < j \le i+1), \tag{8.24} \end{aligned}$$

because, given that we always have customers, the moments of completion are a Poisson process with intensity μ, and its increment for x appears in Eq. (8.23). In this case, from Eq. (8.24) it follows that p_{ij} depends only on the differences in indices, i.e.,

$$p_{ij} = \mathbf{P}(V_{n+1} = i - j + 1) = \beta_{i-j+1} \quad (0 < j \le i+1).$$

The sum of elements in the rows of matrix $\boldsymbol{\Pi}$ is equal to 1; consequently,

$$p_{i0} = 1 - \sum_{j=1}^\infty p_{ij} = 1 - \sum_{j=1}^{i+1} \beta_{i-j+1} = 1 - \sum_{k=0}^i \beta_k.$$

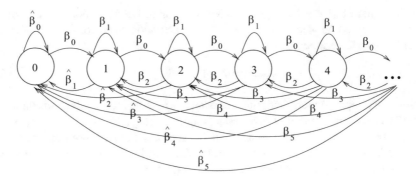

Fig. 8.9 Embedded CTMC of the G/M/1 queue

The system behavior before customers arrive is depicted in Fig. 8.9, and the one-step
state-transition probability matrix is

$$
\Pi = \begin{pmatrix}
1 - \beta_0 & \beta_0 & 0 & 0 & \cdots \\
1 - \beta_0 - \beta_1 & \beta_1 & \beta_0 & 0 & \cdots \\
1 - \beta_0 - \beta_1 - \beta_2 & \beta_2 & \beta_1 & \beta_0 & \cdots \\
\vdots & \vdots & \vdots & \vdots & \ddots
\end{pmatrix},
$$

where

$$
\beta_k = \int_0^\infty \frac{(\mu x)^k}{k!} e^{-\mu x} a(x)\, dx. \tag{8.25}
$$

Consider the Markov chain defined by the preceding matrix. Let $P_{k\ell}(n)$ be the
probability of the event that the system for n steps from state k arrives at state ℓ, and
let us introduce the following notations:

$$
\hat{\beta}_k = 1 - \sum_{i=0}^k \beta_i = \sum_{i=k+1}^\infty \beta_i, \qquad Q(z) = \sum_{i=0}^\infty \beta_i z^i, \qquad C(z) = \frac{Q(z)}{z},
$$

$$
P_\ell(n, z) = \sum_{k=0}^\infty P_{k\ell}(n) z^k, \qquad P_\ell(t, z) = \sum_{n=0}^\infty P_\ell(n, z) t^n, \qquad P_{0\ell}(t) = P_{0\ell}(n) t^n.
$$

Let us fix the final state ℓ and write the inverse Kolmogorov equations for transition
probabilities for $n + 1$ steps:

$$
P_{k\ell}(n + 1) = \sum_{i=0}^k \beta_i P_{k+1-i,\ell}(n) + \hat{\beta}_k P_{0\ell}(n), \qquad k = 0, 1, 2, \ldots
$$

Let us multiply these equations by z^k and sum up by k:

$$\sum_{k=0}^{\infty} P_{k\ell}(n+1)z^k = \sum_{k=0}^{\infty}\sum_{i=0}^{k} \beta_i P_{k+1-i,\ell}(n)z^k + \sum_{k=0}^{\infty} \hat{\beta}_k P_{0\ell}(n)z^k$$

$$= \frac{1}{z}Q(z)[P_\ell(n,z) - P_{0\ell}(n)] + P_{0\ell}(n)\sum_{k=0}^{\infty} \hat{\beta}_k z^k. \quad (8.26)$$

Since

$$\sum_{k=0}^{\infty} \hat{\beta}_k z^k = \sum_{k=0}^{\infty}(1 - \beta_0 - \ldots - \beta_k)z^k$$

$$= (1 - \beta_0) + (1 - \beta_0 - \beta_1)z + (1 - \beta_0 - \beta_1 - \beta_2)z^2 + \ldots$$

$$= (1 - \beta_0)(1 + z + z^2 + \ldots) - \beta_1(z + z^2 + \ldots) - \beta_2(z^2 + z^3 + \ldots) - \ldots$$

$$= (1 - \beta_0)\frac{1}{1-z} - \beta_1\frac{z}{1-z} - \beta_2\frac{z^2}{1-z} - \ldots = \frac{1 - Q(z)}{1-z},$$

from Eq. (8.26)

$$P_\ell(n+1, z) = \frac{1}{z}Q(z)\left(P_\ell(n, z) - P_{0\ell}(n)\right) + P_{0\ell}(n)\frac{1 - Q(z)}{1-z}.$$

Multiplying this equation by t^n and summing up by n

$$\sum_{n=0}^{\infty} P_\ell(n+1)t^n = \frac{1}{z}Q(z)\sum_{n=0}^{\infty}\left(P_\ell(n, t)t^n - P_{0\ell}(n)t^n\right) + \frac{1 - Q(z)}{1-z}\sum_{n=0}^{\infty} P_{0\ell}(n)t^n,$$

i.e.,

$$\frac{1}{t}[P_\ell(t, z) - P_\ell(0, z)] = \frac{1}{z}Q(z)P_\ell(t, z) + P_{0\ell}(t)\frac{z - Q(z)}{z(1-z)},$$

or (using the initial value)

$$P_\ell(t, z) = \left(1 - t\frac{Q(z)}{z}\right)^{-1}\left(z^\ell + P_{0\ell}(t)t\frac{z - Q(z)}{z(1-z)}\right). \quad (8.27)$$

This expression contains the unknown generating function $P_{0\ell}(t)$, which will be determined [since $P_\ell(t, z)$ is analytical function] by means of the roots of the equation

Fig. 8.10 The case where
$C'(1) < 0$

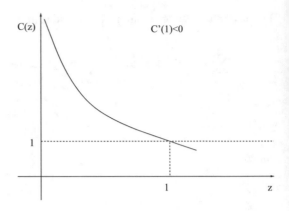

$$1 - t\frac{Q(z)}{z} = 0$$

in $(0, 1)$. We will need the following results.

Lemma 8.2 *If $C'(1) \leq 0$, then $C(z)$ is a continuous function on $(0,1]$, and it is decreasing from $+\infty$ to 1.*

Proof See Fig. 8.10.

$$C'(z) = \sum_{i=0}^{\infty}(i-1)\beta_i z^{i-1}, \qquad C''(z) = \sum_{i=0}^{\infty}(i-1)(i-2)\beta_i z^{i-3}.$$

$C''(z) > 0$ on the interval $(0, 1]$, so $C'(z)$ is monotonically increasing on $(0, 1]$ and on the open interval $C'(z) < C'(1) \leq 0$. Since $C(z)$ is continuous on $(0, 1]$, it decreases from $+\infty$ to $C(1) = 1$. $\square$

Lemma 8.3 *If $0 < C'(1) < +\infty$, then there exists $\theta \in (0, 1)$ such that $C(z)$ monotonically decreases on $(0, \theta]$ from $+\infty$ to $C(\theta) < 1$ and on $[\theta, 1]$ is continuous and monotonically increases from $C(\theta)$ to $C(1) = 1$.*

Proof See Fig. 8.11. $C''(z) > 0$ on $(0, 1)$, so $C'(z)$ monotonically increases from $-\infty$ to $C'(1) > 0$. Consequently, there is one and only one value $\theta \in (0, 1)$ for which $C'(\theta) = 0$. On $(0, \theta)$, $C'(z) < 0$ and $C(z)$ is monotonically decreasing; on $(\theta, 1)$ $C'(z) > 0$ and $C(z)$ is monotonically increasing, $C(\theta) < C(1) = 1$. $\square$

Corollary 8.3 *Under the conditions of lemma there exists one and only one $z^* \in (0, \theta)$ such that on $(0, z^*)$ $C(z)$ monotonically decreases from $+\infty$ to 1, on $[z^*, \theta]$ monotonically decreases from 1 to $C(\theta)$ and on $(\theta, 1)$ monotonically increases from $C(\theta)$ to 1.*

Fig. 8.11 The case where
$C'(1) > 0$

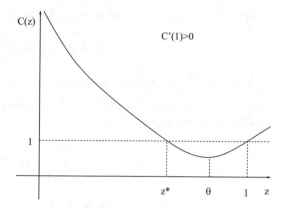

Proof The existence and uniqueness of z^* follow from the fact $C(z)$ is monotonically decreasing function on $(0, z^*]$, and $C(0+) = +\infty$ and $C(\theta) < 1$.

Let $z(t)$ be the root of the equation

$$\frac{1}{t} = C(z(t))$$

on the interval $(0, 1)$. Substituting it into the numerator of right-hand side of Eq. (8.27),

$$z^\ell(t) + P_{0\ell}(t)t\frac{1 - C(z(t))}{1 - z(t)} = z^\ell(t) + P_{0\ell}(t)t\frac{1 - \frac{1}{t}}{1 - z(t)}$$

$$= z^\ell(t) + P_{0\ell}(t)\frac{t - 1}{1 - z(t)} = 0,$$

whence

$$P_{0\ell}(t) = z^\ell(t)\frac{1 - z(t)}{1 - t}.$$

The chain is irreducible and aperiodic, so the equilibrium distribution does not depend on the initial state. By the Tauberian theorem [16]

$$P_\ell = \lim_{t \to 1}(1 - t)P_{0\ell}(t) = z^{*\ell}(1 - z^*), \tag{8.28}$$

where z^* is the root of the equation $C(z) = 1$ on the interval $(0, 1)$. If $C'(1) > 0$, then z^* lies between 0 and 1, so we get a nondegenerate distribution. In the case $C'(1) \leq 0$, we have $z^* = 1$ and the distribution is degenerate.   □

Remark 8.2 If $A(x)$ denotes the distribution function of interarrival times, then the generating function of the number of customers served is $Q(z) = A^\sim(\mu - \mu z)$, and from $C(z^*) = 1$ follows that z^* is the only root of

$$z^* = A^{\sim}(\mu - \mu z^*)$$

in the interval $0 < z^* < 1$ (see Kleinrock [9]).

Remark 8.3 The condition $C'(1) > 0$ can easily be expressed with the help of generating function $Q(z)$ since

$$\left. \frac{Q'(z)z - Q(z)}{z^2} \right|_{z=1} = Q'(1) - 1 > 0,$$

from which the stability condition is $Q'(1) > 1$. It has a simple meaning, namely, the mean value of number of customers served between the entries of two successive customers must be more than one. This condition is equivalent to the inequality $\mu/\lambda > 1$.

Remark 8.4 We have examined the G/M/1 system at moments just before the arrivals and found the stability condition for these points. We mention that in contrast to the M/G/1 system, this distribution does not hold for the inner points.

8.6   Exercises

Exercise 8.1 There is an M/G/1 queue. The arrival intensity is λ and the service time is exponentially distributed with parameter μ_2 with probability $1 - p$ and it is the sum of two independent exponentially distributed random variable with parameters μ_1 and μ_2 with probability p.

- Compute the utilization of the server.
- Compute the coefficient of variation of the service time.
- Compute the mean system time of customers.
- Compute the mean number of customers in the buffer.

Exercise 8.2 Customers arrive to a dentist according to a Poisson process with intensity λ. Arriving customers enter the dentist's surgery if it is idle, otherwise they wait in the waiting room. At the dentist's surgery there is a registration of time D (deterministic). With probability p the patient is directed to the dentist for treatment which takes an exponentially distributed time with parameter μ, with probability $1 - p$ the patient is rejected.

- Compute the mean time of customers in the waiting room.
- Compute the probability that an arriving customer has to wait.
- Compute the mean waiting time.

Exercise 8.3 $F_A(t)$ is the interarrival distribution in an G/M/1 queue whose service rate is μ. $N(t)$ is the number of customers in the system at time t and $T_1, T_2, \ldots$ denote the arrival instants of the first, second, etc. customers. The mean of the stationary number of customers is $\bar{N} = \lim_{t\to\infty} E(N(t))$ and the mean of the stationary number of customers at arrival instants is $\check{N} = \lim_{n\to\infty} E(N(T_n-))$. Compute the relation of $\bar{N}$ and $\check{N}$ if

- interarrival distribution is hyper-exponential ($F_A(t) = 1 - pe^{\lambda_1 t} - (1-p)e^{\lambda_2 t}$),
- interarrival distribution is deterministic,
- interarrival distribution is exponential.

Exercise 8.4 Find the mean value of number of customers in the system and in the waiting queue in the M/G/1 system. Consider the cases of of M/M/1 and M/D/1 systems.

Exercise 8.5 By using the Pollaczek–Khinchin transform equation show that in the M/M/1 system the equilibrium distribution is geometrical.

Exercise 8.6 Let us consider the M/G/1 system with bulk arrivals, an arriving group with probability g_i consists of i customers. Show that the generating function of number of customers entering during time t is $e^{-\lambda t(1-G(z))}$, where λ is the intensity of arrivals and $G(z) = \sum_{i=1}^{\infty} g_i z^i$.

Exercise 8.7 Show that in the M/G/1 system with bulk arrivals the generating function of number of customers arriving for the service time of a customer is $b^\sim(\lambda(1 - G(z)))$, where $b^\sim(s)$ is the Laplace–Stieltjes transform of distribution function of this service time.

8.7 Solutions

Solution 8.1

- Let S be the service time and S_i be exponentially distributed with parameter μ_i. The mean service time and the utilization are

$$\mathbf{E}(S)=(1-p)\mathbf{E}(S_1)+p\,(\mathbf{E}(S_1 + S_2)) = (1-p)\frac{1}{\mu_2}+p\left(\frac{1}{\mu_1} + \frac{1}{\mu_2}\right), \quad \rho = \lambda\mathbf{E}(S).$$

- The second moment and the coefficient of variation of the service time are

$$\mathbf{E}(S^2) = (1 - p)\mathbf{E}(S_1^2) + p\,(\mathbf{E}((S_1 + S_2)^2))$$
$$= (1 - p)\mathbf{E}(S_1^2) + p\,(\mathbf{E}(S_1^2) + 2\mathbf{E}(S_1)\mathbf{E}(S_2) + \mathbf{E}(S_2^2))$$
$$= (1 - p)\frac{2}{\mu_2^2} + p\left(\frac{2}{\mu_1^2} + \frac{2}{\mu_1\mu_2} + \frac{2}{\mu_2^2}\right)$$

$$CV(S) = \frac{E(S^2)}{E(S)^2} - 1.$$

- The mean system time of customers is

$$E(T) = E(S) \left(1 + \frac{\rho}{1 - \rho} \frac{1 + CV(S)}{2} \right).$$

- The mean waiting time and the mean number of customers in the buffer are

$$E(W) = E(S) \frac{\rho}{1 - \rho} \frac{1 + CV(S)}{2}, \quad E(L_W) = \lambda E(W).$$

An alternative solution of the exercise is to recognize that the service time is PH distributed with representation

$$\beta = \{p, 1 - p\}, \boldsymbol{B} = \begin{array}{|c|c|} \hline -\mu_2 & \mu_2 \\ \hline & -\mu_1 \\ \hline \end{array}.$$

Based on the PH representation of the service time we can apply the analysis of the M/PH/1 queue for which closed form expressions are available.

Solution 8.2

- Let S be the service time and S_T be the treatment time which is exponentially distributed with parameter μ. The mean service time and the utilization are

$$E(S) = D + pE(S_T) = D + p\frac{1}{\mu}, \quad \rho = \lambda E(S).$$

The second moment and the coefficient of variation of the service time are

$$E(S^2) = D^2 + pE(S_T^2) = D^2 + p\frac{2}{\mu^2}$$

$$CV(S) = \frac{E(S^2)}{E(S)^2} - 1.$$

Based on these quantities the mean waiting time is

$$E(W) = E(S) \frac{\rho}{1 - \rho} \frac{1 + CV(S)}{2}.$$

- The probability that an arriving customer has to wait can be computed from the utilization of the system as follows

$$P(\text{waiting}) = 1 - \rho.$$

- The waiting time is the time a customer spends in the waiting room.

Solution 8.3 $\bar{N}$ denotes the mean number of customers in a G/M/1 queue at a random time instant and $\check{N}$ denotes the mean number of customers right before an arrival instant. The arrival process is not a Poisson process and consequently PASTA property does not hold. That is, the distribution of the number of customers at a random time instant and the distribution of number of customers right before an arrival instant are different (in general). From (8.28) we have that the number of customers right before arrivals, $N(T_n-)$, is geometrically distributed with parameter z^*, that is $P_\ell = \mathbf{P}(N(T_n-) = \ell) = z^{*\ell}(1 - z^*)$, where z^* is the solution of $z^* = A^\sim(\mu - z^*\mu)$ and $A^\sim(s)$ is the Laplace Stieltjes transform of $F_A(t)$, $A^\sim(s) = \int_t e^{-st} dF_A(t)$.

The $N(t)$ process, number of customers in the G/M/1 queue at time t, is a Markov regenerative process with regeneration instants at customer arrivals. The relation of the distribution at a regeneration instant and at a random time instant is provided by (4.29). The stationary distribution (at random time) can be computed from the embedded distribution as:

$$\pi_k = \frac{\sum_j P_j \tau_{jk}}{\sum_j P_j \tau_j}. \tag{8.29}$$

where τ_j is the mean time to the next embedded instant starting from state j, and τ_{jk} is the mean time spent in state k before the next embedded instant starting from state j as it is indicated in the following time diagram.

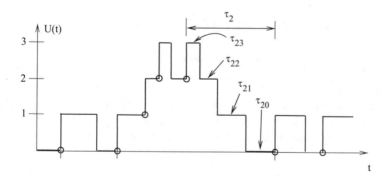

Denoting the mean arrival intensity by $\bar{\lambda}$, we have $\bar{\lambda} = 1/\int_t 1 - F_A(t) dt$. In a G/M/1 queue the embedded instants are the arrival instants, and $\tau_i = 1/\bar{\lambda}$ for all $i \geq 0$ is the mean interarrival time. The following diagram details the stochastic process between arrival instants in order to compute τ_{ik}.

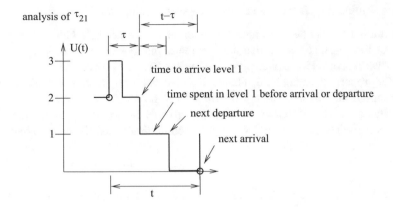

analysis of τ_{21}

In the figure τ is the sum of $i+1-k$ service times and it has an Erlang$(i+1-k, \mu)$ distribution. Using that we have

$$\tau_{ik} = \int_{t=0}^{\infty} \int_{\tau=0}^{t} \int_{x=0}^{t-\tau} e^{-\mu x} \, dx \, f_{Erl(i+1-k)}(\tau) \, d\tau \, dA(t)$$

and

$$\tau_{i0} = \int_{t=0}^{\infty} \int_{\tau=0}^{t} (t - \tau) \, f_{Erl(i+1-k)}(\tau) \, d\tau \, dA(t).$$

Note the level independent behavior of $\tau_{i,k}$, that is $\tau_{i,k} = \tau_{i+j,k+j}$, $\forall j \geq 0$ and $\forall k > 0$.

Computing τ_{ik} and substituting into (8.29) result in

$$\pi_0 = 1 - \rho \quad \text{and} \quad \pi_k = \rho(1 - z^*)z^{*k-1}, \quad k \geq 1.$$

where $\rho = \bar{\lambda}\frac{1}{\mu}$.

Based on the stationary behavior at random instant we can compute the mean number of customers in the queue, $\bar{N} = L$, the mean system time, T, the mean number of waiting customers, L_W, and the mean waiting time

$$\bar{N} = L = \sum_{i=0}^{\infty} i\pi_i = \frac{\rho}{1 - z^*}, \quad T = \frac{L}{\bar{\lambda}} = \frac{1}{\mu}\frac{1}{1 - z^*}.$$

$$L_W = \sum_{i=1}^{\infty} (i - 1)\pi_0 = \frac{\rho \, z^*}{1 - z^*}, \quad W = \frac{L_W}{\bar{\lambda}} = \frac{1}{\mu}\frac{z^*}{1 - z^*}.$$

and the mean number of customers in the queue right before arrivals is

$$\check{N} = \sum_{i=0}^{\infty} i P_i = \frac{z^*}{1 - z^*}$$

Special G/M/1 queues

- exponentially distributed interarrival time—M/M/1 queue:

$$A^\sim(s) = \frac{\lambda}{s + \lambda}$$

$$z^* = A^\sim(\mu - z^*\mu) = \frac{\lambda}{\mu - z^*\mu + \lambda}$$

and its valid (inside the unit disk) solution is $z^* = \frac{\lambda}{\mu} = \rho$. $z^* = 1$ is also a solution of the equation but it is not valid.

- Erlang(λ, 2) distributed interarrival time—E_2/M/1 queue:

$$A^\sim(s) = \left(\frac{\lambda}{s + \lambda}\right)^2$$

$$\rho = \frac{\lambda}{\mu} = \frac{\lambda}{2\mu}, \quad z^* = 2\rho + \frac{1}{2} - \sqrt{2\rho + \frac{1}{4}}$$

- deterministic interarrival time—D/M/1 queue:

$$A^\sim(s) = e^{-sD}$$

$$\rho = 1/\mu D, \quad z^* = A^\sim(\mu - z^*\mu) = e^{-\mu D(1 - z^*)}$$

- hyper-exponentially distributed interarrival time—H_2/M/1 queue:

$$A^\sim(s) = \frac{p_1 \lambda_1}{s + \lambda_1} + \frac{p_2 \lambda_2}{s + \lambda_2}$$

Assuming $p_1 = p_2 = 0.5$, and $\lambda_1 = 2\lambda_2 = \lambda = 1$, we have $\bar{\lambda} = \frac{2\lambda}{3}$,

$$\rho = \frac{\bar{\lambda}}{\mu} = \frac{2\lambda}{3\mu} \text{ and } z^* = \frac{9\rho}{8} + \frac{1}{2} - \sqrt{\frac{9\rho^2}{64} + \frac{1}{4}}$$

Solution 8.4 The mean value of number of customers is provided in (8.5). For the mean value of number of waiting customers we have

$$\sum_{k=1}^{\infty}(k-1)p_k = \sum_{k=1}^{\infty} kp_k - \sum_{k=1}^{\infty} p_k = \rho + \frac{\lambda^2 \mathbf{E}(Y^2)}{2(1 - \rho)} - \rho = \frac{\lambda^2 \mathbf{E}(Y^2)}{2(1 - \rho)}.$$

In case of the M/M/1 system the second moment of service time is $\mathbf{E}(Y^2) = \frac{2}{\mu^2}$. By using this value the main queue length is

$$\rho + \frac{\lambda^2 \mathbf{E}(Y^2)}{2(1 - \rho)} = \frac{\lambda}{\mu} + \frac{\lambda^2 \cdot \frac{2}{\mu^2}}{2\left(1 - \frac{\lambda}{\mu}\right)} = \frac{\lambda}{\mu - \lambda}.$$

The main number of waiting customers is

$$\sum_{k=1}^{\infty}(k-1)p_k = \sum_{k=1}^{\infty}kp_k - \sum_{k=1}^{\infty}p_k = \rho + \frac{\lambda^2\mathbf{E}(Y^2)}{2(1-\rho)} - \rho = \frac{\lambda^2\mathbf{E}(Y^2)}{2(1-\rho)}$$

$$= \frac{\lambda^2}{\mu(\mu-\lambda)} = \frac{\rho^2}{1-\rho}.$$

These values can be computed knowing that for the M/M/1 system the stationary distribution is geometrical. The mean value of number of customers is

$$\sum_{k=1}^{\infty}k\cdot(1-\rho)\rho^k = (1-\rho)\rho\frac{1}{(1-\rho)^2} = \frac{\rho}{1-\rho}.$$

The mean value of waiting customers

$$\sum_{k=1}^{\infty}(k-1)(1-\rho)\rho^k = (1-\rho)\rho\left(\sum_{k=1}^{\infty}k\rho^{k-1} - \sum_{k=1}^{\infty}\rho^{k-1}\right)$$

$$= (1-\rho)\rho\left[\frac{1}{(1-\rho)^2} - \frac{1}{1-\rho}\right] = \frac{\rho^2}{1-\rho}.$$

Let the service time be equal to T in the M/D/1 system, then $\rho = \lambda T$, $\mathbf{E}(Y^2) = T^2$. The mean value of number of customers in the system is

$$\rho + \frac{\lambda^2\mathbf{E}(Y^2)}{2(1-\rho)} = \lambda T + \frac{\lambda^2 T^2}{2(1-\lambda T)} = \frac{2\lambda T - \lambda^2 T^2}{2(1-\lambda T)} = \frac{\rho(2-\rho)}{2(1-\rho)},$$

the mean value of waiting customers

$$\frac{\lambda^2\mathbf{E}(Y^2)}{2(1-\rho)} = \frac{\lambda^2 T^2}{2(1-\lambda T)} = \frac{\rho^2}{2(1-\rho)}.$$

Solution 8.5 Use the fact that for the distribution of service time

$$b^{\sim}(s) = \int_0^{\infty} e^{-sx}\mu e^{-\mu x}\mathrm{d}x = \frac{\mu}{s+\mu}.$$

Solution 8.6 Let $P_i(k)$ denote the probability of event that in i groups together appear k customers. For the generating function of entering customers we have

$$
e^{-\lambda t} + \sum_{i=1}^{\infty} \sum_{k=i}^{\infty} \frac{(\lambda t)^i}{i!} e^{-\lambda t} P_i(k) z^k = \sum_{i=0}^{\infty} \frac{(\lambda t)^i}{i!} e^{-\lambda t} \sum_{k=i}^{\infty} P_i(k) z^k
$$

$$
= \sum_{i=0}^{\infty} \frac{(\lambda t)^i}{i!} e^{-\lambda t} G^i(z) = \sum_{i=0}^{\infty} \frac{[\lambda t G(z)]^i}{i!} e^{-\lambda t} = e^{-\lambda t[1 - G(z)]}.
$$

Solution 8.7 Let $P_i(k)$ has the same meaning as in the previous exercise. We have

$$
\int_0^{\infty} e^{-\lambda x} \, dB(x) + \int_0^{\infty} \sum_{i=1}^{\infty} \sum_{k=i}^{\infty} \frac{(\lambda x)^i}{i!} e^{-\lambda x} P_i(k) z^k \, dB(x)
$$

$$
= \int_0^{\infty} \sum_{i=0}^{\infty} \frac{(\lambda x)^i}{i!} e^{-\lambda x} \sum_{k=i}^{\infty} P_i(k) z^k \, dB(x)
$$

$$
= \sum_{i=0}^{\infty} \frac{(\lambda x)^i}{i!} e^{-\lambda x} G^i(z) dB(x) = \int_0^{\infty} \sum_{i=0}^{\infty} \frac{[\lambda x G(z)]^i}{i!} e^{-\lambda x} \, dB(x)
$$

$$
= b^{\sim}(\lambda(1 - G(z))) = \sum_{j=0}^{\infty} c_j z^j.
$$

References

1. Borovkov, A.A.: Stochastic Processes in Queueing Theory. Applications of Mathematics. Springer, New York (1976)
2. Borovkov, A.A.: Asymptotic Methods in Queueing Theory. Wiley, New York (1984)
3. Ceric, V., Lakatos, L.: Measurement and analysis of input data for queueing system models used in system design. Syst. Anal. Model. Simul. **11**, 227–233 (1993)
4. Cox, D.R.: The analysis of non-Markovian stochastic processes by the inclusion of supplementary variables. Proc. Camb. Philos. Soc. **51**, 433–440 (1955)
5. Gnedenko, B.V., Kovalenko, I.N.: Introduction to Queueing Theory, 2nd edn. Birkhauser Boston Inc., Cambridge (1989)
6. Henderson, W.: Alternative approaches to the analysis of the M/G/1 and G/M/1 queues. J. Operat. Res. Soc. Jpn. **15**, 92–101 (1972)
7. Kendall, D.G.: Stochastic processes occurring in the theory of queues and their analysis by the method of the imbedded Markov chain. Ann. Math. Stat. **24**, 338–354 (1953)
8. Khinchin, A.: Mathematisches über die Erwartung vor einem öffentlichen Schalter. Rec. Math. **39**, 72–84 (1932). Russian, with German summary
9. Kleinrock, L.: Queuing Systems, Volume 1: Theory. Wiley Interscience, New York (1975)
10. Lakatos, L.: A note on the Pollaczek-Khinchin formula. Ann. Univ. Sci. Budapest. Sect. Comput. **29**, 83–91 (2008)
11. Lindley, D.V.: The theory of queues with a single server. Math. Proc. Camb. Philos. Soc. **48**, 277–289 (1952)

12. Medgyessy, P., Takács, L.: Probability Theory. Tankönyvkiadó, Budapest (1973). In Hungarian
13. Palm, C.: Methods of judging the annoyance caused by congestion. Telegrafstyrelsen **4**, 189–208 (1953)
14. Takács, L.: Investigation of waiting time problems by reduction to Markov processes. Acta Math. Acad. Sci. Hung. **6**, 101–129 (1955)
15. Takács, L.: The distribution of the virtual waiting time for a single-server queue with Poisson input and general service times. Oper. Res. **11**, 261–264 (1963)
16. Takács, L.: Combinatorial Methods in the Theory of Stochastic Processes. Wiley, New York (1967)
17. Tijms, H.: Stochastic Models, An Algorithmic Approach. Wiley, New York (1994)

Chapter 9
Queueing Systems with Structured Markov Chains

In the previous chapters we have studied queueing systems with different interarrival and service time distributions. Chapter 7 is devoted to the analysis of queueing systems with exponential interarrival and service time distributions. The number of customers in these queueing systems are characterized by CTMCs with a generally nonhomogeneous birth and death structure. In contrast, Chap. 8 is devoted to the analysis of queueing systems with nonexponential interarrival and service time distributions. It turned out that far more complex analysis approaches are required for the analysis of queues with nonexponential interarrival and service time distributions. In this chapter we introduce queueing systems whose interarrival and service time distributions are nonexponential, but they can be analyzed with CTMCs. Indeed in this chapter we demonstrate the use of the results of Chap. 5 for the analysis of queueing systems with PH distributed interarrival and service times and/or with arrival and service processes which are MAPs. The main message of this chapter is that in queueing models the occurrence of PH, MAP, or BMAP processes instead of exponential distributions results in a generalization of the underlying CTMCs from birth and death processes to QBDs or from the embedded Markov chain of a G/M/1 or an M/G/1 queue to G/M/1 type or M/G/1 type Markov chain [1, 3, 5].

9.1 PH/M/1 Queue

One of the simplest queueing systems with nonexponentially distributed interarrival time distribution is the PH/M/1 queue. We study this queue in detail in order to demonstrate the elementary steps to construct the matrix block structure of the CTMC describing the behavior of the queue.

© Springer Nature Switzerland AG 2019
L. Lakatos et al., *Introduction to Queueing Systems with Telecommunication Applications*, https://doi.org/10.1007/978-3-030-15142-3_9

We consider the queueing system whose arrival process is composed by independent and identically PH distributed interarrival periods which are characterized by initial probability vector τ and transient generator matrix T. Consequently, the arrival process is a PH renewal process with representation (τ, T). The service time is exponentially distributed with parameter μ. The queue has one server and an unlimited buffer. The service discipline is FIFO.

9.1.1   The QBD of the PH/M/1 Queue

This queueing system can be analyzed as a G/M/1 queue using the results of the previous chapter, because the PH distributed interarrival time is a special case of general, nonexponential interarrival time distribution. But using the fact that the PH distribution is characterized by a background Markov chain we can also analyze the PH/M/1 queue as a compound CTMC $\{N(t), J(t)\}$, where $N(t)$ is the number of customers in the queue and $J(t)$ is the state (phase) of the Markov chain characterizing the PH distributed arrivals at time t.

If at time t the state of the compound CTMC is $(N(t), J(t)) = (n, j)$, then the following state transitions are possible.

- There might be a phase transition from phase j to k ($k \neq j$) in the background Markov chain of the PH distribution without an arrival. The rate of this transition from (n, j) to (n, k) is T_{jk}.
- The Markov chain of the PH distribution might move to the absorbing state and generate an arrival. In this case the number of customers in the system increases by one and a new PH distributed interval starts according to the initial phase distribution τ. Let t be the column vector containing the transition rates to the absorbing state, $t = -T\mathbb{1}$. The transition rate from (n, j) to $(n + 1, k)$ due to these steps is $t_j \tau_k$.
- If $n > 0$, then there is a customer in the server, which is served with an exponentially distributed service time with parameter μ. When the service completes, the number of customers in the system decreases by one and, due to the independence of the arrival and the service processes, the service completion does not affect the phase arrival process. This way the transition rate from (n, j) to $(n - 1, j)$ is μ and from (n, j) to $(n - 1, k)$ ($k \neq j$) is 0.

These possible transitions define all nondiagonal elements of the generator matrix of the CTMC. In case of a PH arrival process with 2 phases the generator matrix has the form

$$Q = \begin{bmatrix} \bullet & T_{12} & t_1\tau_1 & t_1\tau_2 & 0 & 0 & 0 & 0 \\ T_{21} & \bullet & t_2\tau_1 & t_2\tau_2 & 0 & 0 & 0 & 0 \\ \mu & 0 & \bullet & T_{12} & t_1\tau_1 & t_1\tau_2 & 0 & 0 \\ 0 & \mu & T_{21} & \bullet & t_2\tau_1 & t_2\tau_2 & 0 & 0 \\ 0 & 0 & \mu & 0 & \bullet & T_{12} & t_1\tau_1 & t_1\tau_2 \\ 0 & 0 & 0 & \mu & T_{21} & \bullet & t_2\tau_1 & t_2\tau_2 \\ & & & & \ddots & \ddots & \ddots & \ddots \\ & & & & \ddots & \ddots & \ddots & \ddots \end{bmatrix} \cdot$$

The diagonal elements are determined by the nondiagonal elements due to the fact that the row sum of the generator matrix is zero. This matrix already indicates that the generator matrix has a regular structure on the matrix block level highlighted by the horizontal and vertical lines. The matrix blocks are closely related to the vectors and the matrix characterizing the PH arrival process

$$Q = \begin{bmatrix} T & t\tau & & & \\ \mu I & T-\mu I & t\tau & & \\ & \mu I & T-\mu I & t\tau & \\ & & \mu I & T-\mu I & t\tau \\ & & & \ddots & \ddots \end{bmatrix}.$$

The nondiagonal elements of this generator matrix defined by matrix blocks are readily identical with the ones of detailed generator matrix. For the validity of the diagonal element we evaluate the row sum of the elements in a row of matrix blocks. In case of $n = 0$ it is $T\mathbb{1} + t\tau\mathbb{1} = T\mathbb{1} + t = \mathbf{0}$, because $\tau\mathbb{1} = 1$ and $t = -T\mathbb{1}$. In case of $n > 0$ it is $\mu I\mathbb{1} + (T - \mu I)\mathbb{1} + t\tau\mathbb{1} = T\mathbb{1} + t = \mathbf{0}$.

The block level structure of the generator matrix shows that $\{N(t), J(t)\}$ is a QBD process with regular level 0. The forward, local, backward, and level 0 local matrices of this QBD process are $F = t\tau$, $L = T - \mu I$, $B = \mu I$ and $L' = T$.

9.1.2   Condition of Stability

From the G/M/1 interpretation of the PH/M/1 queue we already know that the queue is stable as long as the mean interarrival time is greater than the mean service time, that is $\tau(-T)^{-1}\mathbb{1} > 1/\mu$. Now we analyze the relation of this condition with the stability condition of the QBD process. The phase process of the regular levels is a CTMC with generator $B + L + F = T + t\tau$. Let α be the stationary distribution of the phase process (i.e., the solution of $\alpha(F + L + B) = \mathbf{0}, \alpha\mathbb{1} = 1$).

Theorem 9.1

$$\alpha = \frac{\tau(-T)^{-1}}{\tau(-T)^{-1}\mathbb{1}}$$

Proof The normalizing condition obviously holds

$$\frac{\tau(-T)^{-1}}{\tau(-T)^{-1}\mathbb{1}}\mathbb{1} = 1.$$

For the product of the stationary solution vector and the generator of the phase process we have

$$\tau(-T)^{-1}(B + L + F) = \tau(-T)^{-1}(T - T\mathbb{1}\tau) = -\tau + \tau = 0$$

where we neglect the normalizing constant $1/\tau(-T)^{-1}\mathbb{1}$.   □

Based on the stationary distribution of the phase process the condition of stability of the QBD is $\alpha B\mathbb{1} > \alpha F\mathbb{1}$ where $\alpha B\mathbb{1} = \alpha\mu I\mathbb{1} = \mu$ and

$$\alpha F\mathbb{1} = \frac{\tau(-T)^{-1}}{\tau(-T)^{-1}\mathbb{1}}(-T\mathbb{1}\tau)\mathbb{1} = \frac{1}{\tau(-T)^{-1}\mathbb{1}}.$$

9.1.3   *Performance Measures*

The main performance measures of PH/M/1 queues are based on the stationary distribution of the $\{N(t), J(t)\}$ QBD process. According to Theorem 5.4 the row vector of the stationary probabilities with n customers can be computed as $\pi_n = \pi_0 R^n$ where matrix R is the solution (the only one whose eigenvalues are inside the unit disk) of

$$F + RL + R^2 B = 0$$

and vector π_0 is the solution of linear system

$$\pi_0(L' + RB) = 0, \quad \pi_0(I - R)^{-1}\mathbb{1} = 1.$$

Below we compute the main performance measures assuming that the matrix geometric stationary distribution is known.

9.1.3.1 Utilization

The only server of the queueing system is busy when the number of customers in the system is at least 1. Consequently, the utilization is

$$\rho = \lim_{t \to \infty} \mathbf{P}(N(t) \geq 1) = \sum_{n=1}^{\infty} \boldsymbol{\pi}_n \mathbb{1} = 1 - \boldsymbol{\pi}_0 \mathbb{1}.$$

9.1.3.2 Number of Customers

The distribution of the stationary number of customers in the queue is

$$p_n = \lim_{t \to \infty} \mathbf{P}(N(t) = n) = \boldsymbol{\pi}_n \mathbb{1} = \boldsymbol{\pi}_0 R^n \mathbb{1}.$$

The mean number of customers can be computed as

$$\mathbf{E}(N) = \lim_{t \to \infty} \mathbf{E}(N(t)) = \sum_{n=0}^{\infty} n p_n = \sum_{n=0}^{\infty} n \boldsymbol{\pi}_0 R^n \mathbb{1} = \sum_{n=0}^{\infty} \sum_{k=0}^{n} \boldsymbol{\pi}_0 R^n \mathbb{1}$$

$$= \boldsymbol{\pi}_0 \sum_{k=0}^{\infty} \sum_{n=k}^{\infty} R^n \mathbb{1} = \boldsymbol{\pi}_0 \sum_{k=0}^{\infty} R^k (I - R)^{-1} \mathbb{1} = \boldsymbol{\pi}_0 (I - R)^{-2} \mathbb{1}.$$

The distribution of the stationary number of customers right before a customer arrival is defined as $q_n = \lim_{k \to \infty} \mathbf{P}(N(T_k-) = n)$, where T_k denotes the arrival instant of the kth customer and $N(T_k-)$ refers to the left limit at T_k. For q_n, we have

$$q_n = \frac{\text{stationary arrival rate from level } n}{\text{stationary customer arrival rate}} = \frac{\boldsymbol{\pi}_n t}{\sum_{i=0}^{\infty} \boldsymbol{\pi}_i t} = \frac{\boldsymbol{\pi}_0 R^n t}{\boldsymbol{\pi}_0 (I - R)^{-1} t},$$

and similarly

$$\mathbf{E}(N_A) = \lim_{k \to \infty} \mathbf{E}(N(T_k-)) = \sum_{n=0}^{\infty} n q_n = \frac{\boldsymbol{\pi}_0 (I - R)^{-2} t}{\boldsymbol{\pi}_0 (I - R)^{-1} t}.$$

It is worth mentioning that the arrival process is not a Poisson process (in general) and the distribution of the stationary number of customers and the stationary number of customers at arrivals differ.

9.1.3.3 System Time

If a customer arrives to the queue when there are n customers in front of it, then its waiting time is the sum of the remaining service time of the customer in the server, if any (which is exponentially distributed with parameter μ), and the total service time of the customers waiting in front of the newly arrived one, if any (which is also exponentially distributed with parameter μ). The system time (T, also referred to as response time) is the sum of the waiting time (W) and the service time (S). All together, if a customer arrives when there are n other customers in the queue, then its waiting time is the sum of n independent exponentially distributed random variable with parameter μ and its system time is the sum of $n + 1$ independent exponentially distributed random variable with parameter μ. We describe the Laplace transform of the system time, because the sum of independent random variables has a simple product form in Laplace domain. The Laplace transform of the exponentially distributed service time with parameter μ is $\mathbf{E}(e^{sS}) = \frac{\mu}{\mu+s}$ and

$$
f_T^*(s) = \mathbf{E}\left(e^{-sT}\right) = \sum_{n=0}^{\infty} q_n \left(\frac{\mu}{\mu+s}\right)^{n+1} = \sum_{n=0}^{\infty} \frac{\pi_0 R^n t}{\pi_0 (I - R)^{-1} t} \left(\frac{\mu}{\mu+s}\right)^n \left(\frac{\mu}{\mu+s}\right)
$$

$$
= \left(\frac{\mu}{\mu+s}\right) \pi_0 \frac{1}{\pi_0(I-R)^{-1}t} \left(I - \frac{\mu}{\mu+s} R\right)^{-1} t
$$

$$
= \frac{\mu}{\pi_0(I-R)^{-1}t} \pi_0 \left(sI + \mu(I-R)\right)^{-1} t
$$

Using the relation of the Laplace transform and the moments, the mean system time can be computed as

$$
\mathbf{E}(T) = -\frac{d}{ds} f_T^*(s)|_{s=0} = -\frac{\mu}{\pi_0(I-R)^{-1}t} \pi_0 \frac{d}{ds} \left(sI + \mu(I-R)\right)^{-1}\Big|_{s=0} t
$$

$$
= \frac{\mu}{\pi_0(I-R)^{-1}t} \pi_0 \left(sI + \mu(I-R)\right)^{-2}\Big|_{s=0} t
$$

$$
= \frac{1}{\mu \pi_0(I-R)^{-1}t} \pi_0 (I-R)^{-2} t.
$$

Further performance measures, like the parameters of the waiting time, can be computed in a similar manner.

9.2 M/PH/1 Queue

In this section we analyze the other simple queueing system with underlying QBD. It is the M/PH/1 queue, where the arrival process is a Poisson process with rate λ, the service time is PH distributed with representation τ, T of size J, there is a single server and an infinite buffer. Similarly, column vector t contains the transition rates to the absorbing state ($t = -T\mathbb{1}$). The most important difference between the PH/M/1 queue and the M/PH/1 queue is the structure of the underlying QBD. It was a QBD with regular level zero in the previous section and it will be a QBD with irregular level zero in this section. Another useful feature of the M/PH/1 queue, which is unique among the queueing systems with underlying QBD, is that matrix R can be expressed in closed from.

9.2.1 The QBD of the M/PH/1 Queue

Similar to the PH/M/1 queue the behavior of the M/PH/1 queue is characterized by a compound CTMC $\{N(t), J(t)\}$, where $N(t)$ is the number of customers in the queue and $J(t)$ is the state (phase) of the Markov chain characterizing the PH distributed service time at t. One of the main differences between the PH/M/1 and the M/PH/1 queue comes from the fact that the service process is inactive (does not exist) when there is no customer in the queue. Consequently, level 0 of the underlying QBD has a different structure than the higher levels. The QBD has a single phase at level 0 and it has J phases at higher levels. Accordingly, the structure of the transitions from and to level 0 is different from the regular ones. If at time t the state of the QBD is $(N(t), J(t)) = (n, j)$, then the following state transitions are possible.

- If $n \geq 1$, then there might be a phase transition from (n, j) to (n, k) ($k \neq j$) in the background Markov chain of the PH distribution without a departure with rate T_{jk}.
- If $n \geq 2$, then the Markov chain of the PH distribution might move to the absorbing state with rate t_j, which represents the service completion of the customer in the server and the departure of this customer. In this case the number of customers in the system decreases by one and a new PH distributed service time starts according to the initial phase distribution τ. The transition rate from (n, j) to $(n - 1, k)$ is $t_j \tau_k$.
- If $n = 1$ and the PH distribution moves to the absorbing state with rate t_j, then the only customer leaves the queue. In this case the queue becomes idle and the service process becomes inactive. As a result there might be a transition from $(1, j)$ to $(0, 1)$ with rate t_j.
- If $n \geq 1$, then there is one customer in the server and the service process is active. In this case an arrival with rate λ increases the number of customers in the queue and maintains the phase of the service process. This way the transition rate from (n, j) to $(n + 1, j)$ is λ and from (n, j) to $(n + 1, k)$ ($k \neq j$) is 0.
- If $n = 0$, then the arrival of a new customer with rate λ initiates the service of the newly arrived customer according to the initial phase distribution τ. In this case the transition rate from $(0, 1)$ to $(1, k)$ is $\lambda \tau_k$.

These transitions define all nondiagonal elements of the generator matrix. In case of $J = 2$ we have

$$Q = \begin{array}{|ccc|cc|cc|cc|}
\bullet & \lambda\tau_1 & \lambda\tau_2 & 0 & 0 & 0 & 0 & 0 & 0 \\
t_1 & \bullet & T_{12} & \lambda & 0 & 0 & 0 & 0 & 0 \\
t_2 & T_{21} & \bullet & 0 & \lambda & 0 & 0 & 0 & 0 \\
0 & t_1\tau_1 & t_1\tau_2 & \bullet & T_{12} & \lambda & 0 & 0 & 0 \\
0 & t_2\tau_1 & t_2\tau_2 & T_{21} & \bullet & 0 & \lambda & 0 & 0 \\
0 & 0 & 0 & t_1\tau_1 & t_1\tau_2 & \bullet & T_{12} & \lambda & 0 \\
0 & 0 & 0 & t_2\tau_1 & t_2\tau_2 & T_{21} & \bullet & 0 & \lambda \\
 & & & & \ddots & \ddots & \ddots & \ddots & \\
 & & & & \ddots & \ddots & \ddots & \ddots & \\
\end{array} \ ,$$

and on the level of matrix blocks the generator matrix of the QBD is

$$Q = \begin{array}{|c|c|c|c|c|}
-\lambda & \lambda\tau & & & \\
t & T-\lambda I & \lambda I & & \\
 & t\tau & T-\lambda I & \lambda I & \\
 & & t\tau & T-\lambda I & \lambda I \\
 & & & & \ddots \quad \ddots \\
\end{array} \ .$$

That is, $F = \lambda I$, $L = T - \lambda I$, $B = t\tau$ and the special matrix blocks at the zero level are $F' = \lambda\tau$, $L' = -\lambda$, $B' = t$. Using $t = -T\mathbb{1}$ it is easy to check that the row sum of each row is zero.

The condition of the stability of this QBD can be computed in a very similar way as in the case of PH/M/1 queue. The QBD is stable if $\lambda < \frac{1}{\tau(-T)^{-1}\mathbb{1}}$.

9.2.2   Closed Form Solution of the Stationary Distribution

Let $\pi = \{\pi_0, \pi_1, \pi_2, \ldots\}$ be the partitioned stationary probability vector of the QBD. The partitioned form of the set of stationary equations $\pi Q = 0$, is

$$- \pi_0 \lambda + \pi_1 t = 0, \tag{9.1}$$

$$\pi_0 \lambda\tau + \pi_1 (T-\lambda I) + \pi_2 t\tau = 0, \tag{9.2}$$

$$\pi_{n-1}\lambda I + \pi_n (T-\lambda I) + \pi_{n+1} t\tau = 0 \quad \forall n \geq 2. \tag{9.3}$$

The solution of this set of equations can be expressed in a closed matrix geometric form.

Theorem 9.2 *For $n \geq 1$*

$$\pi_n = \pi_0 \tau R^n,$$

where $\pi_0 = 1 - \lambda \tau (-T)^{-1} \mathbb{1}$ and $R = \lambda(\lambda I - T - \lambda \mathbb{1}\tau)^{-1}$.

Proof Substituting (9.1) into (9.2) gives:

$$\pi_1(t\tau + T - \lambda I) + \pi_2 t\tau = 0 .$$

Multiplying this expression with $\mathbb{1}$ from the right we obtain $\pi_1 \lambda \mathbb{1} = \pi_2 t$. Now we take (9.3) with $n = 2$, multiply it with $\mathbb{1}$ from the right and substitute $\pi_1 \lambda \mathbb{1} = \pi_2 t$. It results in $\pi_2 \lambda \mathbb{1} = \pi_3 t$. Recursively multiplying (9.3) with $\mathbb{1}$ and substituting the previous result we obtain:

$$\lambda \pi_n \mathbb{1} = \pi_{n+1} t \quad \forall n \geq 1.$$

Substituting this expression into the third term of (9.3) gives:

$$\lambda \pi_{n-1} + \pi_n (T - \lambda I) + \lambda \pi_n \mathbb{1}\tau = 0 \quad \forall n \geq 2,$$

from which

$$\pi_n = \pi_{n-1} \underbrace{\lambda(\lambda I - T - \lambda \mathbb{1}\tau)^{-1}}_{R} \quad \forall n \geq 2.$$

Additionally from (9.2) we have $\pi_1 = \pi_0 \tau R$. π_0, the probability that the server is idle can be obtained from the Little's law when it is applied for the server itself. It says that $\mathbf{E}(N_S) = \lambda \mathbf{E}(S)$ (the mean number of customers in the server equals with the arrival rate times the mean service time). In our case $\mathbf{E}(S) = \tau(-T)^{-1}\mathbb{1}$. In a single server queue $\mathbf{E}(N_S)$ is the probability that the server is busy, i.e., $\mathbf{E}(N_S) = 1 - \pi_0$, indeed, it is the utilization in this case.   □

9.2.3 Performance Measures

The computation of the main performance measures follow the same pattern as the ones of the PH/M/1 queue, but in case of M/PH/1 queues we can utilize the closed form of the stationary distribution.

9.2.3.1 Number of Customers

The distribution of the stationary number of customers in the queue is

$$p_n = \lim_{t \to \infty} \mathbf{P}(N(t) = n) = \pi_n \mathbb{1} = \pi_0 \tau R^n \mathbb{1} \quad n \geq 1,$$

and $p_0 = \pi_0 = 1 - \lambda \tau (-T)^{-1} \mathbb{1}$. The mean number of customers can be computed in a similar way as in case of the PH/M/1 queue

$$\mathbf{E}(N) = \lim_{t \to \infty} \mathbf{E}(N(t)) = \sum_{n=0}^{\infty} n p_n = \sum_{n=1}^{\infty} n \pi_0 \tau R^n \mathbb{1} = \pi_0 \tau (I - R)^{-2} \mathbb{1}.$$

The distribution of the stationary number of customers right before a customer arrival is

$$q_n = \frac{\text{stationary arrival rate from level } n}{\text{stationary customer arrival rate}} = \frac{\pi_n \mathbb{1} \lambda}{\sum_{i=0}^{\infty} \pi_i \mathbb{1} \lambda} = \pi_n \mathbb{1} = p_n.$$

The Poisson arrival process ensures that the distribution of the stationary number of customers and the stationary number of customers at arrivals are identical (which is referred to as PASTA property).

9.2.3.2 System Time

The Laplace transform of the PH distributed service time is $\mathbf{E}(e^{sS}) = \tau (sI - T)^{-1} t$. Using this the Laplace transform of the system time is

$$f_T^*(s) = \mathbf{E}(e^{sT}) = \sum_{n=0}^{\infty} q_n \left(\tau (sI - T)^{-1} t \right)^n = \sum_{n=0}^{\infty} \pi_0 \tau R^n \mathbb{1} \left(\tau (sI - T)^{-1} t \right)^n =$$

$$= \pi_0 \tau \left(I - \tau (sI - T)^{-1} t R \right)^{-1} \mathbb{1}.$$

The mean system time can be computed from the Laplace transform as

$$\mathbf{E}(T) = -\frac{d}{ds} f_T^*(s)|_{s=0} = \pi_0 \tau \left. \frac{d}{ds} \left(I - \tau (sI - T)^{-1} t R \right)^{-1} \right|_{s=0} \mathbb{1}$$

$$= \pi_0 \tau \left. \left(I - \tau (sI - T)^{-1} t R \right)^{-2} \tau (sI - T)^{-2} t R \right|_{s=0} \mathbb{1}$$

$$= \pi_0 \tau (I - R)^{-2} \tau (-T)^{-1} R \mathbb{1},$$

where we utilized $t = -T \mathbb{1}$ in the last step.

9.3 Further Queues with Underlying QBD

9.3.1 MAP/M/1 Queue

The difference between the PH/M/1 queue and the MAP/M/1 queue is minor. We focus our attention mainly to the extension from PH/M/1 queue to MAP/M/1 queue. Let the arrival process be MAP with representation D_0, D_1, and the service time be exponentially distributed with parameter μ. The MAP/M/1 queue has a single server and an infinite buffer. These possible transitions of the $(N(t), J(t)))$ CTMC in case of a MAP arrival process with 2 phases are as follows

$$
Q = \begin{bmatrix}
\bullet & D_{012} & D_{111} & D_{112} & 0 & 0 & 0 & 0 \\
D_{021} & \bullet & D_{121} & D_{122} & 0 & 0 & 0 & 0 \\
\mu & 0 & \bullet & D_{012} & D_{111} & D_{112} & 0 & 0 \\
0 & \mu & D_{021} & \bullet & D_{121} & D_{122} & 0 & 0 \\
0 & 0 & \mu & 0 & \bullet & D_{012} & D_{111} & D_{112} \\
0 & 0 & 0 & \mu & D_{021} & \bullet & D_{121} & D_{122} \\
& & & & \ddots & \ddots & \ddots & \ddots \\
& & & & \ddots & \ddots & \ddots & \ddots
\end{bmatrix},
$$

from which the block level structure is

$$
Q = \begin{bmatrix}
D_0 & D_1 & & & \\
\mu I & D_0 - \mu I & D_1 & & \\
& \mu I & D_0 - \mu I & D_1 & \\
& & \mu I & D_0 - \mu I & D_1 \\
& & & \ddots & \ddots
\end{bmatrix},
$$

where the row sum is zero due to $(D_0 + D_1)\mathbb{1} = 0$. Comparing the QBD process of the PH/M/1 queue and the MAP/M/1 queues we have that a PH/M/1 queue is a special MAP/M/1 queue with $D_0 = T$ and $D_1 = t\tau$.

9.3.2 M/MAP/1 Queue

The consecutive inter event times of a MAP are correlated (in general). That is, the consecutive service times of the M/MAP/1 queue are correlated (in general) and it is independent of the fact if a departure left the queue idle or not. Due to this property the phase of the service process is carried on also when the queue is idle. Consequently, the zero level of the QBD contains the same number of phases as the higher level. This feature of the M/MAP/1 queue is similar to the one of the MAP/M/1 queue, but significantly differs from the one of the M/PH/1 queue.

If the arrival process is a Poisson process with rate λ and the service process is MAP with representation S_0, S_1, then the block level structure of the QBD is

$$
Q =
\begin{array}{|c|c|c|c|c|c|}
\hline
-\lambda I & \lambda I & & & & \\
\hline
S_1 & S_0 - \lambda I & \lambda I & & & \\
\hline
& S_1 & S_0 - \lambda I & \lambda I & & \\
\hline
& & S_1 & S_0 - \lambda I & \lambda I & \\
\hline
& & & & \ddots & \ddots \\
\hline
\end{array}
.
$$

The zero level of this matrix indicates that the service process is "switched off" at the zero level, but the phase of the service MAP is maintained while the queue is idle and the service MAP resumes its evolution from the same phase when a customer arrives at the system.

In case of PH/M/1 queue with (τ, T) and MAP/M/1 queue with $D_0 = T$ and $D_1 = t\tau$ the QBD of the PH/M/1 and the MAP/M/1 queues are identical, because both of them contain J phases at the zero level. The sizes of the zero levels of M/PH/1 queue and M/MAP/1 queues differ.

Fortunately, the representation of the M/PH/1 queue with (τ, T) (zero level with one phase) as a special M/MAP/1 queue with $S_0 = T$ and $S_1 = t\tau$ (zero level with J phases) remains valid with respect to all queue related parameters computed from the two different QBDs. The behavior of the M/MAP/1 queue with $S_0 = T$ and $S_1 = t\tau$ (zero level with J phases) can be interpreted as the customer who leaves the system idle decides the initial phase of the next service time (independent of the fact that the queue becomes idle), and this phase is preserved by the QBD during the idle time of the queue. In summary, we emphasize that PH arrival and service processes can always be represented as special MAPs.

9.3.3 MAP/PH/1 Queue

If both the arrival and the service processes are characterized by a background Markov chain, then the $(N(t), J(t))$ QBD can still be used for the analysis of the queueing system, but the phase process $J(t)$ has to represent the phase of both background Markov chains. That is, the phase process of the QBD is the Cartesian product of the phase processes of the arrival and the service processes. The Markov chain describing the independent evolution of the arrival and the service processes can be expressed by Kronecker operators. If the arrival process is a MAP with representation D_0, D_1, and the service time is PH distributed with representation τ, T ($t = -T\mathbb{1}$), then the structure of the generator matrix is

$$Q = \begin{array}{|c|c|c|c|c|} \hline D_0 & D_1 \otimes \tau & & & \\ \hline I \otimes t & D_0 \oplus T & D_1 \otimes I & & \\ \hline & I \otimes t\tau & D_0 \oplus T & D_1 \otimes I & \\ \hline & & I \otimes t\tau & D_0 \oplus T & D_1 \otimes I \\ \hline & & & \ddots & \ddots \\ \hline \end{array}.$$

That is, $F = D_1 \otimes I$, $L = D_0 \otimes I + I \otimes T = D_0 \oplus T$, $B = I \otimes t\tau$ and $F' = D_1 \otimes \tau$, $L' = D_0$, $B' = I \otimes t$.

9.3.4   MAP/MAP/1 Queue

Similarly, if the arrival process is a MAP with representation D_0, D_1, and the service process is a MAP with representation S_0, S_1, then the structure of the generator matrix is

$$Q = \begin{array}{|c|c|c|c|} \hline D_0 \oplus I & D_1 \otimes I & & \\ \hline I \otimes S_1 & D_0 \oplus S_0 & D_1 \otimes I & \\ \hline & I \otimes S_1 & D_0 \oplus S_0 & D_1 \otimes I \\ \hline & & \ddots & \ddots \\ \hline \end{array}.$$

That is, $F = D_1 \otimes I$, $L = D_0 \otimes I + I \otimes S_0 = D_0 \oplus S_0$, $B = I \otimes S_1$ and $L' = D_0 \oplus I$.

9.3.5   MAP/PH/1/K Queue

Finally, we demonstrate that the analysis of finite QBD processes can be used for the analysis of finite buffer queues. For example, if the arrival process is a MAP with representation D_0, D_1, the service time is PH distributed with representation τ, T and at most K customers can be present in the queue, then the structure of the QBD process describing the queue behavior is

$$Q = \begin{array}{|c|c|c|c|c|} \hline L' & F' & & & \\ \hline B' & L & \ddots & & \\ \hline & B & \ddots & F & \\ \hline & & \ddots & L & F \\ \hline & & & B & L'' \\ \hline \end{array},$$

where $F = D_1 \otimes I$, $L = D_0 \oplus T$, $B = I \otimes t\tau$, $F' = D_1 \otimes \tau$, $L' = D_0$, $B' = I \otimes T$ and $L'' = (D_0 + D_1) \oplus T$.

9.4   Queues with Underlying G/M/1 and M/G/1 Type Processes

This section discusses some queueing systems whose analyses are based on G/M/1 or M/G/1 type processes.

9.4.1   BMAP/MAP/1 Queue

In a BMAP/MAP/1 queue, customers arrive according to a BMAP $(D_0, D_1, D_2, \ldots)$ process and are served according to a MAP (S_0, S_1) process. There is an infinite buffer and a single server, which serves the customers in FIFO order.

Due to the possibility of batch arrivals the number of customers in the queue can increase by more than one at a time and the $\{N(t), J(t)\}$ CTMC characterizing the process behavior is M/G/1 type, where $N(t)$ is the number of customers in the queue and $J(t)$ is the phase of the Markov environment characterizing the arrival and service processes. More precisely, $J(t)$ is a vector variable $\{J_A(t), J_S(t)\}$, where $J_A(t)$ is the phase of the BMAP characterizing the arrivals and $J_S(t)$ is the phase of the MAP characterizing the service. The block structure of its generator matrix is

$$Q = \begin{array}{|c|c|c|c|c|} \hline L' & F_1 & F_2 & F_3 & \cdots \\ \hline B & L & F_1 & F_2 & \\ \hline & B & L & F_1 & \ddots \\ \hline & & B & L & \ddots \\ \hline & & & \ddots & \ddots \\ \hline \end{array},$$

where the matrix blocks are $L = D_0 \oplus S_0$, $L' = D_0 \otimes I$, $B = I \otimes S_1$, and $F_n = D_n \otimes I$ for $n \geq 1$.

9.4.1.1   Condition of Stability

We can define the condition of stability based on the queue load and based on the stability of the M/G/1 type process with generator Q. Focusing on the queue load the mean arrival rate, λ, given in (5.8), has to be less than the mean service rate of the MAP(S_0, S_1) service process, μ defined in (5.5).

For the stability of the M/G/1 type process with generator Q condition (5.47) needs to hold, which can be written as

$$d = \gamma \sum_{n=1}^{\infty} n \, F_n \mathbb{1} - \gamma B \, \mathbb{1} < 0,$$

where γ is the solution of $\gamma (B + L + \sum_{n=1}^{\infty} F_n) = 0$, with normalizing condition $\gamma \mathbb{1} = 1$. Substituting the structure of the B, L, F_n matrices into the linear equation we have

$$0 = \gamma \left(I \otimes S_1 + I \otimes S_0 + D_0 \otimes I + \sum_{n=1}^{\infty} D_n \otimes I \right) = \gamma \left(\sum_{n=0}^{\infty} D_n \oplus (S_0 + S_1) \right),$$

whose solution is $\gamma = \alpha \otimes \sigma$, where α is the solution of $\alpha \sum_{n=0}^{\infty} D_n = 0$, $\alpha \mathbb{1} = 1$, and σ is the solution of $\sigma(S_0 + S_1) = 0$, $\sigma \mathbb{1} = 1$. Substituting it into the drift expression we have

$$d = \underbrace{\alpha \otimes \sigma}_{\gamma} \left(\underbrace{\sum_{n=1}^{\infty} n \, D_n \otimes I}_{F_n} - \underbrace{I \otimes S_1}_{B} \right) \underbrace{\mathbb{1}_A \otimes \mathbb{1}_S}_{\mathbb{1}} = \alpha \left(\sum_{n=1}^{\infty} n D_n \right) \mathbb{1}_A - \sigma S_1 \, \mathbb{1}_S = \lambda - \mu,$$

where $\mathbb{1}_A$ and $\mathbb{1}_S$ are the column vector of ones whose size is identical with the cardinality of the arrival and the service process. The last expression verifies that the stability condition by the queue load is identical with the stability condition by the M/G/1 type process.

9.4.1.2   Performance Measures

The stationary distribution of the BMAP/MAP/1 queue can be obtained from the analysis of the M/G/1 type process with generator Q, which is provided in Theorem 5.16. Having the stationary solution the distribution of the number of customers in the stationary system, N, is $\mathbf{P}(N = n) = \pi_n \mathbb{1}$.

Let δ_n be the probability vector whose ith element is the probability that after an arrival the process is in state (n, i), i.e., there are n customers in the systems and the background process is in phase i. For δ_n we have

$$\delta_n = \frac{\sum_{\ell=0}^{n-1} \pi_\ell F_{n-\ell}}{\sum_{k=1}^{\infty} \sum_{\ell=0}^{k-1} \pi_\ell F_{k-\ell} \mathbb{1}} = \frac{1}{\lambda} \sum_{\ell=0}^{n-1} \pi_\ell F_{n-\ell},$$

where the numerator contains the stationary arrival rate to state (n, i) for all phases and the denominator is the sum of stationary arrival rates to all levels and all phases. In the second step, we utilized that the denominator is indeed the arrival rate of the BMAP.

Based on the distribution right after arrivals we can compute the distribution of the busy period, B, using the $G^\sim(s)$ matrix of the M/G/1 type process, which is the minimal nonnegative solution of

$$sG^\sim(s) = B + LG^\sim(s) + \sum_{i=1}^{\infty} F_i G^\sim(s)^{i+1}. \tag{9.4}$$

For the Laplace transform of B we have

$$\mathbf{E}\left(e^{-sB}\right) = \frac{\delta_1}{\delta_1 \mathbb{1}} G^\sim(s)\mathbb{1},$$

where the first term characterizes the phase distribution at the beginning of the busy period and the second term, $G^\sim(s)$, gives the phase dependent distribution of the busy period.

The stationary sojourn time distribution can also be computed from δ_n. The service time of a customer is characterized by the MAP (S_0, S_1) service process, that is the Laplace transform of the service time starting from phase i and starting the consecutive service in phase j, $S(s)_{ij} = \mathbf{E}(e^{-sS} \mathscr{I}_{\{J(S)=j\}} | J(0) = i))$, is such that

$$S(s) = (sI - I \otimes S_0)^{-1} I \otimes S_1,$$

from which the Laplace transform of the stationary sojourn time distribution is

$$\mathbf{E}\left(e^{-sT}\right) = \sum_{n=1}^{\infty} \delta_n \, (S(s))^n \, \mathbb{1}.$$

9.4.2 MAP/BMAP/1 Queue

In a MAP/BMAP/1 queue, customers arrive according to a MAP(D_0, D_1) process and are served according to a MAP($S_0, S_1, S_2, \ldots$) process. There is an infinite buffer and a single server, which serves the customers in FIFO order, but at a service instant more than one customer can leave the system as it is determined by the service BMAP process. The number of customers who leave the system cannot be larger than the number of customers in the system, which results in special service rates to level zero. The $\{N(t), J(t)\}$ CTMC characterizing the process behavior is G/M/1 type, where $N(t)$ is the number of customers in the queue and $J(t)$ is the phase of the Markov environment characterizing the arrival and service processes. The block structure of its generator matrix is

$$Q = \begin{array}{|c|c|c|c|c|c|c|}
\hline
L' & F & & & & & \\
\hline
B'_1 & L & F & & & & \\
\hline
B'_2 & B_1 & L & F & & & \\
\hline
B'_3 & B_2 & B_1 & L & F & & \\
\hline
B'_4 & B_3 & B_2 & B_1 & L & F & \\
\hline
\vdots & & & & \ddots & \ddots & \ddots \\
\hline
\end{array}$$

where the matrix blocks are $L = D_0 \oplus S_0$, $L' = D_0 \otimes I$, $B_n = I \otimes S_n$ and $B'_n = I \otimes \sum_{k=n}^{\infty} S_k$ for $n \geq 1$, and $F = D_1 \otimes I$.

9.4.2.1   Condition of Stability

Similar to the BMAP/MAP/1 case, we can define the condition of stability based on the queue load and based on the stability of the G/M/1 type process with generator Q. Considering the queue load the mean arrival rate, λ, of MAP(D_0, D_1) (c.f., (5.5)), has to be less than the mean service rate of the BMAP($S_0, S_1, S_2, \ldots$) service process, μ (c.f., (5.8)).

For the stability of the G/M/1 type process with generator Q condition (5.45) needs to hold, which can be written as

$$d = \gamma F \mathbb{1} - \gamma \sum_{n=1}^{\infty} n \, B_n \, \mathbb{1} < 0,$$

where γ is the solution of $\gamma \left(\sum_{n=1}^{\infty} B_n + L + F \right) = 0$, with normalizing condition $\gamma \mathbb{1} = 1$. Substituting the structure of the B_n, L, F matrices into the linear equation we have

$$0 = \gamma \left(\sum_{n=0}^{\infty} I \otimes S_n + \sum_{n=0}^{1} D_n \otimes I \right) = \gamma \left(\sum_{n=0}^{1} D_n \oplus \sum_{n=0}^{\infty} S_n \right),$$

whose solution is $\gamma = \alpha \otimes \sigma$, where α is the solution of $\alpha \sum_{n=0}^{1} D_n = 0$, $\alpha \mathbb{1} = 1$, and σ is the solution of $\sigma \sum_{n=0}^{\infty} S_n = 0$, $\sigma \mathbb{1} = 1$. Substituting it into the drift expression we have

$$d = \underbrace{\alpha \otimes \sigma}_{\gamma} \left(\underbrace{D_1 \otimes I}_{F} - \underbrace{\sum_{n=1}^{\infty} n \, I \otimes S_n}_{B_n} \right) \underbrace{\mathbb{1}_A \otimes \mathbb{1}_S}_{\mathbb{1}} = \alpha D_1 \mathbb{1}_A - \sigma \left(\sum_{n=1}^{\infty} n S_n \right) \mathbb{1}_S = \lambda - \mu,$$

which represents the identity of the stationary conditions also for the MAP/BMAP/1 queue.

9.4.2.2 Performance Measures

The stationary distribution of the G/M/1 type process with generator Q is matrix geometric, $\pi_n = \pi_0 R^n$, where R is the minimal nonnegative solution of (5.46). Based on this matrix geometric stationary solution the distribution of the number of customers in the stationary system, N, is $P(N = n) = \pi_n \mathbb{1} = \pi_0 R^n \mathbb{1}$.

For the probability vector describing the distribution right after an arrival, δ_n, we have

$$\delta_n = \frac{1}{\lambda}\pi_{n-1}F = \frac{1}{\lambda}\pi_0 R^{n-1}F.$$

The main difficulty of the analysis of the MAP/BMAP/1 queue is that the departure process is level dependent. For that reason we need to adopt a detailed analysis based on the block matrices introduced in Sect. 5.3.3.

For the Laplace transform of the busy period, B, we write

$$\mathbf{E}\!\left(e^{-sB}\right) = \frac{\delta_1}{\delta_1 \mathbb{1}}\left(V^*(s)B_1' + V^*(s)FV^*(s)B_2' + \ldots\right)\mathbb{1}$$

$$= \frac{\pi_0 R}{\pi_0 RF\mathbb{1}}\sum_{n=1}^{\infty}\underbrace{\left(FV^*(s)\right)^n}_{R^*(s)}B_n' = \frac{\pi_0 R}{\pi_0 RF\mathbb{1}}\sum_{n=1}^{\infty}R^*(s)^n B_n',$$

where the coefficient in front of the bracket is the initial probability vector of the busy period, the first term in the bracket describes that the process stays at level 1 at time t and it returns to level zero by a transition in matrix B_1', the next term describes the case that the process stays at level 1 at time τ it modes to level 2 at time τ, it stays at level 2 at time t such that it does not visit level 1 between τ and t, and move to level 0 at time t by a transition in matrix B_2', etc. This set of events to reach level 0 is complete and mutually exclusive. Matrix $R^*(s)$ is the characteristic matrix of the G/M/1 type process, and it is the minimal nonnegative solution of

$$sR^*(s) = F + R^*(s)L + \sum_{n=1}^{\infty}R^*(s)^n B_n.$$

The service process is a BMAP($S_0, S_1, S_2, \ldots$) process. Let $G_n(s)$, be the matrix Laplace transform of the time to serve n customers, when there are n customers in the queue. $G_n(s)$ satisfies

$$G_n(s) = (sI - I \otimes S_0)^{-1}\left(\underbrace{I \otimes S_n'}_{B_n'} + \sum_{k=1}^{n-1}\underbrace{I \otimes S_k}_{B_k}\,G_{n-k}(s)\right), \qquad (9.5)$$

and it can be computed starting from $G_1(s)$ with consecutive increasing indices. From δ_n and $G_n(s)$, the Laplace transform of the stationary sojourn time distribution is

$$\mathbf{E}\left(e^{-sT}\right) = \sum_{n=1}^{\infty} \delta_n G_n(s) \mathbb{1}$$

$$= \sum_{n=1}^{\infty} \frac{1}{\lambda} \pi_0 R^{n-1} F (sI - I \otimes S_0)^{-1} \left(B_n' + \sum_{k=1}^{n-1} B_k G_{n-k}(s) \right) \mathbb{1}$$

$$= \sum_{n=1}^{\infty} \frac{1}{\lambda} \pi_0 R^{n-1} F (sI - I \otimes S_0)^{-1} B_n' \mathbb{1}$$

$$+ \sum_{k=1}^{\infty} \sum_{n=k+1}^{\infty} \frac{1}{\lambda} \pi_0 R^{n-1-k} R^k F (sI - I \otimes S_0)^{-1} B_k G_{n-k}(s) \mathbb{1},$$

where the structure of the infinite summations is similar to, e.g., $\sum_{n=1}^{\infty} R^*(s)^n B_n$, but there is a constant matrix $(F (sI - I \otimes S_0)^{-1})$ in between the parameter dependent matrices. In some queueing models one of the parameter dependent matrices can be interchanged with the constant matrix in between [2] due to special commutativity properties. If it is not the case the infinite summation with the constant matrix in between can be computed by Kronecker expansion [6].

A kind of duality of the MAP/BMAP/1 and the BMAP/MAP/1 queues is visible from the sojourn time distribution of the two queues. In case of the BMAP/MAP/1 queue, π_n is matrix geometric and $G_n(s)$ needs to be computed by a recursive relation starting from $G_1(s)$, while in case of the MAP/BMAP/1 queue, π_n needs to be computed by a recursive relation starting from π_0 and $G_n(s) = (S(s))^n$ is matrix geometric. Furthermore the recursive relations (5.48) and (9.5) have the same structure.

9.4.3   BMAP/G/1 Queue

The BMAP/G/1 queue, where the customers arrive according to a BMAP with representation $(D_0, D_1, D_2, \ldots)$, the service times are i.i.d random variables with CDF $H(t) = \mathbf{P}(S < t)$, there is a single server and an infinite buffer, is analyzed in [4], which we summarize below. We adopt the BMAP notations from Sect. 5.2.3:

- $D = \sum_{i=0}^{\infty} D_i$, $D(z) = \sum_{i=0}^{\infty} D_i z^i$, the arrival intensity is $\lambda = \alpha \sum_{k=1}^{\infty} k D_k \mathbb{1}$, where α is the solution of $\alpha D = 0, \alpha \mathbb{1} = 1$
- The number of arrivals in $(0, t)$ is $N(t)$, whose distribution is $\hat{P}(t, z) = e^{D(z)t}$ (c.f. (5.11)), where $P(t, n)_{ij} = \mathbf{P}(N(t) = n, J(t) = j \mid J(0) = i)$.

The utilization of the server is $\rho = \lambda E(S)$, where $E(S)$ is the mean service time and we assume that $\rho < 1$.

Let T_i be the departure instant of the ith customer. Then the embedded DTMC right after departure instants, $(U(T_i^+), J(T_i^+))$, is an M/G/1 type Markov chain with block structure

$$
P = \begin{array}{|cccccc|}
\hline
B_0 & B_1 & B_2 & B_3 & \ldots \\
A_0 & A_1 & A_2 & A_3 & \ldots \\
& A_0 & A_1 & A_2 & \ldots \\
& & A_0 & A_1 & \ldots \\
& & & \ddots & \ddots \\
\hline
\end{array}, \tag{9.6}
$$

where $U(t)$ is the number of customers in the queue and $J(t)$ is the phase process of the BMAP at time t and the matrix blocks of P has the following stochastic interpretation

- A_{nij} is the probability that the phase moves from i to j and there are n arrivals during a service,
- B_{nij} is the probability that the phase moves from i to j and there are $n+1$ arrivals during an arrival and a service.

Based on its definition

$$
B_n = (-D_0)^{-1} \sum_{k=0}^{n} D_{k+1} A_{n-k},
$$

because $(-D_0)^{-1} \sum_{k=0}^{n} D_{k+1}$ describes the probability that $k+1$ customers arrive at the idle system and A_{n-k} describes the probability that $n-k$ customer arrived during the first service of the busy period. For $A(z)$ and $B(z)$, we have

$$
A(z) = \sum_{n=0}^{\infty} z^n A_n = \sum_{n=0}^{\infty} z^n \int_{t=0}^{\infty} P(t, n) \mathrm{d}H(t) = \int_{t=0}^{\infty} \hat{P}(t, z) \mathrm{d}H(t) = \int_{t=0}^{\infty} e^{D(z)t} \mathrm{d}H(t),
$$

$$
B(z) = \sum_{n=0}^{\infty} z^n B_n = \sum_{n=0}^{\infty} z^n (-D_0)^{-1} \sum_{k=0}^{n} D_{k+1} A_{n-k} \tag{9.7}
$$

$$
= (-D_0)^{-1} \sum_{k=0}^{\infty} z^k D_{k+1} \sum_{n=k}^{\infty} z^{n-k} A_{n-k} = (-D_0)^{-1} [D(z) - D_0] z^{-1} A(z).
$$

Utilizing the regular block structure of P, the ith block of the stationary equation, $\pi P = \pi$, can be written as

$$\pi_i = \pi_0 B_i + \sum_{k=1}^{i+1} \pi_k A_{i+1-k}, \quad i \geq 0. \tag{9.8}$$

Multiplying the ith equation with z^i and summing up give

$$\pi(z) = \pi_0 B(z) + z^{-1}(\pi(z) - \pi_0) A(z),$$

from which the queue length distribution at departure can be computed as

$$\pi(z)(zI - A(z)) = \pi_0 (zB(z) - A(z)) = \pi_0(-D_0)^{-1} D(z) A(z), \tag{9.9}$$

and finally

$$\pi(z) = \pi_0(-D_0)^{-1} D(z) A(z)(zI - A(z))^{-1}.$$

For this last expression we note that $D(z)$, $A(z)$ and $(zI - A(z))^{-1}$ commute by their definition and the matrix inverse exists only for $z < 1$. For $z = 1$, $A(1)\mathbb{1} = \int_{t=0}^{\infty} e^{Dt} dH(t)\mathbb{1} = \mathbb{1}$ and $(I - A(1))\mathbb{1} = 0$, from which $I - A(1)$ is singular.

Let $G(n)$ be the G *matrix* of the discrete-time M/G/1 type process characterized by the transition probability matrix in (9.6). Its i, j element is defined by $G_{ij}(n) = P(J(\gamma_0) = j, \gamma_0 = n \mid J(0) = i, U(0) = 1)$, where γ_0 is the random time to reach level zero. Additionally, we define $\widehat{G}(z) = \sum_{n=0}^{\infty} z^n G(n)$. Based on the results of Sect. 5.5 for the first visit to level zero starting from level 1 we have

$$\widehat{G}(z) = z \sum_{k=0}^{\infty} A_k \widehat{G}^k(z), \quad \text{and at } z = 1, \ \ G = \sum_{k=0}^{\infty} A_k G^k. \tag{9.10}$$

For the first visit to level zero starting from level zero we have

$$K(z) = z \sum_{k=0}^{\infty} B_k \widehat{G}^k(z), \quad \text{and at } z = 1, \ \ K = \sum_{k=0}^{\infty} B_k G^k.$$

Substituting (9.7), we obtain

$$K(z) = z \sum_{n=0}^{\infty} (-D_0)^{-1} \sum_{k=0}^{n} D_{k+1} A_{n-k} \widehat{G}^n(z)$$

$$= (-D_0)^{-1} \sum_{k=0}^{\infty} D_{k+1} z \underbrace{\sum_{n=k}^{\infty} A_{n-k} \widehat{G}^{n-k}(z)\, \widehat{G}^k(z)}$$

$$= (-D_0)^{-1} \sum_{k=0}^{\infty} D_{k+1} \widehat{G}^{k+1}(z) = (-D_0)^{-1}(D(\widehat{G}(z)) - D_0),$$

where $D(\widehat{G}(z)) = \sum_{k=0}^{\infty} D_k \widehat{G}^k(z)$ and the braced expression is the right-hand side of (9.10).

The state transition probability matrix of the restricted M/G/1 type process on level zero is K and π_0 satisfies the linear system $\pi_0 K = \pi_0$. For finding an associated normalizing equation we introduce the mean return time from level zero to level zero as vector κ^*, from which $\pi_0 \kappa^* = 1$ is the normalizing equation for π_0. For κ^* we write

$$\kappa^* = \frac{d}{dz} K(z) \mathbb{1} \Big|_{z=1} = (-D_0)^{-1} \sum_{k=1}^{\infty} D_k \frac{d}{dz} \widehat{G}^k(z) \mathbb{1}|_{z=1}$$

$$= (-D_0)^{-1} \sum_{k=1}^{\infty} D_k \sum_{j=0}^{k-1} G^j \widehat{G}'(1) \underbrace{G^{k-j-1} \mathbb{1}}_{\mathbb{1}}$$

where the procedure from (5.25) can be used for computing the finite sum $\sum_{j=0}^{k-1} G^j$. Using (9.10), for the last term we write

$$\widehat{G}'(1)\mathbb{1} = \frac{d}{dz} z \sum_{k=0}^{\infty} A_k \widehat{G}^k(z) \mathbb{1} \Big|_{z=1} = \underbrace{\sum_{k=0}^{\infty} A_k G^k \mathbb{1}}_{\mathbb{1}} + \sum_{k=1}^{\infty} A_k \frac{d}{dz} \widehat{G}^k(z) \mathbb{1}|_{z=1}$$

$$= \mathbb{1} + \sum_{k=1}^{\infty} A_k \sum_{j=0}^{k-1} G^j \widehat{G}'(1) G^{k-j-1} \mathbb{1} = \mathbb{1} + \sum_{k=1}^{\infty} A_k \sum_{j=0}^{k-1} G^j \widehat{G}'(1) \mathbb{1},$$

from which

$$\widehat{G}'(1)\mathbb{1} = \left(I - \sum_{k=1}^{\infty} A_k \sum_{j=0}^{k-1} G^j \right)^{-1} \mathbb{1}.$$

Having κ^* and then π_0 one can compute the consecutive $\pi_1, \pi_2, \ldots$ stationary probability vectors from (9.8), which gives the stationary behavior of the queue at departure instants. Our next task is to obtain the time homogeneous stationary distribution (ψ_i). The $(N(t), J(t))$ process of a MAP/G/1 queue is a Markov regenerative process with embedded regenerative points at departures. We have the stationary distributions at embedded regenerative instants (π_i) and compute the time homogeneous distribution using (4.29).

The mean time the $(N(t), J(t))$ process spends in a state during the regenerative period, $(0, T_1)$, is

$$T_{ij}(k, \ell) = \mathbf{E}\left(\int_0^{T_1} \mathscr{I}_{\{N(t)=\ell, J(t)=j\}} dt \mid N(0) = k, J(0) = i \right).$$

$T(k, \ell) = 0$ for $\ell < k$. For $\ell = k = 0$, we have $T(0, 0) = \int_{t=0}^{\infty} e^{D_0 t} \, dt = (-D_0)^{-1}$.
For $0 < k \leq \ell$, we have

$$T(k, \ell) = \int_{t=0}^{\infty} P(t, \ell - k) \, (1 - H(t)) \, dt \, ,$$

and for $k = 0, \ell > 0$ it is

$$T(0, \ell) = \sum_{k=1}^{\ell} \underbrace{(-D_0)^{-1} D_k}_{\text{1st arrival}} \int_{t=0}^{\infty} P(t, \ell - k) \, (1 - H(t)) \, dt \, .$$

Applying (4.29), the time homogeneous stationary distribution is

$$\psi_\ell = \frac{\displaystyle\sum_{k=0}^{\ell} \pi_k T(k, \ell)}{\displaystyle\sum_{k=0}^{\infty} \pi_k \sum_{n=k}^{\infty} T(k, n) \mathbb{1}} = \lambda \sum_{k=0}^{\ell} \pi_k T(k, \ell) \, , \tag{9.11}$$

where the denominator is the mean time of a regenerative period, i.e., mean inter-departure time. Assuming a stable queue, it equals to the mean inter-arrival time, $1/\lambda$, because in a stable queue the customer departure rate and the customer arrival rate (λ) are identical and the mean regenerative period is the mean time between departures.

For $\ell = 0$, we have $\psi_0 = \lambda \pi_0 (-D_0)^{-1}$ which satisfies $\psi_0 \mathbb{1} = 1 - \rho$, since $\pi_0 (-D_0)^{-1} \mathbb{1}$ is the mean idle time in a regeneration period. For $\ell > 0$, we multiply (9.11) with z^ℓ and sum up from 1 to infinity and get

$$\psi(z) - \psi_0 = \lambda \left(\pi_0 (-D_0)^{-1} D(z) + \pi(z) \right) \int_{t=0}^{\infty} \hat{P}(t, z) \, (1 - H(t)) \, dt \, ,$$

where

$$\int_{t=0}^{\infty} \hat{P}(t, z)(1 - H(t)) \, dt = \int_{t=0}^{\infty} e^{D(z)t} \, dt - \int_{t=0}^{\infty} e^{D(z)t} H(t) \, dt \tag{9.12}$$

$$= \int_{t=0}^{\infty} e^{D(z)t} \, dt - \int_{t=0}^{\infty} (-D(z))^{-1} e^{D(z)t} \, dH(t) = (-D(z))^{-1} (I - A(z)) \, ,$$

and we used that $D(z)$ and $A(z)$ commute. Using this, we further have

$$\psi(z) - \psi_0 = \lambda\left(\pi_0(-D_0)^{-1}D(z) + \pi(z)\right)(-D(z))^{-1}(I - A(z))$$

$$= \lambda\left(-\pi_0(-D_0)^{-1} + \pi(z)(-D(z))^{-1}\right)(I - A(z))$$

$$\underbrace{\phantom{-\pi_0(-D_0)^{-1} + \pi(z)(-D(z))^{-1}}}_{-\psi_0}$$

$$= -\psi_0 + \lambda\pi_0(-D_0)^{-1}A(z) + \lambda\pi(z)(-D(z))^{-1}(I - A(z)).$$

Adding ψ_0 to both sides, substituting $\pi_0(-D_0)^{-1}D(z)A(z)$ from (9.9), and using that $D(z)$ and $A(z)$ commute give

$$\psi(z) = \lambda\pi(z)(zI - A(z))(D(z))^{-1} + \lambda\pi(z)(-D(z))^{-1}(I - A(z)),$$

which can be written as

$$\psi(z)D(z) = \lambda(z - 1)\pi(z).$$

The inverse transformation of this transform expression gives a simple and elegant recursive formula for the time homogeneous stationary probability vector

$$\psi_{\ell+1} = \left(\sum_{k=0}^{\ell} \psi_k D_{\ell+1-k} - \lambda(\pi_\ell - \pi_{\ell+1})\right)(-D_0)^{-1}.$$

9.4.3.1 Performance Measures

The stationary queue length distribution is given by $\psi(z)$ in transform domain, since $\mathbf{E}(z^N) = \psi(z)\mathbb{1}$.

The Laplace transform of the sojourn time a customer spends in the system is

$$\mathbf{E}(e^{-sT}) = \frac{1}{\lambda}\psi(h^*(s))(D(h^*(s)) - D_0)\mathbb{1},$$

because the probability that after a customer arrival there are n customers in the system is $(\sum_{\ell=0}^{n-1} \psi_\ell D_{n-\ell}\mathbb{1})/\lambda$, whose moment generating function is

$$\frac{1}{\lambda}\sum_{n=1}^{\infty} z^n \sum_{\ell=0}^{n-1} \psi_\ell D_{n-\ell}\mathbb{1} = \frac{1}{\lambda}\sum_{\ell=0}^{\infty} z^\ell \psi_\ell \sum_{n=\ell+1}^{\infty} z^{n-\ell} D_{n-\ell}\mathbb{1} = \frac{1}{\lambda}\psi(z)(D(z) - D_0)\mathbb{1}$$

from which using a $z = h^*(s)$ substitution we have

$$E\left(e^{-sT}\right) = \frac{1}{\lambda}\sum_{n=1}^{\infty} h^*(s)^n \sum_{\ell=0}^{n-1} \psi_\ell D_{i-\ell} \mathbb{1} = \frac{1}{\lambda}\psi(h^*(s))(D(h^*(s)) - D_0)\mathbb{1}.$$

One of the main results presented in [4] is associated with matrix $D(\widehat{G}(z)) = \sum_{i=0}^{\infty} D_i \widehat{G}(z)^i$, which appeared in the derivation of $K(z)$. It was shown in [4] that $D(\widehat{G}(z))$ and $\widehat{G}(z)$ commute, as well as, G and $D(G)$ with $D(G) = \sum_{i=0}^{\infty} D_i G^i$. We mention this here, because for the analysis of BMAP/G/1 queue we have not utilized this property, but in the next section we utilize its counterpart for the analysis of G/BMAP/1 queue.

9.4.4 G/BMAP/1 Queue

The G/BMAP/1 queue, where the customers arrive according to a renewal process with interarrival time distribution $H(t) = \mathbf{P}(A < t)$, and they are served according to a BMAP with representation $(S_0, S_1, S_2, \ldots)$, there is a single server and an infinite buffer. For the BMAP service process we have:

- $S = \sum_{i=0}^{\infty} S_i$, the service intensity is $\mu = \sigma \sum_{k=1}^{\infty} k S_k \mathbb{1}$, where σ is the solution of $\sigma S = 0, \sigma \mathbb{1} = 1$.
- With sufficiently large number of customers, the number of served customers in $(0, t)$ is $N(t)$, whose distribution is $P_{ij}(t, n) = \mathbf{P}(N(t) = n, J(t) = j \mid J(0) = i)$, where $J(t)$ refers to the phase of the BMAP at time t.

The utilization of the server is $\rho = \frac{\lambda}{\mu} = \frac{1}{\mu \mathbf{E}(A)}$, where $\mathbf{E}(A)$ is the mean interarrival time and $\lambda = 1/\mathbf{E}(A)$ is the arrival rate.

$\{U(t), J(t)\}$ denotes stochastic process composed by the number of customers in the system and the phase of the BMAP service process. Let T_i be the arrival instant of the ith customer. Then the embedded DTMC right before arrival instants, $\{U(T_i^-), J(T_i^-)\}$, is a G/M/1 type Markov chain with block structure

$$P = \begin{array}{|c|c|c|c|c|c|}
\hline
B_0 & A_0 & & & & \\
\hline
B_1 & A_1 & A_0 & & & \\
\hline
B_2 & A_2 & A_1 & A_0 & & \\
\hline
B_3 & A_3 & A_2 & A_1 & A_0 & \\
\hline
B_4 & A_4 & A_3 & A_2 & A_1 & A_0 \\
\hline
\vdots & & & \ddots & \ddots & \ddots \\
\hline
\end{array}$$

where the matrix blocks of P has the following stochastic interpretation

- A_{nij} is the probability that the phase moves from i to j and there are n customers served during an interarrival period, i.e., $A_n = \int_{t=0}^{\infty} P(t, n) \mathrm{d}H(t)$
- B_{nij} is the probability that the phase moves from i to j and the departure BMAP serves more than n customers during an interarrival period, i.e., $B_n = \sum_{k=n+1}^{\infty} A_k$.

For $A(z) = \sum_{n=0}^{\infty} z^n A_n$ and $B(z) = \sum_{n=0}^{\infty} z^n B_n$, we have

$$A(z) = \sum_{n=0}^{\infty} z^n \int_{t=0}^{\infty} P(t, n) \mathrm{d}H(t) = \int_{t=0}^{\infty} \hat{P}(t, z) \mathrm{d}H(t) = \int_{t=0}^{\infty} e^{S(z)t} \mathrm{d}H(t)$$

$$B(z) = \sum_{n=0}^{\infty} z^n B_n = \sum_{n=0}^{\infty} z^n \sum_{k=n+1}^{\infty} A_k = \sum_{k=1}^{\infty} A_k \sum_{n=0}^{k-1} z^n = \sum_{k=1}^{\infty} A_k \frac{1 - z^k}{1 - z}$$

$$= \frac{1}{1 - z} (A(1) - A(z)).$$

Utilizing the regular block structure of P, the ith block of the stationary equation, $\pi P = \pi$, can be written as

$$\pi_i = \sum_{k=0}^{\infty} \pi_{i-1+k} A_k, \quad i \geq 1. \tag{9.13}$$

whose solution is matrix geometric, $\pi_i = \pi_0 R^i$, where R is the solution of

$$R = \sum_{i=0}^{\infty} R^i A_i$$

and π_0 is the solution of the linear system $\pi_0 = \pi_0 \sum_{i=0}^{\infty} R^i B_i$, with normalizing equation $\pi_0 (I - R)^{-1} \mathbb{1} = 1$. To compute the infinite sum in the linear equation we note that $\hat{R}(z)$, defined in (5.43), satisfies

$$\hat{R}(z) = z \sum_{k=0}^{\infty} \hat{R}^k(z) A_k, \tag{9.14}$$

and for the first visit to level zero starting from level zero we have

$$K(z) = z \sum_{k=0}^{\infty} \hat{R}^k(z) B_k, \quad \text{and at } z = 1, \quad K = \sum_{k=0}^{\infty} R^k B_k.$$

Substituting $B_n = \sum_{k=n+1}^{\infty} A_k$, we obtain

$$K(z) = z \sum_{n=0}^{\infty} \widehat{R}^n(z) \sum_{k=n+1}^{\infty} A_k = z \sum_{k=0}^{\infty} \sum_{n=0}^{k-1} \widehat{R}^n(z) A_k = z \sum_{k=0}^{\infty} \left(I - \widehat{R}(z)\right)^{-1} \left(I - \widehat{R}^k(z)\right) A_k$$

$$= \left(I - \widehat{R}(z)\right)^{-1} \left(z \sum_{k=0}^{\infty} A_k - z \underbrace{\sum_{k=0}^{\infty} \widehat{R}^k(z) A_k}\right) = \left(I - \widehat{R}(z)\right)^{-1} \left(zA(1) - \widehat{R}(z)\right)$$

where the braced expression is $\widehat{R}(z)$. At $z = 1$ we obtain

$$K = K(1) = (I - R)^{-1} (A(1) - R) = (I - R)^{-1} \left(\int_{t=0}^{\infty} e^{St} dH(t) - R\right).$$

For computing the distribution of the sojourn time (T), we write

$$P(T > t) = \sum_{i=0}^{\infty} \pi_i \sum_{k=0}^{i} P(t, k)\mathbb{1},$$

which means that if a tagged customer arrives to the system at time 0 such that after its arrival there are $i + 1$ customers in the system and at most i are served in the interval $(0, t)$, then at time t the tagged customer is still in the system and its sojourn time is larger than t. For π_i and $P(t, k)$, we can further write

$$P(T > t) = \sum_{i=0}^{\infty} \pi_i \sum_{k=0}^{i} P(t, k)\mathbb{1} = \pi_0 \sum_{i=0}^{\infty} R^i \sum_{k=0}^{i} P(t, k)\mathbb{1}$$

$$= \pi_0 \sum_{k=0}^{\infty} \sum_{i=k}^{\infty} R^{i-k} R^k P(t, k)\mathbb{1} = \pi_0 (I - R)^{-1} \underbrace{\sum_{k=0}^{\infty} R^k P(t, k)\, \mathbb{1}}$$

$$= \pi_0 (I - R)^{-1} W(t)\mathbb{1}.$$

To compute $W(t) = \sum_{k=0}^{\infty} R^k P(t, k)$, we first note that a generator matrix (e.g., Q) and the associated state transition probability matrix (e^{Qt}) commute, from which, according to (5.10), we can also write

$$\frac{d}{dt} P(t, n) = \sum_{k=0}^{n} P(t, n - k) S_k = \sum_{k=0}^{n} S_k P(t, n - k).$$

Multiplying both sides by R^n we get

$$\frac{d}{dt}W(t) = \sum_{n=0}^{\infty} R^n \frac{d}{dt} P(t,n) = \sum_{n=0}^{\infty} R^n \sum_{k=0}^{n} S_k P(t,n-k)$$

$$= \sum_{k=0}^{\infty} \sum_{n=k}^{\infty} R^{n-k} R^k S_k P(t,n-k) = \sum_{n=0}^{\infty} R^n \underbrace{\sum_{k=0}^{\infty} R^k S_k} P(t,n).$$

In [7], Sengupta showed that $R(S) = \sum_{k=0}^{\infty} R^k S_k$ and R commute, due to which

$$\frac{d}{dt}W(t) = \sum_{n=0}^{\infty} R^n R(S) P(t,n) = R(S) \sum_{n=0}^{\infty} R^n P(t,n) = R(S)W(t),$$

with initial condition $W(t) = I$. That is $W(t) = e^{R(S)t}$ and the CDF of the sojourn time distribution is

$$P(T < t) = 1 - \pi_0 (I - R)^{-1} e^{R(S)t} \mathbb{1}.$$

This matrix exponential distribution function refers for ME sojourn time distribution. Additionally, all π_0, R, $(I - R)^{-1}$ are nonnegative and $R(S)$ is a transient generator matrix, hence the sojourn time is PH distributed with representation $PH(\pi_0(I - R)^{-1}, R(S))$. We further note that

$$R = \sum_{n=0}^{\infty} R^n A_n = \sum_{n=0}^{\infty} R^n \int_{t=0}^{\infty} P(t,n) dH(t) = \int_{t=0}^{\infty} \sum_{n=0}^{\infty} R^n P(t,n) dH(t)$$

$$= \int_{t=0}^{\infty} W(t) dH(t) = \int_{t=0}^{\infty} e^{R(S)t} dH(t). \tag{9.15}$$

To compute the time homogeneous stationary distribution we introduce the mean time the $(U(t), J(t))$ process spends in a state during the regenerative period, $(0, T_1)$, as

$$T_{ij}(k,\ell) = \mathbf{E}\left(\int_0^{T_1} \mathscr{I}_{\{U(t)=\ell, J(t)=j\}} dt \mid N(0) = k, J(0) = i\right).$$

For $0 < \ell \le k + 1$, we have

$$T(k,\ell) = \int_{t=0}^{\infty} P(t, k+1-\ell)(1 - H(t)) dt.$$

and for $k \ge 0$, it is

$$T(k,0) = \sum_{\ell=k+1}^{\infty} \int_{t=0}^{\infty} P(t,\ell)(1 - H(t)) dt.$$

Applying (4.29), for $\ell > 0$, the time homogeneous stationary distribution is

$$\psi_\ell = \frac{\displaystyle\sum_{k=\ell-1}^{\infty} \pi_k T(k, \ell)}{\displaystyle\sum_{n=1}^{\infty}\sum_{k=n-1}^{\infty} \pi_k T(k, n) + \sum_{k=1}^{\infty} \pi_k T(k, 0)} = \lambda \sum_{k=\ell-1}^{\infty} \pi_k T(k, \ell), \qquad (9.16)$$

where the denominator is the mean time of a regenerative period, i.e., mean inter arrival time, which is the inverse of the arrival rate, λ. Substituting π_k and $T(k, \ell)$, we further get

$$\psi_\ell = \lambda \pi_0 R^{\ell-1} \sum_{k=\ell-1}^{\infty} R^{k+1-\ell} \int_{t=0}^{\infty} P(t, k+1-\ell)\,(1 - H(t))\,dt$$

$$= \lambda \pi_0 R^{\ell-1} \sum_{k=0}^{\infty} R^k \int_{t=0}^{\infty} P(t, k)\,(1 - H(t))\,dt = \lambda \pi_0 R^{\ell-1} \int_{t=0}^{\infty} W(t)\,(1 - H(t))\,dt$$

$$= \lambda \pi_0 R^{\ell-1} \underbrace{\int_{t=0}^{\infty} e^{R(S)t}\,(1 - H(t))\,dt}.$$

Using (9.15) and following the same steps as in (9.12), for the braced expression we can write $(-R(S))^{-1}(I - R)$ by which

$$\psi_\ell = \lambda \pi_0 R^{\ell-1}(-R(S))^{-1}(I - R) = \lambda \pi_0 (-R(S))^{-1}(I - R)\,R^{\ell-1},$$

where we used again that R and $R(S)$ commute [7]. As a result, ψ_ℓ form a matrix geometric sequence for $\ell \geq 1$.

9.5 Exercises

Exercise 9.1 Define a MAP representation of the departure process of an M/M/1/2 queue with arrival rate λ and service rate μ.

Exercise 9.2 Define a MAP representation of the departure process of a MAP/M/1/1 queue with arrival MAP $(\hat{D}_0, \hat{D}_1)$ and service rate μ.

Exercise 9.3 Define a MAP representation of the customer loss process of a MAP/M/1/1 queue with arrival MAP $(\hat{D}_0, \hat{D}_1)$ and service rate μ.

Exercise 9.4 Compute the generator of the CTMC which describes the number of customers and the phase of the arrival PH distribution in a $PH/M/1$ queue if the

representation of the PH distributed interarrival time is (α, A), with $\alpha = (1, 0)$ and
$A = \begin{pmatrix} -\alpha & \alpha/2 \\ 0 & -\gamma \end{pmatrix}$ and the service rate is μ.

Exercise 9.5 Compute the generator of the CTMC which describes the number of customers and the phase of the service PH distribution in a M/PH/1 queue if the arrival rate is λ and the representation of the PH distributed service time is (β, B), with $\beta = (1/3, 2/3)$ and $B = \begin{pmatrix} -\mu & \mu \\ 0 & -\gamma \end{pmatrix}$.

Exercise 9.6 The packet transmission is performed in two phases in a slotted time communication protocol. The first phase is the resource allocation and the second is the data transmission. The times of both phases are geometrically distributed with parameters q_1 and q_2. In every time slot one packet arrives with probability p (and no packet arrives with probability $1 - p$). Compute the probability of packet loss if at most 2 packets can be in the system.

Exercise 9.7 Requests arrive to a computer according to a Poisson process with rate λ. The service of these requests requires first a processor operation for an exponentially distributed amount of time with parameter μ_1. After this processor operation the request leaves the system with probability p or requires a consecutive disk operation with probability $1 - p$. The time of the disk operation is exponentially distributed with parameter μ_2. After the disk operation the request requires a processor operation as it was a new one. There can be several loops of processor and disk operations. The processor is blocked during the disk operation and one request is handled at a time.

Compute the efficient utilization of the processor, and the request loss probability if there is no buffer in the system.

Compute the efficient utilization of the processor, and the system time of the requests if there is an infinite buffer in the system.

9.6 Solutions

Solution 9.1

$$D_0 = \begin{pmatrix} -\lambda & \lambda & \\ & -\lambda-\mu & \lambda \\ & & -\mu \end{pmatrix} \qquad D_1 = \begin{pmatrix} & & \\ \mu & & \\ & \mu & \end{pmatrix}$$

Solution 9.2

$$D_0 = \begin{array}{|c|c|} \hline \hat{D}_0 & \hat{D}_1 \\ \hline \mu I \ \hat{D}_0 & \hat{D}_0 + \hat{D}_1 - \mu I \\ \hline \end{array}, \qquad D_1 = \begin{array}{|c|c|} \hline 0 & 0 \\ \hline \mu I & 0 \\ \hline \end{array}.$$

Solution 9.3

$$D_0 = \begin{array}{c|c} \hat{D}_0 & \hat{D}_1 \\ \hline \mu I & \hat{D}_0 - \mu I \end{array}, \quad D_1 = \begin{array}{c|c} 0 & 0 \\ \hline 0 & \hat{D}_1 \end{array}.$$

Solution 9.4

$$\begin{array}{cc|cc|cc|cc}
-\alpha & \alpha/2 & \alpha/2 & 0 & 0 & 0 & 0 & 0 \\
0 & -\gamma & \gamma & 0 & 0 & 0 & 0 & 0 \\
\hline
\mu & 0 & -\alpha-\mu & \alpha/2 & \alpha/2 & 0 & 0 & 0 \\
0 & \mu & 0 & -\gamma-\mu & \gamma & 0 & 0 & 0 \\
\hline
0 & 0 & \mu & 0 & -\alpha-\mu & \alpha/2 & \alpha/2 & 0 \\
0 & 0 & 0 & \mu & 0 & -\gamma-\mu & \gamma & 0 \\
\hline
& & & & \ddots & \ddots & \ddots & \ddots \\
& & & & \ddots & \ddots & \ddots & \ddots \\
\end{array}.$$

Solution 9.5 Solution 1: If the idle state of the queue is represented with a single state of the Markov chain, then the generator is

$$\begin{array}{c|cc|cc|cc}
-\lambda & \lambda/3 & 2\lambda/3 & 0 & 0 & 0 & 0 \\
\hline
0 & -\lambda-\mu & \mu & \lambda & 0 & 0 & 0 \\
\gamma & 0 & -\lambda-\gamma & 0 & \lambda & 0 & 0 \\
\hline
0 & 0 & 0 & -\lambda-\mu & \mu & \lambda & 0 \\
0 & \gamma/3 & 2\gamma/3 & 0 & -\lambda-\gamma & 0 & \lambda \\
\hline
0 & 0 & 0 & \ddots & \ddots & \ddots & \ddots \\
0 & 0 & 0 & \ddots & \ddots & \ddots & \ddots \\
\end{array}.$$

Solution 2: If the idle state of the queue is represented with two states of the Markov chain, then the generator is

$$\begin{array}{cc|cc|cc|cc}
-\lambda & 0 & \lambda & 0 & 0 & 0 & 0 & 0 \\
0 & -\lambda & 0 & \lambda & 0 & 0 & 0 & 0 \\
\hline
0 & 0 & -\lambda-\mu & \mu & \lambda & 0 & 0 & 0 \\
\gamma/3 & 2\gamma/3 & 0 & -\lambda-\gamma & 0 & \lambda & 0 & 0 \\
\hline
0 & 0 & 0 & 0 & -\lambda-\mu & \mu & \lambda & 0 \\
0 & 0 & \gamma/3 & 2\gamma/3 & 0 & -\lambda-\gamma & 0 & \lambda \\
\hline
0 & 0 & 0 & 0 & \ddots & \ddots & \ddots & \ddots \\
0 & 0 & 0 & 0 & \ddots & \ddots & \ddots & \ddots \\
\end{array}.$$

Solution 9.6 The service time distribution is indeed a discrete PH distribution with representation

$$\boldsymbol{\beta} = (1, 0) \text{ and } \boldsymbol{B} = \begin{pmatrix} 1 - q_1 & q_1 \\ 0 & 1 - q_2 \end{pmatrix}.$$

and the queueing system is a discrete-time M/PH/1/2 queue. The following DTMC characterizes its behavior,

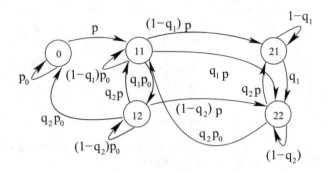

where $p_0 = 1 - p$ and state i, j indicates that there are i customers in the system and the service process of the customer in service is in phase j. The probability of packet loss is

$$P_{\text{loss}} = \frac{p_{2,1} p + p_{2,2} (1 - q_2) p}{p} = p_{2,1} + p_{2,2} (1 - q_2),$$

where $p_{i,j}$ denotes the stationary probability of state i, j.

Solution 9.7 Similar to the previous exercise the service time distribution is a continuous PH distribution with representation

$$\boldsymbol{\beta} = (1, 0) \text{ and } \boldsymbol{B} = \begin{pmatrix} -\mu_1 & \mu_1 (1 - p) \\ \mu_2 & -\mu_2 \end{pmatrix}.$$

and the queueing system is a (continuous-time) M/PH/1/1 queue if there is no buffer and an M/PH/1 queue if there is an infinite buffer. The related CTMCs are as follows.

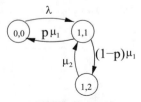

M/PH/1/1 queue

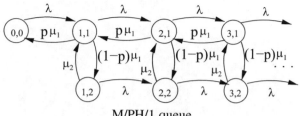

M/PH/1 queue

State i, j indicates that there are i customers in the system and the service process of the customer in service is in phase j and $p_{i,j}$ denotes the stationary probability of state i, j.

In case of no buffer the effective utilization of the server and the loss probability are

$$\rho = p_{1,1} \text{ and } p_{\text{loss}} = p_{1,1} + p_{1,2},$$

and in case of infinite buffer the effective utilization of the server and the loss probability are

$$\rho = \sum_{i=1}^{\infty} p_{i,1} \text{ and } p_{\text{loss}} = 0.$$

References

1. He, Q.M.: Fundamentals of Matrix-Analytic Methods. Springer, New York (2014)
2. Horváth, G., Van Houdt, B., Telek, M.: Commuting matrices in the queue length and sojourn time analysis of MAP/MAP/1 queues. Stoch. Model. **30**(4), 554–575 (2014)
3. Latouche, G., Ramaswami, V.: Introduction to Matrix Analytic Methods in Stochastic Modeling. SIAM Press, Philadelphia (1999)
4. Lucantoni, D.M.: New results on the single server queue with a batch Markovian arrival process. Commun. Stat. Stoch. Models **7**(1), 1–46 (1991)
5. Neuts, M.F.: Matrix Geometric Solutions in Stochastic Models. Johns Hopkins University Press, Baltimore (1981)
6. Razumchik, R., Telek, M.: Delay analysis of a queue with re-sequencing buffer and Markov environment. Queueing Syst. **82**(1), 7–28 (2016)
7. Sengupta, B.: The semi-Markovian queue: theory and applications. Stoch. Models **6**(3), 383–413 (1990)

Chapter 10
Queueing Networks

10.1 Introduction of Queueing Networks

Up to now, we have overviewed the main methods for the analysis of individual queueing systems. However, the analysis of large telecommunication systems or computer systems executing complex inter-related tasks (e.g., transaction processing systems and web server farms) requires the application of systems models which contain several servers (potentially of different kinds) and the customers are traveling among these servers for consecutive services.

The queueing network models are commonly used for the analysis of these kinds of systems. A queueing network is a graph with directed arcs, whose nodes represent such kinds of queueing systems as we studied till now. The arcs of the graph describe the potential transitions of the customers among these queueing systems.

It is a commonly applied modeling assumption in queueing networks that the transition of a customer from one node to the next one is memoryless and independent of the network state. That is, it is independent of the past history of the network, the current number of customers at the network nodes, and the status of the servers. After getting served at a network node a customer chooses the next node according to the weight (probability) associated with the outgoing arcs of the given node.

There are two main classes of queueing networks, they are the classes of open and closed queueing networks. In the closed queueing networks there is a given fixed number of customers circulating in the network, and there is no arrival/departure from/to the environment. In the open queueing networks customers arrive from the environment, obtain a finite number of services at the nodes of the network (nodes are potentially visited more than once), and leave the network eventually.

Queueing networks are classified also based on the structure of the directed arcs. The queueing networks without a loop (series of directed arcs forming a loop) are referred to as acyclic or feed-forward queueing networks, and the ones with a loop are referred to as cyclic or feed-back queueing networks. Acyclic networks are

© Springer Nature Switzerland AG 2019
L. Lakatos et al., *Introduction to Queueing Systems with Telecommunication Applications*, https://doi.org/10.1007/978-3-030-15142-3_10

meaningful only in case of open queueing networks. The nodes of acyclic networks can be numbered such that arcs are directed always from a node with lower index to a node with higher index or to the environment. From now on we assume that the nodes of acyclic networks are numbered this way.

10.2   Burke's Theorem

It is possible to analyze a class of open acyclic queueing networks based on the following theorem.

Theorem 10.1 ([2])   *The customer departure process of a stable M/M/m queue is a Poisson process with the same rate as the arrival process of the queue.*

Proof The number of customers in an M/M/m is a *reversible* Markov chain (see Sect. 3.3.6). The time reverse of the process is stochastically identical (according to all finite dimensional joint probabilities) with the original process. This way the departure instants of the original process (which are the arrival instants of the reverse process) are stochastically identical with the arrival instants of the original process (which are the departure instants of the reverse process) which is a Poisson process.   □

Apart from the general proof of the theorem by reversibility argument, we provide a specific one for M/M/1 queue based on stationary remaining departure time analysis.

Proof Let $D^*(s)$ be the Laplace transform of the time till the next departure, $A^*(s)$ the Laplace transform of the interarrival time distribution, $B^*(s)$ the Laplace transform of the service time distribution, and p the probability that in equilibrium the queue is idle, then

$$D^*(s) = p \, B^*(s) + (1 - p) \, A^*(s) \, B^*(s).$$

Using that $B^*(s) = \frac{\mu}{s+\mu}$, $A^*(s) = \frac{\lambda}{s+\lambda}$, $p = \frac{\lambda}{\mu}$ we have

$$D^*(s) = \frac{\lambda}{\mu} \frac{\mu}{s+\mu} + \frac{\mu - \lambda}{\mu} \frac{\lambda}{s+\lambda} \frac{\mu}{s+\mu},$$

and after some algebra

$$D^*(s) = \frac{\mu}{s+\mu} \frac{s\lambda + \lambda^2 + \mu\lambda - \lambda^2}{\mu(s+\lambda)} = \frac{\lambda}{s+\lambda}.$$

                                                                                                                               □

This specific proof indicates that complex exponentially distributed random variables might lead to very simple result.

10.3 Tandem Network of Two Queues

The simplest queueing network is the open tandem network composed by two M/M/1 queues in which the customers arrive from the environment attend queue 1 and after getting served in queue 1 they attend to queue 2 from where after getting served they depart to the environment. Let the arrival rate from the environment to queue 1 be λ and the service rate at queue 1 and 2 be μ_1 and μ_2, respectively (Fig. 10.1).

From Burke's theorem we have that the arrival intensity to both queues is λ, and this way the condition of stability is

$$\frac{\lambda}{\mu_1} < 1 \qquad \frac{\lambda}{\mu_2} < 1$$

that is

$$\lambda < \min(\mu_1, \mu_2).$$

Let us consider the Markov chain describing the number of customers in both queues. We identify the states of this Markov chain with the vector of the number of customers in the first queue and the second queue. That is, state $\{i, j\}$ refers to the state when there are i customers in the first and j customers in the second queue. The transition rates of this Markov chain are as follows

$$\begin{aligned}
\{i, j\} &\to \{i + 1, j\} &&: \lambda, \\
\{i, j\} &\to \{i - 1, j + 1\} &&: \mu_1 \text{ when } i \geq 1, \\
\{i, j\} &\to \{i, j - 1\} &&: \mu_2 \text{ when } j \geq 1.
\end{aligned}$$

We denote the stationary probability of state $\{i, j\}$ by $p_{i,j}$. The balance equations of the Markov chains are

$$\begin{aligned}
\lambda p_{0,0} &= \mu_2 p_{0,1}, \\
(\lambda + \mu_2) p_{0,j} &= \mu_1 p_{1,j-1} + \mu_2 p_{0,j+1} && \text{when } j \geq 1, \\
(\lambda + \mu_1) p_{i,0} &= \lambda p_{i-1,0} + \mu_2 p_{i,1} && \text{when } i \geq 1, \\
(\lambda + \mu_1 + \mu_2) p_{i,j} &= \lambda p_{i-1,j} + \mu_1 p_{i+1,j-1} + \mu_2 p_{i,j+1} && \text{when } i, j \geq 1.
\end{aligned}$$

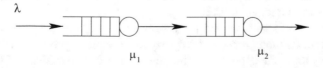

λ

μ_1　　　　μ_2

Fig. 10.1 Tandem network of two nodes

According to Burke's theorem in equilibrium the arrival Process of queue 2 is a Poisson process with rate λ. Using this fact the stationary state probabilities are

$$p_{i,j} = p_i^{(1)} p_j^{(2)} = \left(1 - \frac{\lambda}{\mu_1}\right) \left(\frac{\lambda}{\mu_1}\right)^i \left(1 - \frac{\lambda}{\mu_2}\right) \left(\frac{\lambda}{\mu_2}\right)^j,$$

where $p_i^{(1)}$ and $p_j^{(2)}$ are the stationary distributions of the corresponding M/M/1 queues.

The stationary solutions of this kind are referred to as *product form solution*, because the joint distribution is the product of two marginal distributions. It is important to note that in spite of the product form stationary distribution the number of customers in the two queues is not independent. There is a very strong correlation between those processes, namely that a departure from the first queue results in an arrival in the second queue.

Based on the stationary distribution we can easily determine the important performance indices. For example the mean number of customers in the system, the mean time spent in the network, and the mean waiting time spent in the network are

$$\mathbf{E}(N) = \sum_i \sum_j (i + j) p_{i,j} = \sum_i i p_i^{(1)} + \sum_j j p_j^{(2)} = \frac{\frac{\lambda}{\mu_1}}{1 - \frac{\lambda}{\mu_1}} + \frac{\frac{\lambda}{\mu_2}}{1 - \frac{\lambda}{\mu_2}},$$

$$\mathbf{E}(T) = \frac{\mathbf{E}(N)}{\lambda} = \frac{\frac{1}{\mu_1}}{1 - \frac{\lambda}{\mu_1}} + \frac{\frac{1}{\mu_2}}{1 - \frac{\lambda}{\mu_2}} = \frac{1}{\mu_1 - \lambda} + \frac{1}{\mu_2 - \lambda},$$

$$\mathbf{E}(W) = \mathbf{E}(T) - \frac{1}{\mu_1} - \frac{1}{\mu_2},$$

where we used Little's law to obtain the last two quantities.

10.4   Acyclic Queueing Networks

Acyclic queueing networks are queueing networks without directed loop. Consequently, the nodes can be indexed such that the outgoing arcs of the nodes are directed toward nodes with higher index or to the environment and a customer visits each node at most once (Fig. 10.2).

Based on Burke's theorem and the results on the superposition and the filtering of independent Poisson processes (Property (h) of Poisson processes in Sect. 2.7.3) we can apply the same approach as the one applied for the analysis of the Tandem queueing network. Namely, we can (explicitly) compute the arrival rate to each node of the network and we can assume that the arrival process at the given node is a Poisson process with that arrival rate. Based on these assumptions the product form

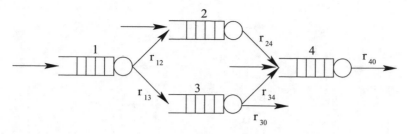

Fig. 10.2 Acyclic queueing network

solution remains valid, that is

$$p_{k_1, k_2, \cdots, k_N} = \prod_{i=1}^{N} p_{k_i}^{(i)},$$

where $p_{k_i}^{(i)}$ is the stationary probability of the k_i state of an M/M/1 queue with
Poisson arrival process with parameter λ_i and exponentially distributed service time
with parameter μ_i, which is

$$p_{k_i}^{(i)} = \left(1 - \frac{\lambda_i}{\mu_i}\right) \left(\frac{\lambda_i}{\mu_i}\right)^{k_i}.$$

10.5 Open, Jackson Type Queueing Networks

In the previous subsections we had acyclic queueing networks and based on Burke's
theorem we assumed that the arrival processes of the queues are independent
Poisson processes. Based on these assumptions we obtained product form solutions.
From now on we consider cyclic queueing networks and consequently we cannot
apply Burke's theorem any more due to the dependences in the arrival processes of
customers to a queue (Fig. 10.3).

The main results on this kind of queueing networks were published by J.R.
Jackson [5], in 1963. Since that these kind of networks are often named as Jackson
type networks. He considered the following queueing network model:

- The network is composed by N nodes,
- there are m_i servers at node i,
- the service time distribution at node i is exponentially distributed with parameter
  μ_i,
- from the environment customers arrive to node i according to a Poisson process
  with rate γ_i,

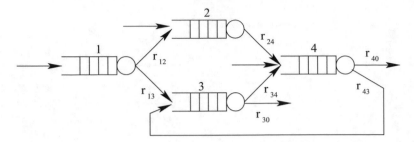

Fig. 10.3 Jackson type queueing network

- a customer getting served at node i goes to node j with probability $r_{i,j}$ ($i, j = 1, 2, \cdots , N$) and the probability that the customer departs from the network is

$$r_{i,0} = 1 - \sum_{k=1}^{N} r_{i,k} \quad i, j = 1, 2, \cdots , N .$$

10.5.1   *Stability Condition of Jackson Type Queueing Networks*

The following *traffic equations* define the traffic rate at the nodes of the network

$$\lambda_i = \gamma_i + \sum_{j=1}^{N} \lambda_j \, r_{j,i} \quad i = 1, 2, \cdots , N. \tag{10.1}$$

The left-hand side of the equation represents the aggregate traffic intensity arriving to node i. Due to the stability of the network nodes it is identical with the departing traffic intensity from node i. The right-hand side of the equation lists the traffic components arrive to node i. γ_i is the traffic component that arrive from the environment, and $\lambda_j \, r_{j,i}$ is the traffic component that departs from node j and goes to node i.

Introducing the row vector $\lambda = \{\lambda_i\}$ and $\gamma = \{\gamma_i\}$ and matrix $R = \{r_{ij}\}$ the traffic equation can be written in the following vector from

$$\lambda = \gamma + \lambda R$$

from which

$$\lambda = \gamma (I - R)^{-1}$$

if $(I - R)$ is non-singular.

The elements of matrix $(I - R)^{-1}$ have got a well-defined physical interpretation according to the following theorem. Let L_{ij} denote the number of visits to node j (before departing to the environment) by a customer arrived to node i.

Theorem 10.2

$$\left[(I - R)^{-1}\right]_{i,j} = E(L_{i,j}),$$

where the left-hand side denotes the i, j element of matrix $(I - R)^{-1}$.

Proof The number of visits to node j satisfies the following equation

$$E(L_{i,j}) = \delta_{i,j} + \sum_{k=1}^{N} r_{i,k} E(L_{k,j}),$$

where $\delta_{i,j}$ is the Kronecker delta, that is $\delta_{i,j} = 1$ if $i = j$, 0 otherwise. Introducing matrix L whose i, j element is $E(L_{i,j})$ we can rewrite the above equation in matrix form

$$L = I + RL,$$

from which the statement comes.   □

The theorem gives a condition for the nonsingularity of matrix $(I - R)$. $(I - R)$ is non-singular if all customers leave the queueing network after a finite number of visits at the nodes of the network.

The queueing network is said to be stable if all queues are stable, which holds when

$$\lambda_i < m_i \mu_i, \quad i = 1, 2, \cdots, N.$$

10.5.2   Stationary Distribution of Jackson Type Queueing Networks

According to the properties of Jackson type queueing networks the number of customers at the nodes of the network is a continuous-time Markov chain. Let k_i denote the number of customers at node i and let us introduce the following notations

$$\begin{aligned}
\mathbf{N} &= (k_1, \cdots, k_i, \cdots, k_j, \cdots, k_N), \\
\mathbf{N_{i,0}} &= (k_1, \cdots, k_i + 1, \cdots, k_j, \cdots, k_N), \\
\mathbf{N_{0,j}} &= (k_1, \cdots, k_i, \cdots, k_j - 1, \cdots, k_N), \\
\mathbf{N_{i,j}} &= (k_1, \cdots, k_i + 1, \cdots, k_j - 1, \cdots, k_N),
\end{aligned}$$

where in the last two cases $k_j \geq 1$. Using these notations we can describe the possible transitions of the Markov chains representing the number of customers at the nodes of the network.

- $\mathbf{N_{0,j}} \rightarrow \mathbf{N}$:
  a new customer arrives to node j from the environment and it increases the number of customers at node j from $k_j - 1$ to k_j. It happens at rate γ_j.
- $\mathbf{N_{i,0}} \rightarrow \mathbf{N}$:
  a customer departs to the environment from node j and it decreases the number of customers at node j from $k_j + 1$ to k_j. It happens at rate $r_{i,0}\alpha_i(k_i + 1)\mu_i$.
- $\mathbf{N_{i,j}} \rightarrow \mathbf{N}$:
  a customer gets served at node i and goes to node j. This transition decreases the number of customers at node i from $k_i + 1$ to i_j and increases the number of customers at node j from $k_j - 1$ to k_j. It happens at rate $r_{i,j}\alpha_i(k_i + 1)\mu_i$.

In the above expressions $\alpha_i(k_i) = \min\{k_i, m_i\}$ defines the coefficient of the service rate of node i when there are k_i customers at the node. When there are more customers at the node than servers, then all servers are working and the service rate is $m_i\mu_i$, and when there are less customers than servers, then there are idle servers and the service rate is $k_i\mu_i$.

Theorem 10.3 *The Markov chain characterized by the above defined state transitions has a product form stationary distribution, that is*

$$p_{\mathbf{N}} = p_{k_1,\cdots,k_N} = p_{k_1}^{(1)} p_{k_2}^{(2)} \cdots p_{k_N}^{(N)} \tag{10.2}$$

where $p_{k_i}^{(i)}$ is the stationary distribution of the M/M/m_i queue with Poisson arrival process with rate λ_i and exponentially distributed service time with parameter μ_i. The stationary probabilities of such queue are given as a function of $p_0^{(i)}$

$$p_{k_i}^{(i)} = \begin{cases} p_0^{(i)} \left(\dfrac{\lambda_i}{\mu_i}\right)^{k_i} \dfrac{1}{k_i!} & 0 \leq k_i \leq m_i \\[2em] p_0^{(i)} \left(\dfrac{\lambda_i}{\mu_i}\right)^{k_i} \dfrac{1}{m_i!} m_i^{m_i - k_i} & k_i \geq m_i \end{cases} \tag{10.3}$$

and $p_0^{(i)}$ can be obtained from the normalizing equation $\sum_{k_i=0}^{\infty} p_{k_i}^{(i)} = 1$.

Proof Based on the possible state transitions of the Markov chain the balance equation of state $\mathbf{N}$ is the following

$$p_N \left(\sum_{i=1}^{N} \gamma_i + \sum_{i=1}^{N} \alpha_i(k_i)\, \mu_i \right) = \sum_{i=1}^{N} p_{N_{i,0}} \alpha_i(k_i+1)\, \mu_i\, r_{i,0} + \sum_{j=1}^{N} p_{N_{0,j}}\, \gamma_j\, \mathcal{I}_{\{k_j>0\}}$$

$$+ \sum_{i=1}^{N} \sum_{j=1}^{N} p_{N_{i,j}}\, \alpha_i(k_i+1)\, \mu_i\, r_{i,j}\, , \tag{10.4}$$

where $\mathcal{I}_{\{k_j>0\}}$ is the indicator of $k_j > 0$, i.e., $\mathcal{I}_{\{k_j>0\}} = 1$ if $k_j > 0$ and $\mathcal{I}_{\{k_j>0\}} = 0$ otherwise.

The left-hand side of the equation is the rate at which the process departs from state $\mathbf{N}$ in equilibrium. It contains the state transitions due to a new customer arrival from the environment and due to a service completion. The right-hand side of the equation is the rate at which the process moves to from state $\mathbf{N}$ in equilibrium. It can happen due to a service of a queue from which the customer leaves the network, or due to the arrival of a new customer from the environment or due to a service completion at node i from where the customer moves to node j.

If $\gamma_i > 0$ and $\mu_i > 0$, then the Markov chain is irreducible, the solution of the stationary equation is unique, and it is enough to show that the product form solution (10.2) satisfies the balance Eq. (10.4). First we substitute the product form solution into the right-hand side of the balance equation and use that fact that from (10.3) we have $p_{k_i+1}^{(i)} = p_{k_i}^{(i)} \frac{\lambda_i}{\mu_i \alpha_i(k_i+1)}$ and $p_{k_i-1}^{(i)} = p_{k_i}^{(i)} \frac{\mu_i \alpha_i(k_i)}{\lambda_i}$. We obtain that

$$\sum_{i=1}^{N} p_{k_1}^{(1)} \cdots p_{k_i+1}^{(i)} \cdots p_{k_N}^{(N)}\, \alpha_i(k_i+1)\, \mu_i\, r_{i,0} + \sum_{j=1}^{N} p_{k_1}^{(1)} \cdots p_{k_j-1}^{(j)} \cdots p_{k_N}^{(N)}\, \gamma_j\, I_{k_j>0}$$

$$+ \sum_{i=1}^{N} \sum_{j=1}^{N} p_{k_1}^{(1)} \cdots p_{k_i+1}^{(i)} \cdots p_{k_j-1}^{(j)} \cdots p_{k_N}^{(N)}\, \alpha_i(k_i+1)\, \mu_i\, r_{i,j}$$

$$= p_{k_1}^{(1)} \cdots p_{k_N}^{(N)} \left(\sum_{i=1}^{N} \lambda_i\, r_{i,0} + \sum_{j=1}^{N} \frac{\mu_j \alpha_j(k_j)}{\lambda_j}\, \gamma_j + \sum_{i=1}^{N} \sum_{j=1}^{N} \frac{\mu_j \alpha_j(k_j)}{\lambda_j}\, \lambda_i\, r_{i,j} \right)$$

$$= p_{k_1}^{(1)} \cdots p_{k_N}^{(N)} \left(\sum_{i=1}^{N} \lambda_i\, r_{i,0} + \sum_{j=1}^{N} \frac{\mu_j \alpha_j(k_j)}{\lambda_j}\, \gamma_j + \sum_{j=1}^{N} \frac{\mu_j \alpha_j(k_j)}{\lambda_j} \underbrace{\sum_{i=1}^{N} \lambda_i\, r_{i,j}}_{\lambda_j - \gamma_j} \right)$$

$$= p_{k_1}^{(1)} \cdots p_{k_N}^{(N)} \left(\sum_{i=1}^{N} \lambda_i\, r_{i,0} + \sum_{j=1}^{N} \mu_j \alpha_j(k_j) \right)$$

$$= p_{k_1}^{(1)} \cdots p_{k_N}^{(N)} \left(\sum_{i=1}^{N} \gamma_i + \sum_{j=1}^{N} \mu_j \alpha_j(k_j) \right) . \tag{10.5}$$

In the third step of the derivation we used the traffic equation of queue j, (10.1) and in the fourth step we utilized that the intensity of customer arrival from the environment $\sum_{i=1}^{N} \gamma_i$ is identical with the intensity of the customer departure to the environment, $\sum_{i=1}^{N} \lambda_i \, r_{i,0}$ in equilibrium.

The obtained expression is the left-hand side of the balance equation assuming the product form solution of the stationary distribution.   □

There might be loops in a Jackson type queueing network from which the arrival processes of the nodes are not independent Poisson processes and the Burke's theorem is not applicable. Consequently, in this case we obtain product form solution in spite of dependent input processes of the queues. Worth noting that the reverse reasoning cannot be applied. The product form solution does not have implication on the dependencies of the arrival processes of the queues.

10.5.3   Traffic Theorem for Open Queueing Networks

Jackson type queueing networks enjoy a traffic property similar to the PASTA (Sect. 6.5) property of queueing systems with Poisson arrival process.

Theorem 10.4 *The distribution of the number of customers in the queues at the arrival instants of node j is identical with the stationary distribution of the number of customers in the queues.*

Proof We define an extended queueing network which contains one additional single server node, node 0, with respect to the original queueing network. The traffic matrix is also similar to the original one. It is modified only such that the customers going to node j is driven to node 0 and from node 0 to node j. The rest of the traffic matrix is unchanged. The extended queueing network is also of Jackson type and consequently its stationary distribution is product form $p_{N'} = p_{k_0}^{(0)} p_{k_1}^{(1)} p_{k_2}^{(2)} \cdots p_{k_N}^{(N)}$.

The service rate of node 0 is μ_0. As $\mu_0 \to \infty$ the behavior of the extended queueing network gets identical with the one of the original and the arrival instants of node j are the instants when there is one customer in node 0. This way the distribution of the customers at an arrival instant of node j is

$$\mathbf{P}(K_1 = k_1, \cdots, K_N = k_N | K_0 = 1) = \frac{\mathbf{P}(K_0 = 1, K_1 = k_1, \cdots, K_N = k_N)}{\mathbf{P}(K_0 = 1)} = p_N$$

□

This theorem is important for computing the delays in the queueing system.

10.6   Closed, Gordon-Newell Type Queueing Networks

The analysis of the closed queueing network counterpart of Jackson type queueing networks was first published by Gordon and Newell in 1967 [4]. Since that this kind of queueing networks are often named after them. The node behavior of the Gordon-Newell type queueing networks is identical with the one of Jackson type networks. At node i there are m_i servers with exponentially distributed service time with parameters μ_i and an infinite buffer.

In contrast to the Jackson type networks there is no arrival from and departure to the environment in closed queueing networks. This way the number of customers in the network is constant, denoted with K. If k_i denotes the number of customers at node i, then in each state of the network we have

$$\sum_{i=1}^{N} k_i = K \ .$$

Similar to the Jackson type network the number of customers at the nodes of the network form a Markov chain. In a closed queueing network the only possible state transition in this Markov chain is the $\mathbf{N_{i,j}} \to \mathbf{N}$ transition, that is a customer gets served at node i and moves to node j, the transition rate of this state transition is $\alpha_i(k_i + 1)\mu_i r_{i,j}$. This state transition decreases the number of customers at node i from $k_i + 1$ to k_i and increases the number of customers at node j from $k_j - 1$ to k_j.

The aggregate arrival rate of the nodes are characterized by the traffic Eq. (10.6)

$$\lambda_i = \sum_{j=1}^{N} \lambda_j \, r_{j,i} \quad i = 1, 2, \cdots, N \tag{10.6}$$

describes that customers arrive to node i are the customers that departed from node j and are directed to node i with probability r_{ij}. In a closed queueing network $\sum_{j=1}^{N} r_{ij} = 1$, since there is no departure to the environment. The solution of the traffic equation of closed queueing networks is not unique. Multiplying an arbitrary solution with a constant gives another solution of the traffic equation.

Theorem 10.5 *The stationary distribution of the number of customers in a Gordon-Newell type queueing network has product form. That is*

$$p_{\mathbf{N}} = p_{k_1, \cdots, k_N} = \frac{1}{G} \prod_{i=1}^{N} h_{k_i}^{(i)}, \tag{10.7}$$

where λ_i is an arbitrary nonzero solution of the traffic equation,

$$h_{k_i}^{(i)} = \begin{cases} \left(\dfrac{\lambda_i}{\mu_i}\right)^{k_i} \dfrac{1}{k_i!} & 0 \le k_i \le m_i \\[12pt] \left(\dfrac{\lambda_i}{\mu_i}\right)^{k_i} \dfrac{1}{m_i!} \, m_i^{m_i-k_i} & k_i \ge m_i \end{cases} \tag{10.8}$$

and $G = \sum_{\mathbf{N}} \prod_{i=1}^{N} h_{k_i}^{(i)}$.

Proof The proof follows the same pattern as the one for the Jackson type network. The balance equation for **N** is

$$p_{\mathbf{N}} \left(\sum_{i=1}^{N} \alpha_i(k_i)\,\mu_i \right) = \sum_{i=1}^{N} \sum_{j=1}^{N} p_{\mathbf{N}_{i,j}}\,\alpha_i(k_i + 1)\,\mu_i\,r_{i,j}, \tag{10.9}$$

where the left-hand side of the equation is the rate at which state **N** is left and the right-hand side is the rate at which state **N** is entered in equilibrium. Due to the irreducibility of the Markov chain we assume a unique solution of the balance equations (together with the normalizing equation, $\sum_{\mathbf{N} \in \mathscr{S}} p_{\mathbf{N}} = 1$) and we only show that the product form satisfies the balance equation.

Substituting the product form into the right-hand side of the balance equation gives

$$\sum_{i=1}^{N} \sum_{j=1}^{N} p_{k_1}^{(1)} \cdots p_{k_i+1}^{(i)} \cdots p_{k_j-1}^{(j)} \cdots p_{k_N}^{(N)}\,\alpha_i(k_i+1)\,\mu_i\,r_{i,j}$$

$$= p_{k_1}^{(1)} \cdots p_{k_N}^{(N)} \left(\sum_{i=1}^{N} \sum_{j=1}^{N} \frac{\mu_j \alpha_j(k_j)}{\lambda_j}\,\lambda_i\,r_{i,j} \right)$$

$$= p_{k_1}^{(1)} \cdots p_{k_N}^{(N)} \left(\sum_{j=1}^{N} \frac{\mu_j \alpha_j(k_j)}{\lambda_j} \underbrace{\sum_{i=1}^{N} \lambda_i\,r_{i,j}}_{\lambda_j} \right) \tag{10.10}$$

$$= p_{k_1}^{(1)} \cdots p_{k_N}^{(N)} \left(\sum_{j=1}^{N} \mu_j \alpha_j(k_j) \right),$$

which is identical with the left-hand side of the balance equation when the product form solution is assumed. The normalizing constant, G, ensures that the normalizing equation is satisfied. □

The main difficulties of the analysis of closed queueing networks are that the solution of the traffic equation is not unique and that the normalizing constant cannot be computed in a node-based manner only for the whole network. The computation of G requires the evaluation of all systems states which gets very high even for reasonable small networks. When there are N nodes and K customers in the network the number of system states is $\binom{N+K-1}{K}$ (e.g., for $N = 10, K = 25$ there are 52,451,256 states).

The commonly applied solution of the first problem is to add an additional equation to the set of traffic equations, $\lambda_1 = 1$, which makes its solution to be unique.

The second problem, the computation of the normalizing constant, G, is a real research challenge. There are a lot of proposals to compute the normalizing constant efficiently. Here we summarize the convolution algorithm [3] and the MVA (mean value analysis) algorithm [6].

10.6.1 Convolution Algorithm

The convolution algorithm was first published by Buzen [3]. In the original paper the nodes have single server, but it is easy to extend the algorithm for Gordon-Newell type queueing networks where the node i has m_i ($m_i \geq 1$) servers and an infinite buffer. We present the more general version of the algorithm.

Assuming that there are n nodes and k customers in the network let the assumed normalizing constant be

$$g(k, n) = \sum_{(k_1,\ldots,k_n),\sum_j k_j=k} \prod_{i=1}^n h_{k_i}^{(i)},$$

and $g(0, n) = 1$. When $g(k, n)$ is known we obtain the normalizing constant of the network with N nodes and K customers as $G = \sum_{\mathbf{N}} \prod_{i=1}^N h_{k_i}^{(i)} = g(K, N)$.

The following formula allows to determine $g(k, n)$ in a recursive manner

$$g(k, n) = \begin{cases} h_k^{(1)} & \text{ha } n = 1, \\ \sum_{j=0}^k h_j^{(n)} g(k-j, n-1) & \text{ha } n > 1. \end{cases} \tag{10.11}$$

In case of one node ($n = 1$) and $k \geq 1$ customers the recursive formula gives $h_k^{(1)}$ and in case of more than one nodes we have

$$g(k, n) = \sum_{(k_1,\ldots,k_n),\sum_j k_j=k} \prod_{i=1}^n h_{k_i}^{(i)}$$

$$= \sum_{(k_1,\ldots,k_n),\sum_j k_j=k,k_n=0} h_0^{(n)} \prod_{i=1}^{n-1} h_{k_i}^{(i)} + \ldots + \sum_{(k_1,\ldots,k_n),\sum_j k_j=k,k_n=k} h_k^{(n)} \prod_{i=1}^{n-1} h_{k_i}^{(i)}$$

$$= h_0^{(n)} g(k, n-1) + \ldots + h_k^{(n)} g(0, n-1).$$

This expression relates the normalizing constant of a network with n nodes with the normalizing constant of a network with $n - 1$ nodes.

The convolution algorithm starts from $n = 1, k = 1, \ldots, K$ and increases n to N step by step according to (10.11). The computational complexity of this algorithm is proportional with N and with K^2 (denoted as $O(NK^2)$) and its memory complexity is proportional with K (denoted as $O(K)$).

Another benefit of the convolution algorithm is that some interesting performance parameters are closely related to the $g(k, n)$ parameters. For example the probability that there are ℓ customers in queue k is

$$\mathbf{P}(k_\ell = k) = \sum_{(k_1,\ldots,k_n),\sum_j k_j=K,k_\ell=k} \frac{1}{G} \prod_{i=1}^{n} h_{k_i}^{(i)} = h_k^{(\ell)} \frac{g(K-k, N-1)}{g(K, N)},$$

and from this the utilization of node ℓ is

$$U_\ell = 1 - \mathbf{P}(k_\ell = 0) = 1 - h_0^{(\ell)} \frac{g(K, N-1)}{g(K, N)}.$$

10.6.2   Traffic Theorem for Closed Queueing Networks

The MVA algorithm is based on the traffic theorem for closed queueing networks. This way we first present the theorem.

Theorem 10.6 *In a closed Gordon-Newell type queueing network containing K customers the distribution of the number of customers at a customer arrival to node j is identical with the stationary distribution of the same network with $K - 1$ customers.*

Proof The proof is practically identical with the one provided for the open queueing network. We extend the network with a single server node 0 and redirect all customers going to node j to node 0 and from node 0 all customers go to node j. The rest of the network is left unchanged. The extended network is of Gordon-Newell type as well and this way it has a product form stationary distribution,

$$P_{k_0,k_1,\ldots,k_N,\sum_{i=0}^{N} k_i=K} = \frac{1}{G'} \prod_{i=0}^{N} h_{k_i}^{(i)}.$$

The service rate of node 0 is μ_0. As $\mu_0 \to \infty$ the behavior of the extended network and the original networks are identical and the arrival instants of node j are the instants when the number of customers in node 0 is 1. This way

$$\mathbf{P}\left(K_1 = k_1, \cdots, K_N = k_N, \sum_{i=0}^{N} k_i = K \mid K_0 = 1\right)$$

$$= \frac{\mathbf{P}\left(K_0 = 1, K_1 = k_1, \cdots, K_N = k_N, \sum_{i=0}^{N} k_i = K\right)}{\mathbf{P}(K_0 = 1)}$$

$$= \mathbf{P}\left(K_1 = k_1, \cdots, K_N = k_N, \sum_{i=1}^{N} k_i = K - 1\right).$$

□

10.6.3   MVA Algorithm

In the convolution algorithm the number of nodes is increasing in an iteration of the algorithm. The MVA algorithm is a kind of counterpart of the convolution algorithm in the sense that the MVA algorithm is also an iterative algorithm, but in this case the number of customers increases in an iteration step. According to this approach we analyze the involved quantities as a function of the number of customers in the network.

In contrast with the convolution algorithm the applicability of the MVA algorithm is limited to the case of single servers at the network nodes, i.e., $m_i = 1, i = 1, \ldots, N$ and the algorithm results mean performance measures. That is where its name comes from.

The mean time a customer spends at node i during a visit to node i is

$$\mathbf{E}(T_i(K)) = (1 + \mathbf{E}(N_i^*(K)))\frac{1}{\mu_i},$$

where $\mathbf{E}(N_i^*(K))$ denotes the mean number of customers present at node i at the arrival of an observed customer. According to the traffic theorem $\mathbf{E}(N_i^*(K))$ is identical with the stationary number of customers at node i when the number of customers in the network is $K - 1$, i.e., $\mathbf{E}(N_i(K-1))$, from which

$$\mathbf{E}(T_i(K)) = (1 + \mathbf{E}(N_i(K-1)))\frac{1}{\mu_i}.$$

On the other hand, the mean number of customer at node i in equilibrium is

$$\mathbf{E}(N_i(K)) = K\frac{\lambda_i \mathbf{E}(T_i(K))}{\sum_{j=1}^{N} \lambda_j \mathbf{E}(T_j(K))},$$

because the arrival rate to node i is proportional to an arbitrary nonzero solution of the traffic equation $\hat{\lambda}_i = \lambda_i c$, according to the Little's law $\mathbf{E}(N_i(K)) = \hat{\lambda}_i \mathbf{E}(T_i(K))$ and

$$K \frac{\lambda_i \mathbf{E}(T_i(K))}{\sum_{j=1}^{N} \lambda_j \mathbf{E}(T_j(K))} = K \frac{\hat{\lambda}_i \mathbf{E}(T_i(K))}{\sum_{j=1}^{N} \hat{\lambda}_j \mathbf{E}(T_j(K))} = K \frac{\mathbf{E}(N_i(K))}{\sum_{j=1}^{N} \mathbf{E}(N_j(K))}$$

$$= K \frac{\mathbf{E}(N_i(K))}{K} = \mathbf{E}(N_i(K)) .$$

Applying the Little's law for another time we obtain

$$\hat{\lambda}_i = \frac{\mathbf{E}(N_i(K))}{\mathbf{E}(T_i(K))} = K \frac{\lambda_i}{\sum_{j=1}^{N} \lambda_j \mathbf{E}(T_j(K))} .$$

By these we have all ingredients of the iterative algorithm:
  Initial value:

$$\mathbf{E}(N_i(0)) = 0,$$

  Iteration step:

$$\mathbf{E}(T_i(K)) = (1 + \mathbf{E}(N_i(K-1))) \frac{1}{\mu_i},$$

$$\mathbf{E}(N_i(K)) = K \frac{\lambda_i \mathbf{E}(T_i(K))}{\sum_{j=1}^{N} \lambda_j \mathbf{E}(T_j(K))},$$

  Closing step:

$$\hat{\lambda}_i = \frac{\mathbf{E}(N_i(K))}{\mathbf{E}(T_i(K))} .$$

The computational and the memory complexity of the algorithm are $O(KN^2)$ and $O(N)$. Compared to the convolution algorithm the MVA is more efficient when K is larger than N.

10.7   BCMP Networks: Multiple Customer and Service Types

The Jackson and the Gordon-Newell type queueing networks exhibit product form stationary distribution. This way efficient computational methods are applicable for the analysis of systems modeled by this kind of networks. The performance analysis and the development of efficient computer systems were based on this kind of simple and computable models for a long time. The analysis of more and more complex system behavior required the introduction of more complex queueing behavior and the analysis of the obtained queueing network models. It resulted in a fertile research to find the most general set of queueing networks with product form stationary distribution. The results of this effort are summarized in [1, 7] and the set of most general queueing networks with product form solution is commonly referred to as BCMP networks, which abbreviation comes from the initial of the authors of [1] Baskett, Chandy, Muntz, and Palacios.

The set of BCMP networks generalizes the previous queueing networks toward two main directions. In the previously discussed queueing networks the customers are indistinguishable and the service discipline is first come first served (FCFS). In BCMP networks the customers belong to customer classes which is distinguished by the system, because customers of different classes might arrive from the environment to the nodes at different rates, might obtain different service (service time distribution and service discipline) at the nodes, and might follow a different traffic routing probability upon a service completion. Still the customers of the same class are indistinguishable.

The arrival of class r customers to node i occurs at rate γ_{ir}. Upon a service of a class r customer at node i the customer attends the queue at node j as a class s customer with probability $P_{ir,js}$. That is, the customers might change their class right after a service completion. Let the number of customer classes be C. Then

$$\sum_{j=0}^{N}\sum_{s=1}^{C} P_{ir,js} = 1, \quad \forall i = 1,\ldots, N,\ r = 1,\ldots, C,$$

$P_{ir,0s}$ denotes the probability of departure to the environment.

A wide range of traffic models can be defined with appropriate setting of arrival rate γ_{ir} and traffic routing probability $P_{ir,js}$. We only list some examples below.

- The customer classes are independent and some classes behave as in open queueing networks and others as in closed queueing networks:

  $P_{ir,js} = 0$ if $r \neq s$, i.e., there is no class change. $\gamma_{ir} = 0$ if $r \leq C_z$ and for all $r > C_z$ there exists i such that $\gamma_{ir} > 0$, i.e., the first C_z classes of customers behave as in closed queueing networks and the rests as in open ones. Obviously the probability of departure to the environment should be related. $P_{ir,0s} = 0$ for $r \leq C_z$ and for all $r > C_z$ there exists i such that $P_{ir,0s} > 0$.
- Background traffic at a subset of the network:

  Let $\gamma_{ir} = 0$, if $i > N_z, r \leq C_z$ and $P_{ir,js} = 0$ if $i \leq N_z, j > N_z, r, s \leq C_z$. In this case the class $r \leq C_z$ customers load only node $i \leq N_z$ and form a kind of background traffic for customers of class $r > C_z$ in that part of the network.
- Multiple service at a node:

  Customer classes can be used to obtain a fixed number of services, u, at node i during a single visit to node i by customers of class v. For example, if for $r = v,\ldots, v+u-2$ let $P_{ir,js} = 1$ if $s = r+1,\ j = i$ and $P_{ir,js} = 0$ otherwise, and for $r = v + u - 1$ let $P_{ir,js} \geq 0$ if $s = r,\ j \neq i$ and $P_{ir,js} = 0$ otherwise, then we have the following behavior. A class v customer arrives to node i and gets served first as class v than as class $v + 1$ and so on, while it departs as a class $v + u - 1$ customer from node i and goes to node j as a class v customer.

The service disciplines at a node of a BCMP network can be one of the following disciplines:

1. FCFS (first come first served)

    The customers get to the server in the same order as they arrived to the node. With this service discipline the service time of all customers are exponentially distributed with the same parameter, which is common for all customer classes. The service intensity might depend on the number of all customers at the node.

2. PS (processor sharing)

    In this case the service capacity of the server is divided into as many equal parts as many customers are at the node and each part of the server capacity is assigned with a customer. That is, when there are n customers at the node all of them are served by $1/n$ portion of the full service capacity. In this case (if there are n customers at the node during the complete service of a customer) the service time of the customer is n times longer than it was if the full service capacity is assigned to this customer. With this service discipline the service time distribution of different customer classes might be different and can be more general than exponentially distributed. Service time distributions with rational Laplace transform (matrix exponential distributions) are allowed in this case.

3. LCFS-PR (last come first served preemptive-resume)

    The server serves one customer at a time, but such a way that the last arrived customer interrupts the service of the customer currently under service (if any) and starts having service. If during its service time a new customer arrives it gets interrupted and waits while all of the customers arrive later get served. At this point it gets to the server again and resumes the service process from the point it was interrupted.

    Similar to the PS case, with this service discipline the service time distribution of different customer classes might be different and can be more general than exponentially distributed. Service time distributions with rational Laplace transform (matrix exponential distributions) are allowed with this service discipline.

4. IS (infinite server)

    There are infinitely many servers in this service discipline, and this way all arriving customers attend an idle server at arrival. Similar to the PS and LCFS-PR cases, with this service discipline the service time distribution of different customer classes might be different and can be more general than exponentially distributed. Service time distributions with rational Laplace transform (matrix exponential distributions) are allowed with this service discipline.

With the introduction of customer classes the traffic equation only slightly modifies,

$$\lambda_{ir} = \gamma_{ir} + \sum_{j=1}^{N} \sum_{s=1}^{C} \lambda_{js} \, P_{js,ir} \, , \tag{10.12}$$

but in order to describe the product form solution of BCMP networks we need to introduce further cumbersome notations. To avoid it we restrict our attention to exponentially distributed service times instead of matrix exponentially distributed ones, but we allow all other generalizations of BCMP service disciplines.

Let N_{ir} denote the number of class r customers at node i and define the vectors $\mathbf{N_i} = \{N_{i1}, \ldots, N_{iC}\}$ and $\mathbf{N} = \{\mathbf{N_1}, \ldots \mathbf{N_N}\}$. This way vector $\mathbf{N}$ defines the distribution of the different classes of customers at the nodes of the network. With this notation the stationary distribution has the form

$$p_{\mathbf{N}} = \frac{1}{G} \prod_{i=1}^{N} h_{\mathbf{N_i}}^{(i)}, \qquad (10.13)$$

where

$$
h_{\mathbf{N_i}}^{(i)} =
\begin{cases}
\dfrac{N_i!}{\mu_i^{N_i}} \displaystyle\prod_{r=1}^{C} \frac{1}{N_{ir}!} \lambda_{ir}^{N_{ir}} & \text{if node } i \text{ is FCFS type,} \\[3ex]
N_i! \displaystyle\prod_{r=1}^{C} \frac{1}{N_{ir}!} \left(\frac{\lambda_{ir}}{\mu_{ir}} \right)^{N_{ir}} & \text{if node } i \text{ is PS or IS type,} \\[3ex]
\displaystyle\prod_{r=1}^{C} \frac{1}{N_{ir}!} \left(\frac{\lambda_{ir}}{\mu_{ir}} \right)^{N_{ir}} & \text{if node } i \text{ is LCFS-PR type,}
\end{cases}
$$

and $N_i = \sum_{r=1}^{C} N_{ir}$. μ_{ir} denotes the service rate of class r customer at node i.

10.8 Non-product Form Queueing Networks

In spite of the fact that BCMP networks allow a wide range of node behavior there are practical examples whose stationary solution do now exhibit product form solution. The most common reasons for not having a product form solution are

- non-Poisson customer arrival process,
- different exponentially distributed service time at FCFS type node for different customer classes,
- non-exponentially distributed service time at FCFS type node,
- non-matrix exponentially distributed service time,
- queueing nodes with finite buffer.

In general queueing networks the stochastic behavior of the number of (different classes of) customers at the nodes is not a Markov chain (e.g., in case of general interarrival or service time distribution). There are also cases when the number of (different classes of) customers at the nodes is a Markov chain but the stationary solution of this Markov chain does not enjoy product form (e.g., in case of Poisson arrival process and exponentially distributed service time distribution and finite capacity FCFS type nodes). In these cases no exact analysis methods are available and we need to resort to approximate analysis methods.

The majority of the approximate analysis methods are somewhat based on a product form solution. They analyze the system as its solution would be of product form and adjust the result obtained from the product form assumptions to better fulfill system equations.

From the set of approximate analysis methods of queueing networks we summarize traffic-based decomposition.

10.9   Traffic-Based Decomposition

The computation of the product form solution can be interpreted as the network nodes are independently analyzed based on the traffic load given by the solution of the traffic equation and the known service process (discipline and service time) of the node.

The traffic-based decomposition is an iterative procedure which analyzes the nodes of the network independently and the traffic load of the node under evaluation is determined based on the departure processes of the network nodes previously analyzed.

The advantages of the procedure are its flexibility and low computational cost, while its disadvantages are the potential inaccuracy of the results and the lack of evidence about the convergence of the procedure. In spite of the disadvantages this is a very often applied approximate analysis method in practice, because in the majority of the cases it converges and gives a reasonable agreement with simulation results.

The traffic-based decomposition procedure iteratively goes through all nodes of the network and performs the following steps for all nodes:

- traffic aggregation:
    aggregates the traffic coming from the environment and from the departure processes of the other nodes (based on the preceding iterations).
- node analysis and departure process computation:
    it is a single queueing system analysis step in which the parameters of the departure process are also computed.
- departure process filtering:
    computation of the traffic components going to the other nodes of the network.

The complexity of an iteration step and the accuracy of the results depend on the applied traffic descriptors. The flexibility of the procedure is due to the wide range of potentially applicable traffic descriptors. The most commonly used traffic descriptor is the average intensity of the traffic, such that a Poisson arrival process is assumed with the given intensity. Using this traffic model with more than one traffic classes results in a nontrivial analysis problem itself. If a more sophisticated traffic model is applied with, e.g., the higher moments or correlation parameters of the interarrival time distribution are considered, then the complexity of the analysis steps increase and the accuracy improves in general.

10.10 Exercises

Exercise 10.1 In the depicted queueing network the requests of input A are forwarded towards output B according to the following traffic routing probabilities $p = 0.3$, $q_1 = 0.2$, $q_2 = 0.5$, $q_3 = 0.3$.

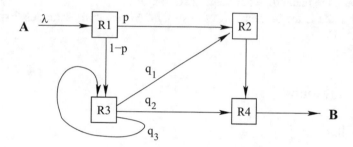

The requests from input A arrive according to a Poisson process with rate $\lambda = 50$. The service times are exponentially distributed in nodes R1, R2, and R3 with parameters $\mu_1 = 90$, $\mu_2 = 35$, and $\mu_3 = 100$, respectively. The service time in R4 is composed of two phases. The first phase is exponentially distributed with parameter $\mu_4 = 400$ and the second phase is deterministic with $D = 0.01$.

- Compute the traffic load of the nodes.
- Compute the mean and the coefficient of variation of the service time at node R4.
- Compute the system time at each node.
- Compute λ_{max} at which the system is at the limit of stability.

Exercise 10.2 In the depicted queueing network the requests of input A are forwarded towards output B according to the following traffic routing probabilities $p_{12} = 0.3$, $p_{13} = 0.7$.

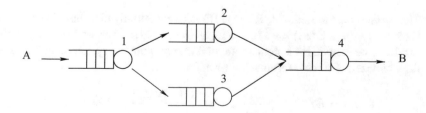

The requests from input A arrive according to a Poisson process with rate $\lambda = 50$. In nodes R1, R2, and R3 there are single servers, infinite buffers and the service times are exponentially distributed with parameters $\mu_1 = 80$, $\mu_2 = 45$, and $\mu_3 = 50$, respectively. There are two servers and two additional buffer at note R4. Both of the servers can serve requests with exponentially distributed service time with parameter $\mu_4 = 40$.

- Characterize the nodes with the Kendall's notation.
- Compute the traffic load of the nodes.
- Compute the system time at each node.
- Compute the utilization of the servers at Node R4.
- Compute the packet loss probability.
- Compute the mean time of a request from A to B.
- Which node is the bottleneck of the system? Which node saturates first when λ increases?

10.11   Solutions

Solution 10.1

- The following traffic equations characterize the traffic load of the nodes.

$$\lambda_1 = \lambda; \ \lambda_2 = p\lambda_1 + q_1\lambda_3; \ \lambda_3 = (1 - p)\lambda_1 + q_3\lambda_3; \ \lambda_4 = \lambda_2 + q_2\lambda_3.$$

With the given probabilities the solution of the traffic equations is

$$\lambda_1 = \lambda; \ \lambda_2 = \lambda/2; \ \lambda_3 = \lambda; \ \lambda_4 = \lambda.$$

- The service time at node R4 is the sum of independent random variables, $S_4 = X + D$, where X is exponentially distributed and D is deterministic.

$$\mathbf{E}\,(S_4) = \mathbf{E}\,(X + D) = \mathbf{E}\,(X) + \mathbf{E}\,(D) = \frac{1}{400} + \frac{1}{100} = \frac{1}{80},$$

$$\mathbf{CV}(S_4) = \frac{\mathrm{Var}(S_4)}{\mathbf{E}\,(S_4)^2} = \frac{\mathrm{Var}(X) + \mathrm{Var}(D)}{\mathbf{E}\,(X + D)^2} = \frac{(1/400)^2 + 0}{(1/400 + 1/100)^2} = \frac{1}{25}.$$

- The service time at node R1, R2, and R3 is exponentially distributed with mean $\mathbf{E}\,(S_1) = 1/\mu_1$, $\mathbf{E}\,(S_2) = 1/\mu_2$, and $\mathbf{E}\,(S_3) = 1/\mu_3$, respectively. Since the service time at node R4 is nonexponential we compute the system time based on the Pollaczek-Khinchin mean value formulae

$$\mathbf{E}\,(T_i) = \mathbf{E}\,(S_i) + \frac{\rho_i}{1 - \rho_i}\,\mathbf{E}\,(S_i)\,\frac{1 + \mathbf{CV}(S_i)}{2}, \quad i = 1, 2, 3, 4,$$

where $\rho_i = \lambda_i \mathbf{E}\,(S_i)$ and the coefficient of variation of the exponential service time is $\mathbf{CV}(S_1) = \mathbf{CV}(S_2) = \mathbf{CV}(S_3) = 1$.

- The utilization of the nodes as a function of λ is

$$\rho_1 = \lambda_1 \mathbf{E}\,(S_1) = \frac{\lambda}{90}, \rho_2 = \lambda_2 \mathbf{E}\,(S_2) = \frac{\lambda/2}{35} = \frac{\lambda}{70},$$

$$\rho_3 = \lambda_3 \mathbf{E}\,(S_3) = \frac{\lambda}{100}, \rho_4 = \lambda_4 \mathbf{E}\,(S_4) = \frac{\lambda}{80}.$$

Consequently the limit of stability is $\lambda_{max} = 70$, because node R2 gets instable at that load.

Solution 10.2

- There are M/M/1 queueing systems at Node R1, R2, and R3, and an M/M/2/4 at Node R4.

-

$$\lambda_1 = \lambda; \ \lambda_2 = p_{12}\lambda_1; \ \lambda_3 = p_{13}\lambda_1; \ \lambda_4 = \lambda_2 + \lambda_3.$$

With the given probabilities the solution of the traffic equations is

$$\lambda_1 = \lambda; \ \lambda_2 = 0.3\lambda; \ \lambda_3 = 0.7\lambda; \ \lambda_4 = \lambda.$$

- The system time at the M/M/1 type nodes is

$$\mathbf{E}\,(T_i) = \mathbf{E}\,(S_i) + \frac{\rho_i}{1 - \rho_i}\,\mathbf{E}\,(S_i), \quad i = 1, 2, 3,$$

where $\rho_i = \lambda_i \mathbf{E}\,(S_i) = \lambda_i/\mu_i$. The systems time at the M/M/2/4 type node can be computed based on the stationary solution of the following CTMC (denoted as p_i).

The system time at Node R4 is

$$\mathbf{E}\,(T_4) = L_4/\bar{\lambda}, \text{ where } L_4 = \sum_{i=0}^{4} i p_i, \bar{\lambda} = \lambda(1 - p_4).$$

- The utilization of the servers at Node R4 is $1 - p_0$.
- There is no packet loss at Node R1, R2, and R3. Packets are only lost at Node R4 with probability p_4. Due to the fact that all packets go to Node R4, the overall packet loss probability is p_4 as well.

- A packet take the path through nodes R1, R2, and R4 with probability p_{12} and through nodes R1, R3, and R4 with probability p_{13}, consequently

$$\mathbf{E}\left(T\right) = p_{12}\left(\mathbf{E}\left(T_1\right) + \mathbf{E}\left(T_2\right) + \mathbf{E}\left(T_4\right)\right) + p_{13}\left(\mathbf{E}\left(T_1\right) + \mathbf{E}\left(T_3\right) + \mathbf{E}\left(T_4\right)\right).$$

- Node R4 never saturates because it has finite buffer. The utilization of the other 3 nodes is

$$\rho_1 = \lambda_1 \mathbf{E}\left(S_1\right) = \frac{\lambda}{80},$$

$$\rho_2 = \lambda_2 \mathbf{E}\left(S_2\right) = \frac{0.3\lambda}{45} = \frac{\lambda}{150},$$

$$\rho_3 = \lambda_3 \mathbf{E}\left(S_3\right) = \frac{0.7\lambda}{50} \sim \frac{\lambda}{71.4}.$$

Node R3 is the bottleneck which saturates first at around $\lambda = 71.4$.

References

1. Baskett, F., Chandy, K.M., Muntz, R.R., Palacios, F.G.: Open, closed and mixed networks of queues with different classes of customers. J. ACM **22**, 248–260 (1975)
2. Burke, P.J.: The output of a queuing system. Oper. Res. **4**, 699–704 (1956)
3. Buzen, J.: Computational algorithms for closed queueing networks with exponential servers. Commun. ACM **16**, 527–531 (1973)
4. Gordon, W.J., Newell, G.F.: Closed queueing systems with exponential servers. Oper. Res. **15**, 254–265 (1967)
5. Jackson, J.R.: Jobshop-like queueing systems. Manag. Sci. **10**, 131–142 (1963)
6. Reiser, M., Lavenberg, S.S.: Mean value analysis of closed multi-chain queueing networks. J. ACM **27**, 313–322 (1980)
7. Leonid Ponomarenko, Che Soong Kim, and Agassi Melikov. *Performance analysis and optimization of multi-traffic on communication networks*. Springer Science & Business Media, 2010.

Chapter 11
Applied Queueing Systems

11.1 Bandwidth Sharing of Finite Capacity Links with Different Traffic Classes

Traditional telephone networks were designed to implement a single type of communication service, i.e., the telephone service. Today's telecommunication networks implement a wide range of communication services. In this section we introduce Markov models of communication services which compete for the bandwidth of a finite capacity communication link.

11.1.1 Traffic Classes

There are several important features of the traffic sources of communication services which allow their classification. Here assume that the traffic sources require the setting up of a connection for a finite period of time during which data communication is carried on between the parties of the communication service. We classify the traffic sources based on the bandwidth of the data transmission during a connection. The simplest case is when data is transmitted with a fixed bandwidth during a connection. This case is commonly referred to as constant bit rate (CBR). A more general traffic behavior is obtained when the bandwidth of data transmission varies during a connection. This case is commonly referred to as variable bit rate (VBR). The most common way of bandwidth variation is when the bandwidth alternates between 0 and a fixed bandwidth. These VBR sources are referred to as ON-OFF sources. We restrict our attention to the ON-OFF case. The most complex traffic sources adjust their bandwidth according to the available capacities of the network resources. There are two classes of this kind of sources.

© Springer Nature Switzerland AG 2019
L. Lakatos et al., *Introduction to Queueing Systems with Telecommunication Applications*, https://doi.org/10.1007/978-3-030-15142-3_11

- *Adaptive* traffic sources set up a connection for a given period of time and transmit data according to the available bandwidth in the network. If the network resources are occupied during the connection of an adaptive traffic source, then the source transmits data with a low bandwidth and the overall transmitted data during a connection is low.
- *Elastic* traffic sources set up a connection for transmitting a given amount of data. The bandwidth of the data transmission depends on the available bandwidth in the network. This way if the network resources are occupied during the connection of an elastic traffic source, then the period of the connection is extended in such a way that the source transmits the required amount of data.

In this section we assume that the traffic sources follow a memoryless time homogeneous stochastic behavior and consequently, the arrival processes are Poisson processes and the connection times are exponentially distributed except for the elastic class where the amount of data to transmit is exponentially distributed. Additionally, the traffic sources are characterized by their bandwidth parameters. In case of CBR and ON-OFF VBR sources, the bandwidth parameter is the bandwidth of the active period. In case of adaptive and elastic sources, the bandwidth parameters are the minimal and maximal bandwidth at which the source can transmit data.

Consequently, in case of the different kind of traffic sources a class k traffic source is characterized by the following parameters:

- constant bit rate connection
  connection arrival intensity λ_k, bandwidth requirement c_k, parameter of the exponentially distributed connection holding time μ_k,
- variable bit rate connection
  connection arrival intensity λ_k, bandwidth requirement in the ON state c_k, parameters of the exponentially distributed connection holding time, ON time and OFF time, μ_k, α_k, β_k, respectively,
- adaptive connection
  connection arrival intensity λ_k, minimal bandwidth $c_{min}^{(k)}$, maximal bandwidth $c_{max}^{(k)}$, parameter of the exponentially distributed connection holding time μ_k,
- elastic connection
  connection arrival intensity λ_k, minimal bandwidth $c_{min}^{(k)}$, maximal bandwidth $c_{max}^{(k)}$, parameter of the exponentially distributed amount of transmitted data δ_k.

These parameters define the arrival process and the bandwidth needs of traffic sources, but does not define completely the service procedure as the common resource (the finite capacity link) is shared among the traffic types and classes. In case of traditional telephone services the procedure at a new telephone call is obvious: accept as many calls as possible with the given finite capacity link. In case of different traffic classes more complex procedures are required to properly utilize the resource and to provide the desired service features to each traffic class. The set of rules concerning the acceptance or rejection of a new connection is referred to as call admission control (CAC). CAC defines the acceptance or rejection of a new connection of all types in case of all possible traffic conditions. We are going to see some typical CAC and their properties.

The most common performance parameters of interest in these kinds of traffic models are

- per class connection dropping probability (at arrival connection arrival),
- VBR connections dropping probabilities (during ongoing connection at an OFF to ON transition),
- per class mean bandwidth of adaptive and elastic connections,
- sojourn time of elastic connections.

Different dimensioning methods apply for different traffic classes. In the following subsections we investigate the simple Markov models of these traffic classes which are the bases of the complex dimensioning methods used in practice.

11.1.2   Bandwidth Sharing by CBR Traffic Classes

One of the first generalizations of traditional telecommunication models is due to the coexistence of communication services with different bandwidth requirements. When a link is utilized by different kinds of CBR connections with the above detailed Markovian properties, then the overall system behavior can be described by a CTMC. The main problem of analyzing the performance parameters through this CTMC is the potentially very high number of states. If a finite capacity link of bandwidth C is utilized by I different kinds of CBR connections, then a state of the CTMC should represent the number of ongoing connections of each class and the number of states is proportional with the product $\prod_{i=1}^{I} \left(\frac{C}{c_i} + 1 \right)$.

To overcome this practical problem an efficient numerical procedure was proposed by two researchers independently [17, 30], which is often named as the Kaufman-Roberts method. The procedure is based on the fact that the large CTMC, which represents the number of ongoing connections of each class, satisfies the local balance equations

$$\lambda_i \, p(n_1, \ldots, n_i - 1, \ldots, n_I) = n_i \mu_i \, p(n_1, \ldots, n_i, \ldots, n_I),$$

where $p(n_1, \ldots, n_i, \ldots, n_I)$ denotes the stationary probability of the state where the number of class i connections is n_i for $i = 1, \ldots, I$. The local balance equation represents that the stationary state transition rate due to an arriving class i connection is in balance with the stationary state transition rate due to a departing class i connection. The main idea of the Kaufman-Roberts method is to unify those states of the large Markov chain which represent the same bandwidth utilization of the link. In state $(n_1, n_2, \ldots, n_I)$ the bandwidth utilization is $c = \sum_{i=1}^{I} n_i c_i$.

Summing up the local balance equations for the states where the bandwidth utilization on the right-hand side is c we have

$$\sum_{i=1}^{I} \lambda_i \, P(c - c_i) \, \mathscr{I}_{\{c_i \geq c\}} = \sum_{i=1}^{I} n_i \mu_i \, P(c),$$

$$\sum_{i=1}^{I} \frac{\lambda_i c_i}{\mu_i} \, P(c - c_i) \, \mathscr{I}_{\{c_i \geq c\}} = \underbrace{\sum_{i=1}^{I} n_i c_i \, P(c),}_{c}$$

$$\sum_{i=1}^{I} \frac{\lambda_i c_i}{\mu_i c} \, P(c - c_i) \, \mathscr{I}_{\{c_i \leq c\}} = P(c),$$

where $P(c)$ denotes the sum of the stationary probabilities of the states where the bandwidth utilization is c, that is $P(c) = \sum_{(n_1, n_2, \ldots, n_I): \sum_{i=1}^{I} n_i c_i = c}$ $p(n_1, n_2, \ldots, n_I)$. The last equation is the core of the Kaufman-Roberts method, which computes the relative (nonnormalized) probabilities of the link utilization levels first and normalizes probabilities after that as follows.

1. Let $\tilde{P}(0) = 1$ and for $c = 1, 2, \ldots, C$ compute

$$\tilde{P}(c) = \sum_{i} \frac{\lambda_i c_i}{\mu_i c} \, \tilde{P}(c - c_i) \, \mathscr{I}_{\{c_i \leq c\}} \, .$$

2. Compute $\tilde{P} = \sum_{c=0}^{C} \tilde{P}(c)$.
3. Normalize the probabilities by $P(c) = \tilde{P}(c)/\tilde{P}$.

There is an implicit technical assumption which is necessary for the application of the Kaufman-Roberts method. It is required that there is a bandwidth unit such that each c_i is an integer multiple of this bandwidth unit. (The method remains applicable if C is not an integer multiple of the bandwidth unit.) Fortunately, in important applications such bandwidth unit exists.

Having the stationary probabilities of the utilization levels we can compute the loss probabilities. If the CAC allows all connections entering the link as long as the available bandwidth is not less than the bandwidth of the entering connection, then the loss probability of class i connections is

$$b_i = \sum_{c > C - c_i} P(c).$$

It is a straightforward consequence of the CAC that connections with higher bandwidth requirements have higher loss probability. If a kind of fairness is required among the different classes such that each class experiences the same loss probability, then the CAC needs to be modified. Let us assume that the traffic class with the highest bandwidth is class I. If the CAC is modified such that each incoming connection is rejected when the available bandwidth is less than c_I, then the distribution of the link utilization changes, but each classes are accepted and rejected at the same time at the different link utilization levels and consequently they have the same loss probability. If the CAC depends only on the link utilization level (like in the case of the modified CAC with identical dropping probabilities), then the Kaufman-Roberts method remains applicable. In this case the main iteration step of the procedure changes to

$$\tilde{P}(c) = \sum_i \frac{\lambda_i c_i}{\mu_i c} \, \tilde{P}(c - c_i) \, \text{CAC}(i, c - c_i) \,,$$

where $\text{CAC}(i, c)$ is one if a class i connection is accepted at link utilization c, and zero otherwise. The link utilization level dependent CAC can also be generalized to probabilistic CAC. In this case the main iteration step of the procedure remains the same as for the deterministic one and $\text{CAC}(i, c)$ indicates the probability that a class i connection is accepted at link utilization c.

11.1.3   Bandwidth Sharing with VBR Traffic Classes

When a link is utilized by different kinds of VBR connections and each of them is characterized by the above described Markovian properties, then the overall system behavior can be described by a CTMC. The states of this CTMC represent the number of ongoing VBR connections of each classes and the number of connections in ON phase, $(n_1, m_1, n_2, m_2, \ldots, n_I, m_I)$. Note that $n_i \geq m_i, i = 1, \ldots, I$, and $\sum_{i=1}^{I} m_i c_i \leq C$, where the second inequality means that the utilized bandwidth should not exceed the link capacity. The state space of this CTMC is even larger than the one for CBR connections, which represents only the number of ongoing connections of each classes, but unfortunately there is no more efficient computation method available for this model than to solve the CTMC. It is due to the fact that this CTMC does not fulfill the local balance equations. Anyhow, the numerical solution of this CTMC is still possible for a limited number of VBR classes and connections.

Generally, the CAC for VBR connections is more complex than the one for CBR connections. A conservative CAC does not allow more VBR connections than the link can serve assuming that all VBR connections are in the ON state. That is, $\sum_{i=1}^{I} n_i c_i \leq C$ holds for each state. Unfortunately, a conservative CAC results in a very low resource utilization, especially when the length of the ON period is short with respect to the length of the OFF period. In these cases, it is worth allowing more VBR connections than a conservative CAC in order to increase the link utilization. The drawback of nonconservative CAC is that accepted ongoing VBRs can be dropped due to insufficient capacity with positive probability at an OFF to ON phase transmission. In practice, it is usually required that the dropping probability of ongoing VBR connections is much lower than the newly arriving ones.

The possible state transitions of the Markov chains are

(a) $(n_1, m_1, \ldots, n_i, m_i, \ldots, n_I, m_I) \rightarrow (n_1, m_1, \ldots, n_i + 1, m_i + 1, \ldots, n_I, m_I)$
    with rate λ_i if $\text{CAC}(i, \{n_1, m_1, \ldots, n_i, m_i, \ldots, n_I, m_I\}) = 1$,

(b) $(n_1, m_1, \ldots, n_i, m_i, \ldots, n_I, m_I) \rightarrow (n_1, m_1, \ldots, n_i - 1, m_i - 1, \ldots, n_I, m_I)$
    with rate $m_i \mu_i$,

(c) $(n_1, m_1, \ldots, n_i, m_i, \ldots, n_I, m_I) \rightarrow (n_1, m_1, \ldots, n_i - 1, m_i, \ldots, n_I, m_I)$
    with rate $(n_i - m_i)\mu_i$,

(d) $(n_1, m_1, \ldots, n_i, m_i, \ldots, n_I, m_I) \rightarrow (n_1, m_1, \ldots, n_i, m_i - 1, \ldots, n_I, m_I)$
    with rate $m_i \alpha_i$,

(e) $(n_1, m_1, \ldots, n_i, m_i, \ldots, n_I, m_I) \rightarrow (n_1, m_1, \ldots, n_i, m_i + 1, \ldots, n_I, m_I)$
    with rate $(n_i - m_i)\beta_i$ if $\displaystyle\sum_{j=1}^{I} m_j c_j + c_i \leq C$,

(f) $(n_1, m_1, \ldots, n_i, m_i, \ldots, n_I, m_I) \rightarrow (n_1, m_1, \ldots, n_i - 1, m_i, \ldots, n_I, m_I)$
    with rate $(n_i - m_i)\beta_i$ if $\displaystyle\sum_{j=1}^{I} m_j c_j + c_i > C$,

where $\text{CAC}(i, \{n_1, m_1, \ldots, n_I, m_I\})$ denotes the CAC decision in state $(n_1, m_1, \ldots, n_I, m_I)$ for an incoming class i connection and state transitions with rate zero are impossible. According to the bandwidth limit of the link

$$\text{CAC}(i, \{n_1, m_1, \ldots, n_I, m_I\}) = 0 \text{ if } \sum_{j=1}^{I} m_j c_j + c_i > C.$$

The transitions represent the following events:

(a) new class i connection arrival,
(b) departure of a class i connection which is in phase ON,
(c) departure of a class i connection which is in phase OFF,
(d) a class i connection switches from phase ON to phase OFF,
(e) a class i connection switches from phase OFF to phase ON,
(f) a class i connection is lost due to insufficient bandwidth for phase OFF to phase ON transition.

Having the stationary probabilities of this CTMC, denoted by $p(n_1, m_1, \ldots, n_I, m_I)$, the dropping probability of class i incoming and ongoing connections can be computed as follows.

$$b_i^{new} = \frac{\text{number of class } i \text{ incoming connections dropped at arrival}}{\text{number of class } i \text{ incoming connections}}$$
$$= \sum_{n_1, m_1, \ldots, n_I, m_I} p(n_1, m_1, \ldots, n_I, m_I)(1 - \text{CAC}(i, \{n_1, m_1, \ldots, n_I, m_I\})),$$

$$b_i^{ongoing} = \frac{\text{number of class } i \text{ dropped ongoing connections}}{\text{number of class } i \text{ incoming connections}}$$
$$= \sum_{\mathscr{S}_i} \frac{(n_i - m_i)\beta_i}{\lambda_i} p(n_1, m_1, \ldots, n_I, m_I),$$

where $\mathscr{S}_i$ denotes the set of states for which $\sum_{j=1}^{I} m_j c_j + c_i > C$. The link utilization is

$$\rho = \sum_{n_1, m_1, \ldots, n_I, m_I} p(n_1, m_1, \ldots, n_I, m_I) \sum_{j=1}^{I} m_j c_j.$$

If the state space of the CMTC is such that the stationary analysis is feasible, then the computation of the performance parameters is straight forward, but the inverse problem, the design of a CAC which fulfills blocking probability constraints and maximizes the link utilization, is still an interesting research problem.

11.1.4   Bandwidth Sharing with Adaptive Traffic Classes

In case of adaptive traffic classes the connections can adopt their bandwidth to the available bandwidth of the link between the class specific bandwidth limits $c_{min}^{(i)}$ and $c_{max}^{(i)}$. If the link is not completely utilized, then each connection receives its maximal bandwidth. If the sum of the maximal bandwidth needs is larger than the link capacity, then the link is completely utilized and a bandwidth reduction affects the bandwidth of all classes according to the following rule. If the actual bandwidth of a class i connection is c and c is less than $c_{max}^{(i)}$, then for any other class j the bandwidth is c if $c \leq c_{max}^{(j)}$ or $c_{max}^{(j)}$ if $c > c_{max}^{(j)}$. It means that the bandwidth of each class is reduced to the same level c, if the class specific maximal bandwidth $c_{max}^{(j)}$ is not less than c. Consequently the main features of the bandwidth sharing with adaptive traffic classes are as follows

- the departure rate of connections are proportional to the number of active connections and it is independent of the instantaneous bandwidth of the connections,
- bandwidth of the connections varies according to the link capacity and the number of active connections,
- an arriving class i connection is rejected when the minimal required bandwidth $c_{min}^{(i)}$ cannot be granted,
- the transmitted data of a connection depends on the instantaneous bandwidth during the connection.

Example 11.1 We demonstrate the behavior of adaptive connections on a finite capacity link in case of a single adaptive class with link bandwidth $C = 3$ Mbps, bandwidth limits $c_{min} = 1$ Mbps, $c_{max} = 2$ Mbps, connection arrival and departure rate λ [1/s] and μ [1/s], respectively. Due to the memoryless arrival and departure processes the number of active connections $X(t)$ is a Markov chain and it is depicted in Fig. 11.1. The figure also indicates the bandwidth of the ongoing connections. If

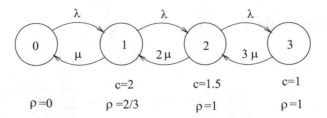

Fig. 11.1 Markov chain of the number of adaptive connections on a finite capacity link

there are 3 ongoing connections the arriving connections are rejected, because in case of 4 connections the common bandwidth $c = 3/4$ Mbps would be smaller than the minimal bandwidth requirement $c_{min} = 1$ Mbps.

The main performance measures of this system are the mean bandwidth of connections

$$\bar{c} = \sum_{i=0}^{3} ic(i)p_i = 2p_1 + 2 \cdot 1.5 p_2 + 3 \cdot 1 p_3,$$

the link utilization

$$\rho = \sum_{i=0}^{3} \rho_i p_i = 2/3 p_1 + 1 p_2 + 1 p_3,$$

and the blocking probability

$$b = p_3,$$

where p_i, ρ_i, and $c(i)$ denote the stationary probability, the utilization, and the bandwidth of a connection in state i, respectively.

Example 11.2 The approach applied for the single class model can be used for the analysis of models with multiple adaptive classes. In case of two adaptive classes with link bandwidth $C = 5$, bandwidth limits $c_{min}^{(1)} = 1.5$, $c_{max}^{(1)} = 3$, $c_{min}^{(2)} = 1$, $c_{max}^{(2)} = 2$, connection arrival and departure rate λ_1, λ_2 and μ_1, μ_2 the Markov chain describing the number of active connections of class 1 and 2 is depicted in Fig. 11.2. The figure indicates the bandwidth of the ongoing connections with bold characters. Arriving connections of both classes are rejected in states $(0, 5)$, $(1, 2)$, $(2, 1)$, and $(3, 0)$, and additionally arriving connections of class 1 are rejected in states $(0, 3)$ and $(0, 4)$. Considering only the minimal bandwidth constraints and the link bandwidth, state $(1, 3)$ would be feasible $(1.5 + 3 \cdot 1 < 5)$, but the identical bandwidth sharing of the classes makes this state infeasible, because it violates the minimal bandwidth requirement of class 1 $(5/4 < c_{min}^{(1)})$. In contrast, in state $(1, 1)$

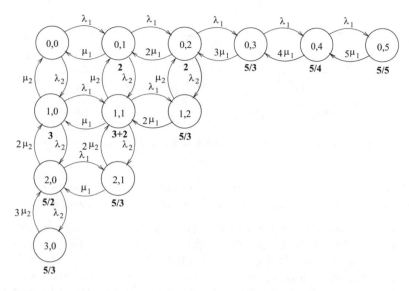

Fig. 11.2 Markov chain of the number of adaptive connections with two adaptive classes

the bandwidth is unevenly divided. It is possible because the class 2 connection obtains its maximal bandwidth and all the remaining bandwidth is utilized by the class 2 connection. The performance measures can be computed in a similar way as in the case of a single adaptive class.

11.1.5   *Bandwidth Sharing with Elastic Traffic Classes*

In case of elastic traffic classes the connections can adopt their bandwidth to the available bandwidth similar to the adaptive class, but the amount of data transmitted through a connection is fixed. This way during the period when the bandwidth is low the sojourn time of the elastic connections is longer. The bandwidth of elastic connections is also bounded by class specific bandwidth limits $c_{min}^{(i)}$ and $c_{max}^{(i)}$, and the bandwidth sharing between traffic classes follows the same role as in case of adaptive connections. The main features of the bandwidth sharing with elastic traffic classes are as follows

- the departure rate of a connection of class i depends on the instantaneous bandwidth of the class i connections, this way the length of the connections varies according to the link capacity and the number of active connections,
- an arriving class i connection is rejected when the minimal required bandwidth $c_{min}^{(i)}$ cannot be granted.

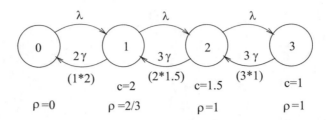

Fig. 11.3 Markov chain of the number of elastic connections on a finite capacity link

- the transmitted data of a connection is a class specific random variable and it does not depend on the instantaneous bandwidth during the connection.

Example 11.3 We demonstrate the behavior of elastic connections with the same model as in Example 11.2 but assuming that the connections are elastic. That is, the link bandwidth is $C = 3$ Mbps, the bandwidth limits are $c_{min} = 1$ Mbps, $c_{max} = 2$ Mbps, the connection arrival rate is λ [1/s], and the amount of transmitted data of an elastic connection is exponentially distributed with parameter $\gamma [1/Mb]$. Due to the memoryless arrival process and the exponential distribution of the amount of transmitted data of the elastic connections the number of active connections $X(t)$ is a Markov chain (Fig. 11.3). The figure indicates the bandwidth of the ongoing connections (parameter c) and the computation of the departure rate of connections in brackets. For example in state two there are two ongoing connections with bandwidth 1.5 Mbps. The rate at which one of them completes the data transmission is 1.5 Mb/s $\times \gamma$ 1/Mb = 1.5 1/s and the sum of the two identical departure rates is 2×1.5 1/s = 3 1/s. Apart from these differences the bandwidth sharing, the link utilization, and the rejection of arriving connections are the same as in the case of adaptive connections.

11.1.6   *Bandwidth Sharing with Different Traffic Classes*

In the previous subsections we discussed the bandwidth sharing of a finite capacity link by traffic classes of the same type. All of the discussed traffic classes have got a memoryless stochastic behavior and this way the performance of the models can be analyzed by CTMCs. Theoretically it is a solved problem, but practical limitations arises when the size of the state space gets large, which is often the case in practically interesting situations. The case when only CBR type connections are present at the link allows an efficient analysis method referred to as Kaufman-Roberts method. If any other type of connections appear this method is not applicable any more. The Markov chain based framework of the previous subsections is also applicable for the analysis of the bandwidth sharing by traffic classes of different types. Interested readers might find further details in [25, 27, 28, 31].

11.2   Packet Transmission Through Slotted Time Channel

In this section we focus on a peculiar detail of modeling slotted time systems with DTMCs. It is the definition of time slots, more precisely the positioning of the beginning of the time slot on the continuous time axes. The modeler has some freedom in this respect and consequently different DTMC models can be obtained for describing the same system behavior. It turns out that these different models result in the same performance parameters if the performance parameters are independent of the slot definition, which is the case with the majority of the practically important queueing parameters. Below we evaluate two models of a simple packet transmitter, which can be seen as discrete-time counterparts of M/M/1 queue.

Consider the packet transmitter with the following properties.

- Packet arrival process: In each time slot 1 packet arrives with probability p and 0 packet arrives with probability $1 - p$ independent of the past history.
- Service (packet transmission) process: If there is at least one packet to transmit, then 1 packet is transmitted with probability q and 0 packet is transmitted with probability $1 - q$ independent of the past history, it means that the service time of a packet is geometrically distributed with parameter q (Pr service time $= k = (1 - q)q^{k-1}$).
- Service discipline: $FIFO$.
- Buffer size: infinite.

Let X_n denote the number of packets in the system at the beginning of the nth time slot. X_n is a DTMC with a special birth and death structure and infinite state space. Depending on the definition of the beginning of a time slot the following two cases arise.

- **Case I**: A time slot starts with packet transmission (if any) and after that packet arrival can happen.
- **Case II**: A time slot starts with packet arrival (if any) and after that packet transmission can happen.

These two cases result in different Markov chains as it is detailed below.

- **Case I** In case I, the X_n Markov chain can be described by the following evolution equation

$$X_{n+1} = (X_n - V_{n+1})^+ + Y_{n+1},$$

where r. v. Y_{n+1} is the number of packets arrivals during time slot $n + 1$ and r. v. V_{n+1} is the number of packets which can be transmitted during time slot $n + 1$. Y_n and V_n are Bernoulli distributed with parameter p and q, respectively. The state transition graph of this Markov chain is

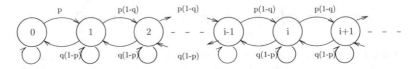

The stationary distribution of this Markov chain is

$$p_0 = \frac{q-p}{q}, \quad p_i = \left(\frac{p(1-q)}{q(1-p)}\right)^i \frac{q-p}{(1-q)q}; \quad i \geq 1.$$

The denominator of the stationary probabilities already indicates that the condition of stability is $p < q$. This result can also be obtained from the evolution equation $\mathbf{E}(V) > \mathbf{E}(Y) \to p < q$ and from the Foster criterion (Theorem 3.15) $q(1-p) > p(1-q) \to p < q$. The basic performance measures can be computed from the stationary distribution. The utilization is

$$\rho = 1 - p_0 = 1 - \left(1 - \frac{p}{q}\right) = \frac{p}{q},$$

the mean of the stationary number of packets in the queue is

$$\mathbf{E}(X) = \sum_{i=1}^{\infty} i p_i = \frac{p(1-p)}{q-p},$$

and the mean of the stationary system time of a packet is

$$\mathbf{E}(T) = \frac{1}{q} p_0 + \sum_{i=1}^{\infty} p_i \left(\frac{i+1}{q} - 1\right) = \frac{1-p}{q-p}.$$

- **Case II** In this case the evolution equation has the form

$$X_{n+1} = (X_n - V_{n+1} + Y_{n+1})^+,$$

and the transition graph of the Markov chain is

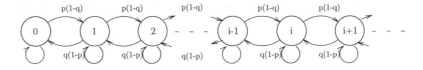

Due to the different transition probabilities around state 0 we have a different stationary distribution

$$p_i = \left(\frac{p(1-q)}{q(1-p)}\right)^i \frac{q-p}{(1-p)q}; \quad i \geq 0.$$

The computation of some performance measures is identical in this case. For example, the condition of stability is $\mathbf{E}(V) > \mathbf{E}(Y) \to p < q$ based on the evolution equation and $q(1 - p) > p(1 - q) \to p < q$ based on the Foster criterion. The computation of some other performance measures is different in case II. For example the utilization is computed as

$$\rho = 1 - p_0(1 - p) = 1 - \frac{(q - p)(1 - p)}{(1 - p)q} = \frac{p}{q},$$

because the server can be utilized by a packet which arrives with probability p when it is idle at the beginning of a time slot. The mean of the stationary system time can be computed as

$$\mathbf{E}(T) = \sum_{i=0}^{\infty} p_i \left(\frac{i + 1}{q} - 1 \right) = \frac{1 - q}{q - p},$$

and the result is identical with the one in case I. In contrast the mean of the stationary number of packets in the queue is

$$\mathbf{E}(X) = \sum_{i=1}^{\infty} i p_i = \frac{p(1 - q)}{q - p},$$

which is different from the results of case I. It reflects the fact that different number of packets are in the system before and after an arrival.

The evaluated performance measures validate the intuitive expectations that there are time slot definition dependent and independent performance measures. A modeler can choose the time slot freely if the required performance measures are time slot definition independent, but the time slot definition should be related to the required performance measures otherwise.

11.3 Performance Analysis of Switching Fabric with Fixed Size Packets

11.3.1 Traffic Model with Fixed Size Packets

In this section we consider the behavior of a switching fabric with N input and N output ports and set up a traffic model of this behavior. We assume that the input and output ports work in a slotted synchronized manner. The length of a time slot is the transmission time of a packet of fixed size, where we assume that the packets of different sizes are transmitted in fragments of identical size, which we refer to as packet for simplicity.

We assume that the arrival processes of packets to the input ports of the switch are independent and memoryless. The packet arrival processes to the input ports are characterized by a vector $q = \{q_i\}$, $i = 1, \ldots, N$, where q_i is the probability that 1 packet arrives to input port i in a time slot. Consequently, the probability that 0 packet arrives to input port i in a time slot is $1 - q_i$.

An incoming packet is directed to one of the N output ports. We assume that a packet from input port i is directed to output port j with probability w_{ij} independent of the past and the state of the system. The matrix composed by these probabilities $W = \{w_{ij}\}$ is referred to as traffic matrix. The traffic is a stochastic matrix, that is, $w_{ij} \geq 0$ and $\sum_j w_{ij} = 1$.

The bandwidth of the input and the output ports is identical. If more than one packet is directed to a given output port in a given time slot, then only one of them can be transmitted and the others are buffered. As it is quantified below the location of the buffers where the colliding packets are stored has a significant effect on the performance of the switch. We consider two cases: buffering at the input ports and buffering at the output ports. Figure 11.4 depicts the structure of these cases.

Real systems contain buffers both at the input and at the output ports. The performance characteristics and the design of the switch determine the proper model of the system. In the worst case one packet arrives to each input port and each packet is directed to the same output. If the switch is designed such that it can transfer all of the N packets to the buffer of the given output port in a single time slot, then the output buffer model describes properly the system. If the switch is designed such that it can transfer only one of the conflicting packets to the output buffer and the remaining $N - 1$ are left at the input buffers, then input buffer model is the proper model of the system.

Between the input and the output buffer models the output buffer seems to provide better performance, because in case of input buffer model it can happen that a given input port is blocked due to the conflict of the first packet in the queue, while other packets waiting in the buffer are directed to idle output ports (see Fig. 11.5).

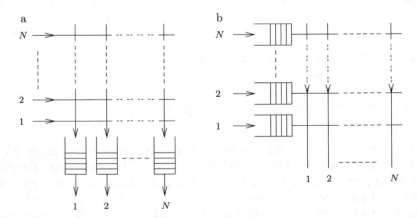

Fig. 11.4 Input and output buffering at packet switching. (**a**) Output queueing. (**b**) Input queueing

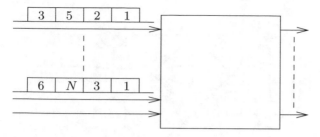

Fig. 11.5 Head of line blocking with input buffering

This phenomenon is often referred to as head of line blocking. This very intuitive qualitative comparison of the two buffer models will be quantified below for some special symmetric configurations.

11.3.2   Input Buffering

In this section we consider the simplest input buffering case when $N = 2$. If two packets at the head of the two input ports are directed to the same output port, then one of them is chosen with even probability $(1/2)$ and the chosen one is transferred to the output port and the other one is left in the buffer of the input port. With the above modeling assumptions the number of packets in the two input buffers is a DTMC. We assume that the time slots are such that if a packet arrives to an idle buffer and it does not collide with any other packet, then it leaves the input port in the same time slot.

Due to the fact that the system state is described by two discrete variables (the number of packets at the two input buffers), it is worth to depict the state space as a two-dimensional one as it is in Fig. 11.6. The state space can be divided into four parts: both queues are idle, queue 1 is idle and queue 2 is busy, queue 2 is idle and queue 1 is busy, and both queues are busy. The state transition probabilities follow the same structure in these four parts.

Figure 11.7 shows the environment of $(0,0)$. It is the state where both buffers are idle and the state transitions starting from this state are depicted. In this and the consecutive figures S denotes the probability of conflict. Conflict occurs when two packets from the head of the two buffers are directed to the same output. Its probability is $S = w_{11}w_{21} + w_{12}w_{22}$, where the first term stands for the case when both packets go to input 1 and second term stands for the case when they go to input 2. Starting from $(0,0)$ these are the following cases:

- the next state is $(1,0)$ if there is a conflict and the packet from input 2 is chosen to be transmitted,
- the next state is $(1,0)$ if there is a conflict and the packet from input 1 is chosen to be transmitted,

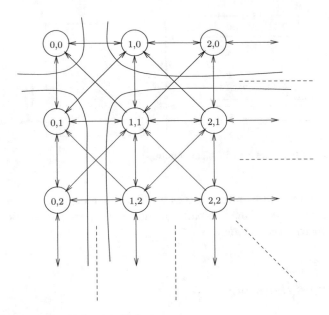

Fig. 11.6 Markov chain with input buffering

Fig. 11.7 Idle buffers

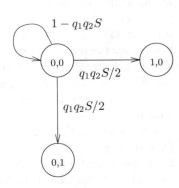

- if there is no conflict (zero or one packet arrives or two packets arrive but the packets are directed to different output) the next state is (0,0).

Figure 11.8 shows the state transitions when buffer 2 is idle and there is at least one packet in buffer 1. Denoting the starting state by $(x, 0)$, $x \geq 1$, the following state transitions can occur:

- $(x, 0) \rightarrow (x - 1, 0)$, if

  - there is no new packet arrival,
  - there is a packet arrival to input 2,

- $(x, 0) \rightarrow (x, 0)$, if

  - there is a packet arrival to input 1 and no packet arrives to input 2,

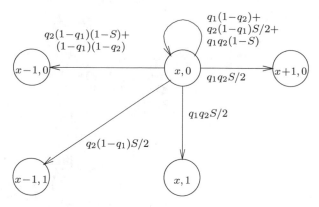

Fig. 11.8 Buffer 2 is idle

- there is a packet arrival to input 2 and no packet arrives to input 1, it is in conflict with the one at the head of buffer 1 and the packet in buffer 2 is chosen to be transmitted.
- there are packet arrivals at both inputs and there is no conflict (the packets at the head of the buffer are directed to different outputs).

- $(x, 0) \rightarrow (x + 1, 0)$, if

  - there are packet arrivals at both inputs, there is a conflict, and the packet in buffer 2 is chosen to be transmitted.

- $(x, 0) \rightarrow (x - 1, 1)$, if

  - there is a packet arrival to input 2 and no packet arrives to input 1, there is a conflict, and the packet in buffer 1 is chosen to be transmitted.

- $(x, 0) \rightarrow (x, 1)$, if

  - there are packet arrivals at both inputs, there is a conflict, and the packet in buffer 1 is chosen to be transmitted.

The states where buffer 1 is idle and buffer 2 is not idle are depicted in Fig. 11.9. The state transitions of these cases follow a similar pattern as the ones in Fig. 11.8 by replacing the role of the buffers.

Figure 11.10 presents a case when there are waiting packets in both buffers. The figure does not show transition $(x, y) \rightarrow (x, y)$ whose probability is 1 minus the sum of the depicted transition probabilities. The following state transitions are possible.

- $(x, y) \rightarrow (x - 1, y - 1)$, if

  - there is no new packet arrival and there is no conflict.

Fig. 11.9 Buffer 1 is idle

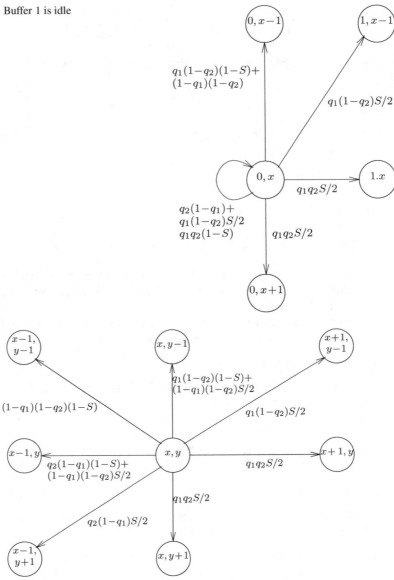

Fig. 11.10 There are packets in both buffers

- $(x, y) \to (x, y - 1)$, if

  - there is a packet arrival to input 1, no packet arrives to input 2, and there is no conflict.
  - there is no new packet arrival, there is a conflict, and the packet in buffer 2 is chosen to be transmitted.

- $(x, y) \rightarrow (x + 1, y - 1)$, if

  - there is a packet arrival to input 1, no packet arrives to input 2, there is a conflict, and the packet in buffer 2 is chosen to be transmitted.

- $(x, y) \rightarrow (x - 1, y)$, if

  - there is a packet arrival to input 2, no packet arrives to input 1, and there is no conflict.
  - there is no new packet arrival, there is a conflict, and the packet in buffer 1 is chosen to be transmitted.

- $(x, y) \rightarrow (x, y)$, if

  - there is a packet arrival to input 1, no packet arrives to input 2, there is a conflict, and the packet in buffer 1 is chosen to be transmitted (with probability $q_1(1 - q_2)S/2$),
  - there is a packet arrival to input 2, no packet arrives to input 1, there is a conflict, and the packet in buffer 2 is chosen to be transmitted (with probability $q_2(1 - q_1)S/2$),
  - new packets arrive to both buffers, and there is no conflict (with probability $q_1q_2(1 - S)$),

- $(x, y) \rightarrow (x + 1, y)$, if

  - new packets arrive to both buffers, there is a conflict, and the packet in buffer 2 is chosen to be transmitted.

- $(x, y) \rightarrow (x - 1, y + 1)$, if

  - there is a packet arrival to input 2, no packet arrives to input 1, there is a conflict, and the packet in buffer 1 is chosen to be transmitted.

- $(x, y) \rightarrow (x, y + 1)$, if

  - new packets arrive to both buffers, there is a conflict, and the packet in buffer 1 is chosen to be transmitted.

11.3.3 Output Buffering

The analytical description of the switch with output buffering is easier than the one with input buffering because in this case the number of packets in a buffer depends only on the properties of the arriving packets and it is independent of the number of packets in the other buffer. Consequently it is possible to analyze one output buffer in isolation.

Figure 11.11 presents the Markov chain of buffer 1 with output buffering. There are two possible state transitions if the buffer is idle:

- $0 \rightarrow 1$ if packets arrive to both input and both packets are directed to output 1.
- $0 \rightarrow 0$ otherwise.

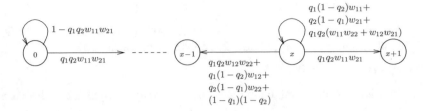

Fig. 11.11 Markov chain of buffer 1 with output buffering

There are three possible state transitions if the buffer is not idle:

- $x \rightarrow x - 1$ if packet arrives to output 1,
- $x \rightarrow x$ if one packet arrives to output 1,
- $x \rightarrow x + 1$ if two packets arrive to output 1.

The probabilities of these state transitions are provided in Fig. 11.11.

11.3.4   Performance Parameters

In this section we compute some performance parameters in case of input and output buffering assuming that the buffers are finite.

11.3.4.1   Mean Number of Packets in the Buffers

Let P_{ij}, $i, j \geq 0$ be the steady state probability of state (i, j) of the Markov chain describing the switch with input buffers. The mean number of packets in buffer 1 and 2 can be computed as

$$E_1 = \sum_{i \geq 0} \sum_{j \geq 0} i P_{ij},$$

$$E_2 = \sum_{i \geq 0} \sum_{j \geq 0} j P_{ij},$$

Similarly, let $P_i^{(1)}$ and $P_i^{(2)}$, $i \geq 1$ be the steady state probability of having i packets in buffer 1 and 2, respectively, in the Markov chains describing the switch with output buffers. The mean number of packets in buffer 1 and 2 can be computed as

$$E_1 = \sum_{i \geq 1} i P_i^{(1)},$$

$$E_2 = \sum_{i \geq 1} i P_i^{(2)},$$

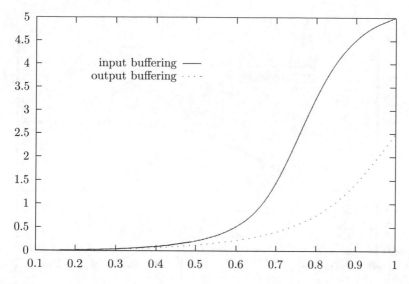

Fig. 11.12 Average number of packets with input and output buffering

Figure 11.12 plots the average buffer content as a function of the arrival probability, $q = q_1 = q_2$, for the input and the output buffer model, when the buffer length is limited to 5, and $w_{11} = w_{21} = 0.5$.

11.3.4.2   Throughtput

The throughput (δ) is the mean number of packets the switch transmits in a time slot. In case of input buffering we can compute the throughput following the same division of the states of the Markov chain. Denoting the stationary probability of the four parts by $P_{00}, P_{x0}, P_{0y}, P_{xy}$ the throughput is

$$\delta = P_{00}[1 \times (q_1(1 - q_2) + q_2(1 - q_1) + q_1q_2S) + 2 \times q_1q_2(1 - S)]$$

$$+ \sum_{x \geq 1} P_{x0}[1 \times ((1 - q_2) + q_2S) + 2 \times q_2(1 - S)]$$

$$+ \sum_{y \geq 1} P_{0y}[1 \times ((1 - q_1) + q_1S) + 2 \times q_1(1 - S)]$$

$$+ \sum_{x \geq 1, y \geq 1} P_{xy}[1 \times S + 2 \times (1 - S)],$$

where we detail the cases with one and with two packets transmission.

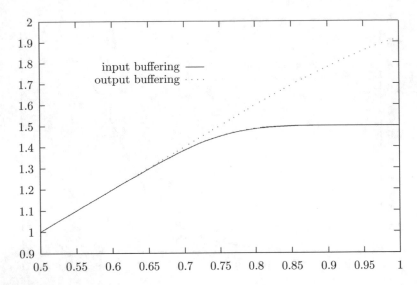

Fig. 11.13 Throughput

In case of output buffering the throughput is

$$\delta = P_0^{(1)}\,(q_1(1 - q_2)w_{11} + q_2(1 - q_1)w_{21} + q_1 q_2(1 - w_{12}w_{22})) + \sum_{x \geq 1} P_x^{(1)}$$

$$+P_0^{(2)}\,(q_1(1 - q_2)w_{12} + q_2(1 - q_1)w_{22} + q_1 q_2(1 - w_{11}w_{21})) + \sum_{x \geq 1} P_x^{(2)}.$$

Figure 11.13 plots the throughput as a function of packet arrival probability, $q = q_1 = q_2$, for input and output buffering when the buffer length is limited to 5, and $w_{11} = w_{21} = 0.5$.

In accordance with intuitive expectations the throughput with input buffering is less than the one with output buffering. As the arrival probability tends to 1 the throughput tends to 1.5 in case of input buffering. A quick intuitive explanation of this property is as follows. If packets arrive in each time slot the buffers are always busy, $\sum_{x \geq 1, y \geq 1} P_{xy}$ tends to 1, and δ tends to $1 \times S + 2 \times (1 - S)$ where $S = 1/2$.

11.3.5 Output Buffering in N × N Switch

Let us consider a single output of an $N \times N$ switch with output buffering and assume that packets arrive to the N input ports according to N independent identical Bernoulli processes. The probability that a packet arrives in a time slot is p. The arriving packets are directed to the output ports according to independent uniform distributions.

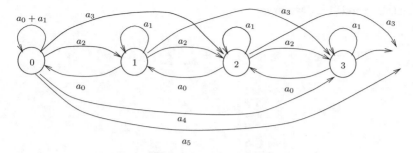

Fig. 11.14 Markov chain modeling the switch with output buffering

The number of packets sent to a tagged output port is binomially distributed.

$$a_i = \mathbf{P}(i \quad \text{packets arrived in the time slot}) = \binom{N}{i}(p/N)^i (1 - p/N)^{N-i}.$$

Figure 11.14 shows the transition probability graph of the Markov chain describing the number of packets in the tagged output port. This Markov chain can also be described by an evolution equation of type II,

$$X_{n+1} = (X_n - 1 + Y_{n+1})^+$$

and its state transition probability matrix is

$$\Pi = \begin{bmatrix} a_0 + a_1 & a_2 & a_3 & a_4 & \cdots & a_{k+1} & \cdots \\ a_0 & a_1 & a_2 & a_3 & \cdots & a_k & \cdots \\ 0 & a_0 & a_1 & a_2 & \cdots & a_{k-1} & \cdots \\ 0 & 0 & a_0 & a_1 & \cdots & a_{k-2} & \cdots \\ \vdots & \vdots & \vdots & \vdots & \vdots & \vdots \end{bmatrix}. \tag{11.1}$$

The stationary probabilities satisfy the following linear equations

$$p_0 = p_0 a_0 + p_0 a_1 + p_1 a_0, \tag{11.2}$$

$$p_k = p_k a_1 + p_{k+1} a_0 + \sum_{i=0}^{k-1} p_i a_{k+1-i} = \sum_{i=0}^{k+1} p_i a_{k+1-i}. \tag{11.3}$$

Let us introduce the following z-transform functions

$$P(z) = \sum_{k=0}^{\infty} p_k z^k, \quad A(z) = \sum_{k=0}^{\infty} a_k z^k.$$

From (11.2) and (11.3) we have

$$P(z) = p_0 a_0 + p_0 a_1 + p_1 a_0 + \sum_{k=1}^{\infty} \sum_{i=0}^{k+1} p_i a_{k+1-i} z^k$$

$$= p_0 a_0 + \sum_{k=0}^{\infty} \sum_{i=0}^{k+1} p_i a_{k+1-i} z^k$$

$$= p_0 a_0 + \sum_{k=0}^{\infty} \sum_{i=1}^{k+1} p_i a_{k+1-i} z^k + \sum_{k=0}^{\infty} p_0 a_{k+1} z^k$$

$$= p_0 a_0 + \sum_{i=1}^{\infty} p_i \sum_{k=i-1}^{\infty} a_{k+1-i} z^k + z^{-1} \sum_{k=0}^{\infty} p_0 a_{k+1} z^{k+1}$$

$$= p_0 a_0 + z^{-1} \sum_{i=1}^{\infty} p_i z^i \sum_{l=0}^{\infty} a_l z^l + z^{-1} p_0 \sum_{m=1}^{\infty} a_m z^m$$

$$= p_0 a_0 + z^{-1}(P(z) - p_0) A(z) + p_0 z^{-1}(A(z) - a_0),$$

from which

$$P(z) = \frac{(1 - z^{-1}) p_0 a_0}{1 - z^{-1} A(z)} = p_0 a_0 \frac{(z - 1)}{z - A(z)}.$$

Considering that $\lim_{z \to 1} P(z) = 1$, and applying the L'Hospital's rule we have

$$1 = p_0 a_0 \frac{1}{1 - A'(z)} \Big|_{z=1},$$

where $A'(1)$, the mean number of packets arriving to the tagged output in a time slot, can be computed.

$$p_0 a_0 = 1 - A'(z) = 1 - p,$$

and using this

$$P(z) = \frac{(1 - p)(z - 1)}{z - A(z)} = \frac{(1 - p)(1 - z)}{A(z) - z}. \tag{11.4}$$

To check the obtained results we set $N = 2$. In this case the probability that i packets arrive to the tagged output post is

$$a_i = \binom{2}{i} \left(\frac{p}{2}\right)^i \left(1 - \frac{p}{2}\right)^{2-i},$$

from which

$$A(z) = \left(1 - \frac{p}{2} + z\frac{p}{2}\right)^2,$$

and

$$P(z) = \frac{(1-p)(1-z)}{\left(1 - \frac{p}{2} + z\frac{p}{2}\right)^2 - z},$$

The probability that the buffer is idle, p_0, is easily obtained from the transform domain expression

$$p_0 = P(z)\,|_{z=0} = \frac{1-p}{\left(1 - \frac{p}{2}\right)^2}.$$

This result can be checked easily by considering that for $N = 2$ the Markov chain has a birth-death structure with forward probability a_2 and backward probability a_0. From this Markov chain we have

$$p_0 = \frac{1}{1 + \sum\limits_{k=1}^{\infty} \left(\frac{a_2}{a_0}\right)^k} = \frac{1-p}{\left(1 - \frac{p}{2}\right)^2}.$$

11.3.6 Throughput of N × N Switch with Input Buffering

The end of Sect. 11.3.4 shows the throughput computation of 2×2 switch with input buffering. In this section we compute the throughput of larger switches, $N > 2$, with input buffering.

We assume that the packets are indistinguishable, the switch chooses one of the packets in conflict with independent uniform distribution, and the packets are directed to output ports in a uniformly distributed manner.

To obtain the maximal throughput of the system we assume that packets are waiting in each buffer of the switch, i.e., none of the input buffers is idle.

We use the following notations:

- let $R_m^{(i)}$ be the number of packets which are at the head of a buffer at time m, are directed to output i, and are not forwarded due to collision,
- let $A_m^{(i)}$ be the number of packets which arrive to the head of a buffer at time m and are directed to output i,

$\boldsymbol{R}_m = \{R_m^{(1)}, \ldots, R_m^{(N)}\}$ is a discrete-time Markov chain. Its evolution is described by the following equation

$$R_m^{(i)} = \max(0, R_{m-1}^{(i)} + A_m^{(i)} - 1), \tag{11.5}$$

where the sum on the right-hand side is reduced by 1 due to the packet which is transmitted to output i. $A_m^{(i)}$, which depends on $\boldsymbol{R}_{m-1}$, follows a binomial distribution

$$\mathbf{P}\left(A_m^{(i)} = k\right) = \binom{F_{m-1}}{k} (1/N)^k (1 - 1/N)^{F_{m-1}-k}, \qquad k = 0, 1, \cdots, F_{m-1},$$

(11.6)

where the number of new packets at the head of the buffer is

$$F_{m-1} = N - \sum_{j=1}^{N} R_{m-1}^{(j)}. \tag{11.7}$$

Equation (11.7) is based on the assumption that none of the input buffers is idle, i.e., the difference between N and $\sum_{j=1}^{N} R_{m-1}^{(j)}$ is the number of input buffers which transmitted a packet at time $m - 1$. Due to the same assumption the number of new packets arrive to the head of the buffers is equal to the number of packets successfully transmitted in a time slot. Consequently the throughput of output i is $\delta^{(i)} = lim_{m\to\infty}\mathbf{E}\left(A_m^{(i)}\right)$.

The parameters of the binomial distribution are F_{m-1} and $1/N$, since there are F_{m-1} new packets at the heads of the buffers and they choose their destination according to a uniform distribution. Here we note that intuitively $lim_{m\to\infty} \mathbf{E}(F_{m-1})$ increases proportionally with N and consequently as N tends to infinity $A_m^{(i)}$ tends to be Poisson distributed.

Having all elements of the evolution equation defined it is possible to compute the stationary distribution of the Markov chain numerically. From the stationary distribution we also have its mean $\mathbf{E}(R^{(i)}) = lim_{m\to\infty}\mathbf{E}\left(R_m^{(i)}\right)$.

Taking the expectation of (11.6) and (11.7) we get

$$\mathbf{E}\left(A^{(i)}\right) = \mathbf{E}(F)/N, \quad \text{and} \quad \mathbf{E}(F) = N - \sum_{i=1}^{N} \mathbf{E}\left(R^{(i)}\right),$$

using the symmetry of the system and introducing $\delta = \delta^{(i)}$, $\mathbf{E}(A) = \mathbf{E}(A^{(i)})$, $\mathbf{E}(R) = \mathbf{E}(R^{(i)})$ we further have

$$\mathbf{E}(F) = N - N\mathbf{E}(R), \quad \text{and} \quad \delta = \mathbf{E}(A) = \mathbf{E}(F)/N = 1 - \mathbf{E}(R).$$

That is, for any finite N, $\mathbf{E}(R)$ needs to be computed numerically and the maximal throughput of one output can be obtained from $\delta = 1 - \mathbf{E}(R)$. Table 11.1 presents these throughput as a function of N. In Sect. 11.3.4, we computed the maximal throughput of the whole switch, not the per output throughput, for $N = 2$.

When $N \to \infty$ the evolution Eq. (11.5) remains valid, $A_m^{(i)}$ tends to be Poisson distributed with parameter δ, and the dependency on the other $R_k^{(i)}$ ($k \neq m$) values

Table 11.1 Per output port
throughput as a function of N

N	Throughput
1	1.0000
2	0.7500
3	0.6825
4	0.6553
5	0.6399
6	0.6302
7	0.6234
8	0.6184
∞	0.5858

vanishes. At the limit we can analyze $R_m^{(i)}$ in isolation, which is a Markov chain
with the same structure as the one in (11.1), with $a_i = \frac{\delta^i}{i!}e^{-\delta}$. Following the same
transform domain analysis leading to (11.4), we have

$$P(z) = \frac{(1 - A'(1))(1 - z)}{A(z) - z},$$

with $A(z) = e^{\delta(z-1)}$, from which $\mathbf{E}(R) = P'(1) = \frac{\delta^2}{2(1-\delta)}$. From the last relation
and $\delta = 1 - \mathbf{E}(R)$ the limiting throughput of the switch is $\delta = 2 - \sqrt{2} \sim 0.5858$.

11.4   Conflict Resolution Methods of Random Access Protocols

One of the main functions of medium access control (MAC) is to share common
resources between randomly arriving customer requests. Different random access
protocols are developed for this purpose. In case of random access protocols several
users try to communicate through a common transmission channel. The users do not
know the activity of the others. In this kind of environments a stable communication
requires the application of a protocol which under a system dependent load level
ensures

- stable communication (with finite mean delay),
- transmission of all packets,
- fairness (users obtain the same service).

These protocols work based on the information available about the state of the
users. The below listed procedures are different members of the set of random access
protocols, which differ

- in the way the users are informed about the status of the common channel, and
  indirectly the activity of the other users, and
- in the design goals to adopt to the alternation of the traffic load.

11.4.1 ALOHA Protocol

The ALOHA protocol [2] is the simplest random access protocol. It was developed for simple radio communication between radio terminals and a central station. It uses two radio channels, one is used by the terminals for communication to the central station and the other one for the communication from the central station to all terminals (Fig. 11.15).

If more than one terminal send message to the central station, the signals interfere and the central station cannot receive any correct message. This case is referred to as collision of the messages. Collision can only happen in the first radio channel, since the second radio channel is used only be the central station. Successfully received messages are acknowledged by the central station. The terminals are informed about the success of their message transmission by these acknowledgments. If no acknowledgment arrives within a given deadline the terminal assumes that the transmission failed. The ALOHA protocol is designed to ensure communication in this system.

ALOHA functions as follows. As soon as a terminal has a new packet to transmit it starts sending it away without any attention to the activity of the other terminals. If there is no acknowledgment arrives within a deadline the terminal assumes that the message collided and it switches message retransmission mode. Terminals in message retransmission mode are referred to as blocked terminals. In this mode the terminal repeats retransmitting the message as long as it gets an acknowledgment about successful transmission.

If after an unsuccessful transmission a terminal would retransmit the message immediately, then there would be multiple collisions among the same set of terminals, as it is demonstrated in Fig. 11.16.

Fig. 11.15 Radio terminal system

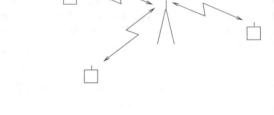

Fig. 11.16 Packet retransmission without random delay

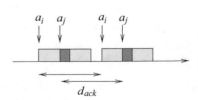

Fig. 11.17 Packet
retransmission with random
delay

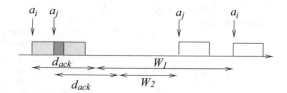

To avoid these multiple collisions among the same set of terminals, the terminals wait for a random amount of time before retransmitting the messages as it is demonstrated in Fig. 11.17. The figures distinguish between the period of collision (dark gray) and the additional time period which is waisted (unavailable for packet transmission) due to the collision (light gray).

The quantitative behavior of the system is straightforward. If the terminals choose large random delay the probability of consecutive collisions with the same set of terminals is low, but the time to the successful transmission is high due to the long delay till retransmissions. In the way around if the delay is short the probability of consecutive collisions is higher, and it could cause that the time to the successful transmission is high due to the high number of repeated transmission attempts. The optimal behavior of the system is somewhere between these extremes.

The modeling and analysis of ALOHA systems is a well-studied area since 1970s. It is still an interesting research area because there are several later introduced random access protocols which contain elements of the basic ALOHA protocol (as it is detailed in the consecutive sections).

There is a wide variety of performance studies. These studies differ in their assumptions about the behavior of the users and the system. It is practically impossible to analyze the simplest ALOHA protocol with all small technical details. To reduce the complexity of the models several simplifying assumptions are used. The obtained simplified models often closely approximate the real system behavior.

In the next sections we introduce some of the simple model of the basic ALOHA system and their analysis.

11.4.1.1   Continuous-Time ALOHA System

We adopt the following modeling assumptions:

- The aggregate arrival process of new and retransmitted messages is a Poisson process with parameter λ.
  This model is not a correct model of the aggregate arrival process (in general) but there are several cases (e.g., the number of blocked terminals is negligible compared to the number of all terminals) when it properly approximates the real system behavior. This kind of models, where the arrivals of the new and the retransmitted messages are considered in an aggregate flow, is referred to as zero order model.
- The length of the messages is fixed and the time of a message transmission is T.

Fig. 11.18 Cycles of busy
and idle periods

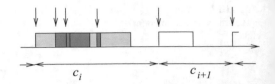

With the zero order model we evaluate which portion of the new and repeated messages, which arrive according to a Poisson process with parameter λ, is transmitted successfully, and what is the related transmission delay and collision probability.

According to Fig. 11.18 we divide the time axes according to the busy and idle periods of the common channel.

The probability of a successful message transmission in a busy period equals with the probability that after the beginning of a busy period the next message arrives later than T. Its probability is $e^{-\lambda T}$.

In order to determine the long-term ratio of idle, successful busy, and unsuccessful busy periods we determine the average length of these periods. The interarrival time in a Poisson process with parameter λ is exponentially distributed with parameter λ. The length of an idle period is the remaining time of an exponentially distributed interarrival time, which is exponential again with the same parameter. This way the mean length of an idle period is $1/\lambda$.

The length of a successful busy period is T. The difficult question is the length of the unsuccessful period. An unsuccessful busy period is composed by $N - 1$ ($N \geq 2$) interarrival intervals shorter than T and a final interval of length T. The case when $N = 1$ is the successful busy period. Due to the memoryless property of Poisson process we can compute the number of colliding messages during the unsuccessful busy period independent of the length of the interarrival times,

$$\mathbf{P}(N = n) = (1 - e^{-\lambda T})^{n-1} e^{-\lambda T}.$$

The CDF of the length of an interarrival interval shorter than T, denoted as U, is

$$F_U(t) = \mathbf{P}(U < t) = \mathbf{P}(\tau < t | \tau < T) = \begin{cases} \dfrac{1 - e^{\lambda t}}{1 - e^{\lambda T}} & \text{if } 0 < t < T, \\ 1 & \text{if } T < t, \end{cases}$$

from which $\mathbf{E}(U) = \dfrac{1 - e^{-\lambda T} - \lambda T e^{-\lambda T}}{\lambda(1 - e^{-\lambda T})}$. Consequently, in a cycle composed by a busy and an idle period

- the mean length of the idle period is $\mathbf{E}(I) = 1/\lambda$,
- the probability of a successful message transmission is $\mathbf{E}(S) = e^{-\lambda T} T$,
- and the mean length of an unsuccessful busy period

$$\mathbf{E}(L) = \sum_{n=2}^{\infty} \mathbf{P}(N = n)\left((n-1)\mathbf{E}(U) + T\right) = \frac{1 - e^{-\lambda T} - \lambda T e^{-\lambda T}}{\lambda e^{-\lambda T}}.$$

The system utilization is characterized by the portion of time associated with successful message transmission

$$\rho = \frac{\mathbf{E}(S)}{\mathbf{E}(I) + \mathbf{E}(S) + \mathbf{E}(L)} = \lambda T e^{-2\lambda T}.$$

It can be seen that the utilization depends only on the λT product. The maximum of the utilization is obtained through the derivative of ρ as a function of λT. The maximum is found at $\lambda T = 0.5$ and it is $\rho = 1/2e \sim 0.18394$. Figure 11.19 shows that the utilization decreases significantly as the load increases above 0.5, consequently these systems should be operated with a load lower than 0.5.

The mean number of arriving messages in a Δ long interval is $\lambda \Delta$. In the same interval the mean time of successful message transmission is $\rho \Delta$. During this interval the mean number of successfully transmitted messages is $\rho \Delta / T$. The ratio between the number of successfully transmitted messages and the number of all message transmission attempts, which is the mean number of transmission attempts per message, is $\mathbf{E}(R) = \lambda T / \rho = e^{2\lambda T}$.

Having the mean number of transmission attempts per message we can compute the message transmission delay

$$\mathbf{E}(D) = \mathbf{E}\left(\sum_{r=1}^{\infty} \mathbf{P}(R = r)\left(rT + \sum_{i=1}^{r-1} d_{ack} + W_i\right)\right)$$
$$= \mathbf{E}(R)T + (\mathbf{E}(R) - 1)(d_{ack} + \mathbf{E}(W)),$$

where d_{ack} is the time a terminal waits for message acknowledgment and W is the random delay spent before message retransmission.

Fig. 11.19 Utilization ρ as a function of the load λT

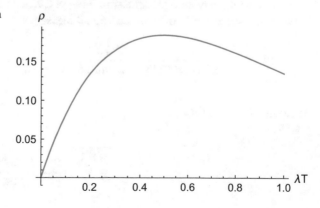

11.4.1.2   Discrete-Time (Slotted) ALOHA System

The main disadvantage of continuous-time ALOHA systems is that the waisted time period at a message collision is large. This phenomenon can be seen in Figs. 11.16 and 11.18. The dark gray period denotes the overlapping intervals of colliding messages, while light gray periods are additional waisted time intervals which cannot be used for useful message transmission.

With a simple modification of the ALOHA system this additional waisted time interval can be avoided. If all terminals work in a synchronized manner, and they initiate message transmission only at the beginning of time slots, then the length while the colliding messages occupy the common channel reduces to T. Naturally in this system the delay of a message retransmission should be an integer multiple of the time slot, T. This system is commonly referred to as slotted ALOHA system (Fig. 11.20).

The zero order model of slotted ALOHA system assumes that the terminals generate a Poisson distributed number of new and repeated messages in a time slot, where the parameter of the Poisson distribution is λT. This model of message arrivals is similar to the zero order model of continuous-time ALOHA system assuming that the messages are generated continuously according to a Poisson process, but the messages generated during a time slot are delayed till the beginning of the next time slot.

With these assumptions utilization of the zero order model of slotted ALOHA system can be computed based on the analysis of a single time slot. Let N be the number of packets generated in a time slot. In this case

$$\rho = \mathbf{P}(\text{successful message transmission}) = \mathbf{P}(N = 1) = \lambda T e^{-\lambda T}$$

The maximum of the utilization is obtained at $\lambda T = 1$ and it is $\rho = 1/e \sim 0.367879$. Compared to the continuous-time ALOHA system the optimal throughput doubles and the aggregated load (new and repeated messages) can be increased to the capacity of the system ($\lambda T = 1$) as it is plotted in Fig. 11.21.

The mean number of retransmission attempts, R, can be computed as the ratio between the successfully transmitted and all messages

$$\mathbf{E}(R) = \frac{\mathbf{E}(N)}{\mathbf{E}(\text{successfully transmitted messages})} = \frac{\lambda T}{\lambda T e^{-\lambda T}} = e^{\lambda T}.$$

Fig. 11.20 Slotted ALOHA system

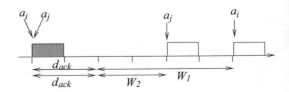

Fig. 11.21 Utilization of the slotted ALOHA system

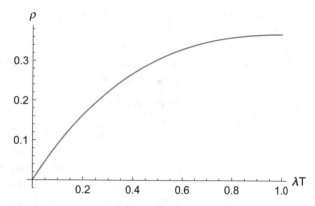

Similar to the continuous-time case the message transmission delay is

$$\mathbf{E}(D) = \mathbf{E}\left(\sum_{r=1}^{\infty} \mathbf{P}(R = r)\left(rT + \sum_{i=1}^{r-1} d_{ack} + W_i\right)\right)$$

$$= \mathbf{E}(R)T + (\mathbf{E}(R) - 1)(d_{ack} + \mathbf{E}(W)).$$

The more complex models of the ALOHA system distinguish the states of the terminals (message generation, new message transmission attempt, waiting random delay, and message retransmission attempt) and characterize the arrival of the new and repeated messages according to those states [8].

In contrast with the terminology of ALOHA systems the general terminology of random access protocols refers to station instead of terminals and packets instead of messages.

11.4.2   *CSMA and CSMA/CD Protocols*

The more advanced random access protocols intend to enhance the channel utilization based on the information available for the stations by the given physical media.

At the introduction of the slotted ALOHA system we have seen that the reduction of the time period while a collision makes the channel unavailable enhances the performance of the protocol. In case of radio terminal systems where the terminals can be outside of each other's propagation range it is hard to further reduce the ineffective time of the channel. In case of wired systems the stations can sense the signals of each other, but it is not immediate, there is a propagation delay of the medium. This direct sensing of the stations can be used to enhance the performance of the multiple access protocol in the following two ways.

- If a station senses that another station is sending a packet when it has a packet to send does not start sending the packet.
- If by accident the packets of two stations collide (because they are sent within the propagation delay) the stations can recognize that the packets collide and finish the useless packet transmission immediately.

The first way is referred to as *carrier sense multiple access* (CSMA), and the second one as *collision detection* (CD).

Figures 11.22 and 11.23 demonstrate the behavior of the CSMA and the CSMA/CD systems. In these systems the time is slotted and the time unit is the maximal propagation delay between the most remote stations, τ. Collision can happen only among packets transmitted within the same slot, because in the next time slot all stations are aware of the busy state of the channel. A station can initiate a packet transmission only if the channel is idle. In case of CSMA without CD the colliding packets are transmitted completely. This way a significant portion of the channel capacity is lost (Fig. 11.22). In case of CSMA with CD the collision is recognized within one time slot and the packet transmission is finished immediately (Fig. 11.23).

11.4.2.1 Performance of Slotted CSMA System

We analyze the zero order model of the system by the analysis of the intervals between consecutive packet transmission attempts. The beginning of these intervals are indicated by the arrows below the time axes in Figs. 11.22 and 11.23. It can also happen that, in contrast with the figures, there is no idle period between two consecutive packet transmission attempts. According to the zero order model of the system we assume that after an idle time slot or the last time slot of a successful packet transmission there are a Poisson distributed number of (new and repeated) packet transmissions initiated with parameter $\tau\lambda$. That is, in contrast with the previously discussed zero order models, the state of the channel affects the arrival

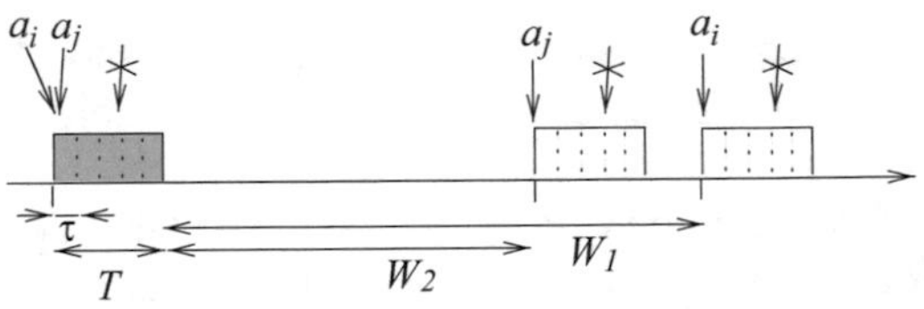

Fig. 11.22 CSMA system

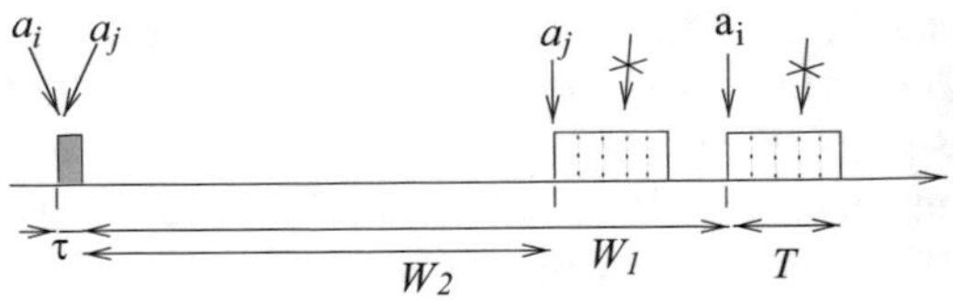

Fig. 11.23 CSMA/CD system

process of the packets. Packets can arrive in the above-mentioned time slots and cannot arrive otherwise. The success of the packet transmission depends on the number of arriving packets, N.

$$\mathbf{P}(\text{successfull packet transmission}) = \mathbf{P}(N = 1 | N \geq 1) = \frac{\lambda \tau e^{-\lambda \tau}}{1 - e^{-\lambda \tau}}.$$

After a successful or colliding packet transmission the channel remains idle until the next packet arrival. Let I denote the number of idle time slots until the next packet arrival. Due to the memoryless property of the arrival process I is geometrically distributed. $\mathbf{P}(I = i) = e^{-\lambda \tau i}(1 - e^{-\lambda \tau})$. Consequently in an interval between consecutive packet transmission attempts

- the mean length of the idle period is $\mathbf{E}(I)\tau = \dfrac{\tau e^{-\lambda \tau}}{1 - e^{-\lambda \tau}}$,
- the mean length of successful packet transmission is $\mathbf{E}(S) = \dfrac{T \lambda \tau e^{-\lambda \tau}}{1 - e^{-\lambda \tau}}$,
- and the mean length of unsuccessful packet transmission is $\mathbf{E}(L) = \dfrac{T (1 - (1 + \lambda \tau)e^{-\lambda \tau})}{1 - e^{-\lambda \tau}}$.

Finally, the utilization is obtained as

$$\rho = \frac{\mathbf{E}(S)}{\mathbf{E}(I) + \mathbf{E}(S) + \mathbf{E}(L)} = \frac{T \lambda \tau e^{-\lambda \tau}}{T(1 - e^{-\lambda \tau}) + \tau e^{-\lambda \tau}}.$$

Figure 11.24 plots the utilization as a function of $\lambda \tau$ when $\tau/T = 0.2$ (dotted line) and when $\tau/T = 0.1$ (short dashed line). It can be seen that the probability of collision is lower and the utilization is higher in case of shorter propagation time ($\tau/T = 0.1$). In any case the utilization reaches an optimum and starts decreasing when the load is increasing. The optimal load level depends on the τ/T ratio.

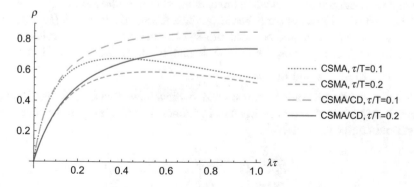

Fig. 11.24 Utilization of slotted CSMA and CSMA/CD systems

11.4.2.2   Performance of Slotted CSMA/CD System

The zero level model of the CSMA/CD system is very similar to the one of the CSMA system. It differs only by the time of the unsuccessful packet transmission, which is shorter due to collision detection $\mathbf{E}(L) = \dfrac{\tau\left(1 - (1 + \lambda\tau)e^{-\lambda\tau}\right)}{1 - e^{-\lambda\tau}}$. As a result the utilization is

$$\rho = \frac{\mathbf{E}(S)}{\mathbf{E}(I) + \mathbf{E}(S) + \mathbf{E}(L)} = \frac{T\,\lambda\tau e^{-\lambda\tau}}{T\lambda\tau e^{-\lambda\tau} + \tau(1 - \lambda\tau e^{-\lambda\tau})}.$$

Figure 11.24 plots the utilization of the CSMA/CD system as a function of the load, $\lambda\tau$, together with the one of the CSMA system. The propagation delay is $\tau/T = 0.2$ (long dashed line) and $\tau/T = 0.1$ (solid line). Also in these cases the shorter propagation delay increases the utilization. In contrast with the CSMA system the utilization is continuously increasing with the load due to the efficient utilization of the channel.

11.4.2.3   Slotted Persistent CSMA/CD System

Up to now we did not discuss the behavior of a station when it has a packet to transmit but the channel is busy. Indeed we implicitly assumed that these stations assumed that their packet collided and delayed the next packet retransmission attempt accordingly. This behavior is referred to as nonpersistent station behavior.

The stations sense the channel and know the history of the channel state from which they can compute when the packet under transmission finishes. Knowing this information the station with a packet to transmit can reduce the packet retransmission time by attempting a packet transmission immediately when the channel becomes idle next. This behavior is referred to as persistent station behavior.

In the zero order model of persistent CSMA system we assume that in each τ long time slot the stations generate a Poisson distributed number of new and repeated packets and those stations which generate packets during a packet transmission attempt to transmit the packet when the channel becomes idle next.

The analysis of this system is based on the analysis of successful (S), colliding (L), and idle (I) intervals, because the system behavior is memoryless at the beginning of these periods. The mean length of these intervals is as follows.
$\mathbf{E}(S) = T$, $\mathbf{E}(L) = \tau$ and $\mathbf{E}(I) = \dfrac{1}{1 - e^{-\lambda\tau}}$.

To compute the utilization we also need to know how often these intervals occur. The following transition probability matrix defines the probability of the occurrence of various consecutive intervals.

$$\boldsymbol{\Pi} = \begin{array}{c} \\ \\ \\ \\ \end{array}\begin{array}{|cc c|c} \multicolumn{1}{c}{S} & \multicolumn{1}{c}{L} & \multicolumn{1}{c}{I} & \\ \hline P(\lambda T, 1) & P(\lambda T, > 1) & P(\lambda T, 0) & S \\ P(\lambda\tau, 1) & P(\lambda\tau, > 1) & P(\lambda\tau, 0) & L \\ P(\lambda\tau, 1) & P(\lambda\tau, > 1) & & \\ \hline P(\lambda\tau, > 0) & P(\lambda\tau, > 0) & 0 & I \end{array}$$

where $P(a, i) = e^{-a}a^i/i!$ and $P(a, > i) = \sum_{j=i+1}^{\infty} P(a, j)$. This is a DTMC whose stationary solution is obtained by the solution of the linear system of equations $\pi \Pi = \pi$, $\pi_S + \pi_L + \pi_I = 1$, where $\pi = \pi_S, \pi_L, \pi_I)$. Having the stationary probabilities π_S, π_L and π_I, the utilization is

$$\rho = \frac{\pi_S \mathbf{E}(S)}{\pi_S \mathbf{E}(S) + \pi_L \mathbf{E}(L) + \pi_I \mathbf{E}(I)}$$

$$= \frac{\lambda \tau \; \lambda T}{1 - \lambda T e^{-\lambda T} + \lambda \tau e^{-\lambda T}\left(1 - \lambda \tau + \lambda T(1 - e^{\lambda \tau})\right) + \lambda \tau \left(e^{\lambda \tau} + \lambda T - 1\right)}.$$

Figure 11.25 plots the utilization as a function of the load, $\lambda \tau$, with propagation delay $\tau/T = 0.1$ (dotted line) and with $\tau/T = 0.2$ (solid line). Similar to the previous cases shorter propagation delays result in lower probability of collision and better utilization. The utilization decreases when the load is high. It is because the probability of collision after a successful packet transmission gets very high due to persistent station behavior.

In summary, the nonpersistent behavior is beneficial when the delay due to the collision at the end of successful packet transmission is less than the normal message retransmission delay. At low load levels the persistent behavior decreases the delay (because the probability of collision of low), while at high load levels the nonpersistent behavior performs better.

There is a continuous transition between the persistent and the nonpersistent behaviors. It is obtained when a station follows persistent behavior with probability p and nonpersistent behavior with probability $1 - p$. This behavior is referred to as p-persistent behavior. Obviously $p = 1$ results the persistent and $p = 0$ the nonpersistent behavior. For a given load level we can optimize the system utilization by setting p to an optimal value.

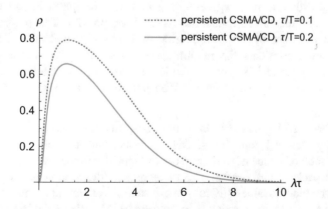

Fig. 11.25 Utilization of slotted persistent CSMA/CD system

11.4.3   IEEE 802.11 Protocol

One of the most commonly used wireless access of computer networks currently is
the wireless fidelity (WF or wifi), which is defined in the IEEE 802.11 standard. The
core of this rather complicated protocol is also an enhanced version of the slotted
ALOHA protocol. The IEEE 802.11 protocol [1] is designed to meet the following
requirements:

- the random delay for packet transmission is bounded,
- the protocol operates in a wide range of traffic load and it adopts the actual level
  of traffic load,
- if large packets are transmitted a priority is given to the completion of the already
  started packets.

According to these requirements the ALOHA protocol is modified as follows
[6, 10].

- At a collision the station draws a random number uniformly distributed between
  1 and M_i and it retransmits the collided packet after the expiration of that many
  slots.
- The upper limit of the uniform distributed delay depends on the number of
  unsuccessful transmission trials. After the first collision this value is $M_1 = 8$
  and it doubles after each consecutive collision $M_i = 8 * 2^{i-1}$ until a predefined
  upper limit, M_{max}, is reached.
- The large packets are transmitted in several small segments. In case of an
  ALOHA system these segments would be transmitted one-by-one and each
  of them has to participate in the contention for the medium. This way the
  packet transmission delay, which is determined by the largest delay of the
  segments, can be very high. IEEE 802.11 reduces the packet transmission delay
  by giving priority to the consecutive segments of a packet under transmission.
  This way only the first segment of a packet participates in the contention
  and other segments are transmitted with high priority. The protocol imple-
  ments this feature by the introduction of two different delays. The stations
  in contention considers the medium available if it is idle for a *distributed
  interframe space* (DIFS) period, while the segment of a packets can be sent
  within a *short interframe space* (SIFS) period which is shorter than a DIFS
  period.

The IEEE 802.11 protocol is built on a so-called basic access method, which
is practically identical with the ALOHA protocol and it combines with various
reservation methods. One of these reservation methods is the mentioned DIFS and
SIFS based packet transmission. The mathematical description of these complex
reservation methods is rather difficult. In this section we present an analytical model
of the basic access method which is introduced in [6]. This model is also based on
a simplifying assumption. It assumes that there are so many independently working
stations in the system that a packet transmission trial is unsuccessful with probability

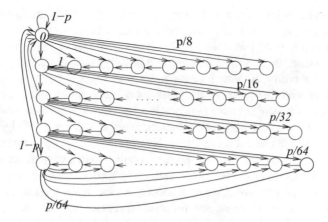

Fig. 11.26 Markov chain describing the basic access method of 802.11

p in each time slot independent of the past history of the system. Furthermore to compute the maximal throughput of we assume that the stations always have packets to transmit.

Having this assumption we can describe the behavior of a station with a DTMC. The state of the Markov chain describes the phase of the actual packet transmission attempt. Figure 11.26 shows the state transition graph of this Markov chain. State 0 indicates that the station just finished transmitting a packet and tries to transmit the next one in the next time slot. If it is unsuccessful, which happen with probability p, then it draws a uniformly distributed random sample between 1 and $M_1 = 8$. Transitions to the right with probability 1 describes that the station waits until the given delay expires. When the chain arrives to the left most state it attempts to transmit the packet again and go back to state 0 or moves to the next row, etc. In this Markov chain the retransmission delay is limited to $M_{max} = 64$. The time between two consecutive visits to state 0 represents the packet transmission delay. The throughput of the station is p_0 and the mean packet transmission delay is $1/p_0$, if p_0 is the stationary probability of state 0 in this Markov chain.

11.5 Priority Service Systems

Priority systems appear in different fields [13, 14, 16, 34]. Several aspects of telecommunication, data management, planning of computer networks, organization of health service, and automatization of production processes could be mentioned. For example, in mobile cellular networks the coverage area is partitioned into cells, and each cell can serve at most c simultaneous communications and use some channels from other cells. There are calls initiated by subscribers from the cell and handover calls from others. Handover calls already use the network resources and they should be prioritized with respect to new calls. Different approaches are possible, e.g., a special channel or a priority queue of handover calls.

The problem may be formulated as follows: customers of different types enter the service system, each of them belongs to a priority class, indexed by a subscript. We assume that customers with a lower index have higher priority in service; this way customers with a lower index can leave the queue earlier than customers with a higher index which were already in the queue at the arrival of customers with a lower index. There are two possible cases: either the entry of customers of higher priority does not interrupt the actual service with lower priority customers or immediately starts its service (in the first case we speak of relative, in the second case of absolute priority). In the second case we have again two choices, whether or not the work up to this moment will be taken into account in the future. With respect to the first possibility, one must complete the residual service, for the second one must complete the whole service later, when the higher priority customers are served. Both of these possibilities occur in computer systems. For example, the results of computations are either regularly saved or not. In the first case results are not lost at a system error. Similar situations appear in other fields. When a disaster occurs, one must first divert the danger and after that to deal with less urgent tasks. For example, a dentist must first see patients who are in pain, other patients can wait.

We will consider service systems with two Poisson arrival processes where the service time will have exponential and general distributions. In the exponential case we will follow the usual method—find the system of differential equations describing the functioning of system, solve it, and at $t \rightarrow \infty$ determine the equilibrium distribution. In the general case, we examine the virtual waiting time by means of the Laplace–Stieltjes transform; the approach is mainly based on the Pollaczek-Khinchin formula concerning waiting time (8.19) (see, e.g., [26]).

11.5.1 Priority System with Exponentially Distributed Service Time

Let us consider the following problem. We have m homogeneous servers and two types of customers. Type i customers arrive to the system according to a Poisson process with parameter λ_i ($i = 1, 2$). If upon the entry of a type 1 customer all servers are occupied, but some servers handle customers of the second type, then a server will change its service and the type 2 customer will be lost. Thus, customers of second type may be lost not only if a type 2 customer arrives and all servers are occupied, but if customers of first type show up as well. First type customers are refused only when there are customers of the same type.

The service times are exponentially distributed with parameters μ_1 and μ_2, respectively.

It is quite clear the service of type 1 customers is denied if all servers were busy and there were no type 2 customers. Thus, the probability of loss of type 1 customers clearly equals

$$P_v = \frac{\frac{\rho_1^m}{m!}}{\sum_{i=0}^{m} \frac{\rho_1^i}{i!}}, \qquad \rho_1 = \frac{\lambda_1}{\mu_1}.$$

Let $p_{ij}(t)$ be the probability of event that at moment t there are i type 1 and j type 2 customers being served ($0 \le i + j \le m$). Furthermore, let

$$p_{i.}(t) = \sum_{j=0}^{m-i} p_{ij}(t) \qquad \text{and} \qquad p_{.j}(t) = \sum_{i=0}^{m-j} p_{ij}(t).$$

The sum $\sum_{i+j=m} p_{ij}(t)$ is the probability of loss of a type 2 customer at moment t. The probability of loss of a type 2 customer during its service is

$$\sum_{i+j=m} p_{ij}(t) - p_{m0}(t).$$

11.5.2 Probabilities $p_{ij}(t)$

The differential equations determining $p_{ij}(t)$ are

$$p'_{00}(t) = -(\lambda_1 + \lambda_2)p_{00}(t) + \mu_1 p_{10}(t) + \mu_2 p_{01}(t); \tag{11.8}$$

if $1 \le i < m$, then

$$p'_{i0}(t) = -(\lambda_1 + \lambda_2 + i\mu_1)p_{i0}(t) + \lambda_1 p_{i-1,0}(t) + (i+1)\mu_1 p_{i+1,0}(t) + \mu_2 p_{i1}(t), \tag{11.9}$$

$$p'_{m0}(t) = -m\mu_1 p_{m0}(t) + \lambda_1[p_{m-1,0}(t) + p_{m-1,1}(t)]; \tag{11.10}$$

in the case of $1 \le j < m$,

$$p'_{0j}(t) = -(\lambda_1 + \lambda_2 + j\mu_2)p_{0j}(t) + \lambda_2 p_{0,j-1}(t) + \mu_1 p_{1j}(t)$$
$$+ (j+1)\mu_2 p_{0,j+1}(t), \tag{11.11}$$

$$p'_{0m}(t) = -(\lambda_1 + m\mu_2)p_{0m}(t) + \lambda_2 p_{0,m-1}(t); \tag{11.12}$$

in the case of $i \geq 1, j \geq 1, i + j < m$,

$$p'_{ij}(t) = -(\lambda_1 + \lambda_2 + i\mu_1 + j\mu_2)p_{ij}(t) + \lambda_1 p_{i-1,j}(t)$$
$$+\lambda_2 p_{i,j-1}(t) + (i+1)\mu_1 p_{i+1,j}(t) + \mu_2 p_{i,j+1}(t); \quad (11.13)$$

in the case of $i > 0, j > 0, i + j = m, i \neq m, j \neq m$,

$$p'_{ij}(t) = -(\lambda_1 + i\mu_1 + j\mu_2)p_{ij}(t) + \lambda_1[p_{i-1,j}(t) + p_{i-1,j+1}(t)] + \lambda_2 p_{i-1,j}(t).$$
$$(11.14)$$

Summing up the Eqs. (11.8), (11.11), and (11.14) by j from 0 to m we obtain

$$p'_{0.}(t) = -\lambda_1 p_{0.}(t) + \mu_1 p_{1.}(t). \quad (11.15)$$

Summing up Eqs. (11.9), (11.13), and (11.14) by j from 0 to m, in the case $1 \leq i < m$,

$$p'_{i.}(t) = -(\lambda_1 + i\mu_1)p_{i.}(t) + \lambda_1 p_{i-1,.}(t) + (i+1)\mu_1 p_{i+1,.}(t). \quad (11.16)$$

Equation (11.10) may be rewritten in the form

$$p'_{m.}(t) = -m\mu_1 p_{m.}(t) + \lambda_1 p_{m-1,.}(t) \quad (11.17)$$

The summation of Eqs. (11.8)–(11.10) by i leads to

$$p'_{.0}(t) = -\lambda_2[p_{.0}(t) - p_{m0}(t)] + \lambda_1 p_{m-1,1}(t) + \mu_2 p_{.1}(t).$$

Summing up Eqs. (11.11), (11.13), and (11.14) by i at $1 \leq j < m$:

$$p'_{.j}(t) = -(\lambda_2 + j\mu_2)[p_{.j}(t) - p_{m-j,j}(t)] + \lambda_2[p_{.,j-1}(t) - p_{m-j+1,j-1}(t)]$$
$$+(j+1)\mu_2 p_{.,j+1}(t) - j\mu_2 p_{m-j,j}(t) - \lambda_1 p_{m-j,j}(t) + \lambda_1 p_{m-j-1,j+1}(t).$$

Equation (11.12) may be written in the form

$$p'_{.m}(t) = -(\lambda_1 + m\mu_2)p_{.m}(t) + \lambda_2[p_{.m}(t) - p_{1,m-1}(t)].$$

From these equations one can see that for type 2 customers the situation is more complicated; type 1 customers play an essential role in the service process.

Let us consider the case $m = 1$. Then Eqs. (11.8)–(11.14) lead to the equations:

$$p'_{00}(t) = -(\lambda_1 + \lambda_2)p_{00}(t) + \mu_1 p_{10}(t) + \mu_2 p_{01}(t),$$
$$p'_{10}(t) = -\mu_1 p_{10}(t) + \lambda_1[p_{00}(t) + p_{01}(t)],$$
$$p'_{01}(t) = -(\lambda_1 + \mu_2)p_{01}(t) + \lambda_2 p_{00}(t).$$

This system may be solved easily; the initial conditions are

$$p_{00}(0) = 1, \qquad p_{10}(0) = 0, \qquad p_{01}(0) = 0.$$

We have

$$p_{10}(t) = \frac{\lambda_1}{\lambda_1 + \mu_1}\left(1 - e^{-(\lambda_1 + \mu_1)t}\right),$$

$$p_{01}(t) = \frac{\lambda_2 \mu_1}{(\lambda_1 + \mu_1)(\lambda_1 + \lambda_2 + \mu_2)} + \frac{\lambda_1 \lambda_2}{(\lambda_1 + \mu_1)(\lambda_2 + \mu_2 - \mu_1)} e^{-(\mu_1 + \mu_2)t}$$
$$- \left(\frac{\lambda_2 \mu_1}{(\lambda_1 + \mu_1)(\lambda_1 + \lambda_2 + \mu_2)} + \frac{\lambda_1 \lambda_2}{(\lambda_1 + \mu_1)(\lambda_2 + \mu_2 - \mu_1)}\right) e^{-(\lambda_1 + \lambda_2 + \mu_2)t},$$

$$p_{00}(t) = \frac{\mu_1(\lambda_1 + \mu_2)}{(\lambda_1 + \mu_1)(\lambda_1 + \lambda_2 + \mu_2)} + \frac{\lambda_1(\mu_2 - \mu_1)}{(\lambda_1 + \mu_1)(\lambda_2 + \mu_2 - \mu_1)} e^{-(\lambda_1 + \mu_1)t}$$
$$+ \frac{\lambda_2}{\lambda_1 + \mu_1}\left(\frac{\mu_1}{\lambda_1 + \lambda_2 + \mu_2} + \frac{\lambda_1}{\lambda_2 + \mu_2 - \mu_1}\right) e^{-(\lambda_1 + \lambda_2 + \mu_2)t}.$$

Consequently,

$$p_{0.}(t) = p_{00}(t) + p_{01}(t) = \frac{\mu_1}{\lambda_1 + \mu_1} + \frac{\lambda_1}{\lambda_1 + \mu_1} e^{-(\lambda_1 + \mu_1)t},$$

$$p_{1.}(t) = p_{10}(t) = \frac{\lambda_1}{\lambda_1 + \mu_1}\left(1 - e^{-(\lambda_1 + \mu_1)t}\right),$$

$$p_{.0}(t) = \frac{\lambda_1(\lambda_1 + \lambda_2 + \mu_2) + \mu_1(\lambda_1 + \mu_2)}{(\lambda_1 + \mu_1)(\lambda_1 + \lambda_2 + \mu_2)} - \frac{\lambda_1 \lambda_2}{(\lambda_1 + \mu_1)(\lambda_2 + \mu_2 - \mu_1)} e^{-(\lambda_1 + \mu_1)t}$$
$$+ \frac{\lambda_2}{\lambda_1 + \mu_1}\left(\frac{\mu_1}{\lambda_1 + \lambda_2 + \mu_2} + \frac{\lambda_1}{\lambda_2 + \mu_2 - \mu_1}\right) e^{-(\lambda_1 + \lambda_2 + \mu_2)t},$$

$$p_{.1}(t) = p_{01}(t) = \frac{\lambda_2 \mu_1}{(\lambda_1 + \mu_1)(\lambda_1 + \lambda_2 + \mu_2)} + \frac{\lambda_1 \lambda_2}{(\lambda_1 + \mu_1)(\lambda_2 + \mu_2 - \mu_1)} e^{-(\lambda_1 + \mu_1)t}$$
$$- \frac{\lambda_2}{\lambda_1 + \mu_1}\left(\frac{\mu_1}{\lambda_1 + \lambda_2 + \mu_2} + \frac{\lambda_1}{\lambda_2 + \mu_2 - \mu_1}\right) e^{-(\lambda_1 + \lambda_2 + \mu_2)t}.$$

The stationary probabilities at $t \to \infty$ are

$$p_{00} = \mu_1(\lambda_1 + \mu_2)/\left((\lambda_1 + \mu_1)(\lambda_1 + \lambda_2 + \mu_2)\right),$$
$$p_{01} = p_{.1} = \lambda_2\mu_1/\left((\lambda_1 + \mu_1)(\lambda_1 + \lambda_2 + \mu_2)\right),$$
$$p_{10} = p_{1.} = \lambda_1/(\lambda_1 + \mu_1),$$
$$p_{0.} = \mu_1/(\lambda_1 + \mu_1),$$
$$p_{.0} = \frac{\lambda_1(\lambda_1 + \lambda_2 + \mu_2) + \mu_1(\lambda_1 + \mu_2)}{(\lambda_1 + \mu_1)(\lambda_1 + \lambda_2 + \mu_2)}.$$

It is interesting to note the probability of the event that a service in process at a given moment will not be interrupted. This happens if during the service no type 1 customers enter, namely,

$$\int_0^\infty e^{-\lambda_1 x} \mu_2 e^{-\mu_2 x}\, dx = \frac{\mu_2}{\lambda_1 + \mu_2},$$

then service will be interrupted with probability $\lambda_1/(\lambda_1 + \mu_2)$.

11.5.3   Priority System with General Service Time

Now we come to priority systems with generally distributed service time. We will consider three cases:

1. If a type 1 customer enters, then the service of a type 2 customer is interrupted and is continued after all type 1 customers have been served. The performed work is taken into account, and the service time is decreased with the work done.
2. The service is realized as above, but when a type 2 customer is served, the performed work will not be taken into account; the service does not decrease with time spent.
3. When a type 1 customer enters, the actual service is interrupted, and the customer is lost.

In all three cases we assume the entering customers constitute Poisson processes with parameters λ_1 and λ_2, the service times are arbitrarily distributed random variables with distribution functions $B_1(x)$ and $B_2(x)$, respectively. The Laplace–Stieltjes transforms of the service time is

$$b_i^\sim(s) = \int_0^\infty e^{-sx}\, dB_i(x), \qquad i = 1, 2.$$

Let us denote the mean values of service times by τ_1 and τ_2, let $V_i(t)$ be the waiting time of a type i customer on the condition that it entered at moment t, and let $\hat{V}_i(t)$ be the time till completion of service. Let

$$F_i(x) = \lim_{t\to\infty} \mathbf{P}(V_i(t) < x), \qquad i = 1, 2, \ldots \qquad (11.18)$$

$$\hat{F}_i(x) = \lim_{t\to\infty} \mathbf{P}\big(\hat{V}_i(t) < x\big), \qquad i = 1, 2, \ldots,$$

be the distribution of the waiting time and the time till service completion and let their Laplace–Stieltjes transforms according to Eq. (11.18) be $f_i^\sim(s)$ and $\hat{f}_i^\sim(s)$.

Type 1 customers are served independently of type 2 customers, so by Eq. (8.19) (if the condition $\lambda_1 \tau_1 < 1$ is fulfilled)

$$\tilde{f_1}(s) = \frac{1 - \lambda_1 \tau_1}{1 - \lambda_1 \frac{1 - \tilde{b_1}(s)}{s}}.$$

The time interval till completion consists of two parts: the waiting time and the service time. They are independent random variables, so for $\hat{V}_1(t)$ we obtain

$$\hat{\tilde{f_1}}(s) = \frac{(1 - \lambda_1 \tau_1) \tilde{b_1}(s)}{1 - \frac{\lambda_1}{s}(1 - \tilde{b_1}(s))}.$$

At the service of type 2 customers the service of type 1 customers may be interpreted as the breakdown of a server. Let L be a random variable denoting the time from the beginning of service of a type 2 customer till the beginning of service of the next one and

$$\tilde{b_L}(s) = \int_0^\infty e^{-sx} \, d\mathbf{P}(L < x).$$

At a fixed moment we have two possibilities: a type 1 customer is absent with probability $1 - \lambda_1 \tau_1$ and present with probability $\lambda_1 \tau_1$, and in its presence according to the service discipline it is being served. If there are no type 1 customers, then by Eq. (8.19) the Laplace–Stieltjes transform of remaining service time is

$$\frac{1 - \lambda_2 \mathbf{E}(L)}{1 - \frac{\lambda_2}{s}(1 - \tilde{b_L}(s))}.$$

In the presence of a type 1 customer, we must first finish the serving existing and entering type 1 customers, then serve type 2 customers (taking into account type 1 customers that enter in the meantime). The Laplace–Stieltjes transform of the service time for existing and entering type 1 customers is

$$\frac{1 - \tilde{b_0}(s)}{s \frac{\tau_1}{1 - \lambda_1 \tau_1}},$$

where $\tilde{b_0}(s)$ is the solution of the functional equation

$$\tilde{b_0}(s) = \tilde{b_1}(s + \lambda_1 - \lambda_1 \tilde{b_0}(s)),$$

i.e., the Laplace–Stieltjes transform of a busy period for the type 1 customers. After having served the type 1 customers we come to the previous situation. Thus, the

Laplace–Stieltjes transform of the time period till the service of type 2 customers entering at a given moment is

$$(1 - \lambda_1 \tau_1) \frac{1 - \lambda_2 \mathbf{E}(L)}{1 - \frac{\lambda_2}{s}(1 - b_L^{\sim}(s))} + \lambda_1 \tau_1 \frac{1 - b_0^{\sim}(s)}{s \frac{\tau_1}{1 - \lambda_1 \tau_1}} \cdot \frac{1 - \lambda_2 \mathbf{E}(L)}{1 - \frac{\lambda_2}{s}(1 - b_L^{\sim}(s))}$$

$$= \frac{(1 - \lambda_1 \tau_1)(1 - \lambda_2 \mathbf{E}(L))[s + \lambda_1(1 - b_0^{\sim}(s))]}{s - \lambda_2(1 - b_L^{\sim}(s))} = f_2^{\sim}(s). \qquad (11.19)$$

In this expression $b_L^{\sim}(s)$ is still unknown, but we will find it for our three models.

1. From the point of view of a type 2 customer we can interpret the system behavior such that the entry of a type customer 1 is a failure and the end of the busy period, generated by this type 1 customer, as maintenance. Based on this interpretation our model can be considered a system with server breakdowns. Thus,

$$b_L^{\sim}(s) = b_2^{\sim}(s + \lambda_1 - \lambda_1 b_0^{\sim}(s)).$$

2. Let us consider the sequences of independent random variables $\{U_n\}$, $\{H_n\}$, and $\{A_n\}$, which have the following meaning:

   U_i: service time of a type 2 customer (with Laplace–Stieltjes transform $b_2^{\sim}(s)$);
   H_i: length of busy period for type 1 customers (the corresponding Laplace–Stieltjes transform is $b_0^{\sim}(s)$);
   A_i: interarrival time for type 1 customers (exponentially distributed random variable with parameter λ_1).

   If $U_1 \leq A_1$, then $L = U_1$ (during the service of type 2 customers no type 1 customers enter, so the type 2 customer leaves after U_1 time from the beginning of service).
   If $A_1 < U_1$ and $U_2 \leq A_2$, then $L = H_1 + A_1 + U_2$ (during the service of a type 2 customer after A_1 time a type 1 customer enters, and for its and the entering customers' service we try time H_1, then the service of a type 2 customer is realized for U_2 without interruption). Similarly, if $A_1 < U_1, A_2 < U_2, \ldots, A_n < U_n, U_{n+1} \leq A_{n+1}$, then $L = A_1 + H_1 + A_2 + H_2 + \ldots + A_n + H_n + U_{n+1}$.
   By the formula of total probability

$$\mathbf{P}(L < x) =$$

$$\sum_{n=0}^{\infty} \mathbf{P}\left(A_i < U_i, 1 \leq i \leq n; U_{n+1} \leq A_{n+1}; \sum_{i=1}^{n} A_i + \sum_{i=1}^{n} H_i + U_{n+1} < x \right).$$

Since

$$\int_0^\infty e^{-sx}\,\mathrm{d}_x P\{A_i < x,\; A_i < U_i\} = \lambda_1 \int_0^\infty e^{-sx}(1 - B_2(x))e^{-\lambda_1 x}\,\mathrm{d}x$$

$$= \frac{\lambda_1}{s+\lambda_1}[1 - b_2^\sim(s+\lambda_1)]$$

and

$$\int_0^\infty e^{-sx}\,\mathrm{d}_x P\{U_i < x,\; U_i \le A_i\} = \int_0^\infty e^{-(s+\lambda_1)x}\,\mathrm{d}B_2(x) = b_2^\sim(s+\lambda_1),$$

we obtain

$$b_L^\sim(s) = \sum_{n=0}^\infty \left\{ \frac{\lambda_1}{s+\lambda_1}[1 - b_2^\sim(s+\lambda_1)]b_0^\sim(s) \right\}^n b_2^\sim(s+\lambda_1)$$

$$= \frac{(s+\lambda_1)b_2^\sim(s+\lambda_1)}{s+\lambda_1 - \lambda_1[1 - b_2^\sim(s+\lambda_1)]b_0^\sim(s)}.$$

3. Using the random variables U_i, H_i, A_i we have

$$L = \begin{cases} U_1, & \text{ha } U_1 \le A_1, \\ A_1 + H_1, & \text{ha } U_1 > A_1. \end{cases}$$

Consequently,

$$b_L^\sim(s) = b_2^\sim(s+\lambda_1) + \frac{\lambda_1}{s+\lambda_1}[1 - b_2^\sim(s+\lambda_1)]b_0^\sim(s).$$

Now let us find the functions $\hat{f}_2^\sim(s)$. In the first two cases the time from the moment t till the end of service is $U_2(t)+L$. They are independent random variables, so in both cases

$$\hat{f}_2^\sim(s) = f_2^\sim(s)b_L^\sim(s).$$

In the third case we can lose the type 2 customer; this happens if during the service of type 2 customer a type 1 customer appears, the probability of losing the type 2 customer is

$$\mathbf{P}(A_1 < U_1) = 1 - b_2^\sim(\lambda_1).$$

Obviously,

$$\hat{U}_2(t) = U_2(t) + \min(A_2, U_1).$$

Since

$$\int_{x=0}^{\infty} e^{-sx}\, d\mathbf{P}(\min(A_1, U_1) < x) = b_2^{\sim}(s + \lambda_1) + \frac{\lambda_1}{s + \lambda_1}[1 - b_2^{\sim}(s + \lambda_1)],$$

then

$$\hat{f}_2^{\sim}(s) = f_2^{\sim}(s)\left\{b_2^{\sim}(s + \lambda_1) + \frac{\lambda_1}{s + \lambda_1}[1 - b_2^{\sim}(s + \lambda_1)]\right\}.$$

These formulas are true if the process has an equilibrium distribution. On the basis of Eq. (11.19) this means that the inequalities $\lambda_1\tau_1 < 1$ and $\lambda_2\mathbf{E}(L) < 1$ hold.

- In the first model $\mathbf{E}(L) = \tau_2/(1-\lambda_1\tau_1)$, from which the condition of equilibrium is $\lambda_2\tau_2 < 1 - \lambda_1\tau_1$.
- In the second model $\mathbf{E}(L) = [1 - b_2^{\sim}(\lambda_1)]/\lambda_1(1 - \lambda_1\tau_1)b_2^{\sim}(\lambda_1)$, from which the condition of equilibrium is $\lambda_2[1 - b_2^{\sim}(\lambda_1)] < \lambda_1(1 - \lambda_1\tau_1)b_2^{\sim}(\lambda_1)$.
- In the third model $\mathbf{E}(L) = [1 - b_2^{\sim}(\lambda_1)]/\lambda_1(1-\lambda_1\tau_1)$, from which the condition of equilibrium is $\lambda_2[1 - b_2^{\sim}(\lambda_1)] < \lambda_1(1 - \lambda_1\tau_1)$.

11.6 Systems with Several Servers and Queues

11.6.1 Multichannel Systems with Waiting and Refusals

Let (X_n, Y_n), $n = 1, 2, \ldots$ be a sequence of i.i.d. random vector variables, where $X_1, X_2, \ldots$ are the interarrival periods of successive customers (the nth one enters at the moment $t_n = X_1 + \cdots + X_n$, $n = 1, 2, \ldots$), and Y_n is the service time of nth customer.

Let us consider the $G/G/m$ system. Introduce the *waiting time vector* of the nth customer:

$$W_n = (W_{n,1}, \ldots, W_{n,m}), \ n = 1, 2, \ldots,$$

where $W_{n,i}$ means the random time interval the nth customer (entering at t_n) has to wait till i servers become free from all earlier (with numbers $1, \ldots, n-1$) customers.

If the initial random vector variable W_0 (on the same probability space) is given, then the sequence W_n, $n \geq 0$, is uniquely determined and a recurrence relation is valid for W_n, i.e., $\{W_n, \ n \geq 0\}$ is a recurrent process. For the arbitrary $x = (x_1, \ldots, x_m) \in \mathbb{R}^m$ let

$$x^+ = (x_1^+, \ldots, x_m^+), \quad \text{where} \quad s^+ = \max(s, 0), \ s \in \mathbb{R},$$

$$R(x) = (x_{i_1}, \ldots, x_{i_m}), \quad x_{i_1} \leq x_{i_2} \leq \cdots \leq x_{i_m},$$

i.e., the function $R(x)$ arranges the components of vector x in increasing order. We introduce the vectors

$$
\begin{array}{cccc}
(1) \ (2) & (m) & (1) & (m) \\
e = (\ 1, \ 0, \ \ldots, \ 0\ ), & & i = (\ 1, \ \ldots, \ 1\ ).
\end{array}
$$

Theorem 11.1 *For the sequence W_n, $n \geq 0$, the recurrence relation*

$$W_{n+1} = R\left([(W_n + Y_n e) - X_n i]^+\right), \quad n \geq 0. \tag{11.20}$$

holds.

Proof Using the definition of included quantities, this is trivial.   □

In the investigation of queueing systems the existence of a limit distribution for the basic characteristics is an important question. Using results from the theory of recurrence processes one can prove a theorem valid in the more general case where instead of total independence we assume the stationarity in a narrower sense and the ergodicity of the process $\{(X_n, Y_n), \ n \geq 1\}$ (see [7]).

Theorem 11.2 *Let $\{(X_n, Y_n), \ n \geq 1\}$ be a sequence of i.i.d. random variables and $E(Y_1 - mX_1) < 0$, $W_0 = 0$; then there exists a stationary, in a narrow sense, process $\{W^{(n)}, \ n \geq 0\}$ satisfying Eq. (11.21), and the distribution function of W_n monotonically converges to the distribution function of $W^{(0)}$.*

11.6.1.1 $G/G/m/0$ Systems with Refusals

Since we are considering a system with refusals, one can speak of waiting only in a virtual sense. Thus, the component $W_{n,i}$ of W_n means the possible waiting time of customers entering at moment t_n till i servers become free from all earlier (with numbers $1, \ldots, n-1$) customers (if $W_{n,1} > 0$, then the nth one will not be served). We can write the recurrence relation

$$W_{n+1} = R\left(\left[(W_n + Y_n e \mathscr{I}_{\{W_{n,1}=0\}}) - X_n i\right]^+\right).$$

The sufficient condition similar to the previous theorem is

$$\mathbf{P}(Y_1 \leq mX_1) > 0, \quad \mathbf{E}(Y_1) < \infty.$$

If (X_n, Y_n), $-\infty < n < \infty$ is not a sequence of independent random variables with the same distribution but a stationary (stationary in a narrower sense), ergodic sequence, even in this case we can give a sufficient condition for the existence of limit distribution, namely,

$$\mathbf{P}(Y_0 \leq X_0 + \ldots + X_{m-1}, \ \ Y_{-1} \leq X_{-1} + X_0 + \ldots + X_{m-2}, \ldots, Y_{-m+1}$$
$$\leq X_{-m+1} + X_{-m+2} + \ldots + X_0) > 0,$$

$$\mathbf{E}(Y_1) < \infty.$$

If we consider instead of the virtual waiting time the queue length L_n at the arrival moment of the nth customer, then it also has a nondegenerate limiting distribution.

11.6.1.2 $G/G/\infty$ System

Now we have an infinite number of servers, so one cannot speak of queueing or losses. In this case the basic characteristic is the queue length: L_k, $k \geq 1$, denotes the number of customers at the arrival moment of the kth customer [at an arbitrary moment t the number of customers present $L(t)$ is left continuous]. Actually, it is the number of occupied servers. At the beginning the system is empty, i.e., $L_1 = 0$.

For the sake of simplicity let X_n, $n \geq 1$, denote the interarrival time of nth and $(n + 1)$st customers, Y_n, $n \geq 1$, the service time of nth customer. Then

$$L_{k+1} = \mathscr{I}_{\{Y_k > X_k\}} + \mathscr{I}_{\{Y_{k-1} > X_{k-1} + X_k\}} + \cdots + \mathscr{I}_{\{Y_1 > X_1 + \cdots + X_k\}}, \quad k \geq 1.$$

Theorem 11.3 *If $\{(X_n, Y_n), \ -\infty < n < \infty\}$ is a sequence of i.i.d. random variables and $0 < E(Y_1) < \infty$ is fulfilled, then*

$$L = \sum_{k \geq 1} \mathscr{I}_{\{Y_{-k} > X_{-k} + \cdots + X_{-1}\}}$$

defines a finite random variable with probability 1, the random variables

$$L_{-n} = \sum_{k=1}^{n} \mathscr{I}_{\{Y_{-k} > X_{-k} + \cdots + X_{-1}\}}, \quad n = 1, 2, \ldots$$

and L_n have the same distribution, and this distribution monotonically converges to the distribution of L.

Proof For the proof it is enough to show that L is finite with probability 1. We need the following lemma; from it with probability 1 follows the finiteness of L. □

Lemma 11.1 *Let* $U_1, U_2, \ldots$ *be a sequence of i.i.d. random variables for which* $\mathbf{P}(U_1 \geq 0) = 1$ *and* $h = \mathbf{E}\left(e^{-U_1}\right) < 1$, *i.e., the distribution of* U_i *is not concentrated at the point 0. Let* V *be an arbitrary random variable (not necessarily independent of* U_i) *for which* $\mathbf{E}(V^+) < \infty$. *Furthermore, let* $\kappa = \frac{1}{2} \log \frac{1}{h}$, $G_V(x) = 1 - \mathbf{P}(V < x)$, $x \in \mathbb{R}$. *Then for arbitrary* $n, N \geq 1$

$$\mathbf{P}(U_1 + \cdots + U_n < V) < e^{-n\kappa} + G_V(n\kappa), \tag{11.21}$$

$$\sum_{n \geq N} \mathbf{P}(U_1 + \cdots + U_n < V) < \frac{1}{1 - e^{-\kappa}} e^{-N\kappa} + \frac{1}{\kappa} \mathbf{E}\left(V \mathscr{I}_{\{N\kappa \leq V\}}\right) \tag{11.22}$$

is true.

Proof For arbitrary $x > 0$

$$\mathbf{P}(U_1 + \cdots + U_n < V)$$
$$= \mathbf{P}(U_1 + \cdots + U_n < V, \ V \leq nx) + \mathbf{P}(U_1 + \cdots + U_n < V, \ nx < V)$$
$$\leq \mathbf{P}(U_1 + \cdots + U_n < nx) + \mathbf{P}(nx \leq V).$$

Using the Markov inequality we obtain

$$\mathbf{P}(U_1 + \cdots + U_n < nx) \leq \mathbf{E}(\exp\{nx - (U_1 + \cdots + U_n)\})$$

$$= e^{nx} \prod_{i=1}^{n} \mathbf{E}\left(e^{-U_i}\right)$$

$$= e^{n(x + \log h)},$$

where at $x = \kappa$ Eq. (11.21) follows.
   Proof of Eq. (11.22): From inequality (11.21)

$$\sum_{n \geq N} \mathbf{P}(U_1 + \cdots + U_n < V) \leq \sum_{n \geq N} \{e^{-n\kappa} + G_V(n\kappa)\}$$

$$= \frac{1}{1 - e^{-\kappa}} e^{-N\kappa} + \sum_{j=0}^{\infty} \mathbf{P}((N + j)\kappa \leq V).$$

Since

$$\mathbf{E}\left(V \mathscr{I}_{\{N\kappa \leq V\}}\right) \geq \sum_{j=0}^{\infty} (N + j)\kappa \mathbf{P}((N + j)\kappa \ \leq \ V \ < \ (N + j + 1)\kappa)$$

$$= (N - 1)\kappa \sum_{j=0}^{\infty} \mathbf{P}((N + j)\kappa \ \leq V \ < \ (N + j + 1)\kappa)$$

$$+ \sum_{j=0}^{\infty} (j+1)\kappa \mathbf{P}((N+j)\kappa \ \leq \ V \ < \ (N+j+1)\kappa)$$

$$= (N-1)\kappa \mathbf{P}(N\kappa \leq V) + \kappa \sum_{j=0}^{\infty} \mathbf{P}((N+j)\kappa \leq V),$$

then

$$\sum_{j=0}^{\infty} \mathbf{P}((N+j)\kappa \leq V) \leq \frac{1}{\kappa} \mathbf{E}\big(V \mathscr{I}_{\{N\kappa \leq V\}}\big) - (N-1)\mathbf{P}(N\kappa \leq V)$$

$$\leq \frac{1}{\kappa} \mathbf{E}\big(V \mathscr{I}_{\{N\kappa \leq V\}}\big).$$

Using Eq. (11.21) we obtain Eq. (11.22).   □

11.6.2   Closed Queueing Network Model of Computers

The queueing network in Fig. 11.27 may be considered as the simplest mathematical model for computers.

In a system there are continuously n customers (tasks) and they can move along the routes indicated in the figure. In front of each service unit there is a waiting buffer of corresponding capacity (for at most $n-1$ customers). On the units the service is realized by the FCFS rule; the service times are independent and on the ith unit have distribution function $F_i(x)$, $0 \leq i \leq M$. After having completed a service on the 0th unit, the customer moves to the ith unit with probability p_i, $0 \leq$

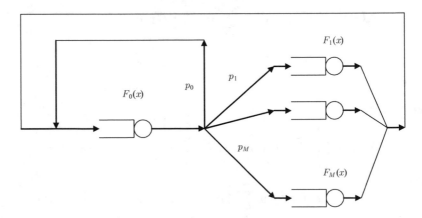

Fig. 11.27   Closed queueing network model

$i \leq M$ ($p_i \geq 0$, $p_0 + \cdots + p_M = 1$), which does not depend on the state of the system or the service time. If the service of a customer is completed on the ith ($1 \leq i \leq M$) unit, then the customer goes to a unit with index 0 with probability 1. The unit 0 plays special role in the network and is called a *central unit*.

It may seem too strict a restriction that the number of customers in a computer is fixed, but this model gives accurate results for several important performance parameters. For example, when we are interested in the maximum performance of a computer, we can assume that the load of the computer is maximal, that is, after completion of a task, a new one enters the system immediately.

If we consider the successive moments when all customers stay at the central unit and the service of a customer has just started, these moments are regeneration points for the network. If the service times have finite pth ($p \geq 1$) order moments, then one can show the pth moment of a regenerative cycle is also finite [32, 33]. It follows that if the mean values of services are finite, then the different characteristics for the system have limiting distributions.

Based only on the structure of the queueing network in Fig. 11.27 the central unit seems to be the decisive element of the system performance. This qualitative judgment might be significantly altered by quantitative analysis. Having the mean waiting times and the mean number of customers at the units, one can better judge the bottleneck of the system. If all units have approximately the same waiting time, then the system is balanced, and improving the performance of a single unit does not increase the overall system performance, which can be measured as the input rate of unit 0, significantly. Instead, when the waiting time (as well as the mean number of customers) of one of the units is much higher than the others, then that unit is the bottleneck of the system, and improving the service capacity of that unit significantly improves the system performance.

11.7   Retrial Systems

11.7.1   Classical Retrial System

If in the case of a phone call the subscriber is occupied, one usually repeats the attempt while the conversation is realized. So the system has two types of requests: primary calls and calls generated by occupied lines. Models constructed for systems with losses do not describe this situation, and they do not take into account repetitions. These problems appeared in Erlang's time, but due to a lack of absence of corresponding theoretical results, these repetitions were considered new arrivals.

Retrial queues constitute a special field of queueing systems; their distinguishing feature is that in the case of a busy server, entering customers leave the service area (go to the orbit) and, after a certain (generally random) time, reinitiate their service.

Let us consider some examples of the retrial phenomenon. The first example is connected with functioning of call centers used by companies to communicate with customers. When a call arrives, it is sent to a call distribution switch. If all agents are busy, then the call center may announce an estimated waiting time. Some customers might decide to wait for a free agent, while some will interrupt the connection immediately or after some time. A portion of these customers will return after some random time.

Random access protocols provide a motivation for the design of communication protocols with retransmission control. If two or more stations transmit packets at the same time, then a collision takes place. Then all packets are destroyed and should be retransmitted. To avoid collision in the next period, this transmission is realized with a certain random delay for each station. This fact motivates the investigation of retrial feature of computer networks.

Two textbooks have been published in this field. The book by Falin and Templeton [9] gives analytical solutions in terms of generating functions and Laplace–Stieltjes transforms, and the one by Artalejo and Gómez-Corral [4] focuses on the application of algorithmic methods studying the M/G/1 and the M/M/m retrial queues and using matrix-analytic techniques to solve some retrial queues with QBD, GI/M/1, and M/G/1 structures.

According to [9] one can divide retrial queues into three large groups: single-channel systems, multi-channel systems, and structurally complex systems (channels with limited availability, link systems, and circuit switched networks). To show the idea of retrial queue we will restrict ourselves to the single-channel system at the simplest conditions and then we will consider the so-called cyclic-waiting systems for both continuous and discrete-time.

Consider a single server queueing system with Poisson arrivals with rate λ. These customers are the primary calls. If the server is free at the arrival moment, its service begins immediately and it leaves after service completion. Otherwise, if the server is busy, the customer becomes the source of repeated calls (goes to the orbit). The pool of sources of repeated calls may be considered as a sort of queue. Each such source produces a Poisson process of repeated calls with intensity μ. If an incoming repeated call finds the server free, it is served and leaves the system after service, the source which generated this call disappears. The service time distribution is $B(x)$ for all calls. The input flow of primary calls, intervals between repetitions, and service times are mutually independent.

The evolution of queueing process is as follows: the $(i-1)$th customer completes its service at η_{i-1} and the server becomes free. The present customers cannot occupy the server immediately, because of their ignorance of server state. The ith customer enters service only after some time interval τ_i during which the server is free though there may be waiting customers. If the number of sources of repeated calls at η_{i-1} is n, then the random variable τ_i has exponential distribution with parameter $\lambda + n\mu$ and it is a primary call with probability $\frac{\lambda}{\lambda+n\mu}$ and repeated call with $\frac{n\mu}{\lambda+n\mu}$. At $\xi_i = \eta_{i-1} + \tau_i$ the ith call's service starts and continues S_i time. All primary calls arriving during S_i will be sources of repeated calls. Repeated calls arriving during this time do not influence the service process. At $\eta_i = \xi_i + S_i$ the ith call completes service and the server becomes free again.

Let $N(t)$ be the number of sources of repeated calls (number of customers on the orbit) and $C(t)$ the number of busy servers (for a single-server system $C(t) = 0$ or 1). At the moment t $(C(t), N(t))$ describes the functioning of the system. If the service time distribution is not exponential, the process $(C(t), N(t))$ is not Markov. In this case one introduces a supplementary variable: if $C(t) = 1$, one defines $\xi(t)$ as the elapsed time of the call being served. Let $B^\sim(s) = \int_0^\infty e^{-sx} dB(x)$ be the Laplace–Stieltjes transform of the service time distribution function, $\beta_k = (-1)^k B^{\sim(k)}(0)$ the kth moment of service time, $\rho = \lambda \beta_1$ the system load due to primary calls, $b(x) = B'(x)/(1 - B(x))$ the instantaneous service time intensity (the probability of event that the service continuing already x time will be finished on $(x, x + \Delta)$), $A(z) = B^\sim(\lambda(1 - z))$.

We investigate the joint distribution of server state and queue length in steady state for both the M/M/1 and the M/G/1 system. The results are obtained under the usual condition $\rho < 1$, and this condition is discussed in [9].

11.7.1.1   M/M/1 System

The distribution function of service time is $B(x) = 1 - e^{-\nu x}$.

Theorem 11.4 *For an M/M/1 retrial queue in the steady state the joint distribution of server state $C(t)$ and queue length $N(t)$, $p_{in} = \mathbf{P}(C(t) = i, N(t) = n)$ is*

$$p_{0n} = \frac{\rho^n}{n! \mu^n} \prod_{i=0}^{n-1} (\lambda + i\mu) \cdot (1 - \rho)^{\frac{\lambda}{\mu}+1}, \qquad (11.23)$$

$$p_{1n} = \frac{\rho^{n+1}}{n! \mu^n} \prod_{i=1}^{n} (\lambda + i\mu) \cdot (1 - \rho)^{\frac{\lambda}{\mu}+1}, \qquad (11.24)$$

where $\rho = \lambda/\nu$. The corresponding partial generating functions are

$$P_0(z) = \sum_{n=0}^{\infty} z^n p_{0n} = (1 - \rho) \left(\frac{1 - \rho}{1 - \rho z} \right)^{\frac{\lambda}{\mu}},$$

$$P_1(z) = \sum_{n=0}^{\infty} z^n p_{1n} = \rho \left(\frac{1 - \rho}{1 - \rho z} \right)^{\frac{\lambda}{\mu}+1}.$$

Proof In this case $(C(t), N(t))$ is a Markov process with state space $\{0, 1\} \times \mathbf{Z}_+$.
From a state $(0, n)$ the following transitions are possible:

- $(1, n)$ with rate λ;
- $(1, n - 1)$ with rate $n\mu$.

The first transition corresponds to the arrival of a primary call, the second one to the arrival of a repeated call.

One can reach the state $(0, n)$ only from the state $(1, n)$ with rate v.

From $(1, n)$ the following transitions are possible:

- $(1, n + 1)$ with rate λ;
- $(0, n)$ with rate v.

The first transition corresponds to the arrival of a primary call, the second to a service completion.

$(1, n)$ can be reached in the following ways:

- $(0, n)$ with rate λ;
- $(0, n + 1)$ with rate $(n + 1)\mu$;
- $(1, n - 1)$ with rate λ.

The equilibrium equations for p_{0n} and p_{1n} have the form

$$(\lambda + n\mu)p_{0n} = v p_{1n}, \tag{11.25}$$

$$(\lambda + v)p_{1n} = \lambda p_{0n} + (n + 1)\mu p_{0,n+1} + \lambda p_{1,n-1}. \tag{11.26}$$

Introduce the generating functions

$$P_0(z) = \sum_{n=0}^{\infty} z^n p_{0n}, \qquad P_1(z) = \sum_{n=0}^{\infty} z^n p_{1n}.$$

Then from (11.25) and (11.26)

$$\lambda P_0(z) + \mu z P_0'(z) = v P_1(z),$$

$$(v + \lambda - \lambda z) P_1(z) = \lambda P_0(z) + \mu P_0'(z).$$

Excluding $P_1(z)$ we obtain the differential equation

$$P_0'(z) = \frac{\lambda \rho}{\mu(1 - \rho z)} P_0(z).$$

Its solution is

$$P_0(z) = \frac{C}{(1 - \rho z)^{\frac{\lambda}{\mu}}},$$

for $P_1(z)$ we obtain

$$P_1(z) = \frac{C \cdot \rho}{(1 - \rho z)^{\frac{\lambda}{\mu}+1}},$$

and the value of the constant may be found from the equality

$$\sum_{n=0}^{\infty}(p_{0n} + p_{1n}) = P_0(1) + P_1(1) = 1,$$

it implies

$$C = (1 - \rho)^{\frac{\lambda}{\mu}+1}.$$

To find the values of p_{0n} and p_{1n} we have to expand the power series $(1 - \rho z)^{-\frac{\lambda}{\mu}}$ and $(1 - \rho z)^{-\frac{\lambda}{\mu}-1}$, and it may be done by the help of the binomial formula in the form

$$(1 + x)^m = \sum_{n=0}^{\infty}\frac{x^n}{n!}\prod_{i=0}^{n-1}(m - i).$$

Since

$$(1 - \rho z)^{-\frac{\lambda}{\mu}} = \sum_{n=0}^{\infty}\frac{(-\rho z)^n}{n!}\prod_{i=0}^{n-1}\left(-\frac{\lambda}{\mu} - i\right) = \sum_{n=0}^{\infty}\frac{(\rho z)^n}{n!\mu^n}\prod_{i=0}^{n-1}(\lambda + i\mu)$$

and

$$(1 - \rho z)^{-\frac{\lambda}{\mu}-1} = \sum_{n=0}^{\infty}\frac{(-\rho z)^n}{n!}\prod_{i=0}^{n-1}\left(-\frac{\lambda}{\mu} - 1 - i\right) = \sum_{n=0}^{\infty}\frac{(\rho z)^n}{n!\mu^n}\prod_{i=1}^{n}(\lambda + i\mu),$$

from the expressions for the generating functions we obtain (11.23) and (11.24).

The stationary distribution of number of sources of repeated calls has the generating function

$$P(z) = P_0(z) + P_1(z) = (1 + \rho - \rho z)\left(\frac{1 - \rho}{1 - \rho z}\right)^{\frac{\lambda}{\mu}+1},$$

from it the mean value and variance are

$$\mathbf{E}(N(t)) = \frac{\rho(\lambda + \rho\mu)}{(1 - \rho)\mu},$$

$$\mathbf{Var}(N(t)) = \frac{\rho(\lambda + \rho\mu + \rho^2\mu - \rho^3\mu)}{(1 - \rho)^2\mu}.$$

The number of customers in the system has the generating function

$$Q(z) = P_0(z) + z P_1(z) = \left(\frac{1-\rho}{1-\rho z} \right)^{\frac{\lambda}{\mu}+1},$$

the corresponding characteristics are

$$\mathbf{E}(K(t)) = \frac{\rho(\lambda + \mu)}{(1-\rho)\mu},$$

$$\mathbf{Var}(K(t)) = \frac{\rho(\lambda + \mu)}{(1-\rho)^2 \mu}.$$

□

For further characteristics see [9].

11.7.1.2   M/G/1 System

The service time has general distribution with distribution function $B(x)$. In the subsequent analysis of the M/G/1 system a supplementary variable (see Sect. 4.5.2) is used.

Theorem 11.5 *For the M/G/1 retrial queue in the steady state the joint distribution of the server state and queue length is*

$$p_{0n} = \mathbf{P}(C(t) = 0, N(t) = n),$$

$$p_{1n}(x) = \frac{\mathrm{d}}{\mathrm{d}x} \mathbf{P}(C(t) = 1, \xi(t) = x, N(t) = n)$$

and we have the partial generating functions

$$P_0(z) = \sum_{n=0}^{\infty} = p_{0n} z^n = (1-\rho) \exp \left\{ \frac{\lambda}{\mu} \int_1^z \frac{1-A(u)}{A(u)-u} \mathrm{d}u \right\},$$

$$P_1(z, x) = \sum_{n=0}^{\infty} p_{1n}(x) z^n = \lambda \frac{1-z}{k(z)-z} P_0(z)[1 - B(x)] e^{-\lambda(1-z)x}.$$

If in the case $C(t) = 1$ we neglect the elapsed service time $\xi(t)$, then for the probabilities $p_{1n} = \mathbf{P}(C(t) = 1, N(t) = n)$ we get

$$P_1(z) = \sum_{n=0}^{\infty} p_{1n} z^n = \frac{1-A(z)}{A(z)-z} P_0(z).$$

Proof For the statistical equilibrium we have the equations

$$(\lambda + n\mu)p_{0n} = \int_0^\infty p_{1n}(x)b(x)dx,$$

$$p'_{1n}(x) = -(\lambda + b(x))p_{1n}(x) + \lambda p_{1,n-1}(x),$$

$$p_{1n}(0) = \lambda p_{0n} + (n+1)\mu p_{0,n+1}.$$

For example, the second equation comes from

$$p_{1n}(x + \Delta) = p_{1n}(x)[1 - (\lambda + b(x))\Delta] + p_{1,n-1}(x)\lambda\Delta + o(\Delta).$$

(There is no change: new customer does not enter and the actual customer on service having passed already x time will not be completed for Δ; or a new customer appears.)

These equations give the following for the generating functions. Since

$$\sum_{n=0}^\infty (\lambda + n\mu)p_{0n}z^n = \lambda P_0(z) + \mu z \sum_{n=1}^\infty np_{0n}z^{n-1},$$

we have

$$\lambda P_0(z) + \mu z \frac{dP_0(z)}{dz} = \int_0^\infty P_1(z, x)b(x)dx;$$

$$\frac{\partial P_1(z, x)}{\partial x} = -(\lambda - \lambda z + b(x))P_1(z, x); \qquad (11.27)$$

$$P_1(z, 0) = \lambda P_0(z) + \mu \frac{dP_0(z)}{dz}. \qquad (11.28)$$

From (11.27)

$$\frac{\partial \ln P_1(z, x)}{\partial x} = -\lambda(1 - z) - b(x),$$

$$\ln P_1(z, x) = -\lambda(1 - z)x - \int \frac{B'(x)}{1 - B(x)}dx + \ln C$$

$$= -\lambda(1 - z)x + \ln[1 - B(x)] + \ln C,$$

$$P_1(z, x) = Ce^{-\lambda(1-z)x}[1 - B(x)],$$

for $z = 0$ we obtain $C = P_1(z, 0)$, so, finally,

$$P_1(z, x) = P_1(z, 0)e^{-\lambda(1-z)x}[1 - B(x)].$$

Since

$$
\int_0^\infty P_1(z, x)b(x)\mathrm{d}x = P_1(z, 0) \int_0^\infty e^{-\lambda(1-z)x} B'(x)\mathrm{d}x = P_1(z, 0)A(z),
$$

we have

$$
\lambda P_0(z) + \mu z \frac{\mathrm{d}p_0(z)}{\mathrm{d}z} = P_1(z, 0)A(z). \tag{11.29}
$$

Eliminating from (11.28) and (11.29) $P_0'(z)$

$$
\lambda P_0(z)(1 - z) = P_1(z, 0)[A(z) - z],
$$

or

$$
P_1(z, 0) = \lambda \frac{1 - z}{A(z) - z} P_0(z),
$$

so

$$
P_1(z, x) = \lambda \frac{1 - z}{A(z) - z} P_0(z)[1 - B(x)]e^{-\lambda(1-z)x}.
$$

Integrating this expression by x and using the formula

$$
\int_0^\infty e^{-sx}[1 - B(x)]\mathrm{d}x = \frac{1 - B^\sim(s)}{s},
$$

we get

$$
\begin{aligned}
P_1(z) &= \lambda \frac{1 - z}{A(z) - z} P_0(z) \int_0^\infty [1 - B(x)]e^{-\lambda(1-z)x}\mathrm{d}x \\
&= \lambda \frac{1 - z}{A(z) - z} P_0(z) \frac{1 - A(z)}{\lambda(1 - z)} \\
&= \frac{1 - A(z)}{A(z) - z} P_0(z).
\end{aligned}
$$

Furthermore, we have

$$
\frac{P_1(1)}{P_0(1)} = \lim_{z\to1} \frac{1 - A(z)}{A(z) - z} = \frac{-A'(1)}{A'(1) - 1} = \frac{-\rho}{\rho - 1},
$$

while

$$
P_0(1) = 1 - \rho \quad \text{and} \quad P_1(1) = \rho.
$$

In order to find the unknown function $P_0(z)$, we use (11.28) and (11.29) and obtain

$$\frac{\lambda}{\mu} \frac{1 - A(z)}{A(z) - z} P_0(z) = \frac{d P_0(z)}{dz}.$$

Let us consider the denominator $f(z) = A(z) - z$. We have

- $f(1) = 0$;
- $f'(z) = -\lambda B^{\sim\prime}(\lambda(1 - z)) - 1$ and, consequently, $f'(1) = -\lambda B^{\sim\prime}(0) - 1 = \rho - 1 < 0$;
- $f''(z) = \lambda^2 B^{\sim\prime\prime}(\lambda(1 - z)) \geq 0$.

So, $f(z)$ is positive decreasing function in $[0, 1]$, it is zero only at $z = 1$ ($\rho < 1$) and

$$z < A(z) \leq 1$$

and

$$\lim_{z \to 1} \frac{1 - A(z)}{A(z) - z} = \frac{A'(1)}{1 - A'(1)} = \frac{\rho}{1 - \rho}.$$

Since $P_0(1) = 1 - \rho$, we can solve this differential equation

$$\ln P_0(z) = \frac{\lambda}{\mu} \int \frac{1 - A(u)}{A(u) - u} du + \ln C,$$

whence

$$P_0(z) = (1 - \rho) \exp \left\{ \frac{\lambda}{\mu} \int_1^z \frac{1 - A(u)}{A(u) - u} du \right\}.$$

Using the generating functions $P_0(z)$ and $P_1(z)$, one can find the characteristics of the system: e.g., the number sources of repeated calls have the generating function

$$P(z) = P_0(z) + P_1(z) = (1 - \rho) \exp \left\{ \frac{\lambda}{\mu} \int_1^z \frac{1 - A(u)}{A(u) - u} du \right\},$$

the mean value is

$$\mathbf{E}(N(t)) = \frac{\lambda^2}{1 - \rho} \left(\frac{\beta_1}{\mu} + \frac{\beta_2}{2} \right);$$

the number of customers in the system has the generating function

$$P_0(z) + z P_1(z) = (1 - \rho) \frac{A(z)(1 - z)}{A(z) - z} \exp \left\{ \frac{\lambda}{\mu} \int_1^z \frac{1 - A(u)}{A(u) - u} du \right\},$$

and the mean value of present customers is

$$\rho + \frac{\lambda^2}{1 - \rho} \left(\frac{\beta_1}{\mu} + \frac{\beta_2}{2} \right).$$

11.7.2   Continuous-Time Cyclic Waiting System

We will consider a model connected with the landing process of airplanes in the case of continuous-time. The model was introduced in [19], and the results for waiting time are contained in [20]. For some numerical results concerning the mean values and optimization of cycle time, see [23, 24]. Let us consider the landing process of airplanes. An airplane appears at an airport ideally positioned for landing. If it is not possible (due to insufficient distance or a waiting queue), it starts circling. The next request for service is possible when it returns to the starting geometrical point on the condition that there are no other airplanes ahead of it.

Similar problems appear at the transmission of optical signals. Signals entering the node must be sent in the order of arrival, but they cannot be stored. They go to delay lines and upon their return can reinitiate their transmission. If all previous signals have already been sent, then the signal is transmitted; in the opposite case they pass through the delay line again, and this process is repeated.

Let us formulate the queueing problem. We investigate a service system where the service may start at the moment of arrival (if the system is available) or at moments differing from it by the multiples of a given cycle time T (in the case of busy server or queue). Service of a customer can be started if all customers who had entered the system earlier have already left (i.e., the FIFO rule works). In such a system the service process is not continuous; during the "busy period" there are idle intervals; these idle intervals are necessary to reach the starting point; for them there is no real service.

Let the service of the nth customer begin at moment t_n, and let us consider the number customers present at the moment just before service begins. Then the number of present customers is determined by the recursive formula

$$N_{n+1} = \begin{cases} \Delta_n - 1, & \text{if} N_n = 0, \\ N_n - 1 + \Delta_n, & \text{if} N_n > 0, \end{cases}$$

where Δ_n is the number of customers appearing for $[t_n, t_{n+1})$. We show that these values constitute a Markov chain.

Let us consider the time intervals during which we record the entering customers. Let $\{Z_i\}$ and $\{Y_i\}$ ($i = 1, 2, \ldots$) be two independent sequences of independent random variables. Z_i means the interarrival time between the ith and $i + 1$st customers (it has exponential distribution with parameter λ), and Y_i the service time of the ith customer (in our case it has an exponential distribution with parameter μ).

Let us assume that at the beginning of service there is one customer in the system. If $Z_i \geq Y_i$, then the time till the beginning of service of the next customer is Z_i (the service of the existing customer will be completed, and the server arrives at a free state and the next customer appears later). If $Z_i < Y_i$, then during the service of first customer a second one appears, and after this moment there will be intervals with length T while we pass the moment of service of the first customer (from the viewpoint of entering customers we are interested in the time from the entry of the second customer till the beginning of its service). The length of this interval is the function of random variables Z_i and Y_i, i.e., a certain $f_1(Z_i, Y_i)$.

If at the beginning of service of the first customer the second one is already present, then the time till the starting moment of its service is determined in the following way. Divide the service time of the first customer into intervals of length T (the last period generally is not full). Since the starting moments for both customers differ by the multiples of T from the moments of arrivals, on each interval of length T there is one point where the service of second customer may start. In reality, this happens at the first moment after the service of the first customer has completed, so the required time period is determined by the service time of the first customer and the interarrival time. Consequently, it will be a certain function of Y_i and Z_i, i.e., $f_2(Y_i, Z_i)$.

Thus, the time intervals for which we consider the number of entering customers are only functions of random variables Y_i and Z_i, consequently they are independent. Taking into account the fact that entering customers form a Poisson process, the quantities Δ_i of these customers are independent random variables, and N_n is a Markov chain.

To describe the functioning of the system we use the embedded Markov chain technique. Our result is formulated in the following theorem.

Theorem 11.6 *Let us consider a service system in which the entering customers form a Poisson process with parameter λ, and the service time is exponentially distributed with parameter μ. The service of a customer may be started at the moment of arrival (in the case of a free system) or at moments differing from it by the multiples of a cycle time T (in the case of busy server or queue); the service discipline is FIFO. Let us define a Markov chain whose states correspond to the number of customers in the system at moments t_k- (t_k is the starting moment of service of the kth customer). The matrix of transition probabilities of this Markov chain has the form*

$$
\begin{bmatrix}
a_0 & a_1 & a_2 & a_3 & \cdots \\
a_0 & a_1 & a_2 & a_3 & \cdots \\
0 & b_0 & b_1 & b_2 & \cdots \\
0 & 0 & b_0 & b_1 & \cdots \\
\vdots & \vdots & \vdots & \vdots & \ddots
\end{bmatrix}
\tag{11.30}
$$

its elements are determined by the generating functions

$$A(z) = \sum_{i=0}^{\infty} a_i z^i = \frac{\mu}{\lambda + \mu} + \frac{\lambda}{\lambda + \mu} z \frac{(1 - e^{-\mu T}) e^{-\lambda(1-z)T}}{1 - e^{-[\lambda(1-z)+\mu]T}}, \qquad (11.31)$$

$$B(z) = \sum_{i=0}^{\infty} b_i z^i =$$

$$= \frac{1}{(1 - e^{-\lambda T})(1 - e^{-[\lambda(1-z)+\mu]T})} \left[\frac{1}{2-z} \left(1 - e^{-\lambda(2-z)T} \right) \left(1 - e^{-[\lambda(1-z)+\mu]T} \right) \right.$$

$$\left. - \frac{\lambda}{\lambda(2-z) + \mu} \left(1 - e^{-[\lambda(2-z)+\mu]T} \right) \left(1 - e^{-\lambda(1-z)T} \right) \right]. \qquad (11.32)$$

The generating function of ergodic distribution of this chain is

$$P(z) = p_0 \frac{B(z)(\lambda z + \mu) - z A(z)(\lambda + \mu)}{\mu[B(z) - z]}, \qquad (11.33)$$

where

$$p_0 = 1 - \frac{\lambda}{\lambda + \mu} \frac{1 - e^{-(\lambda+\mu)T}}{e^{-\lambda T}(1 - e^{-\mu T})}. \qquad (11.34)$$

The ergodicity condition is

$$\frac{\lambda}{\mu} < \frac{e^{-\lambda T}(1 - e^{-\mu T})}{1 - e^{-\lambda T}}. \qquad (11.35)$$

Proof Our original system, where during the busy period there are possible idle intervals, too, is replaced by another one. In it the service process is continuous, and the service time of a customer consists of two parts: the first part is the real service, the second part holds from the end of service till the second one gets to the corresponding position.

For a description of the operation we use an embedded Markov chain; its states are the number of customers in the system at moments t_k-, i.e., we consider it at moments just before starting service. Let us find the transition probabilities for this chain. We have to distinguish two cases: at the starting moment of service the next customer is present or not. First we will consider the second possibility (Fig. 11.28), which happens for the states 0 and 1. Suppose that the service time of the first customer is U, and the second customer enters at V time after the beginning of service. The probability of event $\{U - V < t\}$ is

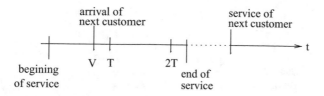

Fig. 11.28 One customer at the beginning of service

$$P(t) = \mathbf{P}(U - V < t)$$

$$= \int_0^t \int_0^U \lambda e^{-\lambda V} \mu e^{-\mu U} dV dU + \int_t^\infty \int_{U-t}^U \lambda e^{-\lambda V} \mu e^{-\mu U} dV dU$$

$$= \frac{\lambda}{\lambda + \mu}\left(1 - e^{-\mu t}\right). \tag{11.36}$$

The time from the entry of the second customer till the beginning of its service equals

$$(I(U - V) + 1)T,$$

where $I(x)$ denotes the integer part of number x/T. This expression is valid for all points excluding multiples of T, but the probability of an event for this time to equal a multiple of T is equal to zero. To determine the transition probabilities, we need the number of customers entering during this period. According to Eq. (11.36) the time from the entry till the beginning of service is equal to iT with probability

$$\frac{\lambda}{\lambda + \mu}\left(e^{-\mu(i-1)T} - e^{-\mu iT}\right),$$

and the generating function of entering customers equals

$$\frac{\lambda}{\lambda + \mu}\sum_{k=0}^\infty \sum_{i=1}^\infty \left(e^{-\mu(i-1)T} - e^{-\mu iT}\right)\frac{(\lambda i T z)^k}{k!}e^{-\lambda iT}$$

$$= \frac{\lambda}{\lambda + \mu}\sum_{i=1}^\infty \left(e^{-\mu(i-1)T} - e^{-\mu iT}\right)e^{-\lambda iT(1-z)} = \frac{\lambda}{\lambda + \mu}\frac{e^{-\lambda(1-z)T}(1 - e^{-\mu T})}{1 - e^{-[\lambda(1-z)+\mu]T}}.$$

This formula is valid if for U at least one customer enters the system, so the desired generating function is

$$A(z) = \frac{\mu}{\lambda + \mu} + \frac{\lambda}{\lambda + \mu}z\frac{(1 - e^{-\mu T})e^{-\lambda T(1-z)}}{1 - e^{-[\lambda(1-z)+\mu]T}},$$

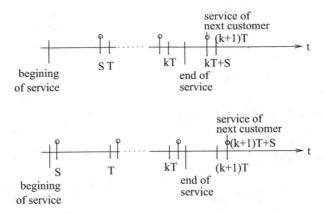

Fig. 11.29 More than one customer at the beginning of service

where $\frac{\mu}{\lambda+\mu} = \int_0^\infty e^{-\lambda x}\mu e^{-\mu x}dx$ is the probability that for the service time no customer appears.

Now we find the transition probabilities for all other states. In this case at the beginning of service the next customer is already present (Fig. 11.29). Let $R = U - I(U)T$ and let S be the *mod T* interarrival time. One can easily see that S has a truncated exponential distribution with distribution function $\frac{1-e^{-\lambda S}}{1-e^{-\lambda T}}$. The time between the starting moments of two successive customers is

$$I(U)T + S \quad \text{if} \quad R \leq S \quad \text{and} \quad (I(U)+1)T + S \quad \text{if} \quad R > S.$$

k customers enter in the two cases with probabilities

$$\frac{(\lambda\{I(U)T+S\})^k}{k!}\exp(-\lambda\{I(U)T+S\}) \tag{11.37}$$

and

$$\frac{(\lambda\{[I(U)+1]T+S\})^k}{k!}\exp(-\lambda\{[I(U)+1]T+S\}). \tag{11.38}$$

Let us fix S and divide the service time of the customer into intervals of length T. S divides each such interval into two parts (the first has length S, the second $T - S$), and the corresponding probability for the first part is Eq. (11.37) and for the second part Eq. (11.38). Let $I(U) = i$. The generating function of the number of entering customers, denoted by N, assuming that the interarrival time *mod T* is equal to S is as follows

$$\mathbf{E}\left(z^N|S\right) = \sum_{k=0}^{\infty}\sum_{i=0}^{\infty}\left(\int_{iT}^{iT+S}\frac{[\lambda(iT+S)z]^k}{k!}e^{-\lambda(iT+S)}\mu e^{-\mu U}\,dU\right.$$

$$\left.+\int_{iT+S}^{(i+1)T}\frac{[\lambda((i+1)T+S)z]^k}{k!}e^{-\lambda((i+1)T+S)}\mu e^{-\mu U}\,dU\right)$$

$$= \frac{1}{1-e^{-[\lambda(1-z)+\mu]T}}\left(e^{-\lambda(1-z)S} - e^{-[\lambda(1-z)+\mu]S}\right.$$

$$\left.+e^{-\lambda(1-z)T}e^{-[\lambda(1-z)+\mu]S} - e^{-\lambda(1-z)S}e^{-[\lambda(1-z)+\mu]T}\right).$$

Multiplying this expression by $\frac{\lambda e^{-\lambda S}}{1-e^{-\lambda T}}$ and integrating by S from 0 to T we obtain the generating function of transition probabilities

$$B(z) = \sum_{i=0}^{\infty}b_i z^i = \frac{1}{(1-e^{-\lambda T})(1-e^{-[\lambda(1-z)+\mu]T})}$$

$$\times\left[\frac{1}{2-z}\left(1-e^{-\lambda(2-z)T}\right)\left(1-e^{-[\lambda(1-z)+\mu]T}\right)\right.$$

$$\left.-\frac{\lambda}{\lambda(2-z)+\mu}\left(1-e^{-[\lambda(2-z)+\mu]T}\right)\left(1-e^{-\lambda(1-z)T}\right)\right].$$

Consider a Markov chain describing the functioning of the system; the matrix of transition probabilities has the form Eq. (11.30). Let us denote the ergodic distribution by p_i ($i = 0, 1, 2, \dots$) and introduce the generating function $P(z) = \sum_{i=0}^{\infty}p_i z^i$. Then

$$p_j = p_0 a_j + p_1 a_j + \sum_{i=2}^{j+1}p_i b_{j-i+1},$$

whence

$$\sum_{j=0}^{\infty}p_j z^j = p_0 A(z) + p_1 A(z) + \sum_{j=0}^{\infty}\sum_{i=2}^{j+1}p_i b_{j-i+1}z^j$$

$$= \frac{1}{z}P(z)B(z) - \frac{1}{z}p_0 B(z) + p_0 A(z) + p_1 A(z) - p_1 B(z),$$

i.e.,

$$P(z) = \frac{p_0\left[zA(z) - B(z)\right] + p_1 z\left[A(z) - B(z)\right]}{z - B(z)}.$$

This expression contains two unknown probabilities—p_0 and p_1—but

$$p_0 = p_0 a_0 + p_1 a_0,$$

i.e.,

$$p_1 = \frac{1 - a_0}{a_0} p_0 = \frac{\lambda}{\mu} p_0.$$

p_0 can be found from the condition $P(1) = 1$,

$$p_0 = \frac{1 - B'(1)}{1 + A'(1) - B'(1) + \frac{\lambda}{\mu}[A' - B'(1)]}.$$

The embedded chain is irreducible, so $p_0 > 0$. Using

$$A'(1) = \frac{\lambda}{\lambda + \mu} \left(1 + \frac{\lambda T}{1 - e^{-\mu T}}\right),$$

$$B'(1) = 1 - \frac{\lambda T e^{-\lambda T}}{1 - e^{-\lambda T}} + \frac{\lambda}{\lambda + \mu} \lambda T \frac{1 - e^{-(\lambda+\mu)T}}{(1 - e^{-\lambda T})(1 - e^{-\mu T})},$$

we obtain

$$\left(1 + \frac{\lambda}{\mu}\right) A'(1) - \frac{\lambda}{\mu} B'(1) = \frac{\lambda}{\lambda + \mu} \lambda T \frac{1 - e^{-(\lambda+\mu)T}}{(1 - e^{-\lambda T})(1 - e^{-\mu T})} > 0,$$

so the condition $1 - B'(1) > 0$ must be fulfilled. This leads to the inequality

$$\frac{\lambda T e^{-\lambda T}}{1 - e^{-\lambda T}} - \frac{\lambda}{\lambda + \mu} \lambda T \frac{1 - e^{-(\lambda+\mu)T}}{(1 - e^{-\lambda T})(1 - e^{-\mu T})} > 0,$$

i.e.,

$$\frac{\lambda}{\lambda + \mu} < \frac{e^{-\lambda T}(1 - e^{-\mu T})}{1 - e^{-(\lambda+\mu)T}}.$$

This is equivalent to Eq. (11.35). Substituting the corresponding values we obtain

$$p_0 = 1 - \frac{\lambda}{\lambda + \mu} \frac{1 - e^{-(\lambda+\mu)T}}{e^{-\lambda T}(1 - e^{-\mu T})}.$$

The theorem is proved.　　　　　　　　　　　　　　　　　　　　　　　　　　□

　During the busy period there are idle intervals, which are necessary to get to the starting position, they alternate between 0 and T. It is clear that if T decreases, then their influence will become increasingly attenuated. In the limit case, the service process becomes continuous, and after having served a customer, we immediately change to the next one.

Theorem 11.7 *The limiting distribution for the system described above as $T \to 0$*

$$P^*(z) = \frac{1-\rho}{1-\rho z} \qquad \left(\rho = \frac{\lambda}{\mu}\right),$$

i.e., it is geometrical with parameter ρ.

Proof We find p_0, $A(z)$, and $B(z)$ as $T \to 0$, and the limiting values are denoted by p_0^*, $A^*(z)$, and $B^*(z)$. On the basis of Eqs. (11.31), (11.32), and (11.34),

$$p_0^* = \lim_{T \to 0} p_0 = \lim_{T \to 0} \left(1 - \frac{\lambda}{\lambda + \mu} \frac{1 - e^{-(\lambda+\mu)T}}{e^{-\lambda T}(1 - e^{-\mu T})}\right) = 1 - \frac{\lambda}{\mu} = 1 - \rho,$$

$$A^*(z) = \lim_{T \to 0} A(z) = \lim_{T \to 0} \left(\frac{\mu}{\lambda + \mu} + \frac{\lambda z}{\lambda + \mu} \frac{e^{-\lambda(1-z)T}(1 - e^{-\mu T})}{1 - e^{-[\lambda(1-z)+\mu]T}}\right)$$

$$= \frac{\mu}{\lambda(1 - z) + \mu},$$

$$B^*(z) = \lim_{T \to 0} B(z) = \frac{1}{(1 - e^{-\lambda T})[1 - e^{-[\lambda(1-z)+\mu]T}]}$$

$$\cdot \left\{ \frac{1}{2 - z}\left(1 - e^{-\lambda(2-z)T}\right)\left(1 - e^{-[\lambda(1-z)+\mu]T}\right) \right.$$

$$\left. - \frac{\lambda}{\lambda(2 - z) + \mu}\left(1 - e^{-[\lambda(2-z)+\mu]T}\right)\left(1 - e^{-\lambda(1-z)T}\right) \right\}$$

$$= \frac{\mu}{\lambda(1 - z) + \mu}.$$

Using these values

$$P^*(z) = (1 - \rho)\frac{(\lambda z + \mu)\frac{\mu}{\lambda(1-z)+\mu} - z(\lambda + \mu)\frac{\mu}{\lambda(1-z)+\mu}}{\mu\left(\frac{\mu}{\lambda(1-z)+\mu} - z\right)} = \frac{1 - \rho}{1 - \rho z}.$$

The preceding formula is the generating function for an M/M/1 system and coincides with the previous results. □

11.7.3   Waiting Time for Continuous Cyclic Waiting System

Let us consider the previously described system. Using Koba's results [18] we determine the distribution of the waiting time. Let t_n denote the moment of arrival of the nth customer; then its service may be started at the moment $t_n + T \cdot X_n$, where X_n is a nonnegative integer. Let $Z_n = t_{n+1} - t_n$, Y_n be the service time of the nth customer. If $X_n = i$, then between X_n and X_{n+1} the following relation holds. If

$$(k - 1)T < iT + Y_n - Z_n \le kT \qquad (k \ge 1),$$

then $X_{n+1} = k$. In this case X_n is a homogeneous Markov chain with transition probabilities p_{ik}, where

$$p_{ik} = \mathbf{P}((k - i - 1)T < Y_n - Z_n \le (k - i)T)$$

if $k \ge 1$, and

$$p_{i0} = \mathbf{P}(Y_n - Z_n \le -iT).$$

Introduce the notations

$$f_j = \mathbf{P}((j - 1)T < Y_n - Z_n \le jT), \tag{11.39}$$

$$p_{ik} = f_{k-i} \quad \text{if} \quad k \ge 1, \qquad p_{i0} = \sum_{j=-\infty}^{-i} f_j = \hat{f}_i. \tag{11.40}$$

The ergodic distribution of the Markov chain satisfies the system of equations

$$p_j = \sum_{i=0}^{\infty} p_i p_{ij} \quad (j \ge 0), \qquad \sum_{j=0}^{\infty} p_j = 1.$$

Theorem 11.8 *Let us consider the system described in Theorem 11.6. Define a Markov chain whose states correspond to the waiting times of customers at the moments of arrivals. The matrix of transition probabilities has the form*

$$\begin{bmatrix} \sum_{j=-\infty}^{0} f_j & f_1 & f_2 & f_3 & f_4 \cdots \\ \sum_{j=-\infty}^{-1} f_j & f_0 & f_1 & f_2 & f_3 \cdots \\ \sum_{j=-\infty}^{-2} f_j & f_{-1} & f_0 & f_1 & f_2 \cdots \\ \sum_{j=-\infty}^{-3} f_j & f_{-2} & f_{-1} & f_0 & f_1 \cdots \\ \vdots & \vdots & \vdots & \vdots & \vdots & \ddots \end{bmatrix} \tag{11.41}$$

and its elements are determined by formulas (11.39)–(11.40). Then the generating function of the waiting time is

$$P(z) = \left[1 - \frac{\lambda}{\mu} \frac{1 - e^{-\lambda T}}{e^{-\lambda T}(1 - e^{-\mu T})}\right]$$

$$\cdot \frac{\dfrac{\mu}{\lambda + \mu} - \dfrac{\mu(1 - e^{-\lambda T})}{\lambda + \mu} \dfrac{z}{z - e^{-\lambda T}}}{1 - \dfrac{\lambda(1 - e^{-\mu T})}{\lambda + \mu} \dfrac{z}{1 - ze^{-\mu T}} - \dfrac{\mu(1 - e^{-\lambda T})}{\lambda + \mu} \dfrac{z}{z - e^{-\lambda T}}}, \qquad (11.42)$$

and the stability condition is

$$\frac{\lambda}{\mu} < \frac{e^{-\lambda T}(1 - e^{-\mu T})}{1 - e^{-\lambda T}}. \qquad (11.43)$$

Proof

$$\mathbf{P}(Z < x) = 1 - e^{-\lambda x}, \qquad \mathbf{P}(Y < x) = 1 - e^{-\mu x}.$$

The distribution function of $Y - Z$ is

$$F(x) = \begin{cases} \frac{\mu}{\lambda + \mu} e^{\lambda x} & \text{if } x \le 0, \\ 1 - \frac{\lambda}{\lambda + \mu} e^{-\mu x} & \text{if } x > 0. \end{cases}$$

We find the transition probabilities. In the case $j > 0$,

$$f_j = 1 - \frac{\lambda}{\lambda + \mu} e^{-\mu(j-1)T} - 1 + \frac{\lambda}{\lambda + \mu} e^{-\mu j T} = \frac{\lambda}{\lambda + \mu}(1 - e^{-\mu T})e^{-\mu(j-1)T},$$

for the negative values ($j \ge 0$)

$$f_{-j} = \frac{\mu}{\lambda + \mu} e^{-\lambda j T} - \frac{\mu}{\lambda + \mu} e^{-\lambda(j+1)T} = \frac{\mu}{\lambda + \mu}(1 - e^{-\lambda T})e^{-\lambda j T},$$

$$p_{i0} = \hat{f}_i = \sum_{j=-\infty}^{-i} f_j = \sum_{j=i}^{\infty} \frac{\mu}{\lambda + \mu}(1 - e^{-\lambda T})e^{-\lambda j T} = \frac{\mu}{\lambda + \mu} e^{-\lambda i T}.$$

Using the matrix of transition probabilities (11.41) we get the system of equations

$$p_0 = p_0 \hat{f}_0 + p_1 \hat{f}_1 + p_2 \hat{f}_2 + p_3 \hat{f}_3 + \dots$$
$$p_1 = p_0 f_1 + p_1 f_0 + p_2 f_{-1} + p_3 f_{-2} + \dots \qquad (11.44)$$
$$p_2 = p_0 f_2 + p_1 f_1 + p_2 f_0 + p_3 f_{-1} + \dots$$
$$\vdots$$

Multiplying the jth equation by z^j, summing up by j from 0 to infinity for the generating function $P(z) = \sum\limits_{j=0}^{\infty} p_j z^j$ we obtain

$$P(z) = P(z)F_+(z) + \sum_{j=1}^{\infty} p_j z^j \sum_{i=0}^{j-1} f_{-i} z^{-i} + \sum_{j=0}^{\infty} p_j \hat{f}_j, \qquad (11.45)$$

where

$$F_+(z) = \sum_{i=1}^{\infty} f_i z^i = \frac{\lambda z}{\lambda + \mu}(1 - e^{-\mu T}) \sum_{i=1}^{\infty} e^{-\mu(i-1)T} z^{i-1}$$

$$= \frac{\lambda(1 - e^{-\mu T})}{\lambda + \mu} \frac{z}{1 - z e^{-\mu T}},$$

$$\sum_{i=0}^{j-1} f_{-i} z^{-i} = \frac{\mu(1 - e^{-\lambda T})}{\lambda + \mu} \sum_{i=0}^{j-1} e^{-\lambda i T} z^{-i} = \frac{\mu(1 - e^{-\lambda T})}{\lambda + \mu} \frac{1 - \left(\frac{e^{-\lambda T}}{z}\right)^j}{1 - \frac{e^{-\lambda T}}{z}},$$

$$\sum_{i=0}^{\infty} p_i \hat{f}_i = \sum_{i=0}^{\infty} p_i \frac{\mu}{\lambda + \mu} e^{-\lambda i T} = \frac{\mu}{\lambda + \mu} P\left(e^{-\lambda T}\right).$$

Using the preceding equations

$$P(z) = P(z)F_+(z) + \sum_{j=1}^{\infty} p_j z^j \frac{\mu(1 - e^{-\lambda T})}{\lambda + \mu} \frac{1 - \left(\frac{e^{-\lambda T}}{z}\right)^j}{1 - \frac{e^{-\lambda T}}{z}} + \frac{\mu}{\lambda + \mu} P\left(e^{-\lambda T}\right)$$

$$= P(z)F_+(z) + \frac{\mu(1 - e^{-\lambda T})}{\lambda + \mu} \frac{z}{z - e^{-\lambda T}} \left[P(z) - P\left(e^{-\lambda T}\right)\right]$$

$$+ \frac{\mu}{\lambda + \mu} P\left(e^{-\lambda T}\right)$$

or

$$P(z)\left[1 - F_+(z) - \frac{\mu(1 - e^{-\lambda T})}{\lambda + \mu} \frac{z}{z - e^{-\lambda T}}\right]$$

$$= P\left(e^{-\lambda T}\right)\left[\frac{\mu}{\lambda + \mu} - \frac{\mu(1 - e^{-\lambda T})}{\lambda + \mu} \frac{z}{z - e^{-\lambda T}}\right].$$

$P(e^{-\lambda T})$ may be computed from the condition $P(1) = 1$,

$$P\left(e^{-\lambda T}\right) = 1 - \frac{\lambda}{\mu} \frac{1 - e^{-\lambda T}}{e^{-\lambda T}(1 - e^{-\mu T})}.$$

So for the generating function we get Eq. (11.42). From it we get the probability of event that the waiting time is equal to zero:

$$p_0 = \left[1 - \frac{\lambda}{\mu} \frac{1 - e^{-\lambda T}}{e^{-\lambda T}(1 - e^{-\mu T})} \right] \frac{\mu}{\lambda + \mu}.$$

In order to have $p_0 > 0$, the inequality

$$\frac{\lambda}{\mu} \frac{1 - e^{-\lambda T}}{e^{-\lambda T}(1 - e^{-\mu T})} < 1$$

must be fulfilled. It leads to condition (11.43) and coincides with the stability condition for the number of customers.   □

11.7.4   Queue Length for the Discrete Cyclic Waiting System

In the previous two paragraphs we have investigated the continuous cyclic-waiting system, and in the following paragraphs we will consider the queue length and waiting time for the discrete-time version in a more general case.

Let us divide the cycle time T into n equal parts called slots. For each slot a new customer may enter with probability r, there is no entry with probability $1 - r$; the service time is i slots with probability q_i. Denoting the interarrival time by Z, the service time by Y, their distributions are

$$\mathbf{P}(Z = i) = (1 - r)^{i-1}r, \qquad \mathbf{P}(Y = i) = q_i \qquad (i \geq 1),$$

i.e., they have geometrical and general distributions, respectively.

As in the case of exponential interarrival and service time distributions we will consider the number of customers N_n at moments just before starting the services of customers. We have seen that these values constitute a Markov chain, we find the transition probabilities of this chain. The reasoning is similar to that in the case of exponential distribution.

Concerning the transition probabilities we have to distinguish two cases: at the moment when the service of a customer begins the next one is present or not. First we find the generating function $A(z)$, corresponding to the case when the next customer is not there yet, then we find the generating function $B(z)$ for the case when the next customer is present, too. These two generating functions are given by the following two lemmas.

Lemma 11.2 (The Generating Function $A(z)$) *If at the beginning of service of a customer there is only one present customer in the system, then the generating function of arriving till the beginning of service of the next customer is*

$$A(z) = \sum_{i=0}^{\infty} a_i z^i \tag{11.46}$$

$$= Q_1 + z \frac{r}{1-r} Q_1 + z \sum_{k=1}^{\infty} (1 - r + rz)^{kn}$$

$$\cdot \left\{ \sum_{i=(k-1)n+2}^{kn+1} q_i + \sum_{i=kn+2}^{\infty} q_i (1-r)^{i-kn-1} - \sum_{i=(k-1)n+2}^{\infty} q_i (1-r)^{i-(k-1)n-1} \right\},$$

Proof This possibility appears at the states 0 and 1. Assume that the service time of first customer is equal to U, and the second customer appears V time after starting its service. The probability of event $\{U - V = \ell\}$ is

$$P\{U - V = \ell\} = \sum_{k=\ell+1}^{\infty} q_k (1-r)^{k-\ell-1} r \qquad (\ell = 1, 2, \ldots).$$

Using these probabilities the generating function of entering customers is $(1 - r + rz)^{kn}$ $(k = 1, 2, \ldots)$ with probability

$$\sum_{i=(k-1)n+2}^{kn+1} q_i + \sum_{i=kn+2}^{\infty} q_i (1-r)^{i-kn-1} - \sum_{i=(k-1)n+2}^{\infty} q_i (1-r)^{i-(k-1)n-1}.$$

Taking into account that the probability of event during the service of a customer a new one does not arrive

$$\sum_{i=1}^{\infty} q_i (1-r)^i = Q_1 = a_0,$$

and the probability of zero waiting time is

$$\sum_{i=1}^{\infty} q_i (1-r)^{i-1} r = \frac{r}{1-r} Q_1,$$

for the generating function $A(z)$ we obtain (11.46).

Its derivative is

$$A'(z) = \frac{r Q_1}{1-r} + \sum_{k=1}^{\infty} (1 - r + rz)^{kn}$$

$$\cdot \left\{ \sum_{i=(k-1)n+2}^{kn+1} q_i + \sum_{i=kn+2}^{\infty} q_i (1-r)^{i-kn-1} - \sum_{i=(k-1)n+2}^{\infty} q_i (1-r)^{i-(k-1)n-1} \right\}$$

$$+ znr \sum_{k=1}^{\infty} k(1 - r + rz)^{kn-1}$$

$$\cdot \left\{ \sum_{i=(k-1)n+2}^{kn+1} q_i + \sum_{i=kn+2}^{\infty} q_i (1 - r)^{i-kn-1} - \sum_{i=(k-1)n+2}^{\infty} q_i (1 - r)^{i-(k-1)n-1} \right\}$$

and at $z = 1$ gives

$$A'(1) = \frac{rQ_1}{1-r} + \sum_{k=1}^{\infty} \sum_{i=(k-1)n+2}^{kn+1} q_i - \frac{Q_2}{1-r} \tag{11.47}$$

$$+ nr \sum_{k=1}^{\infty} k \sum_{i=(k-1)n+2}^{kn+1} q_i - nr \sum_{k=1}^{\infty} \sum_{i=(k-1)n+2}^{\infty} q_i (1 - r)^{i-(k-1)n-1} ,$$

where $Q_k = \sum_{i=k}^{\infty} q_i (1 - r)^i$. □

Lemma 11.3 (The Generating Function $B(z)$**)** *If at the beginning of service of a customer there is present the next customer in the system, too, then the generating function of arriving between the starting moments of services of two customers is*

$$B(z) = \sum_{i=0}^{\infty} b_i z^i \tag{11.48}$$

$$= \sum_{k=0}^{\infty} \sum_{j=1}^{n} q_{kn+j} (1 - r + rz)^{kn+j}$$

$$\cdot \left\{ \frac{r}{1-(1-r)^n} \frac{1-(1-r)^{j-1}(1-r+rz)^{j-1}}{1-(1-r)(1-r+rz)} (1-r+rz)^{n-j+1} \right.$$

$$\left. + \frac{r(1-r)^{j-1}}{1-(1-r)^n} \frac{1-(1-r)^{n-j+1}(1-r+rz)^{n-j+1}}{1-(1-r)(1-r+rz)} \right\} .$$

Proof At the beginning of service of the first customer the second customer is present, too. Let $R = U - \left[\dfrac{U-1}{n} \right] n$ ($[x]$ denote the integer part of x), and let S be the mod T interarrival time ($1 \le S \le n$). The time elapsed between the starting moments of two successive customers is

$$\left[\frac{U-1}{n} \right] n + S \quad \text{if} \quad R \le S \quad \text{and} \quad \left(\left[\frac{U-1}{n} \right] + 1 \right) n + S \quad \text{if} \quad R > S .$$

One can easily see that S has truncated geometrical distribution with probabilities

$$P\{S = \ell\} = \frac{(1-r)^{\ell-1} r}{1 - (1-r)^n} \qquad (\ell = 1, 2, \ldots, n) ,$$

the generating function of entering customer for a time slice is $r(z) = 1 - r + rz$. The generating functions of entering customers depending on the service time and the mod T interarrival time is given by a table containing n rows and a countable number of columns (the rows correspond to the mod T interarrival and the columns to the service times) and consisting of $n \times n$ blocks. The kth ($k \geq 1$) block is

$$
\begin{array}{cccc}
q_{(k-1)n+1}r^{(k-1)n+1}(z) & q_{(k-1)n+2}r^{kn+1}(z) & \cdots & q_{kn-1}r^{kn+1}(z) \; q_{kn}r^{kn+1}(z) \\
q_{(k-1)n+1}r^{(k-1)n+2}(z) & q_{(k-1)n+2}r^{(k-1)n+2}(z) & \cdots & q_{kn-1}r^{kn+2}(z) \; q_{kn}r^{kn+2}(z) \\
q_{(k-1)n+1}r^{(k-1)n+3}(z) & q_{(k-1)n+2}r^{(k-1)n+3}(z) & \cdots & q_{kn-1}r^{kn+3}(z) \; q_{kn}r^{kn+3}(z) \\
\vdots & \vdots & & \vdots \qquad\qquad \vdots \\
q_{(k-1)n+1}r^{kn-1}(z) & q_{(k-1)n+2}r^{kn-1}(z) & \cdots & q_{kn-1}r^{kn-1}(z) \; q_{kn}r^{(k+1)n-1}(z) \\
q_{(k-1)n+1}r^{kn}(z) & q_{(k-1)n+2}r^{kn}(z) & \cdots & q_{kn-1}r^{kn}(z) \;\; q_{kn}r^{kn}(z)
\end{array}
$$

Summing up the elements in the columns, then considering these sums shifted by n [i.e., the sums of columns corresponding to the service times q_j, q_{n+j}, q_{2n+j}, $\cdots$ $(1 \leq j \leq n)$] for a concrete deviation j the generating function equals

$$
\frac{r}{1-(1-r)^n} \; \frac{1-(1-r)^{j-1}(1-r+rz)^{j-1}}{1-(1-r)(1-r+rz)}(1-r+rz)^{n-j+1}\sum_{k=0}^{\infty} q_{kn+j}(1-r+rz)^{kn+j}
$$

$$
+ \frac{r(1-r)^{j-1}}{1-(1-r)^n} \; \frac{1-(1-r)^{n-j+1}(1-r+rz)^{n-j+1}}{1-(1-r)(1-r+rz)}\sum_{k=0}^{\infty} q_{kn+j}(1-r+rz)^{kn+j} ,
$$

and $B(z)$ will be

$$
B(z) = \sum_{k=0}^{\infty}\sum_{j=1}^{n} q_{kn+j}(1-r+rz)^{kn+j}
$$

$$
\cdot \left\{ \frac{r}{1-(1-r)^n} \; \frac{1-(1-r)^{j-1}(1-r+rz)^{j-1}}{1-(1-r)(1-r+rz)}(1-r+rz)^{n-j+1} \right.
$$

$$
\left. + \frac{r(1-r)^{j-1}}{1-(1-r)^n} \; \frac{1-(1-r)^{n-j+1}(1-r+rz)^{n-j+1}}{1-(1-r)(1-r+rz)} \right\} .
$$

Its derivative at $z = 1$ is

$$
B'(1) = \frac{r}{1-(1-r)^n} \sum_{j=1}^{n}\sum_{k=0}^{\infty} q_{kn+j}
$$

$$
\cdot \left\{ [1-(1-r)^n]\frac{[(k+1)n+1]r^2 + r(1-r)}{r^2} - \frac{nr^2(1-r)^{j-1}}{r^2} \right\} .
$$

After some arithmetics we obtain

$$B'(1) = \sum_{k=0}^{\infty} \sum_{j=1}^{n} q_{kn+j}[(k+1)nr+1] - \frac{nr}{1-(1-r)^n} \sum_{k=0}^{\infty} \sum_{j=1}^{n} q_{kn+j}(1-r)^{j-1} .$$

(11.49)

For details see [21].   □

Now we formulate our result.

Theorem 11.9 *Let us consider a discrete queueing system in which the interarrival time has geometrical distribution with parameter r, the service time has general distribution with probabilities q_i $(i = 1, 2, \ldots)$. The service of a customer may start upon arrival or (in case of busy server or waiting queue) at moments differing from it by the multiples of a given cycle time T (equal to n time units) according to the FCFS discipline. Let us define an embedded Markov chain whose states correspond to the number of customers in the system at moments t_k-, where t_k is the moment of beginning of service of the kth one. The matrix of transition probabilities has the form (11.30), and its elements are determined by the generating functions (11.46) and (11.48).*

The generating function of ergodic distribution $P(z) = \sum_{i=0}^{\infty} p_i z^i$ has the form

$$P(z) = \frac{p_0[zA(z) - B(z)] + p_1 z[A(z) - B(z)]}{z - B(z)} ,$$

(11.50)

where

$$p_1 = \frac{1 - a_0}{a_0} p_0 ,$$

$$p_0 = \frac{a_0[1 - B'(1)]}{a_0 + A'(1) - B'(1)} .$$

(11.51)

The ergodicity condition is

$$\sum_{i=1}^{\infty} q_i \left[\frac{i}{n} \right] < \frac{1}{1-(1-r)^n} \sum_{i=1}^{\infty} q_i (1-r)^{i-1 \pmod n} .$$

(11.52)

Proof The matrix of transition probabilities is given in the matrix (11.30), and the generating functions of transition probabilities are (11.46) and (11.48). Denote the ergodic probabilities by p_i $(i = 0, 1, \ldots)$ and introduce the generating function $P(z) = \sum_{i=0}^{\infty} p_i z^i$. According to the theory of Markov chains we have

$$p_j = p_0 a_j + p_1 a_j + \sum_{i=2}^{j+1} p_i b_{j-i+1} \qquad (j \geq 1) ,$$

(11.53)

$$p_0 = p_0 a_0 + p_1 a_0 .$$

(11.54)

From (11.53) and (11.54) one can obtain the expression

$$P(z) = \frac{p_0 \left[z A(z) - B(z) \right] + p_1 z \left[A(z) - B(z) \right]}{z - B(z)} .$$

This expression includes two unknown probabilities p_0 and p_1 from the desired distribution, by (11.54) p_1 can be expressed via p_0,

$$p_1 = \frac{1 - a_0}{a_0} p_0 ,$$

and p_0 can be found from the condition $P(1) = 1$, i.e.,

$$p_0 = \frac{a_0 [1 - B'(1)]}{a_0 + A'(1) - B'(1)} .$$

Using the corresponding (11.47) and (11.49) values, we obtain

$$a_0 + A'(1) - B'(1)$$

$$= \frac{1}{1-(1-r)^n} \left\{ -\sum_{i=1}^{\infty} q_i (1-r)^{i-1 \; (\mathrm{mod} \; n)} [1-(1-r)^{\left\lceil \frac{i}{n} \right\rceil n}] + \sum_{i=1}^{\infty} q_i (1-r)^{i-1 \; (\mathrm{mod} \; n)} \right\}$$

$$= \frac{1}{1-(1-r)^n} \sum_{i=1}^{\infty} q_i (1-r)^{i-1 \; (\mathrm{mod} \; n)} (1-r)^{\left\lceil \frac{i}{n} \right\rceil n} > 0 ,$$

consequently the numerator must be positive, too; so the condition

$$1 - B'(1) > 0$$

must be fulfilled. This leads to the ergodicity condition $B'(1) < 1$, i.e.,

$$\sum_{k=0}^{\infty} \sum_{j=1}^{n} q_{kn+j} [1 + (k+1)nr] - \frac{nr}{1-(1-r)^n} \sum_{k=0}^{\infty} \sum_{j=1}^{n} q_{kn+j} (1-r)^{j-1} < 1 .$$

Taking into account $\sum_{k=0}^{\infty} \sum_{j=1}^{n} q_{kn+j} = 1$ and canceling nr we get an expression which can be written in the form (11.52). $\square$

11.7.5   Waiting Time for the Discrete Cyclic Waiting System

We consider the waiting time for the system described above. We will follow the scheme of Sect. 11.7.3, all reasoning concerning the idea of introduced Markov chain and its transition probabilities (11.39) and (11.40) remain valid. We give the concrete values in the case of discrete system.

Using the notations of Sect. 11.7.3, we have

$$\mathbf{P}(Z = k) = (1 - r)^{k-1} r, \qquad \mathbf{P}(Y = k) = q_k.$$

As we have seen, the transition probabilities are determined by the difference $Y - Z$. Let $Y - Z = j > 0$. Then $Y = Z + j$, and the probability of this event is

$$\hat{q}_j = \sum_{i=1}^{\infty} (1 - r)^{i-1} r q_{i+j} = \frac{r}{(1 - r)^{j+1}} Q_{j+1},$$

where

$$Q_i = \sum_{k=i}^{\infty} q_k (1 - r)^k.$$

If $Y - Z = -j \leq 0 \ (j \geq 0)$, then $Y + j = Z$, and we have

$$\hat{q}_{-j} = \mathbf{P}(Y - Z = -j) = \sum_{i=1}^{\infty} q_i (1 - r)^{i+j-1} r = \frac{r(1 - r)^j}{1 - r} Q_1.$$

Lemma 11.4 *The transition probabilities of the Markov chain are*

$$f_{-j} = \frac{Q_1[1 - (1 - r)^n]}{1 - r} (1 - r)^{jn}$$

for the nonpositive jumps, and

$$f_j = \sum_{i=(j-1)n+2}^{jn} q_i \left[1 - (1 - r)^{i-(j-1)-1} \right] + \sum_{i=jn+1}^{\infty} q_i [1 - (1 - n)^n](1 - r)^{i-jn-1}$$

for the positive jumps.

Proof Using the values $\hat{q}_{-j}$ and $\hat{q}_j$ we obtain

$$f_{-j} = \sum_{k=jn}^{(j+1)n-1} \hat{q}_{-k} = \sum_{k=jn}^{(j+1)n-1} r(1 - r)^{k-1} Q_1 = \frac{rQ_1}{1 - r} \sum_{k=jn}^{(j+1)n-1} (1 - r)^k$$

$$= \frac{Q_1[1 - (1 - r)^n]}{1 - r} (1 - r)^{jn}.$$

Furthermore,

$$\hat{f}_j = \sum_{i=-\infty}^{-j} f_i = \sum_{i=j}^{\infty} \frac{Q_1[1 - (1 - r)^n]}{1 - r} (1 - r)^{in} = \frac{Q_1}{1 - r} (1 - r)^{jn}.$$

Consider the positive jumps. As we have mentioned the probability of $Y - Z = j$ $(j \geq 1)$ is

$$\hat{q}_j = \frac{r}{(1-r)^{j+1}} \sum_{i=j+1}^{\infty} q_i(1-r)^i = r \sum_{i=j+1}^{\infty} q_i(1-r)^{i-j-1}.$$

The transition probabilities f_k are

$$f_k = \hat{q}_{k+1} + \hat{q}_{k+2} + \ldots + \hat{q}_{k+n}$$

$$= \sum_{i=(k-1)n+2}^{kn} q_i \left[1 - (1-r)^{i-(k-1)n-1}\right] + \sum_{i=kn+1}^{\infty} q_i \left[1 - (1-r)^n\right](1-r)^{i-kn-1}.$$

After some arithmetics the corresponding generating function will be

$$F_+(z) = \sum_{k=1}^{\infty} z^k \left\{ \sum_{i=(k-1)n+2}^{kn} q_i + \frac{1}{(1-r)^{kn+1}} \sum_{i=kn+1}^{\infty} q_i(1-r)^i \right.$$

$$\left. - \frac{1}{(1-r)^{(k-1)n+1}} \sum_{i=(k-1)n+2}^{\infty} q_i(1-r)^i \right\}.$$

For the computations there will be required the knowledge of its first derivative at $z = 1$. We obtain this value.

$$F'_+(1) = \sum_{i=2}^{n} q_i + 2 \sum_{i=n+2}^{2n} q_i + 3 \sum_{i=2n+2}^{3n} + \ldots + k \sum_{i=(k-1)n+2}^{kn} q_i + \ldots$$

$$+ \sum_{k=1}^{\infty} k q_{(k-1)n+1} - \sum_{k=1}^{\infty} \frac{1}{(1-r)^{(k-1)n+1}} \sum_{i=(k-1)n+1}^{\infty} q_i(1-r)^i$$

$$= \sum_{k=1}^{\infty} k \sum_{i=(k-1)n+1}^{kn} q_i - \sum_{k=1}^{\infty} \frac{1}{(1-r)^{(k-1)n+1}} \sum_{i=(k-1)n+1}^{\infty} q_i(1-r)^i.$$

The second term of this expression is

$$\sum_{k=1}^{\infty} \sum_{i=(k-1)n+1}^{\infty} q_i(1-r)^{i-(k-1)n-1},$$

or may be written in the form of table

$$q_1 \; q_2(1-r) \; \ldots \; q_n(1-r)^{n-1} \; q_{n+1}(1-r)^n \; \ldots \; q_{2n+1}(1-r)^{2n} \; \ldots$$
$$q_{n+1} \qquad\qquad\qquad \ldots \; q_{2n+1}(1-r)^n \; \ldots$$
$$q_{2n+1} \qquad\qquad\qquad \ldots$$

etc.

From each n columns one can factor out

$$q_i(1-r)^{i-1 \ (\mathrm{mod}\ n)},$$

there remain the powers of $(1-r)^n$, their sums are of the form $\frac{1-(1-r)^{jn}}{1-(1-r)^n}$, i.e., in the first n columns $\frac{1-(1-r)^n}{1-(1-r)^n}$, in the second n columns $\frac{1-(1-r)^{2n}}{1-(1-r)^n}$, etc. The resulting sum will be

$$\sum_{i=1}^{\infty} q_i(1-r)^{i-1 \ (\mathrm{mod}\ n)} \frac{1-(1-r)^{\lceil \frac{i}{n} \rceil n}}{1-(1-r)^n},$$

and, finally,

$$F'_+(1) = \sum_{k=1}^{\infty} k \sum_{i=(k-1)n+1}^{kn} q_i - \sum_{i=1}^{\infty} q_i(1-r)^{i-1 \ (\mathrm{mod}\ n)} \frac{1-(1-r)^{\lceil \frac{i}{n} \rceil n}}{1-(1-r)^n},$$

which can be written in the form

$$F'_+(1) = \sum_{i=1}^{\infty} q_i \left\{ \left\lceil \frac{i}{n} \right\rceil - (1-r)^{i-1 \ (\mathrm{mod}\ n)} \frac{1-(1-r)^{\lceil \frac{i}{n} \rceil n}}{1-(1-r)^n} \right\}.$$

Our result is formulated in the following.

Theorem 11.10 *Let us consider the above described system and introduce a Markov chain whose states correspond to the waiting time (in the sense the waiting time is the number of actual state multiplied by T) at the arrival time of the customers. The matrix of transition probabilities for this chain has the form (11.41), its elements are defined by (11.39) and (11.40). The generating function of the ergodic distribution is*

$$P(z) = \left(1 - \frac{\sum_{i=1}^{\infty} q_i \left(\left\lceil \frac{i}{n} \right\rceil - (1-r)^{i-1 \ (\mathrm{mod}\ n)} \frac{1-(1-r)^{\lceil \frac{i}{n} \rceil n}}{1-(1-r)^n} \right)}{\dfrac{Q_1(1-r)^n}{(1-r)[1-(1-r)^n]}} \right)$$

$$\cdot \frac{\dfrac{Q_1}{1-r} - \dfrac{Q_1[1-(1-r)^n]}{1-r} \dfrac{z}{z-(1-r)^n}}{1-F_+(z) - \dfrac{Q_1[1-(1-r)^n]}{1-r} \dfrac{z}{z-(1-r)^n}}, \qquad (11.55)$$

where

$$F_+(z) = \sum_{j=1}^{\infty} f_j z^j, \qquad Q_1 = \sum_{j=1}^{\infty} q_j(1-r)^j;$$

the condition of existence of ergodic distribution is

$$\frac{\sum_{i=1}^{\infty} q_i \left\{ \left\lceil \frac{i}{n} \right\rceil - (1-r)^{i-1 \pmod{n}} \frac{1 - (1-r)^{\lceil \frac{i}{n} \rceil n}}{1 - (1-r)^n} \right\}}{\frac{Q_1(1-r)^n}{(1-r)[1-(1-r)^n]}} < 1. \qquad (11.56)$$

Proof Using the transition probabilities f_j $(-\infty < j < \infty)$, the ergodic probabilities are the solution of the system of equations (11.44). Multiplying the jth equation by z^j, summing up from zero to infinity, for the generating function $\sum_{j=0}^{\infty} p_j z^j$ we obtain (11.45), where $F_+(z)$ is the generating function of positive jumps.

In this expression

$$\sum_{i=0}^{j-1} f_{-i} z^{-i} = \sum_{i=0}^{j-1} \frac{Q_1[1-(1-r)^n]}{1-r}(1-r)^{in} z^{-i}$$

$$= \frac{Q_1[1-(1-r)^n]}{1-r} \frac{1 - \left(\frac{(1-r)^n}{z}\right)^j}{1 - \frac{(1-r)^n}{z}},$$

$$\sum_{j=1}^{\infty} p_j z^j \sum_{i=0}^{j-1} f_{-i} z^{-i} = \sum_{j=1}^{\infty} p_j z^j \frac{Q_1[1-(1-r)^n]}{1-r} \frac{1 - \left(\frac{(1-r)^n}{z}\right)^j}{1 - \frac{(1-r)^n}{z}}$$

$$= \frac{Q_1[1-(1-r)^n]}{1-r} \frac{z}{z-(1-r)^n} \left[P(z) - P((1-r)^n)\right],$$

$$\sum_{j=0}^{\infty} p_j \hat{f}_j = \sum_{j=0}^{\infty} p_j \frac{Q_1}{1-r}(1-r)^{jn} = \frac{Q_1}{1-r} P((1-r)^n).$$

Substituting these expressions, (11.45) yields

$$P(z)\left[1 - F_+(z) - \frac{Q_1[1-(1-r)^n]}{1-r} \frac{z}{z-(1-r)^n}\right]$$

$$= P((1-r)^n)\left[\frac{Q_1}{1-r} - \frac{Q_1[1-(1-r)^n]}{1-r} \frac{z}{z-(1-r)^n}\right].$$

The value of $P((1-r)^n)$ can be found from the condition $P(1) = 1$. By using L'Hospital's rule

$$P((1-r)^n) = 1 - F'_+(1)\frac{(1-r)[1-(1-r)^n]}{Q_1(1-r)^n},$$

and the generating function takes on the form

$$P(z) = \left[1 - F'_+(1)\frac{(1-r)[1-(1-r)^n]}{Q_1(1-r)^n}\right]\frac{\dfrac{Q_1}{1-r} - \dfrac{Q_1[1-(1-r)^n]}{1-r}\dfrac{z}{z-(1-r)^n}}{1 - F_+(z) - \dfrac{Q_1[1-(1-r)^n]}{1-r}\dfrac{z}{z-(1-r)^n}}. \tag{11.57}$$

From this generating function

$$p_0 = \left[1 - F'_+(1)\frac{(1-r)[1-(1-r)^n]}{Q_1(1-r)^n}\right]\frac{Q_1}{1-r}.$$

It is positive if

$$F'_+(1)\frac{(1-r)[1-(1-r)^n]}{Q_1(1-r)^n} < 1, \tag{11.58}$$

which serves as stability condition. $\qquad\square$

The substitution of this value into (11.57) leads to the expression (11.55), and the condition of ergodicity (11.58) gives (11.56).

11.7.5.1 Coincidence of Ergodicity Conditions

In the case of queue length the ergodicity condition was obtained in the form $B'(1) < 1$, i.e.,

$$\sum_{k=0}^{\infty}\sum_{j=1}^{n} q_{kn+j}[(k+1)nr+1] - \frac{nr}{1-(1-r)^n}\sum_{k=0}^{\infty}\sum_{j=1}^{n} q_{kn+j}(1-r)^{j-1} < 1,$$

which can be written in the form

$$\sum_{k=0}^{\infty}\sum_{j=1}^{n} q_{kn+j}(k+1) < \frac{1}{1-(1-r)^n}\sum_{k=0}^{\infty}\sum_{j=1}^{n} q_{kn+j}(1-r)^{j-1}$$

or

$$\sum_{i=1}^{\infty} q_i\left[\frac{i}{n}\right] < \frac{1}{1-(1-r)^n}\sum_{i=1}^{\infty} q_i(1-r)^{i-1 \ (\mathrm{mod}\ n)}.$$

In the case of waiting time the ergodicity condition was (11.56), i.e.,

$$\sum_{i=1}^{\infty} q_i \left\{ \left\lceil \frac{i}{n} \right\rceil - (1-r)^{i-1 \ (\mathrm{mod}\ n)} \frac{1 - (1-r)^{\left\lceil \frac{i}{n} \right\rceil n}}{1 - (1-r)^n} \right\} < \frac{Q_1(1-r)^n}{(1-r)[1 - (1-r)^n]}.$$

We show they coincide. One has

$$\frac{Q_1(1-r)^n}{(1-r)[1 - (1-r)^n]} = \frac{(1-r)^{n-1}}{1 - (1-r)^n} \sum_{i=1}^{\infty} q_i(1-r)^i$$

$$= \frac{1}{1 - (1-r)^n} \sum_{i=1}^{\infty} q_i(1-r)^{i+n-1}$$

$$= \frac{1}{1 - (1-r)^n} \sum_{i=1}^{\infty} q_i(1-r)^{i-1 \ (\mathrm{mod}\ n)}(1-r)^{\left\lceil \frac{i}{n} \right\rceil n}.$$

Using this value, one gets

$$\sum_{i=1}^{\infty} q_i \left\lceil \frac{i}{n} \right\rceil < \frac{1}{1 - (1-r)^n} \sum_{i=1}^{\infty} q_i(1-r)^{i-1 \ (\mathrm{mod}\ n)},$$

i.e., we come to the same inequality.
    For details see [22].

11.8   Queues with Negative Arrivals

G-networks (generalized or Gelenbe networks, respectively the queues in such networks called G-queues), i.e., networks with negative arrivals were first introduced by Gelenbe [11] to model neural networks; the negative arrivals remove a customer waiting in a queue, so resemble inhibitory signal in neural networks. Regular arrivals correspond to customers who upon arrival join the queue with the intention to get service and to leave the system. The system is affected by a negative arrival if and only if customers are present, in the case of negative arrival a customer is removed from the system. Negative arrivals can be considered as commands to delete some transactions, and they can represent inhibitory and excitatory signals in the mathematical models of neural networks or impatient customers in queueing systems.
    Artalejo [3] presented a survey of the main results and methods used to analyze G-networks (product form solution, stability, quasi-reversibility, etc.); Gelenbe [12] summarized the differences between the research trends related to G-networks and the random neural network and pointed the applications less known to the community. Many researchers extended the idea of negative customer with different service disciplines and removal strategies: analyzed Markov modulated multiserver

G-queue, introduced the concept of "reset" customers which can be used to reset the state of a queue in the network to its steady state distribution whenever the queue may become empty. These results were mainly connected with continuous-time models; discrete-time queues with negative customers received less attention, though they are more appropriate to describe the communication systems. In this area one can mention the early work of Atencia and Moreno [5] or [29].

For the queues with negative customers there are possible different service disciplines, FCFS and LCFS for the positive customers, and killing strategies which remove a customer in the queue: RCE—removal of customer at the end, and the customer in service RCH—removal of customer at the head, if there is any.

Following [15] we give a detailed description of RCH strategy with LCFS preemptive restart and resampling (PRR) discipline.

We consider a M/G/1 queue with Poisson arrivals with rates λ^+ for positive customers and λ^- for negative customers, respectively, $\lambda = \lambda^+ + \lambda^-$. We use the method of supplementary variables from Sect. 4.5.2 including the partial service time in service. Let $N(t)$ be the number of customers in the system at moment t, $Y_0(t)$ the elapsed service time of the customer in service at t, and $Q(z)$ the generating function of the number of customers in the steady state.

Let

$$p_k(t) = \mathbf{P}(N(t) = k),$$

$$p_k(t, t_1)\Delta = \mathbf{P}(N(t) = k, t_1 < Y_0(t) \le t_1 + \Delta),$$

$$p_k = \lim_{t \to \infty} p_k(t),$$

$$p_k(t_1) = \lim_{t \to \infty} p_k(t, t_1),$$

when the limits exist. Introduce the generating functions

$$R(z, t) = \sum_{k=1}^{\infty} p_k(t)z^k \quad \text{and} \quad R(z) = \int_0^{\infty} R(z, t)dt.$$

$r(t)\Delta$ is the probability of event the service ends in the interval $[t, t + \Delta)$ on condition it has not completed till the moment t, i.e.,

$$r(t) = \frac{B'(t)}{1 - B(t)},$$

where $B(x)$ is the service time distribution.

Theorem 11.11 *The generating function $Q(z)$ of equilibrium queue length distribution is*

$$Q(z) = \frac{\lambda^- - \lambda^+ + 2\lambda^+ B^\sim(\lambda)}{\lambda^- + \lambda^+ B^\sim(\lambda) - \lambda^+ z[1 - B^\sim(\lambda)]} = \frac{1 - \rho}{1 - \rho z}.$$

Proof At the moment $t + \Delta$ the system may be in the state either $(k, t + \Delta, t_1 + \Delta)$
$(k > 0)$ or $(k, t + \Delta, 0)$ $(k \geq 0)$.

The state $(k, t + \Delta, t_1 + \Delta)$ can be reached from the state (k, t, t_1) in the case of
neither positive, nor negative arrivals, and nor completion of service is started, i.e.,
with probability $1 - (\lambda^+ + \lambda^- + r(t_1))\Delta + o(\Delta)$:

$$p_k(t + \Delta, t_1 + \Delta) = p_k(t, t_1)[1 - (\lambda + r(t_1))\Delta + o(\Delta).$$

Divide the equation by Δ and take the limits as $t \to \infty$ and $\Delta \to 0$, and this
provides the equation

$$\frac{dp_k(t_1)}{dt_1} = -(\lambda + r(t_1))p_k(t_1), \tag{11.59}$$

its solution is

$$p_k(t) = p_k(0)[1 - B(t)]e^{-\lambda t}.$$

The state $(0, t + \Delta, 0)$ can be reached from the states

- $(1, t, t_1)$ in the case of a negative arrival or a service completion with probability
  $(\lambda^- + r(t_1))\Delta + o(\Delta)$; or
- $(0, t, 0)$ in the case of a negative arrival or no arrival at all.

This leads to the equation

$$\lambda^+ p_0 = \int_0^\infty (\lambda^- + r(t))p_1(t)dt. \tag{11.60}$$

The state $(k, t + \Delta, 0)$ $(k > 0)$ can be reached from the states

- $(k + 1, t, t_1)$ in the cases of a negative arrival or a service completion with
  probability $(\lambda^- + r(t_1))\Delta + o(\Delta)$;
- $(k - 1, t, t_1)$ in the case of a positive arrival with probability $\lambda^+ \Delta + o(\Delta)$.

This gives the equation

$$p_k(0) = \int_0^\infty (\lambda^- + r(t_1))p_{k+1}(t_1)dt_1 + \lambda^+ p_{k-1}. \tag{11.61}$$

The normalizing condition is

$$p_0 + \sum_{k=1}^\infty \int_0^\infty p_k(t)dt = 1.$$

Multiplying (11.60) by z and (11.61) by z^{k+1} and summing up we obtain

$$z\sum_{k=1}^\infty p_k(0)z^k = \int_0^\infty (\lambda^- + r(t))\sum_{k=1}^\infty p_k(t)z^k dt + \lambda^+ z^2 \sum_{k=1}^\infty p_{k-1}z^{k-1} - \lambda^+ z p_0,$$

or

$$zR(z,0) = \int_0^\infty r(t)R(z,0)\mathrm{d}t + \lambda^- R(z) + \lambda^+ R(z) + \lambda^+ z^2 R(z) + \lambda^+ z(z-1)p_0.$$
(11.62)

Multiplying (11.59) by z^k and summing up over k, we get

$$\frac{\partial R(z,t)}{\partial t} = -(\lambda + r(t))R(z,t),$$
(11.63)

whose solution is

$$R(z,t) = R(z,0)[1 - B(t)]e^{-\lambda t},$$
(11.64)

whence

$$\int_0^\infty r(t)R(z,t)\mathrm{d}t = R(z,0)B^\sim(\lambda).$$

It makes possible to express $R(z,0)$ from (11.62), namely,

$$R(z,0) = \frac{\lambda^+ z(z-1)p_0 + (\lambda^+ z^2 + \lambda^-)R(z)}{z - B^\sim(\lambda)}.$$
(11.65)

By integrating (11.64) from 0 to ∞

$$R(z) = R(z,0)\frac{1 - B^\sim(\lambda)}{\lambda}.$$
(11.66)

We eliminate $R(z,0)$ from (11.65) and (11.66), and this provides

$$R(z) = \frac{\lambda^+ z[1 - B^\sim(\lambda)]p_0}{\lambda^- - \lambda^+ z + \lambda^+ B^\sim(\lambda)(1+z)}.$$

By the definition of $R(z)$, $p_0 = 1 - R(1)$, i.e.,

$$p_0 = 1 - R(1) = 1 - \frac{\lambda^+[1 - B^\sim(\lambda)]p_0}{\lambda^- - \lambda^+ + 2\lambda^+ B^\sim(\lambda)},$$

this yields

$$p_0 = 1 - \frac{\lambda^+[1 - B^\sim(\lambda)]}{\lambda^- + \lambda^+ B^\sim(\lambda)} = 1 - \rho,$$

i.e.,

$$\rho = \lambda^+ \frac{1 - B^\sim(\lambda)}{\lambda^- + \lambda^+ B^\sim(\lambda)}.$$

The generating function of present customers is

$$Q(z) = p_0 + R(z) = p_0 \left[1 + \frac{\lambda^+ + \lambda^+ z[1 - B^\sim(\lambda)]}{\lambda^- - \lambda^+ z + \lambda^+ B^\sim(\lambda)(1 + z)} \right]$$

$$= \frac{\lambda^- - \lambda^+ + 2\lambda^+ B^\sim(\lambda)}{\lambda^- + \lambda^+ B^\sim(\lambda) - \lambda^+ z[1 - B^\sim(\lambda)]} = \frac{1 - \rho}{1 - \rho z},$$

consequently,

$$p_k = (1 - \rho)\rho^k.$$

Corollary 11.1

$$B^\sim(\lambda) > (\lambda^+ - \lambda^-)/2\lambda^+.$$

It is the consequence of inequality

$$\rho = \lambda^+ \frac{1 - B^\sim(\lambda)}{\lambda^- + \lambda^+ B^\sim(\lambda)} < 1.$$

Corollary 11.2 *The mean value of queue length is*

$$E(N) = \frac{\rho}{1 - \rho} = \frac{\lambda^+[1 - B^\sim(\lambda)]}{\lambda^- + \lambda^+[2B^\sim(\lambda) - 1]}.$$

In the case of an M/M/1 queue,

$$B^\sim(\lambda) = \int_0^\infty e^{-\lambda t} \mu e^{-\mu t} \, dt = \frac{\mu}{\lambda + \mu} \quad \text{and} \quad \rho = \frac{\lambda^+}{\lambda^- + \mu},$$

this leads to the mean queue length

$$E(N) = \frac{\lambda^+}{\lambda^- + \mu - \lambda^+}.$$

In the M/G/1 system with no negative customers $\lambda^- = 0$ and $\lambda^+ = \lambda$. For the generating function we obtain

$$Q(z) = \frac{2B^\sim(\lambda) - 1}{B^\sim(\lambda) - z[1 - B^\sim(\lambda)]} = \frac{1 - \rho}{1 - \rho z},$$

where

$$\rho = \frac{1 - B^\sim(\lambda)}{B^\sim(\lambda)}.$$

Corollary 11.3 *The equilibrium exists if $B^{\sim}(\lambda) > \frac{1}{2}$. In the M/M/1 case this means* $\lambda < \mu$.

11.9 Exercises

Exercise 11.1 A transmission link with capacity $C = 5\,$MB/s serves two kinds of CBR connections. Type i connections arrive according to a Poisson process with rate λ_i and occupy c_i bandwidth of the link for an exponentially distributed amount of time with parameter μ_i $(i = 1, 2)$, where $c_1 = 1\,$MB and $c_2 = 2\,$MB.

1. Describe the system behavior with a CTMC and compute the loss probability of type 1 customers if $\lambda_2 = 0$.
2. Describe the system behavior with a CTMC when both λ_1 and λ_2 are positive and compute the loss probability of type 1 and type 2 connections and the overall loss probability of connections.
3. Which loss probability is higher, the one of type 1 or the one of type 2 connections? Why?
4. Compute the link utilization factor when both arrival intensities are positive.
5. Compute the link utilization of type 1 and type 2 connections.

Exercise 11.2 There is a transmission link with capacity $C = 13\,$MB/s, which serves adaptive connections. The connections arrive according to a Poisson process with rate λ and their length is exponentially distributed with parameter μ. The minimal and maximal bandwidth of the adaptive connections are $c_{min} = 2\,$MB/s and $c_{max} = 3\,$MB/s, respectively. Compute the average bandwidth of an adaptive connection in equilibrium.

Exercise 11.3 There is a transmission link with capacity $C = 13\,$MB/s, which serves elastic connections. The connections arrive according to a Poisson process with rate λ and during an elastic connection an exponentially distributed amount of data is transmitted with parameter γ. The minimal and maximal bandwidth of the elastic connections are $c_{min} = 2\,$MB/s and $c_{max} = 3\,$MB/s, respectively. Compute the average bandwidth of an elastic connection in equilibrium. Compute the average time of an elastic connection in equilibrium.

Exercise 11.4 A transmission link with capacity $C = 3\,$MB/s serves two kinds of elastic connections. Type 1 connections arrive according to a Poisson process with rate $\lambda_1 = 0.5\,$1/s, transmit an exponentially distributed amount of data with parameter $\gamma_1 = 4\,$1/MB. The minimal and maximal bandwidth of type 1 connections are $\check{c}_1 = 1\,$MB/s and $\hat{c}_1 = 1\,$MB/s. Type 2 connections are characterized by $\lambda_2 = 0.1\,$1/s, $\gamma_1 = 2\,$1/MB, $\check{c}_2 = 1\,$MB/s, and $\hat{c}_2 = 2\,$MB/s.

(a) Describe the system behavior with a CTMC.
(b) Compute the mean number of type 1 and type 2 connections.

(c) Compute the mean channel utilization.
(d) Compute the loss probability of type 1 and type 2 connections.
(e) Compute the average bandwidth of type 2 connections.

Exercise 11.5 A transmission link with capacity $C = 3$ MB/s serves two kinds of connections, an elastic and an adaptive. Type 1 elastic connections arrive according to a Poisson process with rate λ_1 [1/s], transmit an exponentially distributed amount of data with parameter γ_1 [1/MB]. The minimal and maximal bandwidth of type 1 connections are $\check{c}_1 = 0.75$ MB/s and $\hat{c}_1 = 1.5$ MB/s. Type 2 adaptive connections arrive according to a Poisson process with rate λ_2 [1/s] and stay in the system for an exponentially distributed amount of time with parameter μ_2 [1/s]. The minimal and maximal bandwidth of type 2 connections are $\check{c}_2 = 1$ MB/s and $\hat{c}_2 = 2$ MB/s.

(a) Describe the system behavior with a CTMC.
(b) Compute the loss probability of type 1 and type 2 connections.
(c) Compute the average bandwidth of type 1 and type 2 connections.
(d) Compute the mean number of type 1 and type 2 connections on the link.

Exercise 11.6 Compute the mean value of waiting time in the cyclic-waiting system.

Exercise 11.7 Let us consider our cyclic-waiting system in case of discrete-time. Divide the cycle time T into n equal parts and suppose that for an interval T/n a new customer enters with probability r (there is no entry with probability $1 - r$), the service in process for such an interval is continued with probability q and completed with probability $1-q$ (i.e., the service time has geometrical distribution). The service may be started at the moment of arrival or at moments differing from it by the multiples of T.

(a) Show that the number of customers in the system at moments t_k- constitute a Markov chain, find its transition probabilities.
(b) Find the generating function of number of customers in the system in equilibrium and the stability condition.

11.10   Solutions

Solution 11.1

• When $\lambda_2 = 0$ we obtain an M/M/5/5 queueing system with the number of type 1 connections and the loss probability is $\text{loss}_1 = p_{5,0}$.

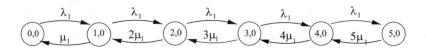

- When $\lambda_2 > 0$ we need to keep track the number of ongoing connections. The states of the Markov chain are identified by the number of ongoing type 1 and type 2 connections.

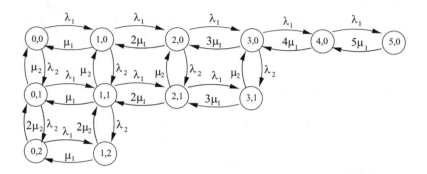

Type 1 connections are lost in states $(1, 2)$, $(3, 1)$, $(5, 0)$, while type 2 connections are lost in states $(1, 2)$, $(3, 1)$, $(5, 0)$, $(0, 2)$, $(2, 1)$, $(4, 0)$. The related loss probabilities are

$$\text{loss}_1 = p_{1,2} + p_{3,1} + p_{5,0}, \ \text{loss}_2 = p_{1,2} + p_{3,1} + p_{5,0} + p_{0,2} + p_{2,1} + p_{4,0}$$

and the overall loss probability is

$$\text{loss} = p_{1,2} + p_{3,1} + p_{5,0} + \frac{\lambda_2}{\lambda_1 + \lambda_2}(p_{0,2} + p_{2,1} + p_{4,0}),$$

where $p_{i,j}$ denotes the stationary probability of state (i, j).
- The loss probability of type 2 connections is higher, because type 2 connections are lost in more states than type 1 connections.
- Link utilization is obtained by weighting utilized bandwidth with the associated state probabilities

$$\rho = \frac{\sum_{i=0}^{5} \sum_{j=0}^{3} p_{i,j} (i \ 1\text{MB} + j \ 2\text{MB})}{5\text{MB}}.$$

- The link utilization of type 1 and 2 connections is

$$\rho_1 = \frac{\sum_{i=0}^{5} \sum_{j=0}^{3} p_{i,j} \ i \ 1\text{MB}}{5\text{MB}}, \quad \rho_2 = \frac{\sum_{i=0}^{5} \sum_{j=0}^{3} p_{i,j} \ j \ 2\text{MB}}{5\text{MB}}.$$

Solution 11.2 The adaptive connections arrive and depart according to the number of customers in an M/M/6/6 queueing system, but the bandwidth of the active connection changes with the arrival and departure of other connections. The

Markov chain indicates the number of active connections as well as the bandwidth utilization.

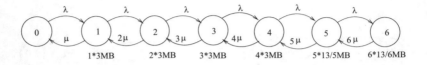

The mean bandwidth of an adaptive connection is

$$\mathbf{E}(S_A) = \sum_{i=1}^{4} p_i \ i \ 3\text{MB} + p_5 \ 5 \ 13/5\text{MB} + p_6 \ 6 \ 13/6\text{MB},$$

where p_i denotes the stationary probability of state i.

Solution 11.3 The elastic connection arrive according to a Poisson process, but their departure rates depend on the bandwidth utilization. The bandwidth of the active connection also changes with the arrival and departure of other connections. The following Markov chain indicates the number of active connections as well as the bandwidth utilization.

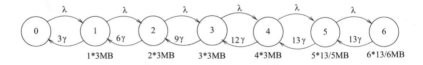

The mean bandwidth of an elastic connection is

$$\mathbf{E}(S_E) = \sum_{i=1}^{4} p_i \ i \ 3\text{MB} + p_5 \ 5 \ 13/5\text{MB} + p_6 \ 6 \ 13/6\text{MB},$$

where p_i denotes the stationary probability of state i. It seems similar to the bandwidth of the adaptive connection in the previous exercise, but the p_i probabilities differ in the two Markov chains.

The average time of an elastic connection is PH distributed with the following representation.

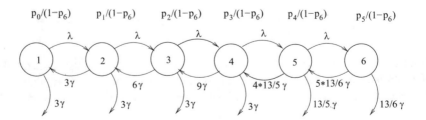

The number above the states indicate the associated initial probabilities and the downward arrows indicate the transitions to the absorbing state. This PH representation describes the sojourn of a randomly chosen tagged customer in the system. The arrival rate is independent of the number of connections. The probability that the tagged connection arrives to the system when there are i ($i = 0, \ldots, 5$) connections is proportional to p_i. The connections arrive when there are 6 active connections are lost and the probabilities are normalized for the states in which incoming connections are accepted. If an arrival occurs when there are i ($i = 0, \ldots, 5$) ongoing connections, then after the arrival there will be $i + 1$ active connections. The Markov chain of the PH distribution describes the behavior of the system up to the departure of the tagged connection. When there are i connections in the system $1/i$ portion of the utilized bandwidth is associated with the tagged connection which terminates with a rate proportional with its instantaneous bandwidth.

Solution 11.4

(a) Describe the system behavior with a CTMC.

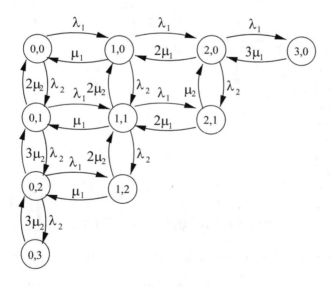

(b) Compute the mean number of type 1 and type 2 connections.

$$E(X_1) = \sum_i \sum_j i p_{ij}, \quad E(X_2) = \sum_i \sum_j j p_{ij}.$$

(c) Compute the mean channel utilization.

$$\rho = 1 - p_{00} - 2/3 p_{10} - 1/3(p_{20} + p_{01})$$

(d) Compute the loss probability of type 1 and type 2 connections.

$$P_{loss} = P_{loss1} = P_{loss2} = p_{03} + p_{12} + p_{21} + p_{30}.$$

(e) Compute the average bandwidth of type 2 connections.

$$\bar{C}_2 = \frac{\text{number of connections and the bandwidth of the connections}}{\text{number of connections}}$$

$$= \frac{1 \cdot 2 \cdot p_{01} + 2 \cdot 1.5 \cdot p_{02} + 3 \cdot 1 \cdot p_{03} + 1 \cdot 2 \cdot p_{11} + 1 \cdot 1 \cdot p_{21} + 1 \cdot 1 \cdot p_{21}}{E(X_2)}.$$

Solution 11.5

(a) The CTMC describing the system behavior:

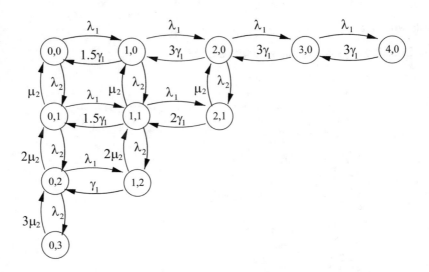

(b) the loss probability of type 1 and type 2 connections:

$$P_{loss1} = p_{40} + p_{21} + p_{12} + p_{03}, \qquad P_{loss2} = p_{40} + p_{30} + p_{21} + p_{12} + p_{03},$$

$$P_{loss} = \frac{(\lambda_1 + \lambda_2)(p_{40} + p_{21} + p_{12} + p_{03}) + \lambda_2 p_{40}}{\lambda_1 + \lambda_2},$$

(c) the average bandwidth of type 1 and type 2 connections:

$$\bar{c}_1 = 1p_{12} + 1.5(p_{10} + p_{11}) + 3(p_{20} + p_{30} + p_{40}),$$

$$\bar{c}_2 = 1p_{21} + 1.5p_{11} + 2(p_{12} + p_{01}) + 3(p_{02} + p_{03}),$$

(d) the mean number of type 1 and type 2 connections on the link:

$$E(X_1) = \sum_i \sum_j i p_{ij} = (p_{10} + p_{11} + p_{12}) + 2(p_{20} + p_{21}) + 3p_{30} + 4p_{40},$$

$$E(X_2) = \sum_i \sum_j j p_{ij} = (p_{01} + p_{11} + p_{21}) + 2(p_{02} + p_{12}) + 3p_{03}.$$

Solution 11.6 The generating function of waiting time (measured in cycles) is

$$P(z) = \left[1 - \frac{\lambda}{\mu} \frac{1 - e^{-\lambda T}}{e^{-\lambda T}(1 - e^{-\mu T})} \right]$$

$$\cdot \frac{\dfrac{\mu}{\lambda + \mu} - \dfrac{\mu(1 - e^{-\lambda T})}{\lambda + \mu} \dfrac{z}{z - e^{-\lambda T}}}{1 - \dfrac{\lambda(1 - e^{-\mu T})}{\lambda + \mu} \dfrac{z}{1 - ze^{-\mu T}} - \dfrac{\mu(1 - e^{-\lambda T})}{\lambda + \mu} \dfrac{z}{z - e^{-\lambda T}}}.$$

Introducing the notations

$$A(z) = \frac{\mu}{\lambda + \mu} - \frac{\mu(1 - e^{-\lambda T})}{\lambda + \mu} \frac{z}{z - e^{-\lambda T}},$$

$$B(z) = 1 - \frac{\lambda(1 - e^{-\mu T})}{\lambda + \mu} \frac{z}{1 - ze^{-\mu T}} - \frac{\mu(1 - e^{-\lambda T})}{\lambda + \mu} \frac{z}{z - e^{-\lambda T}},$$

the mean value of number of cycles is

$$\lim_{z \to 1} \left[1 - \frac{\lambda}{\mu} \frac{1 - e^{-\lambda T}}{e^{-\lambda T}(1 - e^{-\mu T})} \right] \frac{A'B - AB'}{2B^2}.$$

By using twice L'Hospital's rule and taking into account

$$\lim_{z \to 1} \frac{A'B - AB'}{2B^2} = \lim_{z \to 1} \frac{A''B' - A'B''}{2B'^2},$$

$$A'(1) = \frac{\mu e^{-\lambda T}}{(\lambda + \mu)(1 - e^{-\lambda T})},$$

$$A''(1) = -\frac{2\mu e^{-\lambda T}}{(\lambda + \mu)(1 - e^{-\lambda T})^2},$$

$$B'(1) = -\frac{\lambda}{(\lambda + \mu)(1 - e^{-\mu T})} + \frac{\mu e^{-\lambda T}}{(\lambda + \mu)(1 - e^{-\lambda T})},$$

$$B''(1) = -\frac{2\lambda e^{-\mu T}}{(\lambda + \mu)(1 - e^{-\mu T})^2} - \frac{2\mu e^{-\lambda T}}{(\lambda + \mu)(1 - e^{-\lambda T})^2},$$

we finally obtain

$$P'(1) = \frac{\lambda[1 - e^{-(\lambda+\mu)T}]}{(1 - e^{-\mu T})[\mu e^{-\lambda T}(1 - e^{-\mu T}) - \lambda(1 - e^{-\lambda T})]}.$$

Solution 11.7

(a) Similar to the continuous-time case we will consider two possibilities: at the beginning of service there is one customer in the system or there are at least two customers in the system.

   The case of one customer. We begin the service of the customer and after a certain time the second one arrives. Let u be the service time and the second customer appears at time v after the beginning of service. The remaining service time is ℓ ($\ell = 0, 1, 2, \ldots$) with probability

$$P\{u - v = \ell\} = \sum_{k=\ell+1}^{\infty} q^{k-1}(1-q)(1-r)^{k-\ell-1}r = \frac{r(1-q)q^{\ell}}{1 - q(1-r)}.$$

We find the time from the entry of second customer till the beginning of its service. It is 0 if the customer arrives during the last time slice of the first customer's service, n if $u - v$ belongs to the interval $[1, n]$, $2n$ if $u - v \in [n+1, 2n]$, and, generally, in if $u - v \in [(i-1)n+1, in]$. The corresponding probabilities are

$$\sum_{\ell=(i-1)n+1}^{in} \frac{r(1-q)q^{\ell}}{1 - q(1-r)} = \frac{rq}{1 - q(1-r)}\left(q^{(i-1)n} - q^{in}\right).$$

The generating function of number of customers arriving for a time slice is $1 - r + rz$, so the generating function of customers entering for the waiting time is

$$\sum_{i=1}^{\infty} \frac{rq(1-q^n)}{1 - q(1-r)}q^{(i-1)n}(1-r+rz)^{in} = \frac{rq(1-r+rz)^n(1-q^n)}{[1 - q(1-r)][1 - q^n(1-r+rz)^n]}.$$

Taking into account that the first customer obligatorily arrives and the waiting time may be equal to zero for the generating function of entering customers we obtain

$$A(z) = \sum_{i=0}^{\infty} a_i z^i = \frac{(1-r)(1-q)}{1 - q(1-r)} + z\frac{r(1-q)}{1 - q(1-r)}$$

$$+ z\frac{rq(1-r+rz)^n(1-q^n)}{[1 - q(1-r)][1 - q^n(1-r+rz)^n]},$$

where $\dfrac{(1-r)(1-q)}{1-q(1-r)}$ is the probability of event for the service of first customer no further customers arrive.

The case of at least two customers. At the beginning of service of first customer the second customer is present, too. Let $x = u - \left[\dfrac{u-1}{n}\right] n$ ($[x]$ denote the integer part of x), and let y be the mod T interarrival time ($1 \le y \le n$). The time elapsed between the starting moments of two successive customers is

$$\left[\frac{u-1}{n}\right]n + y \quad \text{if} \quad x \le y \quad \text{and} \quad \left(\left[\frac{u-1}{n}\right]+1\right)n + y \quad \text{if} \quad x > y.$$

Let us fix y and consider the cycle $[in + 1, (i + 1)n]$. If the service of first customer ends till y (including y), then the time till the beginning of service of second customer is $in + y$ and the probability of this event is

$$\sum_{j=in+1}^{in+y} q^{j-1}(1-q) = q^{in} - q^{in+y},$$

in case $x > y$ the time is $(i + 1)n + y$ and the probability is

$$\sum_{j=in+y+1}^{(i+1)n} q^{j-1}(1-q) = q^{in+y} - q^{(i+1)n}.$$

i changes from 0 to ∞ (the summation is extended for all possible values of service time), for fixed y the generating functions of entering customers in the two cases will be

$$\sum_{i=0}^{\infty}[q^{in} - q^{in+y}](1 - r + rz)^{in+y} = \frac{(1 - r + rz)^y}{1 - q^n(1 - r + rz)^n}$$

$$- \frac{(1 - r + rz)^y q^y}{1 - q^n(1 - r + rz)^n},$$

$$\sum_{i=0}^{\infty}[q^{in+y} - q^{in+n}](1 - r + rz)^{in+n+y} = \frac{q^y(1 - r + rz)^{n+y}}{1 - q^n(1 - r + rz)^n}$$

$$- \frac{q^n(1 - r + rz)^{n+y}}{1 - q^n(1 - r + rz)^n}.$$

y has truncated geometrical distribution, it takes on the value k ($k = 0, 1, 2, \ldots, n$) with probability $\dfrac{(1-r)^k r}{1 - (1-r)^n}$.

Consequently, the generating function of transition probabilities is

$$B(z) = \sum_{k=1}^{n} \frac{(1-r)^{k-1}r}{1-(1-r)^n} \frac{1}{1-q^n(1-r+rz)^n}$$

$$\cdot [(1-r+rz)^k - (1-r+rz)^k q^k + (1-r+rz)^{n+k} q^k$$

$$- (1-r+rz)^{n+k} q^n]$$

$$= \frac{1-(1-r)^n(1-r+rz)^n}{1-(1-r)(1-r+rz)} \frac{r(1-r+rz)}{1-(1-r)^n}$$

$$+ \frac{1-q^n(1-r)^n(1-r+rz)^n}{1-q(1-r)(1-r+rz)} \frac{rq(1-r+rz)[(1-r+rz)^n-1]}{[1-(1-r)^n][1-q^n(1-r+rz)^n]}.$$

We have seen that, as in the continuous case, the length of interval between two successive starting moments is determined by the service time of first customer and the interarrival time of first and second customers, so they are independent random variables. By using the memoryless property of geometrical distribution, we obtain the number of customers in the system at moments just before the beginning of services constitute a Markov chain.

(b) The system is considered at moments just before starting the services of customers. Let us denote the ergodic distribution by p_i $(i = 0, 1, 2, \ldots)$ and introduce the generating function by $P(z) = \sum_{i=0}^{\infty} p_i z^i$. For p_i we have the system of equations

$$p_j = p_0 a_j + p_1 a_j + \sum_{i=2}^{j+1} p_i b_{j-i+1},$$

from which

$$\sum_{j=0}^{\infty} p_j z^j = p_0 A(z) + p_1 A(z) + \sum_{j=0}^{\infty} \sum_{i=2}^{j+1} p_i b_{j-i+1} z^j$$

$$= \frac{1}{z} P(z) B(z) - \frac{1}{z} p_0 B(z) + p_0 A(z) + p_1 A(z) - p_1 B(z),$$

or

$$P(z) = \frac{p_0[z A(z) - B(z)] + p_1 z[A(z) - B(z)]}{z - B(z)}.$$

Since

$$p_0 = p_0 a_0 + p_1 a_0,$$

we have

$$p_1 = \frac{1 - a_0}{a_0} p_0 = \frac{r}{(1 - r)(1 - q)} p_0.$$

We find p_0 from the condition $P(1) = 1$

$$p_0 = \frac{1 - B'(1)}{1 - B'(1) + A'(1) + \frac{r}{(1-r)(1-q)}[A'(1) - B'(1)]}.$$

The chain is irreducible, so $p_0 > 0$.

Using the values

$$A'(1) = \frac{r}{1 - q(1 - r)} + \frac{nr^2 q}{[1 - q(1 - r)](1 - q^n)},$$

$$B'(1) = 1 - \frac{nr(1 - r)^n}{1 - (1 - r)^n} + \frac{nr^2 q[1 - q^n(1 - r)^n]}{(1 - q^n)[1 - (1 - r)^n][1 - q(1 - r)]},$$

we obtain

$$\left(1 + \frac{r}{(1 - r)(1 - q)}\right) A'(1) - \frac{r}{(1 - r)(1 - q)} B'(1)$$

$$= \frac{nr^2 q}{(1 - q^n)[1 - q(1 - r)]} + \frac{nr^2(1 - r)^n}{(1 - r)[1 - (1 - r)^n][1 - q(1 - r)]} > 0,$$

so the condition $1 - B'(1) > 0$ must be fulfilled. This leads to the expression

$$\frac{nr(1 - r)^n}{1 - (1 - r)^n} - \frac{nr^2 q[1 - q^n(1 - r)^n]}{(1 - q^n)[1 - (1 - r)^n][1 - q(1 - r)]} > 0.$$

From it we obtain the stability condition

$$\frac{rq}{1 - q^n} \frac{1 - q^n(1 - r)^n}{1 - q(1 - r)} < (1 - r)^n.$$

References

1. 802.11: IEEE standard for information technology-telecommunications and information exchange between systems-local and metropolitan area networks-specific requirements - part 11: wireless LAN medium access control (MAC) and physical layer (PHY) specifications (2007). http://ieeexplore.ieee.org/servlet/opac?punumber=4248376

2. Abramson, N.: The ALOHA system - another alternative for computer communications. In: Proceedings Fall Joint Computer Conference. AFIPS Press, Montvale (1970)
3. Artalejo, J.R.: G-networks: a versatile approach for work removal in queueing networks. Euro. J. Oper. Res. **126**, 233–249 (2000)
4. Artalejo, J.R., Gomez-Corral, A.: Retrial Queueing Systems: A Computational Approach. Springer, Berlin (2008)
5. Atencia, I., Moreno, P.: The discrete-time Geo/Geo/1 queue with negative customers and disasters. Comput. Oper. Res. **31**, 1537–1548 (2004)
6. Bianchi, G.: Performance analysis of the IEEE 802.11 distributed coordination function. IEEE J. Sel. Areas Commun. **18**, 535–547 (2000)
7. Borovkov, A.A.: Stochastic processes in queueing theory. In: Applications of Mathematics. Springer, Berlin (1976)
8. Dallos, G., Szabó, C.: Random Access Methods of Communication Channels. Akadémiai Kiadó, Budapest (1984). In Hungarian
9. Falin, G.I., Templeton, J.G.C.: Retrial Queues. Chapman and Hall, London (1997)
10. Foh, C.H., Zukerman, M.: Performance analysis of the IEEE 802.11 MAC protocol. In: Proceedings of European Wireless Conference, Florence (2002)
11. Gelenbe, E.: Random neural networks with negative and positive signals and product form solution. Neural Comp. **1**, 502–510 (1989)
12. Gelenbe, E.: Introduction to the special issue on g-networks and the random neural network. Perform. Eval. **68**, 307–308 (2011)
13. Gnedenko, B.V., Kovalenko, I.N.: Introduction to Queueing Theory, 2nd edn. Birkhauser Boston Inc., Cambridge (1989)
14. Gnedenko, B., Danielyan, E., Dimitrov, B., Klimov, G., Matveev, V.: Priority Queues. Moscow State University, Moscow (1973). In Russian
15. Harrison, P.G., Pitel, E.: The M/G/1 queue with negative customers. Adv. Appl. Probab. **28**, 540–566 (1996)
16. Jaiswal, N.K.: Priority Queues. Academic, New York (1968)
17. Kaufman, J.: Blocking in a shared resource environment. IEEE Trans. Commun. **29**, 1474–1481 (1981)
18. Koba, E.V.: On a retrial queueing system with a FIFO queueing discipline. Theory Stoch. Proc. **8**, 201–207 (2002)
19. Lakatos, L.: On a simple continuous cyclic waiting problem. Annales Univ. Sci. Budapest Sect. Comp. **14**, 105–113 (1994)
20. Lakatos, L.: Cyclic waiting systems. Cybern. Syst. Anal. **46**, 477–484 (2010)
21. Lakatos, L.: On the queue length in the discrete cyclic-waiting system of Geo/G/1 type. CCIS **678**, 121–131 (2016)
22. Lakatos, L.: On the waiting time in the discrete cyclic-waiting system of Geo/G/1 type. CCIS **601**, 86–93 (2016)
23. Lakatos, L., Efrosinin, D.: Some aspects of waiting time in cyclic-waiting systems. CCIS **356**, 115–121 (2014)
24. Lakatos, L., Serebriakova, S.: Number of calls in a cyclic-waiting system. Reliab. Theory Appl. **11**, 37–43 (2016)
25. Massoulie, L., Roberts, J.: Bandwidth sharing: objectives and algorithms. In: Infocom. IEEE, Piscataway (1999)
26. Medgyessy, P., Takács, L.: Probability Theory. Tankönyvkiadó, Budapest (1973). In Hungarian
27. Rácz, S., Telek, M., Fodor, G.: Call level performance analysis of 3rd generation mobile core network. In: IEEE International Conf. on Communications, ICC 2001, vol. 2, pp. 456–461. IEEE, Piscataway (2001)
28. Rácz, S., Telek, M., Fodor, G.: Link capacity sharing between guaranteed- and best effort services on an ATM transmission link under GoS constraints. Telecommun. Syst. **17**(1,2), 93–114 (2001)
29. Rakhee, G.S., Priya, K.: Analysis of G-queue with unreliable server. OPSEARCH **50**, 334–345 (2013)

30. Roberts, J.: A service system with heterogeneous user requirements - application to multi-service telecommunications systems. In: Proceedings of Performance of Data Communications Systems and Their Applications, pp. 423–431. North-Holland Publishing, Amsterdam (1981)
31. Ross, K.W.: Multiservice Loss Models for Broadband Telecommunication Networks. Springer, Berlin (1995)
32. Szeidl, L.: Estimation of the moment of the regeneration period in a closed central-server queueing network. Theory Probab. Appl. **31**, 309–313 (1986)
33. Szeidl, L.: On the estimation of moment of regenerative cycles in a general closed central-server queueing network. Lect. Notes Math. **1233**, 182–189 (1987)
34. Takagi, H.: Queueing Analysis. Elsevier, Amsterdam (1991)

Appendix A
Functions and Transforms

A.1 Nonlinear Transforms

Many theoretical and practical problems can be converted into easier forms if instead of the discrete or continuous distributions their different transforms are applied, which can be solved more readily. In probability theory, numerous transforms are applied. Denote by F the distribution function of a random variable X and by f the density function if it exists. The general form of the most frequently used transform depending on the real or complex parameter w is

$$\mathbf{E}\left(w^X\right) = \int_{-\infty}^{\infty} w^x \mathrm{d}F(x).$$

If the density function exists, then the last Riemann-Stieltjes integral can be rewritten in the form of Riemann integral as follows:

$$\mathbf{E}\left(w^X\right) = \int_{-\infty}^{\infty} w^x f(x) \mathrm{d}x.$$

1. In the general case, setting $w = e^{\Im t}$, $t \in \mathbb{R}$, we have the *characteristic function* (Fourier-Stieltjes transform)

$$\varphi(t) = \mathbf{E}\left(e^{\Im t X}\right) = \int_{-\infty}^{\infty} e^{\Im t x} \mathrm{d}F(x).$$

2. If the random variable X has a discrete distribution with values $0, 1, \ldots$ and probabilities $p_0, p_1, \ldots$ corresponding to them, then setting $z = w$, $|z| < 1$, we get

© Springer Nature Switzerland AG 2019
L. Lakatos et al., *Introduction to Queueing Systems with Telecommunication Applications*, https://doi.org/10.1007/978-3-030-15142-3

$$G(z) = \mathbf{E}(z^X) = \int_{-\infty}^{\infty} z^x \mathrm{d}F(x) = \sum_{k=0}^{\infty} p_k z^k,$$

which is the *generating function* of X.

3. The Laplace-Stieltjes transform plays a significant role when considering random variables taking only nonnegative values (usually we consider this type of random variable in queuing theory), which we obtain with $w = e^{-s}$, $s \geq 0$:

$$F^{\sim}(s) = \mathbf{E}(e^{-sX}) = \int_{0}^{\infty} e^{-sx} \mathrm{d}F(x).$$

For the case of continuous distributions it can be rewritten in the form

$$f^*(s) = \mathbf{E}(e^{-sX}) = \int_{0}^{\infty} e^{-sx} f(x) \mathrm{d}x = F^{\sim}(s),$$

where f^* denotes the Laplace transform of the density function f.

4. The generating function plays a significant role when considering discrete random variables taking only nonnegative values, which we obtain with $w = z$:

$$f(z) = \mathbf{E}(z^X) = \sum_{n=0}^{\infty} f_n z^n.$$

The identical background of the transformations given above determines some identical properties. When considering various problems, the use of separate transforms may be advantageous. For example, in the case of random variables taking nonnegative integer values the z-transform, and in case of general nonnegative random variables the Laplace-Stieltjes or Laplace transform is favorable to apply.

Note that we define the transforms given above for more general classes of functions than the distribution functions.

A.2 z-Transform

Let $f_0, f_1, \ldots$ be a sequence of real numbers and define the power series

$$f(z) = \sum_{n=0}^{\infty} f_n z^n = f_0 + f_1 z + f_2 z^2 + \ldots + f_n z^n + \ldots \tag{A.1}$$

It is known from the theory of power series that if the series (A.1) is not everywhere divergent except the point $z = 0$, then there exists a number $A > 0$ such that the series (A.1) is absolute convergent ($\sum_{n=0}^{K} |f_n z^n| < \infty$) for all $|z| < A$ and divergent for all $|z| > A$. The series (A.1) may be convergent or divergent at the points $z =$

$\pm A$ depending on the values of the parameters f_i, $i = 0, 1, \ldots$ The number A is called the **convergence radius** of the power series (A.1). By the Cauchy-Hadamard theorem, A can be given in the form

$$A = 1/a, \quad \text{where} \quad a = \limsup_{n \to \infty} (|f_n|)^{1/n}.$$

In the last formula we set $A = +\infty$ if $a = 0$ and $A = 0$ if $a = +\infty$. The first relation $A = +\infty$ means that the power series (A.1) is convergent in all points of the real line, and the second one means that Eq. (A.1) is convergent only at the point $z = 0$.

A finite power series $f(z) = \sum_{n=0}^{K} f_n z^n$ (K-order polynomial, which corresponds to the case $f_i = 0$, $i \geq K + 1$) is convergent at all points of the real line.

Definition A.1 Let $f_0, f_1, \ldots$ be a sequence of real numbers satisfying the condition $a = \limsup_{n \to \infty} (|f_n|)^{1/n} < \infty$. Then the power series

$$f(z) = \sum_{n=0}^{\infty} f_n z^n, \ |z| < A = 1/a$$

is called the z-**transform** of the sequence $f_0, f_1, \ldots$

It is clear from this definition that if we use a discrete distribution f_n, $k \geq 0$, $\sum_{k=0}^{\infty} f_k = 1$, then the z-transform of the sequence $f_0, f_1, \ldots$ is identical with the generating function $G(z)$, which was introduced earlier.

A.2.1   Main Properties of the z-Transform

1. *Derivatives.* If the convergence radius A does not equal 0, then the power series $f(z)$ is an anytime differentiable function for all points $|z| < A$ and

$$\frac{d^k}{dz^k} f(z) = \sum_{n=k}^{\infty} n(n-1) \ldots (n-k+1) f_n z^{n-k}, \ k \geq 1.$$

2. *Computing the coefficients of the z-transform.* For all $k = 0, 1, \ldots$ the following relation is true:

$$f_k = \frac{1}{k!} \frac{d^k}{dz^k} f(z) \bigg|_{z=0}, \ k \geq 0. \tag{A.2}$$

It is clear from relation (A.1) that if the condition $A > 0$ holds, then the function $f(z)$ defined by the power series (A.1) and the sequence $f_0, f_1, \ldots$ uniquely determine each other, that is, the z-transform realizes a one-to-one correspondence between the function $f(z)$ and the sequence $f_0, f_1, \ldots$.

3. *Convolutions*. Let $f(z) = \sum_{n=0}^{\infty} f_n z^n$ and $g(z) = \sum_{n=0}^{\infty} g_n z^n$ be two z-transforms determined by the sequences f_n and g_n, respectively. Define the sequence h_n as the convolution of f_n and g_n, that is,

$$h_n = \sum_{k=0}^{n} f_k g_{n-k}, \ n \geq 0.$$

Then the z-transform $h(z) = \sum_{n=0}^{\infty} h_n z^n$ of the sequence $h_0, h_1, \ldots$ satisfies the equation

$$h(z) = f(z) \cdot g(z).$$

A.3   Laplace-Stieltjes and Laplace Transform in General Form

Let $H(x)$, $0 \leq x < \infty$ be a function of bounded variation. A function H is said to be of bounded variation on the interval $[a, b]$, if its total variation $V_H([a, b])$ is bounded (finite). The total variation is defined as

$$V_H([a, b]) = \sup_{P} \sum_{k=1}^{K_P} \left| H(x_{P,k}) - H(P_{,k-1}) \right|$$

where the supremum is taken over the set of all partitions

$$P = \{x_{P,0} = a < x_{P,1} < \ldots < x_{P,K_P} = b\}$$

of the interval $[a, b]$. The function H is of bounded variation on the interval $[0, \infty)$ if $V_H([0, b])$ is bounded by some number V for all $b > 0$.

The function

$$H^{\sim}(s) = \int_0^{\infty} e^{-sx} dH(x). \tag{A.3}$$

is called the **Laplace-Stieltjes transform** of the function H. If the function H can be given in the integral form

$$H(x) = \int_0^x h(u) du, \ x \geq 0,$$

where h is an integrable function (this means that H is an absolute continuous function with respect to the Lebesgue measure), then the Laplace transform of the function h satisfies the equation

$$h^*(s) = \int_0^\infty e^{-sx} h(x) dx = H^\sim(s).$$

Theorem A.1 *If the integral (A.3) is convergent for $s > 0$, then $H^\sim(s)$, $s > 0$ is an analytic function, and for every positive integer k*

$$\frac{d^k}{ds^k} H^\sim(s) = \int_0^\infty e^{-sx} (-x)^k dH(x).$$

The transform $H^\sim$ satisfies the following asymptotic relation [2]. If the integral (A.3) is convergent for $\mathrm{Re}\, s > 0$ and there exist constants $\alpha \geq 0$ and A such that

$$\lim_{x \to \infty} \frac{H(x)}{x^\alpha} = \frac{A}{\Gamma(\alpha + 1)},$$

then the convergence

$$\lim_{s \to 0+} s^\alpha H^\sim(s) = A \qquad (A.4)$$

holds.

Theorem A.2 *Assume that there exists a function $h(x)$, $x \geq 0$, and its Laplace transform $h^*(s)$, $s > 0$; moreover, the function $h(x)$ is convergent as $x \to \infty$, i.e.,* $\lim_{x \to \infty} h(x) = h_\infty$. *Then*

$$\lim_{s \to 0+} s h^*(s) = h_\infty.$$

Proof Denote $H(x) = \int_0^x h(s) ds$, $x \geq 0$. Choosing $\alpha = 1$ we have

$$\lim_{x \to \infty} \frac{H(x)}{x} = \lim_{x \to \infty} \frac{1}{x} \int_0^x h(s) ds = h_\infty = \frac{h_\infty}{\Gamma(1 + 1)},$$

thus by relation (A.4) the assertion of the theorem immediately follows. □

Theorem A.3 *If there exists a Laplace transform f^* of the nonnegative function $f(t)$, $t \geq 0$, and there exists the finite limit $\lim_{x \to 0+} f(x) = f_0$, then*

$$\lim_{s \to \infty} s f^*(s) = f_0.$$

Proof It is clear that

$$s \int_0^\infty e^{-sx} dx = \int_0^\infty e^{-x} dx = 1$$

and

$$s \int_{1/\sqrt{s}}^\infty e^{-sx} dx = \int_{\sqrt{s}}^\infty e^{-y} dy = e^{-\sqrt{s}} = o(1), \ s \to \infty;$$

therefore,

$$sf^*(s) - f_0 = s \int_0^{1/\sqrt{s}} e^{-sx}[f(x) - f_0]dx + s \int_{1/\sqrt{s}}^\infty e^{-sx} f(x)dx + f_0 o(1).$$

Since there exists the finite limit $\lim_{x \to \infty} f(x) = f_0$, with the notation

$$\delta(z) = \sup_{0 < x \le z} |f(x) - f_0| \to 0, \ z \to 0+,$$

we obtain

$$s \int_0^{1/\sqrt{s}} e^{-sx} |f(x) - f_0| \, dx < \delta(1/\sqrt{s}) \int_0^\infty se^{-sx} dx = \delta(1/\sqrt{s}) \to 0, \ s \to \infty.$$

On the other hand, for all $0 < t \le s$ the relation

$$f^*(s) = \int_0^\infty e^{-sx} f(x)dx \le \int_0^\infty e^{-tx} f(x)dx = f^*(t),$$

is true, then

$$s \int_{1/\sqrt{s}}^\infty e^{-sx} f(x)dx \le se^{-(1/2)\sqrt{s}} \int_{1/\sqrt{s}}^\infty e^{-(s/2)x} f(x)dx$$

$$\le se^{-(1/2)\sqrt{s}} \int_0^\infty e^{-(s/2)x} f(x)dx = se^{-(1/2)\sqrt{s}} f^*(s/2)$$

$$\le se^{-(1/2)\sqrt{s}} f^*(1) \to 0,$$

as $s \to \infty$ $(s \ge 2)$. Summing up the results obtained above, the assertion of the theorem follows.   □

A.3.1 Examples for Laplace Transform of Some Distributions

(a) Deterministic distribution ($a > 0, \mathbf{P}(X = a) = 1$):

$$F^{\sim}(s) = \int_0^\infty e^{-sx} dF(x) = e^{-sa}, \quad \mathbf{E}(X) = a :$$

(b) $B(n, p)$ binomial distribution:

$$F^{\sim}(s) = \int_0^\infty e^{-sx} dF(x) = \sum_{k=0}^n \binom{n}{k} p^k (1 - p)^{n-k} e^{-sk}$$

$$= \sum_{k=0}^n \binom{n}{k} (pe^{-s})^k (1 - p)^{n-k} = [1 + p(e^{-s} - 1)]^n,$$

$$\mathbf{E}(X) = npe^{-s}[1 + p(e^{-s} - 1)]^{n-1}\Big|_{s=0} = np.$$

(c) Poisson distribution with parameter λ:

$$F^{\sim}(s) = \int_0^\infty e^{-sx} dF(x) = \sum_{k=0}^\infty e^{-sk} \frac{\lambda^k}{k!} e^{-\lambda}$$

$$= \sum_{k=0}^\infty \frac{1}{k!} (\lambda e^{-s})^k e^{-\lambda} = \exp\{\lambda(e^{-s} - 1)\},$$

$$\mathbf{E}(X) = \lambda e^{-s} \exp\{\lambda(e^{-s} - 1)\}\Big|_{s=0} = \lambda.$$

(d) Uniform distribution on the interval $[a, b]$:

$$F^{\sim}(s) = \int_a^b e^{-sx} \frac{1}{b - a} dx = \begin{cases} \frac{1}{s(b-a)} (e^{-sa} - e^{-sb}), & s > 0, \\ 1, & s = 0. \end{cases}$$

and by the use of the L'Hospital's rule:

$$\mathbf{E}(X) = \frac{1}{b - a} \lim_{s \to 0+} -\frac{1}{s^2} \left([e^{-sa} - e^{-sb}] - [sae^{-sa} - sbe^{-sb}] \right)$$

$$= \frac{1}{b - a} \lim_{s \to 0+} \frac{b^2 se^{-sb} - a^2 e^{-sa}}{2s} = \frac{b^2 - a^2}{2(b - a)} = \frac{a + b}{2}.$$

(e)   Exponential distribution with parameter λ:

$$F^\sim(s) = \int_0^\infty e^{-sx}\lambda e^{-\lambda x}dx = \lambda \int_0^\infty e^{-(s+\lambda)x}dx = \frac{\lambda}{s+\lambda},$$

$$\mathbf{E}(X) = \frac{\lambda}{(s+\lambda)^2}\bigg|_{s=0} = \frac{1}{\lambda}.$$

A.3.2   Sum of a Random Number of Independent Random Variables

Theorem A.4 *Let K be a random variable with nonnegative integer values, and consider the sum of K random variables $Y = \sum_{n=1}^{K} X_n$, where*

(1) The random variables K and $\{X_n,\ n \geq 0\}$ are independent.
(2) The distributions of the random variables X_n are identical with common distribution function $F(x)$.

Denote by $F_X^\sim(s)$ the Laplace-Stieltjes transform of X_n and by $G_K(z)$ the generating function of K. Then the Laplace-Stieltjes transform of random variable Y has the form

$$E(e^{-sY}) = G_K(F_X^\sim(s)).$$

Proof Since

$$\mathbf{E}\left(\exp\left\{-s\sum_{n=1}^{K} X_n\right\}\ \bigg|\ K = k\right) = F_X^\sim(s)^k,$$

then we obtain by the use of the formula of total expected value

$$\mathbf{E}\left(\exp\left\{-s\sum_{n=1}^{K} X_n\right\}\right) = \sum_{k=1}^{\infty}\mathbf{E}\left(\exp\left\{-s\sum_{n=1}^{K} X_n\right\}\ \bigg|\ K = k\right)\mathbf{P}(K = k)$$

$$= \sum_{k=1}^{\infty} F_X^\sim(s)^k\ \mathbf{P}(K = k) = \mathbf{E}(F_X^\sim(s)^K) = G_K(F_X^\sim(s)).$$

$\square$

A.4 Bessel and Modified Bessel Functions of the First Kind

Definition A.2 The nonzero solutions of Bessel's differential equation

$$x^2 u'' + x u' + (x^2 - v^2)u = 0 \tag{A.5}$$

are called v-**order Bessel functions**, where v is a real number.

Definition A.3 The solutions of Bessel's differential equation are called **Bessel functions of the first kind** and denoted by $J_v(x)$, which are nonsingular at the origin $x = 0$.

The v-order Bessel functions of the first kind $J_v(x)$ ($v \geq 0$) can be defined by their Taylor series expansion around $x = 0$ as follows:

$$J_v(x) = \sum_{k=0}^{\infty} \frac{(-1)^k}{\Gamma(k + v + 1)\Gamma(k + 1)} \left(\frac{x}{2}\right)^{2k+v}, \tag{A.6}$$

where $\Gamma(x) = \int_0^{\infty} e^{-t} t^{x-1} dt$ is the gamma function. This formula is valid, providing $v \neq -1, -2, \ldots$. The Bessel functions

$$J_{-v}(x) = \sum_{k=0}^{\infty} \frac{(-1)^k}{\Gamma(k + v + 1)\Gamma(k + 1)} \left(\frac{x}{2}\right)^{2k-v}$$

is given by replacing v in Eq. (A.6) with a $-v$.

An important special case of the Bessel function of the first kind is that of a purely imaginary argument.

Definition A.4 The function

$$I_v(x) = \Im^{-v} J_v(\Im x) = \sum_{k=0}^{\infty} \frac{1}{\Gamma(k + v + 1)\Gamma(k + 1)} \left(\frac{x}{2}\right)^{2k+v}$$

is called a **modified v-order Bessel function of the first kind**.

Both the Bessel functions $J_v(x)$ and $I_v(x)$ can be expressed in terms of the generalized hypergeometric function $_0F_1(v; x)$ as follows [1]:

$$J_v(x) = \frac{1}{\Gamma(v + 1)} \left(\frac{x}{2}\right)^v {}_0F_1(v + 1; -\frac{x^2}{4}),$$

$$I_v(x) = \frac{1}{\Gamma(v + 1)} \left(\frac{x}{2}\right)^v {}_0F_1(v + 1; \frac{x^2}{4}),$$

where

$$_0F_1(v; x) = \sum_{k=0}^{\infty} \frac{\Gamma(v)}{\Gamma(k+v)\Gamma(k+1)} x^k.$$

A.5 Notations

$\mathbb{N}^+$	Set of the nonnegative integer numbers
$\mathbb{R}$	Set of the real numbers ($\mathbb{R} = (-\infty, \infty)$)
$\Im$	Imaginary unit ($\Im^2 = -1$)
δ_{ij}	Kronecker delta function, that is, $\delta_{ij} = 1$, if $i = j$, otherwise it equals 0
a^+	Positive part of a real number a, i.e., $a^+ = \max(a, 0)$
$\overline{A}$	Complementary event of A
$\mathscr{I}_{\{A\}}$	Indicator function of an event A, that is, it equals 1 if the event A occurs and otherwise it equals 0
$\mathbf{P}(A)$	Probability of an event A
$\mathbf{E}(X)$	Expected value of a random variable X
$\mathbf{D}(X)$	Variation of a random variable X
S	State space of a Markov chain
Π	(One-step) transition probability matrix of a discrete-time Markov chain
Q	rate matrix of a continuous-time Markov chain
I	Unit matrix

References

1. Prudnikov, A.P., Brychkov, Y.A., Marichev, O.I.: Integrals and Series, vol. 2. Gordon and Breach Science Publishers, New York (1986). Special functions
2. Takács, L.: Combinatorial Methods in the Theory of Stochastic Processes. Wiley, London (1967)

Index

© Springer Nature Switzerland AG 2019
L. Lakatos et al., *Introduction to Queueing Systems with Telecommunication Applications*, https://doi.org/10.1007/978-3-030-15142-3

Printed in the United States
By Bookmasters